Lecture Notes on Data Engineering and Communications Technologies

Volume 134

Series Editor

Fatos Xhafa, Technical University of Catalonia, Barcelona, Spain

The aim of the book series is to present cutting edge engineering approaches to data technologies and communications. It will publish latest advances on the engineering task of building and deploying distributed, scalable and reliable data infrastructures and communication systems.

The series will have a prominent applied focus on data technologies and communications with aim to promote the bridging from fundamental research on data science and networking to data engineering and communications that lead to industry products, business knowledge and standardisation.

Indexed by SCOPUS, INSPEC, EI Compendex.

All books published in the series are submitted for consideration in Web of Science.

More information about this series at https://link.springer.com/bookseries/15362

Zhengbing Hu · Ivan Dychka ·
Sergey Petoukhov · Matthew He
Editors

Advances in Computer Science for Engineering and Education

 Springer

Editors
Zhengbing Hu
School of Computer Science
Hubei University of Technology
Wuhan, China

Sergey Petoukhov
Mechanical Engineering Research Institute
of the Russian Academy of Sciences
Moscow, Russia

Ivan Dychka
Faculty of Applied Mathematics
National Technical University of Ukraine
"Igor Sikorsky Kiev Polytechnic Institute"
Kiev, Ukraine

Matthew He
Halmos College of Arts and Sciences
Nova Southeastern University
Ft. Lauderdale, FL, USA

ISSN 2367-4512   ISSN 2367-4520 (electronic)
Lecture Notes on Data Engineering and Communications Technologies
ISBN 978-3-031-04811-1   ISBN 978-3-031-04812-8 (eBook)
https://doi.org/10.1007/978-3-031-04812-8

This Springer imprint is published by the registered company Springer Nature Switzerland AG
The registered company address is: Gewerbestrasse 11, 6330 Cham, Switzerland

Preface

Modern engineering and educational technologies provide dramatically new opportunities because of computer science. These include the ability to conduct computer experiments, detailed visualization of the objects being built and analyzed, distant learning, rapid retrieval of information from large databases, and the development of artificial intelligence systems, among other things. As a result, governments and scientific-technological communities are paying close attention to computer science issues and their applications in engineering and education. As a result, higher education institutions are faced with the difficult task of educating new generations of professionals who will be able to properly use and further advance computer science and its applications. International cooperation helps to facilitate and speed the development of acceptable solutions to this critical subject.

For these reasons, the Fifth International Conference on Computer Science, Engineering, and Education Applications (ICCSEEA2022) was jointly organized by the National Technical University of Ukraine "Igor Sikorsky Kyiv Polytechnic Institute," National Aviation University, and the International Research Association of Modern Education and Computer Science on February 21–22, 2022, in Kiev, Ukraine. The ICCSEEA2022 brings together leading scholars from all around the world to share their findings and discuss outstanding challenges in Computer Science, Engineering, and Education Applications.

The best contributions to the conference were selected by the programme committee for inclusion in this book out of all submissions.

February 2022

Zhengbing Hu
Ivan Dychka
Sergey Petoukhov
Matthew He

Organization

Conference Organizers and Supporters

National Technical University of Ukraine "Igor Sikorsky Kyiv Polytechnic Institute", Ukraine
Wuhan University of Technology, China
Nanning University, China
National Aviation University, Ukraine
Huazhong University of Science and Technology, China
Polish Operational and Systems Society, Poland
Wuhan Technology and Business University, China
International Research Association of Modern Education and Computer Science, Hong Kong

Contents

Computer Science for Manage
of Natural and Engineering Processes

Digital Twin-Driven Warehouse Management System for Picking Path Planning Problem

Jingjing Cao[✉] and Ahui Lei

School of Transportation and Logistics Engineering, Wuhan
University of Technology, Wuhan 430063, China
bettycao@whut.edu.cn

Abstract. As an important link in logistics, warehousing management plays an important role in improving the overall service level of logistics. This paper applies digital twin technology to warehouse management, proposes and studies a warehouse management system based on digital twin. According to the bottleneck problems in the actual operation of warehouse management, the application requirements of the system are considered. Then, the system is finished by designing the overall architecture, planning functions and designing the operation mechanism. Finally, this paper designs and constructs the basic digital twin model of the whole operation process in the warehouse, especially the model of the picking path planning. As a result, the system realizes the digital twin model-oriented goods picking path plan and real-time interactive adjustment to form a dynamic interactive picking path optimization method based on digital twin. Based on this system, the actual operation of a warehouse is taken as an example to demonstrate the effect of the digital twin system in the actual picking operation, which proves that the system can reasonably improve the picking efficiency and the overall operation level of the warehouse.

Keywords: Digital twin · Warehouse management · Picking · Dynamic interaction

1 Introduction

With the rapid development of e-commerce, the diversity of customer demand will lead to a rapid increase in the types and quantities of goods in the warehouse. Therefore, warehousing as the key link of logistics faces challenges: higher requirements for the warehouse operation efficiency and service level. Nowadays, warehouse management mainly includes seven links: order processing inbound shelving, replenishment, picking, sorting, quality inspection and outbound operations. Improvement of single and multiple links can play an optimal role in the whole process of warehousing. Particularly, picking operations as an important step in the warehouse operations, the number of operators is the most, accounting for 50% of human resources, the high complexity of the operation makes the time accounted for 30% of the warehouse operations, processing costs accounted for 90% of the overall warehouse processing costs. Therefore, reasonable

© The Author(s), under exclusive license to Springer Nature Switzerland AG 2022
Z. Hu et al. (Eds.): ICCSEEA 2022, LNDECT 134, pp. 3–16, 2022.
https://doi.org/10.1007/978-3-031-04812-8_1

planning of picking operations is particularly important in improving the overall operational efficiency of the warehouse, picking speed directly affects the outbound efficiency, and the correctness of picking affects the overall logistics service level.

Nowadays, most of the warehouses are still manual picking or "semi-manual + semi-automated" mode, and most of the strategies selected are simple-minded traversal-type strategies, with the disadvantages of fixed operation patterns, long time consumption and long picking distances. The biggest problem is the inability to cope with the dynamic changes and uncertainties in actual operations. With the development of information technology, to solve the problem of the high cost of physical experiments, the concept of Digital Twin (DT) was born. Digital twin technology refers to the use of digital technology to describe and model the characteristics, behavior, formation process and performance of physical entity objects to form a digital twin model, which can then be monitored and reversed. The Internet of Things, blockchain, artificial intelligence and other technologies provide data support for the digital twin, which applies virtual simulation and other technologies to map physical scenes to virtual scenes, uses real-time dynamic interaction functions to achieve dynamic planning and scheduling, gets the optimal solution to send back instructions to feedback to logistics activities and related entities, completes control operations and operational guidelines, greatly simplifies the problem of planning and scheduling operations. By applying this technology to warehouse management, it is possible to cope with dynamic changes in warehouse operations such as picking and sorting, eliminate the influence of uncertainties and improve the operational efficiency of the warehouse.

In this paper, the characteristics of virtual reality mapping and interactive fusion of digital twin technology are applied [1]. It proposed and studied a digital twin-based warehouse management system and an example of its application in picking operations. The rest of this paper is organized as follows. Section 2 introduce the application field of data twin technology, especially in logistics and warehousing management to verify the feasibility and necessity of this paper; Sect. 3 introduces the application requirements, architecture design, functional planning and overall operation mechanism of the warehouse management system based on digital twin technology; Sect. 4 introduces the implementation process and simulation case analysis of dynamic interactive picking path planning in the digital twin warehouse management system; Sect. 5 concludes this paper.

2   Related Works

As an emerging technology, digital twin technology is currently applied to areas such as product development and design, manufacturing process, anomaly detection and shop floor management [2–5]. For example, Tingyu Lin et al. proposed an intelligent industrial product development model based on the Evolutionary Digital Twin (EDT), which leverages the self-evolution of machine learning to enable greater flexibility and adaptability of industrial products [2]. Alberto Villalonga et al. proposed a dynamic scheduling decision framework based on the digital twin for physical production systems, using real-time monitoring functions to detect changes in manufacturing processes and make dynamic decisions [3]. Andrea Castellani et al. addressed the industrial anomaly detection problem, use the digital twin to simulate the normal operation of machinery and introduce clustering algorithms to achieve anomaly detection [4]. Cunbo Zhuang et al. developed a DT-based visual monitoring and prediction system (DT-VMPS) for real-time monitoring and 3D visualization of shop floor operation status and Markov chain-based shop floor operation status prediction [5]. It can be seen that the application of digital twin's dynamic interaction and virtual-real mapping can solve problems in various fields, and its application areas are expanding.

Nowadays, the application of digital twin technology in the logistics field is relatively small, and the relevant studies have been made so far include production and logistics system, product and service system. Y. H. Pan et al. proposed a digital twin-driven production logistics synchronization system, mainly for the delivery vehicle routing problem in industrial parks, proposed a digital twin-driven decision-making architecture and designed a real-time dynamic synchronization control mechanism [6]. Digital twin technology is less studied in warehouse operations. Jiewu Leng et al. proposed a digital twin system that fuses real-time data from a large automated high-rise warehouse product service system, maps it to a network model, obtains the best decision through a joint optimization model and feeds it to a semi-physical simulation engine in the digital twin system to implement control [7]. Most of the existing studies have addressed the application of digital twin in highly automated warehouses, which are mostly "semi-manual + semi-automated" mode, so this paper investigates the application of digital twin in general warehouse management and how to achieve the scheduling and planning of critical operations, proposes related approaches.

3 Architecture Design of Digital Twin-Based Warehouse Management System

3.1 Application Requirements of the System

The main process of warehouse operation includes "inbound - replenishment - picking - quality inspection - packaging - outbound", through the digital twin, the entire process of the warehouse is virtually mapped. The complex process is replaced with a model for easy observation and analysis. The analysis and scheduling results are reversed in the real warehouse operation to improve the efficiency of the entire operation and enhance the overall service level of logistics. Because of the diversification of customer needs, the SKU of goods in the warehouse can reach hundreds of thousands of millions, it is difficult to solve many problems in the operations only with conventional warehouse management systems and simple information technology, for example, the picking-path optimization problem. As an emerging technology, establishing a digital twin system can use the characteristics of practical interaction and virtual and real integration to solve the problem of the warehouse operation process at the minimum cost. The requirements of the system include:

3.1.1 Optimizing Internal Layout of the Warehouse

By adjusting the layout of the internal facilities of the warehouse, including the placement of shelves, merchandise display, can make the operation process more efficient and faster. In the actual warehouse, the warehouse is usually designed according to experience.

Once the installation of shelves and equipment is completed, it is difficult to adjust and the optimal layout is uncertain, which is a big problem in the transformation and upgrading of traditional warehouses. On the other hand, for the high shelves of the whole warehouse, usually up to 12–15 m, which requires high forklifts, the location of the goods is particularly important, and a more convenient location for frequently outbound goods. Therefore, by establishing a digital twin system, using the virtual simulation technology combined with the optimization algorithm, to visually display the optimal solution.

3.1.2 Optimizing Picking Paths

Picking is an important part of the warehouse operation. Due to the diversified needs of customers, there are often many different combinations of goods, which is a major difficulty for picking, so how to make a reasonable allocation of orders and optimize the picking path within the constraints of the number of warehouse personnel and time is the key. The simulation optimization model of the digital twin system can integrate the basic information of the warehouse and continuously optimize to find the shortest path.

3.1.3 Realizing Dynamic Demand Forecast and Improving Inventory Strategy

Accurately grasping the needs of each commodity is particularly important for the inventory adjustment of the warehouse. If the inventory is insufficient, it will affect the picking efficiency, the current warehouse master inventory information through manual inventory and inform replenishment personnel to replenishment, there is a time difference. Some warehouses use a management system to display the inventory, which is relatively efficient, but when faced with large orders and urgent orders there will be out of stock. Using the digital twin model for demand forecasting and inventory visualization of commodities to assist the warehouse to complete inventory strategy and replenishment strategy development is feasible.

3.1.4 Achieving the Overall Visualization of the Warehouse Operation, Facilitating Decision-Making and Emergency Response

In the actual operation of the warehouse, there may be an equipment failure, manual misoperation and other situations, which may cause equipment damage or personnel injury. However, the monitoring system in the warehouse is sometimes unable to find the abnormalities of the equipment, and it is difficult to manage so much equipment. Therefore, the realization of the overall visualization of the warehouse operation and emergency response is particularly important. By connecting corresponding sensors, fault detection equipment and emergency alarm devices to the system, the digital twin system realizes visual display and automatic early warning. It is convenient to find on-site operation problems and provide emergency solutions.

3.2 System Architecture Design

The overall architecture of the digital twin system designed based on the above requirements includes five layers: physical layer, IoT layer, data layer, model layer and O&M service layer, as shown in Fig. 1.

1) **Physical layer.**

   This includes entities such as personnel, forklifts, goods and conveyors involved in the actual operational process within the warehouse, as well as the specific operations in the actual operational process, such as shelving, picking and packing. The physical layer is capable of receiving operation instructions from the digital twin system and realizing the instructions, providing guidance for the actual work of personnel and completing the control of equipment status and completing the physical activities within the entire warehouse.

2) **IoT layer.**

   This layer mainly plays the role of connecting the physical layer and the data layer. It is the basis and key link to reflect reality. It completes the collection and transmission of information on logistics activities and related entities through a series of logistics information technology, such as RFID, sensors, high-definition camera, 5G communication technology and so on. The information mainly includes real-time information on goods (location, inventory), operation status of equipment in the warehouse (forklifts, conveyors, packaging equipment), picking path and operators' information. It also needs to transfer the instructions generated by the operation and maintenance service layer, mainly including the digital twin data acquisition module, the network transmission module and the mobile terminal processing module (responsible for receiving the instructions and displaying them on the computer and electronic display).

3) **Data layer.**

   A large amount of data is generated during the operation process in the warehouse, which is the basis for realizing the physical to virtual model and is the core of the digital twin system. The data layer processes, fuses and stores the data collected by the IoT layer, which includes the twin data generated by the operation and maintenance service layer, the basic warehouse information data, cargo data, order data and customer data inside the warehouse management system. It provides data support for the physical scene, virtual scene and digital twin system operation. It is an important carrier for realizing the virtual-to-real conversion.

4) **Model layer.**

   To solve the problems existing in the warehouse operation through the digital twin technology, this layer provides a model for the digital twin-based warehouse management system, which is a digital description of the physical activity. The layer generates a dynamic digital twin model of the warehouse according to the data layer to describe the dynamic information of the entity objects in the system, including the twin of the warehouse environment, the twin of goods and shelf and the digital twin model of picking path optimization.

5) **Operation and maintenance service layer.**

   This layer interacts directly with the data layer, the model layer, the physical warehouse and the virtual warehouse, integrating all the functions and services, is the value of the entire digital twin system. It achieves the transformation from the basic IoT layer, data layer, model layer to a more easily observable visualization level. For example, the visualization of inventory information and sales information of goods in the warehouse is displayed. It uses data mining algorithms, optimization algorithms, etc., to optimize the warehouse operations or carry out value-added services, transmits instructions to the physical layer and completes the feedback from the virtual to the real. For example, this layer can predict real-time orders and optimize the inventory strategy, use path optimization algorithm and virtual simulation technology to guide the pickers to task efficiently.

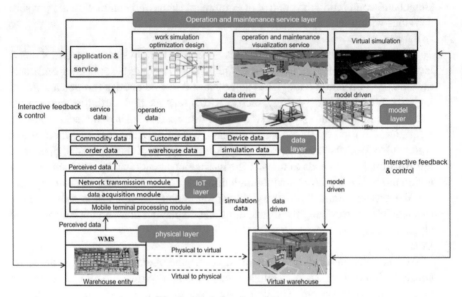

Fig. 1. Digital twin architecture

3.3 System Function Planning

The core of the digital twin-based warehouse management system lies in the full-cycle management of goods, including the entire process from inbound to outbound. Using digital twin technology, we can obtain real-time information data about the flow of goods in the warehouse, realize data integration and visual management of the entire process, map the relevant physical activities and related entities into the twin model and feed the results back to the actual through the simulation and optimization of the twin model. The results are fed back to the actual operation after finishing the simulation and optimization of the twin model, which drives the warehouse management and improves the efficiency of the whole warehouse operation. Therefore, the system functions include basic warehouse information management functions, such as picking personnel management, order management and warehouse equipment management. Secondly, operation status real-time monitoring function includes warehouse temperature monitoring, goods quality monitoring, goods trajectory tracking and equipment status monitoring. Thirdly, digital twin picking path optimization function includes warehouse and goods layout optimization, digital twin model construction, goods stacking stability analysis and dynamic optimization of the picking paths. Finally, value-added services for the warehousing process include the visual query of goods' status and locations equipment operation, the real-time data of warehouse operations and real-time control.

3.4 Operational Mechanism Design Based on Digital Twin Warehouse Management System

The use of digital twin technology can achieve real-time interaction between physical warehousing and virtual warehousing, the integration of the whole data, the whole process, the whole business and the dynamic mapping of reality in the actual operation process. The whole includes the physical scene of the warehouse operation, the virtual scene of the warehouse operation, the warehouse operation twin data and the digital twin warehouse management system.

The physical scene of operations in the warehouse includes the whole process of warehouse management. It will receive feedback from the virtual scene and the control information of the warehouse management system in the actual process. The main scenarios include: *a.* quality inspection and shelves in the warehouse; *b.* a series of operations such as picking and packing, feedback on the status of goods and equipment; *c.* receiving real-time instructions and real-time interaction with the digital twin system.

The virtual scene of operations in the warehouse is the digital expression of operations in the warehouse, which is a collection of many physical models, which can be divided into two parts: **a.** the digital expression of operation physical entities in the warehouse, that is, static twins, including circulating goods, high forklift trucks, picking vehicles, etc.; **b.** the simulation reproduction of cargo status and warehouse operations is dynamic twins, can realize dynamic tracking of goods and real-time interaction with the real scene. The management of the storage process can be carried out efficiently by simulating physical scenes with virtual scenes. It can monitor and manage the status of goods, equipment, personnel and picking paths.

Twin data of the warehouse model is the basis of the operation and interaction of physical scene, virtual scene and digital twin warehouse management system, including physical scene data, virtual scene data and digital twin storage management system operation data. The physical scene data includes the warehouse data (the location and status of the goods), the data of the equipment (the equipment location and operation status), the personnel information (the personnel location and operation information), the result data of the twin model feedback to the entity after simulation optimization, etc.

The role of the digital twin warehouse management system is to acquire tasks, collect data, fuse data, process data, optimize and analyze, evaluate and decide and issue instructions. Its operation mechanism is shown in Fig. 2, as follows.

1) First of all, the system receives receipts, plans the location of the goods and arranges the task for the forklift driver to complete the goods on the shelves. Then it generates the initial task-based storage digital twin model according to the basic attributes of goods, warehouse equipment, existing selection route planning, warehouse layout, etc., realizes the initial virtual warehouse operation.

2) Digital twin warehouse management system evaluates the rationality of the warehouse layout and selection path based on the twin data of 1), receives the simulation

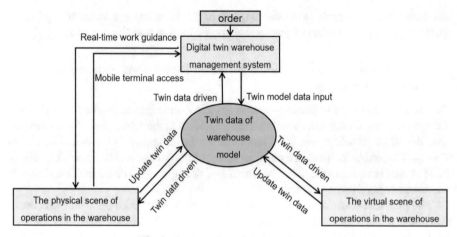

Fig. 2. Design of operation mechanism

and analysis results generated by the virtual scene of the warehouse operation. The system adjusts the layout scheme and updates the warehouse twin data according to the results. Then, it drives the twin model to generate a visual warehouse layout and selection scheme, guides the actual selection operation and drives the operation of the physical warehouse.

3) The physical warehouse operates following the layout and picking method of the virtual warehouse. It also transmits the time of actually selected path and other data into the digital twin system according to the scanning equipment and GPS. The corresponding twin data is updated to ensure the consistency between the virtual twin model and the physical activities of the entity.

4) According to the planning and scheduling of the twin system, the whole cycle management of the commodity storage link is realized. In the actual operation, the real-time data collected from the warehouse information, commodity status, and equipment operation state will be feedback to the digital twin storage management system. The staff scans the handheld terminal to feedback the completed operation to the system update data. The staff can also access the digital twin system through the mobile terminal to obtain information about the processing situation, location and quantity of the goods in the warehouse in the order.

5) The Digital twin storage management system uses the digital twin model to realize the 3-dimensional visual monitoring of the data. For example, the track tracking function is used to visualize the circulation track of goods in the warehouse through the location change data of goods, which is useful for analyzing the operation process and corresponding time of goods in the warehouse. It monitors the operation in the warehouse in real-time through the work data of personnel and equipment, determines whether personnel scheduling, line planning, additional equipment, discontinued equipment, emergency response and other operations are necessary. It provides decision-making services for the operation in the warehouse. The above operation mechanism can optimize the process in time, and its application scenario is shown in Fig. 3.

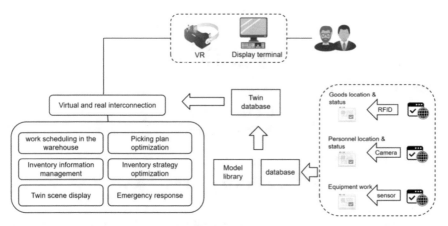

Fig. 3. Application scenarios of warehouse management system based on digital twin.

4 Dynamic Interactive Picking Path Planning Model Design

4.1 Digital Twin Model for Picking Operations

When planning the picking path, the system maps the actual picking process to the twin model firstly. Then, it conducts path planning in the model and feedbacks to the actual picking operation. In the picking stage, the system will divide the order into batches and generate the picking order according to the shortest picking time, give it to the picking personnel to complete the picking operation. The goods are a basic twin, the location of the goods can be described by the location coordinates. By simply selecting the origin coordinates, we can get the location information of all the goods contained in the order. After completing the construction of the digital twin model for virtual picking, we can shorten the operation time, improve the outbound rate and reduce the warehouse operation cost by optimizing the picking path.

As there are too many types of goods in the warehouse, the system needs to complete the classification of goods before the path planning. The system uses ABC classification to improve the location layout of goods and simplify the picking path problem to the TSP problem. After completing the above operation, the system optimizes the digital twin for the picking work of the warehouse picking district, assuming includes: *a.* the total weight of the picked goods is less than or equal to the weight-bearing capacity of the picking trolley; *b.* the layout of the shelves in the warehouse is: all the shelves layout in the warehouse is: all single shelves are discharged horizontally and in double rows side by side; *c.* the picker walks at a uniform speed throughout the process.

Modeling: In the problem of optimizing order distribution and planning picking paths, two main objectives are considered, one is the shortest picking path and the other is the shortest picking time. It is necessary to take into account the changes in dynamic factors, which require full use of the real-time interaction function in the digital twin model, and the model is as follows.

objective function:

$$\min f = \sum_{i=1}^{K} \sum_{j=1}^{K} d_{ij} a_{ij} \tag{1}$$

decision Variables:

$$a_{ij} = \begin{cases} 1, & Choose\ j\ after\ choosing\ i \\ 0, & Choose\ i\ over\ j \end{cases} \tag{2}$$

constraint conditions:

$$\sum_{i=1}^{K} a_{ij} = 1\ (i \neq j, j = 1,2,3,\ldots,K) \tag{3}$$

$$\sum_{j=1}^{K} a_{ij} = 1\ (i \neq j, i = 1,2,3,\ldots,K) \tag{4}$$

$$\sum_{i=1}^{K} m_i \leq M_{max} \tag{5}$$

$$i,j = 1,2,3,\ldots,K \tag{6}$$

where K is the type of goods to be picked by the picker each time, d_{ij} means the shortest distance from good i to good j, a_{ij} is the decision variable, which means the order of picking goods. Formula (1) means the total distance of the picker's path. Formula (3) and (4) mean that the picker only passes once for each kind of goods in the picking process. Formula (5) means that the weight of the goods picked by the picker at one time must be less than or equal to the maximum weight of the picking vehicle.

4.2 Model Optimization Algorithm Design

Nowadays, most scholars use the heuristic algorithm and its improved algorithm to optimize the picking path, which includes the genetic algorithm, A * algorithm, etc. Li-Mei Duan proposed the improved A * algorithm to realize the batch picking path planning of multi-warehouse logistics robots [8]. Li Zhou et al. proposed the V-type warehouse picking path planning based on the Internet of Things, establish the return picking method model [9]. Makusee Masae et al. proposed the optimal routing algorithm based on the graph theory concept, and establish the optimal order selector routing strategy for the glyph warehouse [10]. Because it needs to consider the impact of the actual layout and dynamic factors of the warehouse, it is difficult and is also the focus of scholars' research.

In the optimization of the model, the classical genetic optimization algorithm is used to complete the optimization of the picking path using the digital twin as a carrier. Genetic Algorithms are widely used in optimal scheduling problems [11–13]. The system realizes real-time dynamic interaction and visualizes the optimization scheme of the selection path to the mobile terminal of the picking person. The selector will be guided to quickly complete the selection process with the voice selection function. The Fig. 4 below is the optimization process of the genetic algorithm:

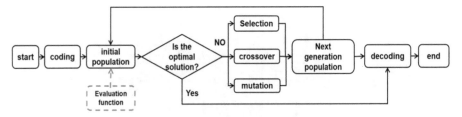

Fig. 4. Genetic algorithm design

Encoding: In the problem of picking path optimization, the goods positions are numbered. The picking order will correspond to multiple numbers and the genes in the chromosome are encoded according to the numbers, with each segment of genes corresponding to one commodity position, for a total of L.

Fitness Function: The larger the population adaptation is the better the overall result, set the adaptation function as $F = 1/(min f + 1)$.

Crossover Operator: Each gene segment represents one cargo, no duplication or omission, and two chromosomes are crossed using a single point.

The Crossover Probability: Takes the value of $pc = 0.8$. For each chromosome, a random number in the interval (0, 1) is generated. If the generated number is less than pc, the crossover operation is performed for that chromosome. The initial path planning is performed mathematically in the initial stage of the model and further optimized by combining the visual digital twin model with the optimization algorithm for route planning adjustments to meet specific requirements.

4.3   Simulation Case Study

Taking warehouse X as an example, a digital twin model is established based on the physical operation of the warehouse and related entities. First, the layout is optimized based on historical data of goods combined with demand forecast results, and the digital twin model design for path planning is constructed and combined with optimization algorithms for picking path optimization. An example is used to introduce how the digital twin system solves the picking path planning problem and provides an optimal solution for warehouse operations.

Taking a pick list containing 21 items as an example, the system optimizes the picking path based on the objective function and constraints in the model. The following figure shows the picking paths under three strategies: the traversal strategy, the improved traversal strategy and the picking path optimized using the genetic algorithm schematic diagram. As shown in Fig. 5, if the traversal strategy is selected, the picking path is 92 units. As shown in Fig. 6, the selected path is 88 units. As shown in Fig. 7, the result of the genetic algorithm modeling optimization is shown as 84 units. The improved crossing strategy uses the midline segmentation to complete the picking on one side first, saving

Fig. 5. Crossing type strategy

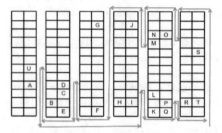

Fig. 6. Improved crossing-type strategy

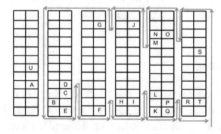

Fig. 7. Genetic algorithm optimization strategy

Table 1. Different measurement measuring mixed state estimation results

Strategy	Total distance	Save distance	Save proportion
Crossing strategy	92	0	0
Improved crossing strategy	88	4	4.34%
Genetic algorithm optimization	84	8	8.69%

partial distance, while the genetic algorithm optimization can make the path the shortest path.

Table 1 shows that the optimal results are obtained by building a digital twin model of the picking operation combined with an optimization algorithm. Picking employees then perform picking through a visualization scheme on mobile terminals. And it is possible to obtain relevant data timely when changes occur in the field operation, such as blockage in the aisle, the location of goods and the insufficient number of personnel by using the real-time interaction feature of the digital twin model. Then the system can complete the adjustment of relevant parameters and re-complete the planning of the picking path, greatly reduce the impact on the efficiency of the picking operation due to unexpected situations. In summary, the use of the digital twin-based warehouse management system combined with algorithms can solve the picking operation path optimization problem, effectively improve the order processing efficiency and enhance the overall service level of logistics.

5   Conclusion

In the actual operation process of the warehouse, dynamic change is an important factor affecting the operational efficiency in the library. It is particularly important to reflect the dynamic change and replanning timely. The digital twin system is well proposed to solve this problem. This paper proposes the warehouse management system based on the digital twin according to the existing problems in warehouse management. The process mainly includes the demanding design, architecture design, functional planning and operation mechanism design, the construction of the digital twin model of the warehouse. The paper takes the long-consuming and highly complex picking job in the warehouse as an example to introduce how to realize the planning and visual presentation of the picking path through the digital twin model. The system realizes the real-time dynamic interaction between the physical scene and the virtual scene. Compared to traditional warehouse management systems, the system realizes responding to the dynamic changes in the warehouse timely by using digital twin technology to obtain real-time data and combined with the optimization algorithm. This paper proves that using the digital twin model combined with the optimization algorithm can greatly improve the operation efficiency by analyzing the relevant case, which verifies the feasibility and effectiveness of the system.

The system designed in this paper is only for general warehouses. It does not apply in some special warehouses, such as cold chain storage, which has different operating procedures and stricter requirements on temperature and operating time. Therefore, it is necessary to further expand the adaptability of the system, so that the system can be applied in a diversified warehouse environment. And it is necessary to further expand the application of digital twin technology in other aspects of logistics.

Acknowledgment. Thanks to the School of Transportation and Logistics Engineering for their support to the research work and thanks to relevant scholars for their research foundation and comments.

References

1. Fang, Y., Peng, C., Lou, P., Zhou, Z., Jianmin, H., Yan, J.: Digital-twin-based job shop scheduling toward smart manufacturing. IEEE Trans. Ind. Informatics **15**(12), 6425–6435 (2019)
2. Lin, T., et al.: Evolutionary digital twin: a new approach for intelligent industrial product development. Adv. Eng. Informatics **47**, 101209 (2021)
3. Villalonga, A., et al.: A decision-making framework for dynamic scheduling of cyber-physical production systems based on digital twin. Annu. Rev. Control. **51**, 357–373 (2021)
4. Castellani, A., Schmitt, S., Squartini, S.: Real-world anomaly detection by using digital twin systems and weakly supervised learning. IEEE Trans. Ind. Informatics **17**(7), 4733–4742 (2021)
5. Zhuang, C., Miao, T., Liu, J., Xiong, H.: The connotation of digital twin, and the construction and application method of shop-floor digital twin. Robotics Comput. Integr. Manuf. **68**, 102075 (2021)
6. Pan, Y.H., Wu, N.Q., Qu, T., Li, P.Z., Zhang, K., Guo, H.F.: Digital-twin-driven production logistics synchronization system for vehicle routing problems with pick-up and delivery in industrial park. Int. J. Comput. Integr. Manuf. **34**(7–8), 814–828 (2021)
7. Leng, J., et al.: Digital twin-driven joint optimisation of packing and storage assignment in large-scale automated high-rise warehouse product-service system. Int. J. Comput. Integr. Manuf. **34**(7–8), 783–800 (2021)
8. Duan, L.-M.: Path planning for batch picking of warehousing and logistics robots based on modified a* algorithm. Int. J. Online Eng. **14**(11), 176–192 (2018)
9. Zhou, L., Liu, J., Fan, X., Zhu, D., Pingyu, W., Cao, N.: Design of v-type warehouse layout and picking path model based on internet of things. IEEE Access **7**, 58419–58428 (2019)
10. Masae, M., Glock, C.H., Vichitkunakorn, P.: Optimal order picker routing in the chevron warehouse. IISE Trans. **52**(6), 665–687 (2020)
11. AlBalas, F., Mardini, W., Bani-Salameh, D.: Optimized job scheduling approach based on genetic algorithms in smart grid environment. Int. J. Commun. Networks Inf. Secur. **9**(2), 172 (2017)
12. Soulegan, N.S., Barekatain, B., Neysiani, B.S.: MTC: minimizing time and cost of cloud task scheduling based on customers and providers needs using genetic algorithm. Int. J. Intell. Syst. Appl. (IJISA) **13**(2), 38–51 (2021). https://doi.org/10.5815/ijisa.2021.02.03
13. Nagar, R., Gupta, D.K., Singh, R.M.: Time effective workflow scheduling using genetic algorithm in cloud computing. Int. J. Inform. Technol. Comput. Sci. (IJITCS) **10**(1), 68–75 (2018). https://doi.org/10.5815/ijitcs.2018.01.08

Analysis of Transfer of Modulated Ink Flows in a Short Printing System of Parallel Structure

Bohdan Durnyak[1], Mikola Lutskiv[1], Petro Shepita[1], Vasyl Sheketa[2], Roman Karpyn[1], and Nadiia Pasyeka[3(✉)]

[1] Ukrainian Academy of Printing, Lviv, Ukraine
durnyak@uad.lviv.ua, lutolen@i.ua
[2] Ivano-Frankivsk National Technical University of Oil and Gas, Ivano-Frankivsk, Ukraine
[3] Vasyl Stefanyk Precarpathian National University, Ivano-Frankivsk, Ukraine
pasyekanm@gmail.com

Abstract. There are model of modulated ink flows and the ink image transfer from the plate to the printed material has been developed, which is expressed by the ink amount per unit of square raster elements area for short ink printing systems of parallel structure of the sixth dimension, the characteristics of inking and tone transfer have been developed and constructed and their properties have been analysed. A graph of modulated ink flows of the printing system has been constructed, which graphically reflects the structure of ink flows, their direction and system parameters, it also allows determining the thickness of ink flows, tone transfer and other variables directly according to the graph.

The functional scheme of the model of modulated ink flows of the printing system in Matlab is developed: Simulink package, which allows one to calculate the system characteristics simultaneously, performs their visualization in the form of image inking, tone transfer, their deviations which are quite informative. The re sults of simulation are presented in the form of nonlinear characteristics of image inking. It has been established that the non-uniformity of the ink thickness in the tone transfer interval is 21.4%. The short printing system of the parallel structure of the sixth dimension enlightens the image on grey tones.

Keywords: Accuracy · Ink amount · Inking · Model · Nonlinearity · Raster · Tone transfer

1 Problem Setting

In offset machines for printing of newspapers and simple printing products, Western companies began to use simple in design short ink units with anilox ink supply device, which have only a few ink rollers. They are simple in design, do not have mechanisms for adjusting the zonal ink supply, so do not provide constant ink thickness on the raster imprint surface on the entire range of tone transfer, which limits their use for printing of quality books and magazines [8, 12, 13, 15, 21, 22]. In the United States, a number of short ink units with an anilox ink supply device of various structures have been patented, which are not implemented in metal, so it is difficult to determine which scheme of the

Z. Hu et al. (Eds.): ICCSEA 2022, LNDECT 134, pp. 17–26, 2022.
https://doi.org/10.1007/978-3-031-04812-8_2

device is the best [9, 19]. Experimental research methods of short printing systems require expensive equipment to measure the ink thickness on rotating rollers and the time and they are inefficient.

In offset printing technology, the reproduction of halftone images is provided by changing the relative areas of blank and printing elements covered with ink, provided that the ink thickness is constant, and depending on the optical density of the image only one parameter is changed – the area of printing elements [6, 11]. The organization of tone reproduction of halftone images for a constant ink thickness, tone reproduction schemes, synthesis and adjustment of tone transfer are known and are reduced to synthesis of areas of raster elements [16, 17, 20], therefore are one-parameter ones, so they are directly inefficient which limit their use for the printing of any book and magazine products [5, 14, 18]. Therefore, there is a problem of modelling and analysis of the transfer of modulated ink flows from the plate to the printed material, expressed by the ink amount per area unit, depending on the tone transfer interval.

2 Literature Review

The tone transfer processes that take place in short printing systems are complex, due to the modulation of the ink flow of the raster printing plate and their transfer to the offset and from it to the printed material. The part of the ink that is not perceived by the blank elements remains on the rollers and creates unmaintained backflows circulating in the system, which causes a decrease in the ink thickness depending on the tone transfer interval. In works [1, 10] mathematical models of short printing systems are developed, on the basis of which the characteristics of coverage of linear raster scales for systems of different dimensions are constructed. The analysis of the accuracy of the ink layer thickness has revealed that it depends on the tone transfer interval and can be 10–20% or more, which does not meet the regulatory requirements for the quality of books and magazines [7, 23]. In the publications of the authors [3, 16], models of inking of raster elements of different shapes for printing systems of sequential structure of different dimensions are constructed. Characteristics of inking are S-shaped curves. The maximum deviation of the characteristic from the linear one depends on the tone transfer interval and it is in the range from −15.6 to +12.6%. However, the properties of short ink printing systems of parallel structure are studied insufficiently.

3 Goal of the Paper

To develop a mathematical model for the transfer of modulated ink flows from the plate to the printed material that describe the tone transfer of a short printing system of parallel structure, to determine and construct the characteristics of tone transfer expressed by the ink amount per area unit for square raster elements and analyse their properties.

4 Presentation of the Main Research Material

To construct a mathematical model of the transfer of modulated ink flows from the plate to the printed material, the following assumptions are made: a mathematical model of the printing system of parallel structure is known, printing elements have a square shape and are described by a relative area, which is the main carrier of information about the image and corresponds to the degree of the plate coverage by printing elements, the image is a linear raster scale. Then, in general, the problem of modelling of the transfer of modulated ink flows from the plate to the printed material, which describe the tone transfer expressed by the ink amount per area unit, is:

$$V = S_k H_a, \quad if \ 0 \leq S_k \leq 1 \tag{1}$$

where S_k – is a relative area of square raster elements.

The ink thickness on the surface of the imprint raster elements is:

$$H_a = F(S_k, H_0) \tag{2}$$

where $F(*)$ – is a function that describes the dependency of the ink thickness on the raster imprints of a given printing system, H_0 – is the thickness of the ink flow at the input of the system model.

To solve this problem, one needs to develop a model of a short printing system and a raster model. Taking into account the novelty and complexity of the problem, one can consider a short printing system of parallel structure of the sixth dimension of the company KVA [1, 9], the scheme of which is presented in Fig. 1.

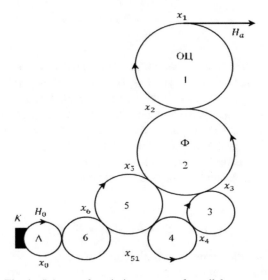

Fig. 1. Scheme of a printing system of parallel structure

The ink supply unit consists of a closed ink chamber K, in which the anilox roller A is installed. The ink fills the raster cells of the roller under pressure. The excess is rolled up with a squeegee, and the metered ink flow is fed to the system input and sequentially unrolled. The printing raster plate Φ modulates the ink flow, which is fed to the offset cylinder OЦ, and from it to the printed material. On the rollers 3 and 5 there is a layer of ink, which is not perceived by the blank elements of the plate, which causes unmaintained backflows, the part is returned back to the chamber, resulting in a decrease in the ink amount on the raster imprint.

The model of ink flow transfer is based on the balance equations of ink flows in the system under certain assumptions [2–4]: printing elements have a raster structure and are evenly distributed on the surface of the plate, the ink printing system is a low-pass filter, a constant thickness of the ink flow is fed to the input of the model, the output of the model is the amplitude value of the ink on the raster elements of the plate. Under these assumptions, the balance equation of ink flows can be replaced by the "thickness balance equation" at the points of contact of the ink rollers and the printing plate, which describe the ink flows for the steady operation. Then, on the basis of Fig. 1, one can write a system of thickness balance equations:

$$x_0 = H_0 + \Upsilon_6 x_6$$
$$l_0 = \Upsilon_0 x_0$$
$$x_6 = \alpha_6 x_0 + R_5 x_{51}$$
$$x_5 = \alpha_5 x_6 + \Upsilon_2 x_2$$
$$x_{51} = \alpha_5 x_5 + R_4 x_4 \tag{3}$$
$$x_4 = \alpha_2 x_5 + \Upsilon_3 x_4$$
$$x_2 = P_3 x_3 + \Upsilon_1 x_1$$
$$x_1 = P_1 x_1$$
$$V = \beta x_1,$$

where x_i – is the average value of the ink flow thickness at the points of contact of the ink rollers, the plate and the offset cylinders, H_0 – is the ink flow thickness at the input of the system model, H_a – is the amplitude value of the ink thickness at the output of the system (on the imprint), I_0 – is the flow thickness that returns back to the ink chamber, α_i, Y_i – are transmission coefficients of direct and back ink flows at the output of the contact points P_i, R_i – is the transfer of modulated and unmaintained flows by raster printing formula, P_1 – is the transfer of the model output.

The transfers of modulated and unmaintained flows are determined by expressions:

$$P_i = \alpha_i S_k$$
$$R_i = 1 - \Upsilon_i S_k \; if \; 0 \le S_k = S^2 \le 1, \tag{4}$$

where S – is a linear raster scale of a printing plate.

The transfer of the model output is:

$$P_1 = \frac{\beta}{S_k} \tag{5}$$

where β – is the ink transfer coefficient from the offset cylinder to the printed material.

If in expressions (3)–(5) one changes S_k according to the quadratic law within the given limits, for the set constant thickness of the ink flow at the model input, then it is possible to calculate and construct the dependency of ink thickness at the model input according to it. To simplify the solution of the problem a graph of modulated ink flows of the printing system of parallel structure is constructed according to the expression (3), shown in Fig. 2.

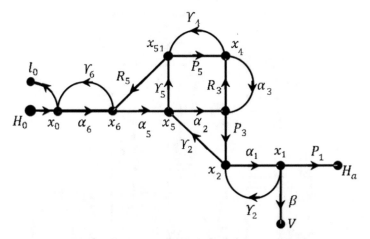

Fig. 2. Graph of modulated ink flows in printing system

The average values of thickness of the ink flows at the contact points of the plate and offset cylinders' rollers correspond to the vertices of the graph. The arcs of the graph are subordinated to the transfer coefficients of the direct and reverse ink flows at the output of contact points. Arrows on the arcs show the direction of the flows. Directly by the graph and on the basis of Maison formula [1, 2, 5] we determined the dependence of the amplitude value of the ink flow thickness at the system output.

$$H_a = \frac{\alpha_6\alpha_5\alpha_2 P_3\alpha_1 P_1(1-\Upsilon_4 P_5) + \alpha_6\alpha_5\Upsilon_5 P_5\alpha_3 P_3\alpha_1 P_1}{\Delta 6} H_0 \qquad (6)$$

The determinant of the graph characterizes the contour part and is defined directly by the graph

$$
\begin{aligned}
\Delta 6 = {} & 1 - \alpha_6\Upsilon_6 - \alpha_5\Upsilon_5 R_5 - \Upsilon_4 P_5 - \alpha_3 R_3 - \alpha_2\Upsilon_2 P_3 - \alpha_1\Upsilon_1 - \alpha_2\alpha_5 R_3\Upsilon_4 R_5 \\
& - \Upsilon_5 P_5\alpha_3 P_3\Upsilon_2 + \alpha_6\Upsilon_6(\Upsilon_4 P_5 + \alpha_3 R_3 + \alpha_2\Upsilon_2 P_3 + \alpha_1\Upsilon_1 + \Upsilon_5 P_5\alpha_3 P_3\Upsilon_2) \\
& + \alpha_5\Upsilon_5 R_5(\alpha_3 R_3 + \alpha_1\Upsilon_1) + \Upsilon_4 P_5(\alpha_2\Upsilon_2 P_3 + \alpha_1\Upsilon_1) + \alpha_3\Upsilon_3\alpha_1\Upsilon_1 \\
& - \alpha_6\Upsilon_6(\Upsilon_4 P_5\alpha_2\Upsilon_2 P_3 + \Upsilon_4 P_4\alpha_1\Upsilon_1 + \alpha_3 R_3\alpha_1\Upsilon_1) - \alpha_5\Upsilon_5 R_5\alpha_3 R_3\alpha_1\Upsilon_1
\end{aligned}
\qquad (7)
$$

Therefore, the amplitude value of the ink thickness at the system output is a nonlinear dependence (6) of the ink printing system parameters. Similarly, the graph can determine the amount of ink on the surface of the raster imprint, which describes the tone transfer of a short printing system. Object-oriented programming in MATLAB: Simulink [1] has been used to simplify the solution of the problem. A block diagram of the ink flows model of the printing system for square shaped raster elements has been developed on the basis of the expressions (3)–(5), the scheme of Fig. 1 and a graph, which is shown in Fig. 3.

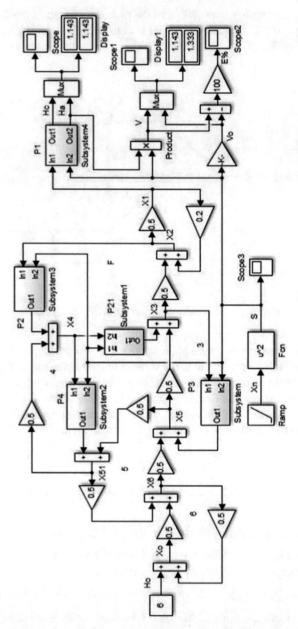

Fig. 3. A block diagram of the ink flows model of the printing system in Simulink

At the top of the figure there is a diagram of the ink flow model for printing system having a parallel structure of the sixth dimension. The arcs of the graph correspond to the Gain blocks in the dialog windows which specify the transfer coefficients α_i, Y_i Adders are subordinated to the vertices of the graph, the direct and inverse ink flows are fed and stacked at the input point, their thickness values x_i. Are obtained in the output point. The transfers P_i and R_i of modulated and unmodulated ink flows are implemented by Simulink by expression (4) and disguised in Subsystem blocks for compactness reason. At the bottom there is a Ramp block that generates a linear raster scale with a single tonal interval of $0 \leq S \leq 1$, which is fed to the input of the Fcn mathematical function block, which determines the relative area of the square-shaped raster elements ($S_k = S^2$) and feeds to the second inputs of subsystem blocks necessary for defining the transfer (4) of modulated and unmodulated ink flows. The Constant block specifies the thickness of the H_0 at the system input. Calculated in the subsystem block (7), the amplitude value of the thickness of H_a is visualized by Scope and Display blocks, and the ton transfer is expressed by the amount of ink per area unit by the Scope1 block. We have adjusted the parameters of the printing system model to the nominal transmission coefficients ($\alpha_i = Y_i = 0{,}5$; $\beta = 0{,}8$; $Y_i = 0{,}2$), set the thickness of the ink flow at the input point of the model $H_0 = 6$ microns and adjusted the parameters of subsystem blocks. The results of the simulation modelling of ink thickness dependence on the tone transfer interval are given in Fig. 4.

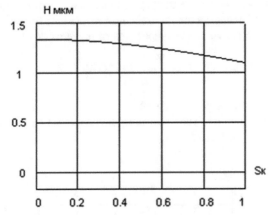

Fig. 4. Dependence of epy ink thickness on the tone transfer interval characteristics of the ink fcoating of the raster scale is a nonlinear curve

At the beginning of the interval, the thickness of the ink is 1,333, it gradually decreases and at the end it is 1,095 microns. The uneven thickness of the ink at the tone transfer interval is 21.4%, which does not meet the regulatory requirements for high-quality book and magazine products. For comparison, we have defined the tone transfer of the printing system under the condition of a sustainable thickness of $H = 1.333$ microns expressed by the amount of ink per area unit

$$V_0 = 1,333 * S_K \tag{8}$$

The results of simulation of tone transfer, expressed by the amount of ink per unit of the raster scale area are given in Fig. 5.

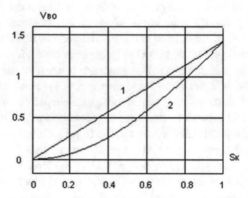

Fig. 5. Characteristics of tone transfer, expressed by the amount of ink per area unit

For comparison, the figure shows the characteristic 1 for the sustainable thickness of the ink H = 1.333 microns, which is located above. The characteristics of the tone transmission are the concave curves. At the beginning of the interval, the characteristics practically coincide, and at the end the amount of ink is 1,741 and 1,43 relative units. To assess the effect of changing the ink thickness on the tone transmission, it has been offered to determine the deviation of the amount of ink at a constant thickness

$$E = \frac{V - V_0}{V_{cp}} * 100\% \qquad (9)$$

where V_{av} - is average value of ink amount.

The results of simulation of tone transmission deviation at a constant ink thickness are shown in Figs. 2, 3, 4, 5 and 6.

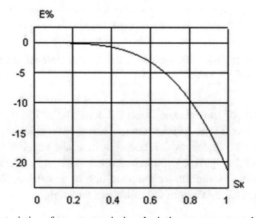

Fig. 6. Characteristics of tone transmission deviation at a constant thickness of ink

The characteristic of tone transmission deviation is a descending curve. Reducing the thickness of the ink at the tone transmission interval significantly reduces the amount of ink on the surface of the screen print, and at the end of the interval the deviation is 21.75%. Therefore, a short, sixth-dimensional printing system with parallel structure significantly illuminates the image at the middle range and especially in the shadow areas.

5   Discussion and Conclusion

A mathematical model of the modulated ink flows transfer from the plate to the printed material has been developed, which describes the tone transmission by the amount of ink per area unit of square shape raster elements for a short printing system with parallel structure of the sixth dimension. The graph of modulated ink flows of the short printing system, which in graphically reflects the structure of ink flows, direction and parameters of the printing system, allows you to directly determine the thickness of ink flows, system ink transfer and other variables, which is convenient for analysis.

The structural diagram of the model of modulated ink flows of the printing system in the Matlab: Simulink package, which makes it possible to simultaneously calculate the characteristics of the system, visualizes the process as the characteristics of thickness of image inking, tone transmission, deviations that are quite informative. The results of simulation modelling are given as ink characteristics that are nonlinear. It has been established that the uneven thickness of the ink at the tone transmission interval is 21.4%, which does not meet the regulatory requirements for high-quality book and magazine products. The short, six-dimensional parallel structure print system illuminates the image in medium to gravy tones that is a disadvantage.

References

1. Durnyak, B., Lutskiv, M., Petriaszwili, G., Shepita, P.: Analysis of raster imprints parameters on the basis of models and experimental research. Proceedings GRID **10**, 379–385 (2020)
2. Durnyak, B., Lutskiv, M., Shepita, P., Hunko, D., Savina, N.: Formation of linear characteristic of normalized raster transformation for rhombic elements. In: Intelligent Information Technologies & Systems of Information Security: CEUR Workshop Proceedings, vol. 2853, pp. 127–133 (2021)
3. Englund, C., Verikas, A.: Ink flow control by multiple models in an offset lithographic printing process. Comput. Ind. Eng. **55**(3), 592–605 (2008). https://doi.org/10.1016/j.cie.2008.01.019. ISSN 0360-8352
4. Friedrich, L., Begley, M.: Printing direction dependent microstructures in direct ink writing. Addit. Manuf. **34**, 101192 (2020). https://doi.org/10.1016/j.addma.2020.101192
5. Jurečić, D., Žiljak, V., Gršić, J.Ž, Rajković, I.: Near infrared spectrography of colorants for offset printing with individualized rasters on drug packaging. Acta Graphica **29**(4), 7–12 (2020)
6. Kusaka, Y., Fukuda, N., Ushijima, H.: Recent advances in reverse offset printing: an emerging process for high-resolution printed electronics. Jpn. J. Appl. Phys. **59**(SG), SG0802 (2020)
7. Kusaka, Y., Kanazawa, S., Ushijima, H.: Design rules for vertical interconnections by reverse offset printing. J. Micromech. Microeng. **28**(3), 035003 (2018)

8. Litunov, S.N., Gusak, E.N., Toshhakova, Y.D.: Numerical study of printing ink structuring. J. Phys. Conf. Ser. **1050**(1), 012045 (2018)
9. Moreira, A., Silva, F.J.G., Correia, A.I., Pereira, T., Ferreira, L.P., De Almeida, F.: Cost reduction and quality improvements in the printing industry. Procedia Manuf. **17**, 623–630 (2018)
10. Nikolov, A., Murad, S., Wasan, D., Wu, P.: How the capillarity and ink-air flow govern the performance of a fountain pen. J. Colloid Interface Sci. **578**, 660–667 (2020)
11. Potts, S.-J., Phillips, C., Jewell, E., Clifford, B., Lau, Y.C., Claypole, T.: High-speed imaging the effect of snap-off distance and squeegee speed on the ink transfer mechanism of screen-printed carbon pastes. J. Coat. Technol. Res. **17**(2), 447–459 (2019)
12. Senkivskyy, V., Babichev, S., Durnyak, B., Zhydetskyy, V., Pikh, I.: Application of optics density-based clustering algorithm using inductive methods of complex system analysis. In: 14th International Scientific and Technical Conference on Computer Sciences and Information Technologies, vol. 1, pp. 169–172. CSIT, Lviv, Ukraine (2019)
13. Shin, S., Kim, S., Cho, Y.T.: A study on the selection of highly flexible blanket for reverse offset printing. J. Korean Soc. Manuf. Process Eng. **20**(5), 121–127 (2021)
14. Takeda, Y., et al.: Organic complementary inverter circuits fabricated with reverse offset printing. Adv. Electron. Mater. **4**(1), 1700313 (2018)
15. Tao, R., et al.: Capillary force induced air film for self-aligned short channel: pushing the limits of inkjet printing. Soft Matter **14**(46), 9402–9410 (2018)
16. Pasyeka, M., Sheketa, V., Pasieka, N., Chupakhina, S., Dronyuk, I.: System analysis of caching requests on network computing nodes. In: 2019 3rd International Conference on Advanced Information and Communications Technologies (AICT), pp. 216–222. Lviv, Ukraine (2019). https://doi.org/10.1109/AIACT.2019.8847909
17. Pasyeka, M., Sheketa, V., Pasieka, N., Chupakhina, S., Dronyuk, I.: System analysis of caching requests on network computing nodes. In: 3rd International Conference on Advanced Information and Communications Technologies, pp. 216–222. AICT (2019)
18. Romanyshyn, Y., Sheketa, V., Pikh, V., Poteriailo, L., Kalambet, Y., Pasieka, N.: Social-communication web technologies in the higher education as means of knowledge. In: 14th International Scientific and Technical Conference on Computer Sciences and Information Technologies, CSIT, pp. 35–38 (2019). https://doi.org/10.1109/STCCSIT.2019.8929753
19. Varepo, L.G., Yu Brazhnikov, A., Volinsky, A.A., Nagornova, I.V., Kondratov, A.P.: Control of the offset printing image quality indices. J. Phys.: Conf. Ser. **858**, 012038 (2017)
20. Wu, Y., Chiu, G.: Modeling height profile for drop-on-demand print of UV curable ink. In: Dynamic Systems and Control Conference, vol. 59155, pp. V002T13A006 (2019)
21. Kaur, S.: An automatic number plate recognition system under image processing. Int. J. Intell. Syst. Appl. **8**(3), 14–25 (2016). https://doi.org/10.5815/ijisa.2016.03.02
22. Tripathy, B.K., Dahiya, R.: Fuzzy clustering of sequential data. Int. J. Intell. Syst. Appl. (IJISA) **11**(1), 43–54 (2019). https://doi.org/10.5815/ijisa.2019.01.05
23. Sun, F., Hea, C.N., Yong, W.S.: Combining multi-feature regions for fine-grained image recognition. Int. J. Image Graphics and Signal Processing (IJIGSP) **14**(1), 15–25 (2022). https://doi.org/10.5815/ijigsp.2022.01.02

Credit Strategy Model of Small, Medium and Micro Enterprises Based on Principal Component Analysis

Bingchan Fan[1] and Shan Wu[1,2(✉)]

[1] Public Course Department, Wuhan Technology and Business University, Wuhan 430065, China
fanbc_sun@yeah.net, hariny@163.com
[2] School of Management, Huazhong University of Science and Technology, Wuhan 430074, China

Abstract. This paper mainly studies the bank credit decisions for small, medium and micro enterprises. Four indicators affecting the loan ability of enterprises are selected from four aspects: cash flow input ratio, transaction failure rate, credibility and VAT. Three principal components of credit risk assessment are obtained based on principal component analysis, and a comprehensive evaluation model of credit risk is established. The credit risk values of 123 enterprises are obtained and ranked. When the total annual credit of the bank is fixed, the turnover rate of enterprises regarded as the loss income. Taking the minimum risk function and the maximum income function of the bank as the objective function, a multi-objective nonlinear programming model is established. By introducing the risk threshold and taking the risk function as the constraint condition, the multi-objective problem is transformed into a single objective nonlinear programming problem. When the bank's loan amount is 50 million, this paper gives the loan strategy of whether to give loans to various enterprises, loan amount and annual interest rate. The results show that only a few enterprises can lend high amount of loans, and even if most enterprises can lend, the amount will not be very high.

Keywords: Principal component analysis · Credit risk · Multi-objective nonlinear programming · Risk threshold

1 Introduction

In the operation process of an enterprise, it borrowings money from banks or other financial institutions in accordance with the specified interest rate and term to meet the needs of production and operation, such behavior is called "credit". Under normal circumstances, the bank will provide loans to enterprises with strong strength and stable loan relationship, and give certain interest rate preference to enterprises with high reputation and low loan risk. But in practice, small and medium-sized enterprises are relatively small in scale and weak in risk resistance. Therefore, banks will evaluate the credit risk of the enterprises according to their strength and reputation, and then decide whether to lend, loan amount, term and interest rate and other credit strategies according to the credit risk and other factors.

The amount of a bank to the lending enterprise is 100–1 million yuan, the annual interest rate is 4–15%, the loan term is 1 year. Based on the bill information of 123 enterprises with credit records from 2017 to 2020 [1], this paper establishes a mathematical model to carry out quantitative analysis of credit risks. When the total annual credit of banks is fixed, the credit strategies of each enterprise are given.

2 Problem Analysis

In this paper, four indexes are selected from four aspects of enterprise's solvency, supply and demand relationship, credibility and profitability, and principal component analysis method is adopted to construct a comprehensive evaluation model of bank's credit to enterprises by the contribution rate and weight of principal components. In this paper, a multi-objective nonlinear programming model is established based on the risk and profit factors. In order to facilitate the processing, the boundary value of risk is introduced, the risk function is taken as the constraint condition, and the single objective linear programming problem is obtained, then the equivalent model of credit strategy can be obtained, so as to determine whether the bank loans to each enterprise and the amount of loan.

3 Model Establishment and Solution

3.1 Establish Credit Risk Indicators

In practice, due to the relatively small scale of small, medium and micro enterprises and the lack of collateral assets, banks usually evaluate their credit risks according to their strength and reputation, and then determine credit strategies based on credit risks and other factors. This paper starts from four aspects of enterprise's repayment ability, supply and demand relationship, credibility and profitability, and selects four indicators that affect enterprise's loan ability: cash flow input ratio, transaction failure rate, credibility and VAT [2–4].

(1) Cash Flow Input Ratio

The bank's credit decisions for small, medium and micro enterprises are closely related to the solvency of enterprises, while cash flow is the inflow and outflow of cash generated in certain economic activities of enterprises, and is a direct manifestation of the solvency of enterprises. The current cash flow is mainly used to purchase products and repay debts. So the cash flow input ratio of enterprises is:

$$C_i = \frac{m_i - n_i}{n_i}, (i = 1, 2, \cdots, 123) \tag{1}$$

Where, C_i is the cash flow input ratio of the enterprise, m_i is the total output price tax of the enterprise, and n_i is the total input price tax of the enterprise.

(2) Transaction Failure Rate

A stable supply-demand relationship is an embodiment of the strength of an enterprise, to some extent, the more stable supply-demand relationship, the stronger the strength of an enterprise. The supply-demand relationship can be reflected by the transaction failure rate of an enterprise.

a. *Failure Rate of Supply Transactions*

Assume that the total number of invalid invoices, negative invoices and valid invoices in the sales invoice of the enterprise are a_i, b_i, c_i, the transaction failure rate of supply can be expressed as

$$G_i = \frac{a_i + b_i}{c_i + a_i}, (i = 1, 2, \cdots, 123) \tag{2}$$

b. *Failure Rate of Demand Transactions*

Assume that the total number of invalid invoices, negative invoices and valid invoices in the input invoice of the enterprise are a_i', b_i', c_i', the transaction failure rate of supply can be expressed as

$$G_i' = \frac{a_i' + b_i'}{c_i' + a_i'}, (i = 1, 2, \cdots, 123) \tag{3}$$

The total transaction failure rate is:

$$G_{总} = G_i + G_i', (i = 1, 2, \cdots, 123) \tag{4}$$

(3) Credibility

Credibility is a non-financial indicator that reflects whether an enterprise can repay the principal and interest after the loan term expires. The higher the credibility, the smaller the credit risk of the enterprise, and the lower the credibility, the higher the credit risk of the enterprise.

a. *The Impact of Default on Credit Rating*

Since some enterprises with the same credit rating may or may not default, the enterprises with the same credit rating will be downgraded as follows: When the credit rating is "A", "B" and "C", the enterprises with the record of default will be downgraded by one level, namely "B", "C" and "D". When the credit rating is the same as "D", because the bank does not lend to the credit rating of "D" enterprises in principle, so such enterprises will not be downgraded.

b. *Quantification of Credit Rating*

Convert the enterprise's credit rating (A, B, C, D) into the corresponding value (4, 3, 2, 1). According to the principle of maximum membership, adopt the larger membership function:

$$\mu(x) = \begin{cases} 0, & 0 \leq x \leq a \\ \frac{1}{2} + \sin\frac{\pi}{b-a}\left(x - \frac{a+b}{2}\right), & a < x < b \\ 1, & x \geq b \end{cases} \tag{5}$$

So the corresponding quantitative value of credit rating (A, B, C, D) is (1, 0.717, 0.283, 0).

(4) **VAT**

The profitability of an enterprise can be measured by the amount of value-added tax (VAT) paid by the enterprise. Then the VAT payable by the enterprise is:

$$ZZS_i = XX_i - JX_i, (i = 1, 2, \cdots, 123) \tag{6}$$

where, XX_i, JX_i, ZZS_i are the total output tax, input tax and VAT of the enterprise.

3.2 Credit Risk Model

The credit risk is mainly obtained by integrating the four indexes of cash flow input ratio, transaction failure rate, credit degree and value-added tax. Principal component analysis method is adopted to calculate the credit risk value of banks to each enterprise, so as to determine the credit risk situation of 123 enterprises [5–9].

Step1. Processing data

Among the four indicators, the transaction failure rate is a very small indicator, while the other three indicators are very large indicators. Now, the transaction failure rate is transformed into a very large indicator. Since the dimensions and magnitude of the four indicators are different, they need to be standardized. Note that the j indicator of the enterprise is a_{ij}, and adopt standard deviation standardization, that is:

$$\tilde{a}_{ij} = \frac{a_{ij} - \mu_j}{s_j}, (i = 1, 2, \cdots, 123; j = 1, 2, 3, 4) \tag{7}$$

where, μ_j, s_j is the sample mean and sample standard deviation of the index j.

Step 2. Calculate the correlation coefficient matrix between the four indicators

As can be seen from Table 1, there is a strong correlation between VAT and cash flow inflow ratio.

Table 1. Correlation coefficient matrix of credit risk indicators

	VAT	Cash flow input ratio	Credibility	Transaction failure rate
VAT	1	0.978290758	−0.054900053	−0.043775903
Cash flow input ratio	0.978290758	1	−0.04814959	−0.000277125
Credibility	−0.054900053	−0.04814959	1	0.095073646
Transaction failure rate	−0.043775903	−0.000277125	0.095073646	1

Step3. Calculate the credit risk value of each enterprise

The contribution rate and cumulative contribution rate are determined by the eigenvalues of the correlation coefficient matrix of credit risk indicators, as shown in Table 2:

Table 2. Principal component analysis results

Number	Contribution rate	Cumulative contribution rate
1	49.6294	49.6294
2	27.2405	76.8699
3	22.6116	99.4815
4	0.5185	100.0000

It can be seen that the cumulative contribution rate of the first three principal components reaches 99%, and the effect of principal component analysis is very good. The first three principal components are selected for comprehensive evaluation of credit risk. With the contribution rate of the three principal components as the weight, the comprehensive credit evaluation model of the bank to the enterprise is constructed as follows:

$$Z = 0.4963y_1 + 0.2724y_2 + 0.2262y_3 \tag{8}$$

For the convenience of subsequent description, Z in the comprehensive credit evaluation model is transformed into credit risk value Z1:

$$Z1 = \max_i\{Z_i\} - Z_i \tag{9}$$

At this time, the smaller the value of Z1, the smaller the credit risk of the enterprise. The credit risk value and comprehensive ranking of 123 enterprises can be obtained.

3.3 Enterprise Credit Strategy Model

From the perspective of investment, banks hope that the lower the risk, the better the return, so this paper from the two perspectives of income and risk to determine whether to lend to enterprises, the amount of loan, interest rate and other credit strategies are discussed.

3.3.1 Model Preparation

The loan amount given by the bank to the enterprise is d_i, the interest rate is l_i, the enterprise loss rate is s_i, the risk rate is f_i, and the bank's loan situation is:

$$y_i = \begin{cases} 0, & not\ lending \\ 1, & lending \end{cases}, (i = 1, 2, \cdots, 123), where, f_i = \frac{Z_i'}{\max(Z_i')} \quad (10)$$

Different credit rating of the enterprise, enterprise turnover rate is different. According to the data in Annex 3, enterprise turnover rate is related to credit rating and bank annual interest rate. The turnover rate of enterprises with different credit ratings can be obtained by polynomial fitting between the turnover rate of enterprises and the interest rate, i.e.

$$S_A(l_i) = -76.41l_i^2 + 21.98l_i - 0.70$$
$$S_B(l_i) = -67.93l_i^2 + 20.21l_i - 0.65 \quad (11)$$
$$S_C(l_i) = 504.72l_i^3 - 207.39l_i^2 - 0.97$$

So, the turnover rate of the enterprise is

$$S_i = \begin{cases} S_A, a_{3i} = a \\ S_B, a_{3i} = b \\ S_C, a_{3i} = c \end{cases} \quad (12)$$

where, a_{3i} is the credibility of the enterprise, a, b, c are the quantified values of credit ratings A, B, C.

3.3.2 Model Building

When lending to enterprises, banks consider both risks and benefits comprehensively. When the total amount of annual credit is fixed, banks lend to enterprises selectively so as to maximize net income and minimize overall risks. This is a multi-objective nonlinear programming problem [10–13].

(1) Risk function: The overall risk function is measured by the biggest risk among all enterprises, i.e.

$$\min \max\{f_i d_i y_i | i = 1, 2, \cdots, 123\} \quad (13)$$

In actual lending, different banks bear different degrees of risk. This paper gives a risk threshold value λ. In principle, banks will not lend to enterprises with a credit rating of "D" and take the minimum risk rate of all enterprises with a credit rating of "D" as the threshold value. In other words, $f_i d_i y_i \leq \lambda \cdot M$, $\lambda = 0.5143$, where, M is the total annual credit of the bank.

(2) Income function: The bank's income mainly comes from the interest earned on loans, but the enterprise turnover rate is positively correlated with the loan profit. In this case, the enterprise turnover rate can be regarded as the profit loss rate, and the income function is:

$$\max \sum_{i=1}^{123} (l_i - S_i) d_i y_i \tag{14}$$

Since it is a multi-objective nonlinear programming problem, this kind of problem is difficult to deal with. Now, a single-objective linear programming problem can be obtained by taking the risk function as the constraint condition through the threshold value of risk [14, 15]. Then, the credit strategy model is as follows.

$$\max \sum_{i=1}^{123} (l_i - S_i) d_i y_i$$

$$s.t. \begin{cases} 10 \le d_i y_i \le 100 \\ 0.04 \le l_i \le 0.15 \\ \sum_{i=1}^{123} d_i y_i = M \\ f_i d_i y_i \le \lambda \cdot M \end{cases} \tag{15}$$

Since the enterprise turnover rate is a discrete piece-wise function, its equivalent model is:

$$\begin{cases} S_i = S_{i1} + S_{i2} + S_{i3} \\ b_{i1} + b_{i2} + b_{i3} = 1, b_{ij} = 0 \, or \, b_{ij} = 1 (j = 1, 2, 3) \\ a - a_{3i} \le 2M(1 - b_{i1}), a_{3i} - a \le 2M(1 - b_{i1}) \\ b - a_{3i} \le 2M(1 - b_{i2}), a_{3i} - b \le 2M(1 - b_{i2}) \\ c - a_{3i} \le 2M(1 - b_{i3}), a_{3i} - c \le 2M(1 - b_{i3}) \\ -2Mb_{i1} \le S_{i1} \le 2Mb_{i1}, S_A - 2M(1 - b_{i1}) \le S_{i1} \le S_A + 2M(1 - b_{i1}) \\ -2Mb_{i2} \le S_{i2} \le 2Mb_{i2}, S_B - 2M(1 - b_{i2}) \le S_{i2} \le S_B + 2M(1 - b_{i2}) \\ -2Mb_{i3} \le S_{i3} \le 2Mb_{i3}, S_C - 2M(1 - b_{i3}) \le S_{i3} \le S_C + 2M(1 - b_{i3}) \end{cases} \tag{16}$$

The bank's credit decision-making strategy model for enterprises is

$$\max \sum_{i=1}^{123} (l_i - S_{i1} - S_{i2} - S_{i3}) d_i y_i \tag{17}$$

The model is subject to constraints, as follows.

$$\text{s.t.} \begin{cases} 10 \le d_i y_i \le 100, 0.04 \le l_i \le 0.15 \\ \sum_{i=1}^{123} d_i y_i = M, f_i d_i y_i \le \lambda \cdot M \\ b_{i1} + b_{i2} + b_{i3} = 1, b_{ij} = 0 \text{ or } b_{ij} = 1 (j = 1, 2, 3) \\ a - a_{3i} \le 2M(1 - b_{i1}), a_{3i} - a \le 2M(1 - b_{i1}) \\ b - a_{3i} \le 2M(1 - b_{i2}), a_{3i} - b \le 2M(1 - b_{i2}) \\ c - a_{3i} \le 2M(1 - b_{i3}), a_{3i} - c \le 2M(1 - b_{i3}) \\ -2Mb_{i1} \le S_{i1} \le 2Mb_{i1}, S_A - 2M(1 - b_{i1}) \le S_{i1} \le S_A + 2M(1 - b_{i1}) \\ -2Mb_{i2} \le S_{i2} \le 2Mb_{i2}, S_B - 2M(1 - b_{i2}) \le S_{i2} \le S_B + 2M(1 - b_{i2}) \\ -2Mb_{i3} \le S_{i3} \le 2Mb_{i3}, S_C - 2M(1 - b_{i3}) \le S_{i3} \le S_C + 2M(1 - b_{i3}) \end{cases} \quad (18)$$

3.3.3 Model Solution

When the loan amount is 50 million, the loan amount and annual interest rate of each enterprise can be solved by lingo. 95 of the 123 enterprises are lent, of which E4, E9 and E7 have the highest loan amount; E28, e72 and E90 have the lowest loan amount. The highest annual interest rate is E3 and the lowest is E110. The data of enterprises being lent is drawn into a statistical graph, as shown in Fig. 1.

As can be seen from Fig. 1, the loan amount of enterprises is mainly distributed in the range of [40,100], of which the loan amount of most enterprises is in the range of [40,70], and only a few enterprises are in the range of [70,100].

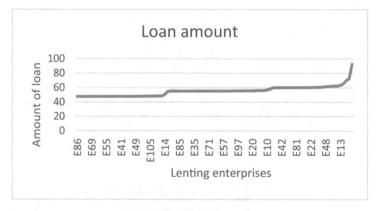

Fig. 1. Corporate loan quota

It can be seen from Table 3, 22.8% of enterprises have not been lent, 30.9% can only lend 500,000–600,000, and only 2.4% can lend more than 700,000. This means that very few companies can borrow at high levels, and most, if at all, do not borrow at high levels.

Table 3. Distribution of enterprise loan amount

Loan amount	0	[40,50)	[50,60)	[60,70)	[70,80)	[80,90)	[90,100]
Enterprise number	28	30	38	24	2	0	1
Ratio	22.8%	24.4%	30.9%	19.5%	1.6%	0%	0.8%

4 Summary and Conclusion

This paper selects four indexes that affect the loan ability of enterprises: cash flow input ratio, transaction failure rate, credibility and VAT. Through principal component analysis, a comprehensive evaluation model of credit risk is established. In this paper, a multi-objective nonlinear programming model is established based on the risk and profit factors. The results show that only a few enterprises can borrow high amount of loans, most enterprises can borrow loans, the amount is not very high.

Acknowledgment. This project is supported by the University-level Scientific Research Projects of China Wuhan Technology and Business University (Grant No. A2018018).

References

1. 2020 Higher Education Society Cup National College Students' mathematical modeling competition [EB/OL].: http://www.mcm.edu.cn/html_cn/node/10405905647c52abfd6377c0311632b5.html
2. Zhou, Z.: On the problems and countermeasures of credit risk management of small and medium-sized enterprises. Finance (Academic Edition) **18**, 57–58 (2017). (in Chinese)
3. Si, S., Sun, Z.: Mathematical Modeling Algorithm and Application. National Defense Industry Press, Beijing (2017). (in Chinese)
4. Chen, F.: Innovative strategies of credit risk management for SMEs in China's commercial banks. Hebei Enterprise (5), 35–36 (2019). (in Chinese)
5. Javed, A.: Face recognition based on principal component analysis. Int. J. Image Graph. Signal Process. **2**, 38–44 (2013)
6. Liu, Y., Chen, H., Ren, H.: Research on supply chain financing model and risk management of SMEs. Econ. Issues (5) (2016)
7. Zhang, A.-L., Tang, H.: Solution and application of linear programming in multiple linear regression. J. Kunming University Sci. Technol. (Natural Science Edition) (1) (2014). (in Chinese)
8. Zhang, J.: DEA method and empirical analysis on efficiency of commercial banks in China from 1997 to 2001. Financ. Res. (3), 11–25 (2003). (in Chinese)
9. Barde, S., Zadgaonkar, A.S., Sinha, G.R.: PCA based multimodal biometrics using ear and face modalities. Int. J. Inf. Technol. Comput. Sci. (5), 43–49 (2014)
10. Zheng, L.M.: Empirical Research on Corporate Credit Risk Assessment of China's Commercial Banks. Shandong University (2018). (in Chinese)
11. Erfanian, H.R., Abdi, M.J., Kahrizi, S.: Solving a linear programming with fuzzy constraint and objective coefficients. Int. J. Intell. Syst. Appl. **7**, 65–72 (2016)
12. Jiang, Q., Xie, J., Ye, J.: Mathematical Models, 4th edn. Higher Education Press, Beijing (2011). (in Chinese)

13. Narwal, S., Kaur, D.: Comparison between minutiae based and pattern based algorithm of fingerprint image. Int. J. Inf. Eng. Electr. Bus. **2**, 23–29 (2016)
14. Mahbub, M.S., Ahmed, S.S., Ali, K.I., Imam, M.T.: A multi-objective optimization approach for solving AUST class timetable problem considering hard and soft constraints. Int. J. Math. Sci. Comput. **5**, 1–14 (2020)
15. Wang, L., Chen, Y.: Diversity based on entropy: a novel evaluation criterion in multi-objective optimization algorithm. Int. J. Intell. Syst. Appl. **10**, 113–124 (2012)

Modelling of Tone Reproduction with Round Raster Elements in a Short Printing System of Parallel Structure

Bohdan Durnyak[1], Mikola Lutskiv[1], Petro Shepita[1], Roman Karpyn[1], Vasyl Sheketa[2], and Mykola Pasieka[2(✉)]

[1] Ukrainian Academy of Printing, Pid Goloskom str. 19, Lviv 7920, Ukraine
`durnyak@uad.lviv.ua`, `lutolen@i.ua`
[2] Ivano-Frankivsk National Technical University of Oil and Gas, Ivano-Frankivsk, Ukraine
`pms.mykola@gmail.com`

Abstract. A mathematical model of two-parameter determination of the optical density of an imprint raster scale with round-shaped elements in a short printing system of parallel structure of the sixth dimension has been developed. The block diagram of the model and the simulator for determining the thickness of the ink layer on the imprint and the optical density for the natural characteristics of rasterization depending on the tone transfer interval have been constructed. The results of the simulation have been presented in the form of characteristics of the scale coating with ink, the optical density and its deviation from the linear one. It has been established that the reduction of the ink thickness on the imprint surface significantly affects the uniformity of the optical density, which must be taken into account when choosing the rule of tone reproduction, in particular in the editorial tone transfer.

Keywords: Model · Tone reproduction · Optical density · Printing system · Block diagram · Graph · Characteristics · Analysis · Properties

1 Introduction

1.1 Problem Statement

Western companies have begun to use simple in design short printing appliance with an anilox block of serve of inks. They do not have mechanisms of adjusting of area of serve of inks, does not give therefore the uniform ink there is a thickness on-the-spot imprint, that limits their use for the seal of high-quality books and magazines [4, 13]. Most patented devices for the short printing systems with the block of serve of aniloksovoy paint not metallic, and only some of them began to be used in the offset printing, in particular in newspaper blocks, to that experience of production operation, adjusting, not enough. And synthesis of transmission of tone [13]. Organization of tone recreation of semitone images for the permanent thickness of paint on an imprint, types of transmission of tone, chart of recreation of tone, the algorithms of synthesis are known

Z. Hu et al. (Eds.): ICCSEEA 2022, LNDECT 134, pp. 37–46, 2022.
https://doi.org/10.1007/978-3-031-04812-8_4

and lighted up in separate publications [1, 3, 9, 18]. Usually in the analysis and synthesis of raster tone transfer, it is assumed that the ink thickness on the surface of a raster imprint is constant and does not depend on the tone transfer interval. Under such conditions, the tone transfer synthesis is reduced to the synthesis of raster elements areas [1, 3, 9]. Systems of automatic zone ink supply to a given circulation are used to ensure the constant ink thickness on the imprint surface [6, 8].

Since traditional methods of tone reproduction synthesis are one-parameter, they cannot be directly applied to the systems of short seal, in which the thickness of paint on the interval of transmission of tone can be diminished to 30% [5, 6]. Therefore, the problem of tone reproduction modelling with round raster elements in short printing systems is relevant.

1.2 Literature Review

An ink device with an anilox inks supply unit is a new class of ink devices of simplified design. The processes that take place in them are complex, due to the circulation direct and reverse streams of inks, modulation of flows by a raster printing plate, so their analysis is very different from traditional objects and image processing systems. The analysis of the ink thickness on the raster imprint has revealed that it depends on the tone transfer interval and it is 20–30% or more, which does not meet the regulatory requirements for the quality of books and magazines [11, 12]. In these publications [4, 5], models of inking of raster elements of various forms which characteristics are S-shaped curves are constructed. Maximal deviation is from linear one is on the middle tones and it is 25%. In the works of the authors [7, 8, 17] the optical density of certainly linear raster scale for the system of short printing. Simulations of normalized raster transformation for round raster elements are presented in [7]. The work [5] proposes formulas for determining the visual density of an ink layers on an imprint, provided that the maximum value of the optical density is known, but it is not related to the area of raster elements.

2 Presentation of the Research Main Material

For the decision of this problems, it is necessary to have two basic models: the model of a short printing system [4] and the model of normalized raster transformation for round elements [7]. The model of the printing plate should describe the dependency of the ink thickness on the imprint surface in relative units on the tone transfer interval, provided that the printing plate has a scale that reproduces the model of normalized raster transformation. Then the two-parameter description of tone reproduction given by the ink amount in relative units will be described by the expression:

$$V = PH, \; if \; 0 \le P \le 1, \; H = \frac{H_a}{H_M} \tag{1}$$

where P is the degree of coverage of the scale with the ink corresponding to the relative area of the normalized raster scale, H_a is value of amplitude of thickness of inks on the raster elements of imprints on the interval of transmission of tone, H_M is the maximum value of thickness of paint is in the initial point of interval.

To determine the visual optical density of the scale image depending on the relative area, we will determine the formula of demodulation (de-rasterization) of Yula-Nichols by analogy [1, 3]:

$$D_{пл} = -nlg[V * 10^{-D_{пл}/n} + (1 - V) * D^{-D_n/n}] \quad (2)$$

where $D_{пл}$ is absorbency of matrix, got on an imprint from the system of short printing, $D_п$ is the optical density of the paper, n is the Yula-Nichols index which depends on a raster line and can be scope $(1.3 \leq n \leq 3.0)$ [1]. To improve the accuracy and quality of calculations in difficult conditions of printing production, the specific value of the optical density of printed prints can be measured with a densitometer.

As an example of solving this problem, we will consider a short printing system of parallel structure of the sixth dimension, the block diagram of which is presented in Fig. 1.

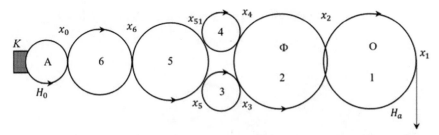

Fig. 1. Block diagram of a short printing system of parallel structure

Anilox roller A is installed in a closed chamber K, the paint, which is under pressure due to hydraulic or pneumatic system, fills the porous cells of anilox (raster cylinder), using a squeegee collects excess ink and the sixth roller is fed a clearly defined amount of ink. Shape the third and fourth cylinders. Modulated with a raster-type printing plate, the ink flow F is transferred to the offset cylinder O, and from there to the printed material. Ink residues that are not transferred to the paper or printed material create backflows of ink on the rollers of printing equipment, an effect of the circulation of backflows of ink that interact with the direct and partially return back to the black tank.

To implement the construction of the printing system model, we assume the following assumption: the peculiarities of operating modes are taken into account, uniform ink flow is fed to the system input, and the printing form used in the printing machine is normalized. At the input of the model, the printing process takes place under stable conditions. With the accepted principles on the basis of the known relations [4, 6, 10] according to the block diagram of Fig. 1, we make a system of equations of the balance of the flow supply and separation in the areas of contact of the rollers and the plate expressed by the ink thickness for stable conditions of the printing system:

$$
\begin{aligned}
x_0 &= H_0 + \Upsilon_6 x_6 & x_4 &= \alpha_4 x_{51} + P_2 x_2 \\
l_0 &= \Upsilon_0 x_0 & x_3 &= \alpha_3 x_5 + P_{21} x_4 \\
x_6 &= \alpha_6 x_0 + \Upsilon_5 x_{51} & x_2 &= \alpha_2 x_3 + \Upsilon_1 x_1 \\
x_5 &= \alpha_5 x_6 + P_3 x_3 & x_1 &= \alpha_1 x_2 \\
x_{51} &= \alpha_5 x_5 + P_4 x_4 & H_a &= P_1 x_1
\end{aligned}
\tag{3}
$$

where x_i is averaging value of thickness of the modulated stream of paint in points where to the contact of offset cylinders, offset plate and rollers that carry and roll a paint, H_0 is a thickness of stream of inks is on included in a model H_a is a value of amplitude of thickness of inks is on the output of the system (on an imprint), l_0 is the thickness of the flow that returns back to the ink chamber, α_i, Y_i is the transmission coefficient of direct and reverse ink flows at the entrance from contact points, P_{21}, P_2, P_3, P_4 there are transmissions of the modulated and reverse streams, P_1 is the transmission of output of model. The transmissions of flows are defined by the expressions (Fig. 2):

$$
P_{21} = \alpha_2 P; \; P_2 = \alpha_3 P; \; P_3 = 1 - \alpha_3 P; \; P_4 = 1 - \alpha_4 P; \; P_1 = \frac{B}{P} \; if \; 0 \le P \le 1 \tag{4}
$$

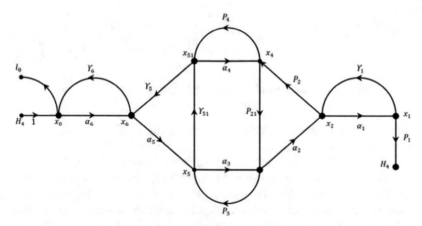

Fig. 2. Scheme of ink flows of the printing system

The vertices of the graph denoted by xi correspond to the average values of the thickness of the flows at the contact points of the rollers, plate and offset cylinders. The arcs of the graph are denoted by αi, Υi and are subject to the transmission coefficients of the direct and reverse ink flows on the output of points of contact. Pointers are on arcs show the direction of flow. By expression (1) and directly by the graph based on Meson formula [6, 7] we determine the two-parameter tone reproduction represented by the ink amount in relative units:

$$
V = \frac{P[\alpha_6 \alpha_5 \alpha_3 \alpha_2 \alpha_1 P_1 (1 + \alpha_4 + P_4) + \alpha_6 \alpha_5 \Upsilon_{51} \alpha_4 P_{21} \alpha_2 \alpha_1 P_1] H_0}{\Delta_6}, \tag{5}
$$

where Δ_6 is therefore determinant of chart, which is bulky, it is not indicated.

Having the ink amount on the scale surface, the expression (2) can determine the visual optical density of the scale that describes the tone reproduction for the short print category system.

The mathematical model of the normalized raster transformation presented in [7, 14] is a double function with the area definition of a closed unit raster square with relative unit constant dimensions. A round-shaped raster element is located in the center of the raster grid. During the process of rasterization, its geometric dimensions change, which are shown by the radius of the circle, the length of which varies from zero to X_H and so, the area, which carries information during rasterization and corresponds to the number of levels of the gray original, change. Gradation conversion function for the first range is the following:

$$S1 = \pi X_H^2, \; if \; 0 \leq X_H \leq 0.5 \tag{6}$$

where X_H – argument (normalized spatial variable), 0.5 half of the side of a unit square.

With a further increase of the radius of the raster element, it loses the shape of a circle, its surface is gradually limited to a unit square, and the radius of the circle approaches its maximum value $X_M = 0.5\sqrt{2} = 0.707$. Then, the surface area is determined by the expression [7]

$$S1 = \pi X_H^2 - 4X_H^2 \sqrt{\frac{X_H^2}{0.5^2} - 1} + 4 * 0.5^2 \sqrt{\frac{X_H^2}{0.5^2} - 1}, \; if \; 0.5 \leq X_H \leq 0.5\sqrt{2} = 0.707 \tag{7}$$

Then, the function of the normalized raster conversion is the following

$$S = S1 + S2 \tag{8}$$

If there are no distortions (dot spread of raster elements) during lighting, manufacturing of the plate, and printing, then, the relative normalized area of the raster element (8) corresponds to the degree of coverage of printing elements with ink on the surface of the imprint (S = P).

To facilitate the tasks in the Matlab application package, a visual programming and modeling environment is used Simulink [2]. On the basis of expressions (1)–(8) and the graph, the structural scheme of the simulator model for defining tone reproduction with a round-shape raster element a parallel structure was developed for short printing systems, it is represented in Fig. 3.

At the top level is a model of various printing systems. Operation units Gain correspond to the arcs of the graph with the transmission ratio α_i, Υ_i. The summators are subordinated to the vertices of the graph, the values of the forward and backward flows of ink are supplied to their inputs. Transfers of modulated and unmaintained ink flows are carried out by Simulink means by expressions (4) which are located in a hidden way in Subsystem units for compactness. At the output of the model, there is an amplitude value of the ink flow thickness H_a and its average value H_c, which are visualized by Scope and Display units. In the dialog box of the mathematical functions Fcn unit according to expression (2) there is a recorded program for calculating the characteristics of tone reproduction (optical density) based on the relative amount of ink V fed to the input unit, which is visualized by Scope1 and Display1 units and its deviation from the linear one by Scope2 unit.

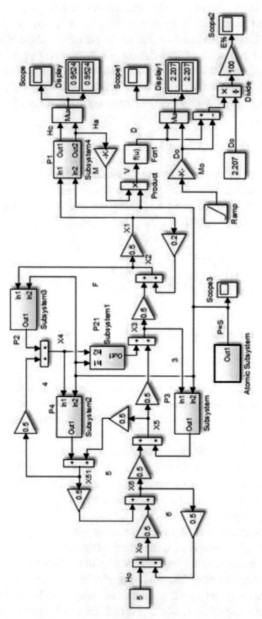

Fig. 3. Structural scheme of the model for defining tone reproduction in a short printing system

In the lower part, there is a scheme of the model of normalized raster conversion, designed on the basis of expressions (6)–(8) and implemented by Simulink means. The Ramp unit generates a normalized geometric size of a raster element. Expressions (6) and (7) are implemented by two units of mathematical functions, which are masked in the Atomic Subsystem unit, the output of which (P = S) is fed to the second inputs of all Subsystem units.

For example, in the Constant block we set the thickness of the paint flow that is fed to the inlet of the model $H_0 = 6$ μm was set. Adjust the model of the printing system to the nominal transfer coefficients ($\alpha_i = \Upsilon_i = 0.5$; $\Upsilon_1 = 0.2$; $\beta = 0.8$). The results of simulation modeling of the dependence of the ink thickness on the print scale on the tone transfer interspace are presented in Fig. 4.

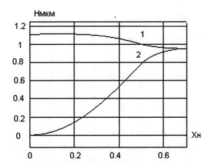

Fig. 4. Dependence of the ink thickness on the tone transfer interval: 1 – amplitude value, 2 – average value

The dependence of the ink thickness on the tone transfer interval is nonlinear. At the starting point of the interval, set the thickness of the paint layer, which is equal to 1.153 μm, it decreases proportionally at the end point of the set interval and is equal to 0.943 μm. The average thickness value gradually increases from zero to 0.943 μm. Therefore, the thickness of the ink does not meet the standard requirements for offset printing. The model of normalized raster conversion (Atomic Subsystem unit) on a round-shaped element was adjusted, the graph of dependence of normalized coverage on the tone transfer interval is shown in Fig. 5.

The characteristics of the normalized raster conversion is a nonlinear S-curved one that causes image distortion, namely, illuminates light areas and darkens gray ones, which must take into account in the synthesis of tone transmission.

Using the expression for demodulation, we determine the optical density of the scale (2) in the unit of mathematical functions Fcn, the values of the optical density of the full tone areas $D_{пл} = 2.0$, and the optical density of the paper $D_п = 0.05$ were set and set the value $n = 3$. The results of simulation of tone reproduction of a raster scale with round-shaped elements in a short printing system are presented by the characteristics of optical density depending on the selected tone transmission interval (Fig. 6).

There is a linear characteristic shown in the figure for comparison. Instead, the optical density characteristic of a short printing system is an S-shaped curve, the maximum value of which at the end of the interval is 1.476. Note, that at simulation, the value of the

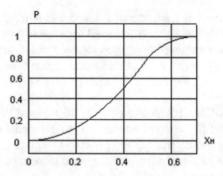

Fig. 5. Graph of the dependence of normalized coverage on the tone transfer interval

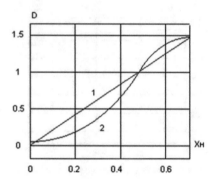

Fig. 6. Characteristics of optical density: 1 – linear, 2 – of a short printing system

optical density of the full tone area was set as 2.0. Reduction of the optical density in the printing system model to 1.476 is due to the reduction of the thickness of the ink from 0.943 to 0.943 µm. To estimate properties of recreation of tone of the system of printing to the shirt, will define deviation of absorbancy from linear.

$$E = D - D_0 \tag{9}$$

where D_0 – is a linear characteristic.

The results of simulation a design of rejection of absorbancy of the printing system is from linear are shown on Fig. 7.

The deviation of the optical density from the linear one for a short printing system is similar to a negative sinuous line. The maximum value of negative deviation is in the light areas and is 17.6%, and the maximum value of +8.9% is on the gray areas.

Let's analyze the obtained results of simulation modeling from the purpose of tone reproduction [3, 13]. If the purpose of the tone reproduction of a black-and-white image is the identical equality of the optical densities of the imprint with the densities of the corresponding parts of the original [3, 9], then, the short printing system is unable to fulfil this. If we are talking about an editorial tone transfer with a compressed interval of optical densities [3, 15, 16], then, S-shaped characteristic of optical density (Fig. 7) for certain types of images can be successfully printed by a short printing system almost

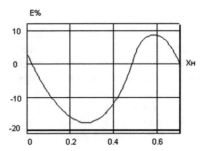

Fig. 7. Deviation of the optical density from the linear one

without additional adjustment. For example, for a photo of a man with a burnt (darkened) face in gray clothes, the S-shaped tone characteristic (7) provides illumination of light areas (face) and darkens men's clothes almost without adjustment (that will improve image quality) at printing.

The main result of the simulation is the made characteristics of optical density and its deviation from the linear, which assess in quantity the toning properties of short printing system, which allows to make correct decisions as to further processing in image processing programs to optimize brightness and contrast, using standard functions including gamma characteristics, "Shadows/Midtones/Light Tones" functions, Gurwes, and others [9].

3   Conclusion

A two-parameter model of tone reproduction of images with raster round-shaped elements for the system of short printing of parallel structure of the sixth order has been developed. A block diagram of the model and a simulator for determining the ink layer thickness on the imprint and the optical density for the natural characteristics of rasterization depending on the tone transfer interval have been constructed. The results of simulation have been presented in the form of characteristics of the coating of the scale with ink, optical density and its deviation from the linear one.

It was found that the characteristics of the optical density of a short printing system is an S-shaped curve and its deviation from the linear one is similar to a negative sinuous line with a maximum negative value in light areas of -17.6%, and in gray ones $+8.9\%$. The main result of the simulation is the made characteristics of optical density and its deviation from the linear one, which assess in quantity the toning properties of a short printing system, which allows making right decisions as to further processing in digital image processing programs to optimize brightness and contrast using standard functions.

References

1. Baranovskyi, I., Yakhymovych, Y.: Polygraphic Processing of Image Information. IZMN Tutorial, Kyiv-L'viv (1998)
2. Hultiaiev, O.: MATLAB 5.2 Simulation Modeling in the WINDOWS Environment: A Practical Guide. Korona Print, S-Pb (1999)

3. Kurka, P.: Characteristics of painting square raster elements in the ink-printing system of the sixth dimension. Comput. Print. Technol. **35**(1), 25–33 (2016)
4. Durnyak, B., Lutskiv, M., Shepita, P., Nechepurenko, V.: Simulation of a combined robust system with a P-fuzzy controller. In: Lytvynenko, V., Babichev, S. (eds.) Intellectual Systems of Decision Making and Problems of Computational Intelligence: Proceedings of the XV International Scientific Conference, vol. 1020, pp. 570–580. Springer, Heidelberg (2019)
5. Lutskiv, M., Makachivskyi, P.: Approximation of the dependence of the optical density on the thickness of the ink layer on the print. Kvalilohiya knyhy **7**, 95–102 (2005)
6. Martyniuk, V.: Fundamentals of Prepress Training of Image Information. University "Ukrayina", Kyiv (2009)
7. Musiiovska, M.: Analysis of the accuracy of raster scale coating with ink in a short printing system of sequential structure. Comput. Print. Technol. **33**(2), 116–124 (2015)
8. Nazar, I., Lazarenko, E.: Roll Offset Printing Parameters. UAP, Lviv (2009)
9. Pashulia, P.: Standardization, Metrology, Compliance, Quality in Printing. UAP, Lviv (2011)
10. Ivaskiv, R., Neroda, T.: Enhancement of conception and embedding the enterprise social network in academy information space. In: Computational Linguistics and Intelligent Systems: CEUR Workshop Proceedings, vol. 2604, pp. 612–621 (2020)
11. Vlah-Vigrinovka, G., Ivanyuk, O., Vigrinovsky, M.: Remote control system for street lighting based on internet of things technologies. Computer Print. Technol. **45**(1), 26–32 (2021)
12. Buben, V.: Sensitivity of the optical reflection density model. Computer Print. Technol. **41**(1), 95–101 (2019)
13. Medykovskyj, M., Pasyeka, M., Pasyeka, N., Tyrchyn, O.: Scientific research life cycle performance of information technology. In: XIIth International Scientific and Technical Conference "Computer Science & Information Technologies" (CSIT'2017), Lviv, Ukraine, 5–8 Sept. 2017, pp. 425–428
14. Pasyeka, N., Mykhailyshyn, H., Pasyeka, M.: Development algorithmic model for optimization of distributed fault-tolerant web-systems. In: IEEE International Scientific Practical Conference "Problems of Infocommunications. Science and Technology", Kharkiv, 9–12 Oct. 2018, pp. 663–669
15. Romanyshyn, Y., Sheketa, V., Poteriailo, L., Pikh, V., Pasieka, N., Kalambet, Y.: Social communication web technologies in the higher education as means of knowledge transfer. In: Proceedings of the IEEE 2019 14th International Scientific and Technical Conference on Computer Sciences and Information Technologies (CSIT), vol. 3, 17–20 Sept. 2019, Lviv, Ukraine, pp. 35–39
16. Prystavka, P., Cholyshkina, O.: Pyramid image and resize based on spline model. Int. J. Image Graph. Signal Process. (IJIGSP) **14**(1), 1–14 (2022). https://doi.org/10.5815/ijigsp.2022.01.01
17. Islam, Md.T., Islam, S.Md.R.: A new image quality index and it's application on MRI image. Int. J. Image Graph. Signal Process. (IJIGSP) **13**(4), 14–32 (2021). https://doi.org/10.5815/ijigsp.2021.04.02
18. Dayang, P, Kouyim Meli, A.S.: Evaluation of image segmentation algorithms for plant disease detection. Int. J. Image Graph. Signal Process. (IJIGSP), **13**(5), 14–26 (2021). https://doi.org/10.5815/ijigsp.2021.05.02

Parametric Optimization of Time-Domain Digital Control System

Igor Golinko and Iryna Galytska[✉]

National Technical University of Ukraine "Igor Sikorsky Kyiv Polytechnic Institute", Peremogy str. 37, Kyiv 03056, Ukraine
irinagalicka@gmail.com

Abstract. The article considers the parametric optimization method for the control system with PID controller modifications, which allows reducing the control system synthesis error due to the plant accurate description in the time domain. When describing the plant model in the control system, the convolution integral is used. For parametric optimization, an integral quality indicator is substantiated, which takes into account the technological process features. It is shown that the digital control system synthesis according to the proposed quality criterion belongs to the one-extremal optimization problems. The proposed method and mathematical models are recommended to be used at the supervisory control systems level as an adviser for setting up and adapting control system. The analysis of PID controller modifications influence on the transient processes quality in control system was carried out. Numerical simulation confirmed the proposed control system optimization method effectiveness. The developed method significant advantage is the digital controller parametric optimization possibility without stage of identifying plant. The proposed mathematical support can be successfully used for the automatic control systems synthesis with controllers' different types.

Keywords: Optimization · Control system · Control plant · Quality criteria · PID controller · PI-D controller · I-PD controller

1 Introduction

Modern technical systems contain controls that are responsible for the system quality as a whole. When developing control systems, the following are used: PID controllers; adaptive controllers [1, 2]; controllers with fuzzy logic [3, 4]; controllers with fractional derivatives [5]; intelligent controllers [6, 7]; neural network controllers and many other controller types. It should be noted that for traditional control systems, where the plant can be linearized, the controller with fuzzy logic, or, for example, the neural network controller does not have a winning quality in relation to PID controller modifications [8].

To adjust the controllers are used: frequency, root and time methods of automatic control systems (ACS) synthesis [9, 10]. Frequency and root methods operate with simplified plant models, which affect the control quality. Time methods for tuning controls are rarely used, since mathematical formalization exists only for simple systems.

© The Author(s), under exclusive license to Springer Nature Switzerland AG 2022
Z. Hu et al. (Eds.): ICCSEEA 2022, LNDECT 134, pp. 47–60, 2022.
https://doi.org/10.1007/978-3-031-04812-8_5

Time methods for ACS synthesis are visual in use, but time-consuming in mathematical implementation.

The time response of plant on an arbitrary waveform input signal can be obtained using the convolution integral. In the system analysis theory and time series identification, this approach is used [11, 12]. The numerical methods use for the convolution integral implementation enables a new time methods evolution for synthesis of controllers without the plant identifying stage by simplified mathematical models. The literature review did not reveal publications in this direction.

In the classical control system used a PID controller. SOFTLOGIC-systems review [13] showed that the PLC software uses in its arsenal PID control law. With the advent of cheap microprocessor automation tools, interest in PID controllers' software implementations has grown [14]. For many use years of the classical PID controllers functioning shortcomings number are revealed. The PID controller has three setting parameters, which in some cases does not allow providing the required control quality. As a result, PID control law modifications began to appear [15, 16]. Among such modifications are PI-D and I-PD controllers, which are widely used by ABB, Honeywell, Fisher-Rosemount, Foxboro, SATT Instruments, Toshiba and others [17].

For these reasons, the digital ACS synthesis with PID controllers parametric optimization according is an urgent task and therefore is necessary to research on three components for ACS synthesis: the criterion selection; the adequate plant mathematical model; the control law type.

2 Research Problem Statement

The publication aim is to develop mathematical support for the optimal ACS synthesis in the time domain. To achieve this aim, it is necessary to analyse and choice the optimization criterion with control influence minimization, develop the plant mathematical model in time domain and provide for PID controller modifications use in the control system.

3 Optimization Criterion Choice

The optimization theory of dynamic systems for the ACS synthesis suggests using the quality criterion integral-quadratic form [18], since such a task has a general analytical solution:

$$I = \mathbf{X}^T \mathbf{S} \mathbf{X}\Big|_{t_f} + \int_0^{t_f} \left(\mathbf{X}^T \mathbf{Q} \mathbf{X} + \mathbf{U}^T \mathbf{R} \mathbf{U} \right) dt, \qquad (1)$$

here, $\mathbf{S}$, $\mathbf{Q}$ are weighted positive semidefinite matrices, $\mathbf{R}$ is a weighted symmetric positive matrix, $\mathbf{X}$ is the controlled plant state vector, $\mathbf{U}$ is the plant control vector, t_f is the final time of observation. The criterion terminal component is the quadratic norm of the actual final state $\mathbf{X}|_{t_f}$ deviation from zero (given) with weight $\mathbf{S}$. The criterion integral component allows assessing the dynamic behaviour of the state vector $\mathbf{X}$ with weight $\mathbf{Q}$ and the control vector $\mathbf{U}$ with weight $\mathbf{R}$.

For digital ACS shown in Fig. 2, the discrete analogue of the quality criterion (1) will take the form:

$$I = T_{KV} \sum_{s=0}^{Nt-1} \left(q e_s^2 + r u_s^2 \right), \text{ if } e_s|_{s \geq Nt} = 0, \tag{2}$$

where, q and r are weight coefficients, Nt is the total number of the transient process observation points with a discrete step T_{KV} $(t_f = Nt \cdot T_{KV})$, s is the current index of discrete time.

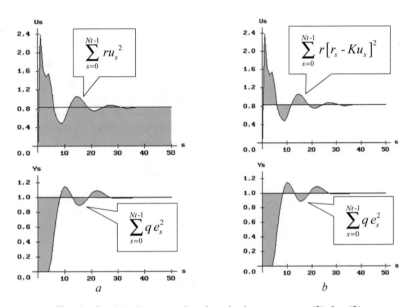

Fig. 1. Graphics interpretation the criterion terms: a – (2); b – (3)

Let's analyse criterion (2). We will consider a controlled plant with positive self-levelling, since most technological equipment have this property. Suppose that ACS is steady and the digital controller settings when changing the setpoint adjustment ensure the condition fulfilment $e_s|_{s \geq Nt} = 0 (e_s = r_s - y_s)$. Thus, the first term sum of criterion (2) over time will acquire a steady value, since $e_s|_{s \geq Nt} = 0$ by condition. To compensate for the control error at the final moment of time digital controller must transfer the control signal from the initial state $u_s|_{s=0} = 0$ (zero initial conditions) to the final $u_s|_{s=Nt} \neq 0$. A graphic interpretation the terms of criterion (2) is shown in Fig. 1, a. If $Nt \to \infty$ and $r \neq 0$, then $\sum_{s=0}^{Nt-1} r u_s^2 \to \infty$, and the quality criterion (2) loses its meaning. Another disadvantage of criterion (2) is the integration of the values e_s and u_s into one numerical value, which are different in physical essence. Even if the signals e_s and u_s are normalized to a unity value, taking into account the weight coefficients q and r, the unit weight of the signal e_s and u_s will be different.

To eliminate the listed disadvantages of criterion (2), it is proposed to consider the following quadratic form [19]

$$I = T_{KV} \sum_{s=0}^{Nt-1} \left(q\, e_s^2 + r[r_s - Ku_s]^2 \right), \tag{3}$$

here, K is the plant transmission coefficient, $K = y_s/u_s|_{s \geq Nt}$. In sum with the second addend in functional (3), there is a control signal u_s, but this signal is "scaled to the dimension" of the signal e_s. For criterion (3), if $Nt \to \infty$ and $r \neq 0$ sum of the second addend $\left(\sum_{s=0}^{Nt-1} r[r_s - Ku_s]^2 \right)$ acquires a steady value, since $y_s|_{s \geq Nt} = Ku_s|_{s \geq Nt}$, and $e_s|_{s \geq Nt} = (r_s - y_s)|_{s \geq Nt} = 0$ by condition. In this case, the physical essence of the addend changes. The addend $r\, u_s^2$ criterion (2) minimizes the value of the control signal with weight r. The addend $\sum_{s=0}^{Nt-1} r[r_s - Ku_s]^2$ criterion (3) minimizes deviation of the control signal u_s from the technologically set value with weight r. A graphic interpretation of the addend in criterion (3) is shown in Fig. 1, b.

As a rule, ACS perform stabilizing functions. Let us take into account the disturbance channel influence in the optimal criterion. To do this, consider the block diagram of the digital ACS is shown in Fig. 2.

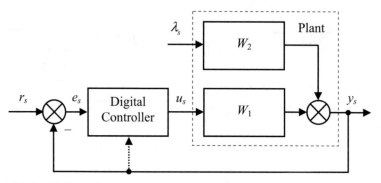

Fig. 2. Digital ACS block diagram: W_1 – plant control channel; W_2 – plant disturbance channel

Using the reasoning considered above according to criteria (2) and (3), we can propose the criterion that takes into account the disturbance channel properties

$$I = T_{KV} \sum_{s=0}^{Nt-1} \left(q\, e_s^2 + r\, [r_s - K_1 u_s - K_2 \lambda_s]^2 \right), \tag{4}$$

where K_1 is the control channel transmission coefficient, $K_1 = y_s/u_s|_{s \geq Nt}$; K_2 is the disturbance channel transmission coefficient, $K_2 = y_s/\lambda_s|_{s \geq Nt}$.

When optimizing ACS, in addition to quadratic criteria, modular ones are used. Considering the above, we modify (4):

$$I = \begin{cases} T_{KV} \sum_{s=0}^{Nt-1} \left[q|e_s| + (1-q)|r_s - K_1 u_s - K_2 \lambda_s| \right], & \text{when } Nvr = 1; \\ T_{KV} \sum_{s=0}^{Nt-1} \left[q e_s^2 + (1-q)(r_s - K_1 u_s - K_2 \lambda_s)^2 \right], & \text{when } Nvr = 2, \end{cases} \tag{5}$$

here Nvr is the quality criterion number ($Nvr = 1$ is the modular criterion, $Nvr = 2$ is the quadratic criterion); q is the weight coefficient, $0 \leq q \leq 1$.

Thus, it is proposed to use functional (5) to optimize digital ACS. The criterion contains two parameters: Nvr is the criterion number; q is the weight coefficient that allows redistributing shares between signals of control and control error. The proposed quality criterion, in contrast to the existing ones, correctly takes into account the control signal, which makes it possible to minimize the consumption of material or energy flows in the ACS dynamic modes.

Optimization problem statement: let for a single-circuit ACS (Fig. 2) the dynamic properties of the exposure channels W_1 and W_2 plant are set in the transient or impulse characteristic forms; by the condition of the problem, the sampling period T_{KV} and the control law are also set; it is necessary to optimize the digital controller settings, which minimizes the quality criterion (5). It is advisable to carry out parametric optimization numerically.

4   Plant Mathematical Model

The analysis of digital ACS involves the numerical methods choice for modelling the reaction of digital controller and controlled plant to the corresponding input arbitrary shape signals. The control plant mathematical model will be based on the use of the convolution integral [20]. For the case under consideration, the discrete representation of plant model will have the form

$$
y_s = \sum_{j=0}^{N} \left(u_j \, g_{s-j}^{W_1} + \lambda_j \, g_{s-j}^{W_2} \right),
\tag{6}
$$

here $s = 0, 1, 2, \ldots Nt$; $g_{s-j}^{W_1}$ is the discrete impulse response of plant, which is the plant response by channel W_1 to a pulse of a single value, the duration of which is T_{KV}; $g_{s-j}^{W_2}$ is the discrete impulse response of plant, which is the plant response by channel W_2. Dependence (6) makes it possible to find the plant response y_s as the sum of the responses on signals u_s and λ_s.

Numerical integration introduces an error in the final result. Consider the mathematical justification for using dependence (6) on the example of the plant control channel W_1. Similarly, the conclusions can be obtained with respect to the plant disturbance channel W_2.

The plant discrete mathematical model can be represented as the serial dynamic units connection: Zero-Order Hold (ZOH) with a transmitting function $W_{ZOH}(p) = \frac{1-e^{-pT}}{p}$; continuous plant which transfer function $W_1(p)$; S_1, S_2, are samplers with the quantization period T_{KV}. For such the system, the block diagram is shown in Fig. 3.

Suppose that the plant dynamic properties are determined by the transient response $h(t)$. ZOH in the time domain can be represented as the difference of single signals shifted by the quantization period T_{KV} (Fig. 4, a):

$$
W_{ZOH}(p) = L\{ 1(t) - 1(t - T_{KV}) \} = L\{ u_{ZOH}(t) \}.
\tag{7}
$$

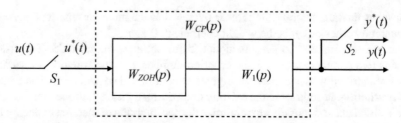

Fig. 3. Serial dynamic units connection for the plant discrete model

In the time domain it is possible to represent the response of the continuous part (CP) (with ZOH and continuous plant, see Fig. 4, *b*) to a single impulse $u^*(t) = \begin{cases} 0, & \text{if } t \neq 0 \\ 1, & \text{if } t = 0 \end{cases}$, as

$$W_{CP}(p) = W_{ZOH}(p) \cdot W_1(p) = L\{h(t) - h(t - T_{KV})\} = L\left\{g^{W_1}(t)\right\}, \tag{8}$$

here $g^{W_1}(t)$ is the CP response to the unit value pulse of duration T_{KV}.

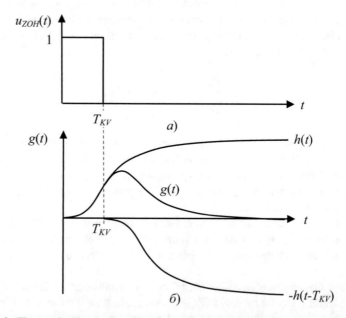

Fig. 4. The continuous plant reaction formation to unit value pulse of duration T_{KV}

If the sequence of pulses $u^*(t)$ arrives at the CP input, then the output signal can be written as:

$$y(t) = u(0) g^{W1}(t) + u(T_{KV}) g^{W1}(t - T_{KV}) + u(2T_{KV}) g^{W1}(t - 2T_{KV}) + \dots. \tag{9}$$

For $t = s \cdot T_{KV}$ ($s = 0, 1, 2, \ldots Nt$) the dependence (9) will take the form:

$$y_s = u_0\, g_s^{W_1} + u_1\, g_{s-1}^{W_1} + \ldots + u_s\, g_0^{W_1} = \sum_{j=0}^{s} u_j\, g_{s-j}^{W_1}. \tag{10}$$

We obtain the z–transformation for both parts of (10), applying the convolution theorem in the time domain $F_1(z)F_2(z) = Z\left\{ \sum_{j=0}^{s} f_1(j) f_2(s-j) \right\}$ [20] and obtain:

$$y(z) = u(z) \cdot W_1(z), \tag{11}$$

here $W_1(z)$ is the plant discrete transfer function, which is determined by the dependence:

$$W_1(z) = \sum_{s=0}^{\infty} g_s^{W_1} z^{-s}. \tag{12}$$

Thus, if the plant dynamic in time domain is described by the impulse response $g^{W_1}(t)$, which is the continuous plant response to unit value pulse of duration T_{KV}, and also for the impulse response $g^{W_1}(t)$ there is a lattice function $g_s^{W_1}$ ($s = 0, 1, 2\ldots$), then the discrete transfer function of the CP can be determined from dependence (12). And dependence (10) can be used to numerically determine the discrete output signal y_s plant response to the sequence of input pulses u_s.

This approach makes it possible not only to reduce the error in the plant modelling by eliminating the step of approximating the transient response by the impulse transfer function, but also to avoid subjectivity when choosing the structure of the approximating dependence.

5   Digital Controller Mathematical Model

The PI-D and I-PD control law mathematical implementation is some-what different from the PID control law. The equations for PI-D, I-PD and PID controllers, used for industrial analog controllers, in operator form obtain as [16, 21]:

$$u(p) = K_R\left(1 + \frac{1}{T_I p}\right) e(p) - K_R \frac{T_D p}{\frac{T_D}{K_F} p + 1}\, y(p), \tag{13}$$

$$u(p) = K_R \frac{1}{T_I p} e(p) - K_R \left(1 + \frac{T_D p}{\frac{T_D}{K_F} p + 1}\right) y(p), \tag{14}$$

$$u(p) = K_R\left(1 + \frac{1}{T_I p} + \frac{T_D p}{\frac{T_D}{K_F} p + 1}\right) e(p), \tag{15}$$

here K_R, T_I, T_D are controller settings parameters (respectively, the transfer coefficient, integration time and differentiation time); K_F is coefficient that determines limiting the filter frequency of the regulator D-component ($K_F = 2\ldots20$) [15], $\frac{T_D}{K_F}$ is the filter time constant; p is the Laplace operator.

We obtain the finite–difference equations of the digital controller for the considered control laws. Consider the integral term for (13)–(15) $I(p) = \frac{1}{T_I p} e(p)$. Let's move from Laplace domain to the original $I(t) = \frac{1}{T_I} \int_0^t e(t)dt$. We use the method of left rectangles for numerical integration

$$I_s = \frac{T_{KV}}{T_I} \sum_{j=0}^{s-1} e_j. \tag{16}$$

Consider the differential term for the PID controller (15) with first-order aperiodic filter $u_d(p) = \frac{T_D p}{\frac{T_D}{K_F} p + 1} e(p)$. Let's move from Laplace domain to the original $\frac{T_D}{K_F} \frac{d u_d(t)}{dt} + u_d(t) = T_D \frac{d e(t)}{dt}$. Replacing the differentials by finite differences, we pass to the difference equation. After simplification, we obtain a recurrent equation to determine influence of the PID controller D-component

$$u_{d,s} = a_d u_{d,s-1} + b_d (e_s - e_{s-1}), \tag{17}$$

here $a_d = \frac{T_D}{T_D + K_F T_{KV}}$, $b_d = a_d K_F$. Taking into account the signal y_s dependence (17) is valid for (13) and (14).

Taking into account (16), (17), discrete analogs of Eqs. (13)–(15) for digital controller will take the form:

$$u_s = K_R \left(e_s + \frac{T_{KV}}{T_I} \sum_{j=0}^{s-1} e_j \right) - K_R \left(a_d u_{d,s-1} + b_d (y_s - y_{s-1}) \right), \tag{18}$$

$$u_s = K_R \frac{T_{KV}}{T_I} \sum_{j=0}^{s-1} e_j - K_R \left(y_s + a_d u_{d,s-1} + b_d (y_s - y_{s-1}) \right), \tag{19}$$

$$u_s = K_R \left(e_s + \frac{T_{KV}}{T_I} \sum_{j=0}^{s-1} e_j + a_d u_{d,s-1} + b_d (e_s - e_{s-1}) \right). \tag{20}$$

As you can see, in contrast to the digital PID (20) control law, PI-D (18) and I-PD (19) control laws for the formation of a control action, in addition to the control error e_s, use the plant output signal y_s (see Fig. 2).

6 Simulation

Based on the above mathematical models (6) and (18)–(20), software is implemented that minimizes the functional (5). When developing the software, it was possible to form an image of the quality index surface (5) by a system of isolines in the parameters space of tuning the PI-controller modifications (or the PID controller with a fixed value of T_d), which makes it possible to get an idea of the surface character for the quality index I.

The digital ACS research example using the developed program is shown below. The plant discrete mathematical model (6) in the form discrete impulse response was formed by numerically differentiating the transient characteristic $h(t)$ for the continuous plant with a transfer function:

$$W_1(p) = W_2(p) = \frac{K}{(T_1p + 1)(T_2p + 1)} e^{-\tau p} \tag{21}$$

according to the method described above. The plant dynamic properties are chosen arbitrarily without reference to a specific equipment: $K = 1{,}2$; $T_1 = 1$; $T_2 = 5$; $\tau = 3$. The modelling results the surface relief for criterion (5) in coordinates $I = f(K_r, T_i)$ with a digital PI-controller ($T_d = 0$, $T_{KV} = 1$) by the channel $r_s \to y_s$ are shown in Fig. 5. The surface topography was modelled using the scanning method. As a result of modelling, it is shown that the synthesis of ACS based on the quality criterion (5) refers to the problems of one-extremal optimization. The surface reliefs of the quality indicator I for ACS with the I-P controller have similar images.

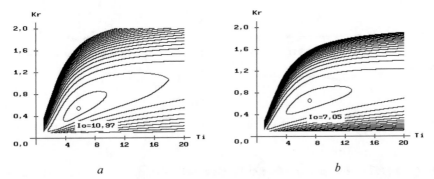

Fig. 5. The surface reliefs of the quality indicator $I = f(K_r, T_i)$ for PI-controller by the channel $r_s \to y_s$: a – modular quality criterion ($Nvr = 1$, $q = 0.5$); b – quadratic quality criterion ($Nvr = 2$, $q = 0.5$)

The next research series demonstrates the influence of criteria (2) and (5) on the quality optimization of ACS with a digital PI-controller. The plant dynamic properties (21) remained unchanged. In Fig. 6 shows the surface reliefs of the quality indicators (2) and (5) for the PI-controller and the corresponding transient processes in the ACS by the channel $r_s \to y_s$. As can be seen from the graphs (Fig. 6, c), the optimization of the ACS according to criterion (2) leads to the PI-controller synthesis, which in its dynamic characteristics approaches the P-controller and tries to find "golden mean" between minimizing the control error e_s and the control u_s. At the same time, the value of the criterion $I_o = 27{,}4$ is rather conditional, given that the criterion numerical integration was carried to $Nt = 50$. With an increase in the observation time of ACS transient process, the numerical value (2) will increase, which indicates its incorrectness. The isolines system (Fig. 6, a) for criterion (2) shows that the optimum is on the right area boundary of the PI-controller parameters, and the PI-controller properties approach the P-control law. For most ACS the P-regulator synthesis is unacceptable. In contrast to criterion (2), the criterion (5) lacks these shortcomings (see Fig. 6, b, d).

Let us consider the criterion parameters (5) influence on the digital ACS optimization result. In subsequent researchers, the Hooke-Jeeves method was used to find the

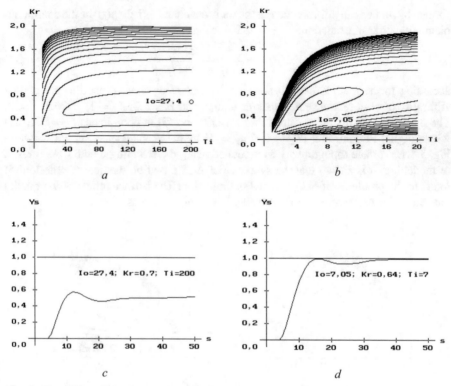

Fig. 6. The ACS optimization results with PI-regulator by the channel $r_s \to y_s$: a – surface relief of the criterion (2); b – surface relief of the criterion (5); c – transition process in ACS by criterion (2) $q = 0.5$; d – transition process in ACS by criterion (5) $Nvr = 2$, $q = 0.5$

local minimum. For simplicity, in all researchers, the plant discrete mathematical model (6) was used in the form discrete impulse response, which corresponds to the transfer function (21). For all researches, the ACS sampling period $T_{KV} = 1$, the differential filter coefficient digital controller $K_f = 5$.

The results of modelling the optimal ACS by the channel $r_s \to y_s$ with the PID controller are shown in Fig. 7. For the proposed criterion (5), the weighting coefficient q effectively effects on optimization result. The ACS optimization by the mean-square control error ($Nvr = 2$, $q = 1$) provides an increased tendency to fluctuations in the transient process compared to the modular criterion ($Nvr = 1$, $q = 1$).

The results of ACS optimization by the channel $\lambda_s \to y_s$ ($Nvr = 1$, $q = 0.5$) are shown in Fig. 8. Curve 1 provided the optimal quality index $I_o = 14{,}24$ (digital controller settings: $K_R = 0.92$; $T_I = 4.77$; $T_D = 1.71$; $K_F = 5$; $T_{KV} = 1$). By the channel $r_s \to y_s$, the obtained digital controller settings provide transient processes in the ACS: curve 2 for the PI-D control law, curve 3 for the I-PD control law, curve 4 for the PI-D control law. Despite the fact that by channel $r_s \to y_s$ the numerical values of the quality criterion for the PID control law modifications are close, the I-PD control law ensures the minimum

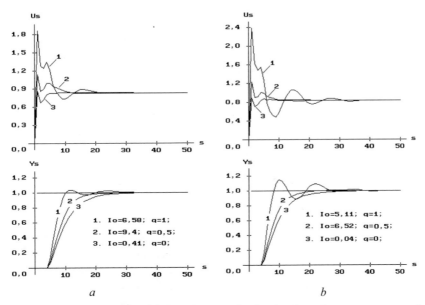

Fig. 7. The ACS optimization with the PID controller by the channel $r_s \rightarrow y_s$: a – modular criterion ($Nvr = 1$); b – quadratic criterion ($Nvr = 2$)

transient process overcontrol. Consequently, in the case of ACS optimization by channel $\lambda_s \rightarrow y_s$ the I-PD control law provides the best ACS dynamics.

The next series of transient processes in the control system is shown in Fig. 9. Here ACS is optimized by channel $r_s \rightarrow y_s$ ($Nvr = 1$, $q = 0.5$). Curve 1 is responsible for the optimal transient process with the PID controller (settings: $K_R = 0.58$; $T_I = 5.61$; $T_D = 0.94$; $K_F = 5$; $T_{KV} = 1$). Curve 2 represents the transient process by the channel $\lambda_s \rightarrow y_s$ with the given PID controller settings. Curve 3 characterizes the optimal transient process with the PI-D controller (digital controller settings: $K_R = 0.62$; $T_I = 6.01$; $T_D = 1.04$; $K_F = 5$; $T_{KV} = 1$). Curve 4 represents the transient process by the channel $\lambda_s \rightarrow y_s$ with the PI-D controller. And accordingly, curve 5 represents the optimal transient process with the I-PD controller (digital controller settings: $K_R = 1$; $T_I = 5.15$; $T_D = 1.53$; $K_F = 5$; $T_{KV} = 1$), and curve 6 characterizes the transient process in the control system by the channel $\lambda_s \rightarrow y_s$ with the I-PD controller. From the presented graphs it can be seen that the PID and PI-D controllers provide the best ACS dynamics by the channel $r_s \rightarrow y_s$, however, the disturbance is not compensated as effectively as the I-PD controller. The ACS dynamics with a PI-D controller is close to the ACS dynamics with a PI controller. The worst dynamics by the channel $r_s \rightarrow y_s$ in the ACS with I-PD regulator, however, the disturbances compensation is worked out best of all.

An important parameter for the digital ACS practical design is the imperfection of the digital controller differential component, which is characterized by the K_F parameter. If in analogy controllers this parameter was present as constructive limitations manifestation, then for digital controller this parameter is programmed to filter high frequency disturbances. Let us consider the imperfection influence of the D-component implementation in the PID controller on the optimal settings value and the optimization result. The

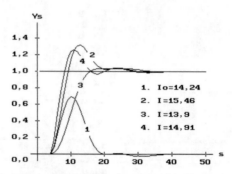

Fig. 8. The results of ACS optimization by channel $\lambda_s \rightarrow y_s$

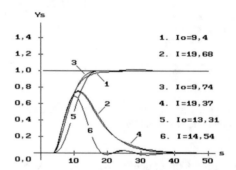

Fig. 9. The results of ACS optimization by channel $r_s \rightarrow y_s$

transients in ACS by channel $r_s \rightarrow y_s$ for the quality criterion (5) ($Nvr = 2$, $q = 0.5$) are shown in Fig. 10. Curve 1 (Fig. 10, a) is the ACS optimal transient characteristic, found when using an ideal PID controller (digital controller settings: $K_R = 0.84$; $T_I = 5.47$; $T_D = 1.87$; $K_F = 10000$; $T_{KV} = 1$). If we take into account the PID controller imperfection, namely, to accept that $K_F = 2$, then with these settings the controller will have a transient characteristic shown by curve 2 (Fig. 10, a).

Curves 1 and 2 differ in their dynamics. In this case, the regulator imperfection impairs the ACS functioning. Let's repeat the search for the optimal setting, taking into account the regulator imperfection.

We get the optimal digital controller: $K_R = 0.76$; $T_I = 5.61$; $T_D = 1.51$; $K_F = 2$; $T_{KV} = 1$; and curve 3 as the optimal ACS transient response. Curve 3 differs from curve 1, but much less than curve 2. In Fig. 10, b shows the transient processes for an experiment similar to the previous one with functional minimization (5) ($Nvr = 1$, $q = 0.5$). Curves 1–3 (Fig. 10, b) do not differ significantly. Obviously, when searching for the optimal digital controller setting, it is necessary to pay attention to the K_F digital controller parameter, the value of which is determined by the disturbance level in useful sensor signal.

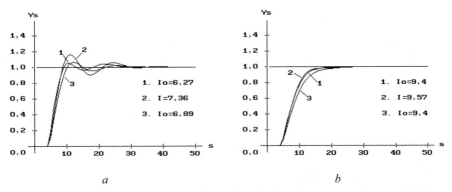

Fig. 10. K_F influence on transients in ACS by channel $r_s \to y_s$: a – quadratic criterion ($Nvr = 2$, $q = 0.5$); b – modular criterion ($Nvr = 1$, $q = 0.5$)

7 Summary and Conclusion

Based on the results of this research, the following conclusions can be drawn.

7.1. The method for control parametric optimization in the time domain without simplifying the plant dynamic properties is proposed. The optimization method basis
is The quality criterion (5), the plant mathematical model (6), the control laws
(18)–(20).

7.2. The quality integral indicator (5) is substantiated, which takes into account ACS
synthesis features in the presence of a control signal. The quality indicator (5) use
makes it possible to evaluate ACS dynamic properties and choose such weight
parameters of the criterion (Nvr, q) for setting up the control system that will meet
technological process requirements.

7.3. To accurately describe the plant model, it is proposed to use the convolution integral. This makes it possible to reduce the ACS modeling error by eliminating the
plant identification stage.

7.4. By the scanning method shows that ACS synthesis based on the quality criterion (5) refers to the one-extremal optimization problems. For these reasons, for
the controller parametric optimization, any numerical method for finding a local
minimum can be used with minor computational resources.

7.5. The transient processes quality dependence on the criterion parameters, influence
channels and PID control law modifications has been research. If ACS performs
stabilizing functions and its main task is to compensate for disturbances, the I-DP
control law should be used. For ACS with change reference in time, the PID control
law provides the best dynamic characteristics. In the control system, PI-D and PID
laws are similar in their properties.

7.6. Modeling confirmed the proposed ACS optimization method effectiveness. The
presented mathematical models can be used at the supervisory control systems top
level as recommendations on setting up a digital controller with modifications to
the PID control law.

7.7. Mathematical support in the quality criterion form (5) and the convolution integral (6) can be successfully used to optimize control systems with controllers different types.

References

1. Singhal, P., Agarwal, S.K., Kumar, N.: Advanced adaptive particle swarm optimization based SVC controller for power system stability. Intelligent Systems and Applications **1**, 101–110 (2015)
2. Puangdownreong, D.: Multiobjective multipath adaptive tabu search for optimal PID controller design. Intelligent Systems and Applications **8**, 51–58 (2015)
3. Misra, Y., Kamath, H.R.: Design algorithm and performance analysis of conventional and fuzzy controller for maintaining the cane level during sugar making process. Intelligent Systems and Applications **1**, 80–93 (2015)
4. Yazdanpanah, A., et al.: Design PID baseline fuzzy tuning proportional derivative coefficient nonlinear controller with application to continuum robot. Intelligent Systems and Applications **5**, 90–100 (2014)
5. Soukkou, A., Belhour, M.C., Leulmi, S.: Review, design, optimization and stability analysis of fractional-order PID controller. Intelligent Systems and Applications **7**, 73–96 (2016)
6. Lakshmi, K.V., Srinivas, P., Ramesh, C.: Comparative analysis of ANN based intelligent controllers for three tank system. Inter. J. Intell. Sys. Appl. **8**(3), 34–41 (2016)
7. Potekhin, V.V., Pantyukhov, D.N., Mikheev, D.V.: Intelligent control algorithms in power industry. EAI Endorsed Trans. Energy Web **3**(11), e5 (2017)
8. Gross, C., Voelker, H.: A comparison of tuning methods for PID-controllers with fuzzy and neural network controllers. Cyber-Phys. Syst. Control, 89–102 (2020)
9. Bubnicki, Z.: Modern Control Theory. Springer, Berlin (2005)
10. Ladanyuk, A.P., Arkhangelskaya, K.S., Vlasenko, L.O.: Automatic Control Theory of Technological Plant. Kiev (2014)
11. Bidyuk, P.I., Romanenko, V.D., Timoshchuk, O.L.: Time-Series Analysis. Kiev (2010)
12. Diligenskaya, A.N.: Control Plant Identification. Samara (2009)
13. Slobodyuk, M., Golinko, I.: Modern means review of programming PC-BASED controllers. In: IX International Scientific and Practical Conference "Modern Problems of En-ergy Scientific Support", p. 304 (2011)
14. Quevedo, J., Escobet, T.: Digital control: past, present and future of PID control. In: Proceedings of the IFAC Workshop. Terrassa, Spain (2000)
15. Äström, K.J., Hägglund, T.: Advanced PID control. ISA (2006)
16. O'Dwyer, A.: Handbook of PI and PID Controller Tuning Rules. Imperial College Press, London (2009)
17. McMillan, G.K.: Tuning and Control Loop Performance, a Practitioner's Guide. Instrumentation Systems (1994)
18. Bensoussan, A., et al.: Representation and Control of Infinite Di-mensional Systems. Birkhäuser, Boston (2007)
19. Sirotynsky, S.O., Golinko, I.M.: Control systems tuning according to integrated quality indicators. In: IX International Scientific and Practical Conference "Modern Problems of Energy Scientific Support", p. 303 (2011)
20. Houpis, C.H., Lamon, G.B.: Digital Control Systems: Theory, Hardware, Software. McGraw-Hill Book Co., Singapore (1992)
21. Johnson, M.A., Moradi, M.H.: PID Control. New Identification and Design Methods. Springer (2005)

Research on Possible Convolution Operation Speed Enhancement via AArch64 SIMD

Andrii Shevchenko[1][✉], Pylyp Prystavka[1], and Vitalii Tymchyshyn[2]

[1] National Aviation University, 1 Lubomir Guzara Avenue, Kyiv, Ukraine
111landreyshevchenko111@gmail.com
[2] Bogolyubov Institute for Theoretical Physics, Kyiv, Ukraine

Abstract. A method of two-dimensional convolution operation (CO) optimization by means of 16-bit SIMD technologies for ARM x64 (aarch64) is proposed. The proposed method is compared with CO implemented in OpenCV by means of statistical analysis. It shown that for the most interesting case of small kernels (2×2 to 7×7) the proposed method is at least x2 faster than OpenCV, the result is statistically significant. We expect that the proposed approach can be employed to accelerate image processing on embedded devices.

Keywords: Convolution · SIMD · Optimization

1 Instruction

Digital image (DI) processing takes an important place in modern computer science and software engineering: video-stream processing (stabilization, filtration, noise reduction, compression, etc.), single image processing, and a number of machine learning (ML) problems are part of a DI processing realm. Despite a wide variety of DI problems, a large number of them can be reduced to or involves some form of convolution operation (CO):

$$p_{i,j} = \sum_{k=0}^{r} \sum_{l=0}^{c} \Gamma_{k,l} P_{k+i-a,l+j-a'}, \tag{1}$$

where $i = a, \ldots, W - (r - a) - 1, j = a', \ldots, H - (c - a') - 1$ are indexing pixels of the destination image p; W and H are width and height of the source P and destination p images (we neglect border effects in the destination at the moment), Γ is the kernel of the convolution (matrix $r \times c$), and a, a' are so-called "anchors" that define relative position of a filtered point within the kernel.

The omnipresence of the operation (1) is what generates an enormous interest in its optimization, for example, the most resource-demanding operation in modern deep neural networks is the CO. Moreover, one may want to run CO on a CPU that supports vectorization, e.g. perform DI processing or running a neural network on different embedded devices. Since modern compilers like GCC and Clang/LLVM support automatic program code vectorization (APCV) based on the SIMD instructions of CPU,

Z. Hu et al. (Eds.): ICCSEEA 2022, LNDECT 134, pp. 61–75, 2022.
https://doi.org/10.1007/978-3-031-04812-8_6

most cases one will use a CO implemented in C/C++ and compiled with appropriate flags that enable APCV. But as was shown in [1–3], it should be possible to get much higher performance than we currently achieve with the APCV.

In current contribution we explore the possibility of CO implementation that is more optimal than the one obtained with the APCV. First, we will show that certain kernels can be quantized and provide a criterion of such reducibility without loss of quality. Then, using quantized kernels, we will provide an implementation of a CO that utilizes ARMx64 SIMD, e.g. aarch64 (NEON64) operations. At last, the new CO will be compared with the OpenCV CO [4] and a statistical analysis of the performances will be provided. It will be shown that the proposed method is at least x2 times faster than the OpenCV implementation and the result is statistically significant.

2   A Brief Overview of Modern Software Optimization

We will perform the overview in a "bottom to top" style - we consider hardware first, then software, and then algorithmic methods of performance enhancement.

2.1   Acceleration by Means of Hardware

It is worth noting that well-chosen hardware architecture is the most influential factor of the software product performance. The first question is whether the task (e.g., CO) allows parallelization of the data flow (or instructions flow). Flynn's taxonomy [5] gives a general perspective on possible solutions.

Today NEON64 (A64 - instruction set; Neon – SIMD for ARM64 CPUs) principles are implemented in both RISC (e.g., Cortex-A53–72/X1 ARM64 CPUs) [6] and CISC (e.g., Intel x64/x86 series) CPUs. Programmers may access the feature (NEON64) through the specific extensions of the assembly language. Similarly, optimization for CISC architecture implies SSE_n and $AVX_{1/2}$ extensions of the assembly language. These extensions provide one with special registers (32 or more), that is above the number of SIMD registers that modern ARM64 CPUs have.

Except using CPU, one can employ co-processor units, e.g. Digital Signal Processor (DSP) like Qualcomm Hexagon. It has been incorporated into Snapdragon-6XX/8XX CPUs to reduce the CPU load up to $\sim 75\%$ and improve audio/video encoding/decoding performance up to ~ 18 times [7, 8]. Moreover, compared to simple NEON64, its performance is ~ 4 times higher. This DSP uses a very long instruction word (VLIW), which means multithreading at the assembler level [9] (as SIMD): during one interruption, three assembly instructions with different inputs are processed.

2.2   Optimization by Means of Software

The software we use (e.g. compiler itself, additional libraries, frameworks) highly influence the product software by employing different optimizations to use the hardware platform capabilities more efficient. In the scope of the current article we are primarily concerned with their ability to perform vectorization without significant loss of the precision and speed. Further we consider three well-known compilers: GNU Compiler Collection (GCC/G++) [10], Clang [11], and nvcc (compiles cu-files for CUDA).

The most popular nowadays is still the GCC compiler developed/supported by the FSF community. First versions of the GCC were a collection of compilers for different programming languages developed by Richard Stallman. Nowadays GCC is no longer a GNU C compiler but a GNU Compiler Collection. GNU is an optimizing compiler produced by the GNU Project that supports various programming languages, hardware architectures, and operating systems.

GCC's main competitor is Clang. For example, Apple already uses it as the basic compiler for its products. Moreover, the UNIX/BSD OS distributives also use it as a default compiler. The Android NDK now uses clang compiler by default instead of previously used GCC. Clang itself is a frontend for different programming languages, e.g. C, C++, Objective-C, Objective-C++, and OpenCL. The actual generation of the binary code and vectorization is performed by the LLVM framework. Both GCC and Clang are performance-oriented, but still they fail compared to the human-made assembly code [1, 2].

One more reasonable approach to achieve performance enhancement of DI processing is the usage of different libraries [12, 13] (proprietary or not) like OpenCV and ARM Compute Library (ACL). Many of them contain NEON64-optimized code for armeaby-v7a and arm64-v8a. Another smart strategy is to use the collection of libraries combined into a single framework so that the advantages of one library compensate drawbacks of the others. OpenCV and ACL [14] are good examples of libraries comprising a wide variety of algorithms including DI processing and DI analysis. Moreover, OpenCV contains modules for CNN training optimized for different CPU architectures, uses SIMD ($AVX_{1/2}$/SSE4, NEON64) and GPU optimized approaches. Also, OpenCV is well-known for its high-quality DI processing. Thus we will consider OpenCV as a reference for comparison.

2.3 Optimization by Means of Metaprogramming Approaches

In our previous paper [2, 3] we provided/used an approach [15, 16] that lead to a huge (over +25%) speed improvement of an algorithm. But the ARM64 architecture was significantly improved compared to the ARMv7-A since then and nowadays the proposed technique is redundant. Moreover, our new research shows that now the loop unrolling yields a 3–5% speed reduction. With IDA we made sure that loops were actually unrolled on ARM64 by the clang (9 versions) compiler. We speculate that the negative effect was cause by redundant comparisons in the CO function. Code for both rolled and unrolled functions (as the ACL lib used) is presented in Fig. 1. The mentioned fact is a good starting point for further research.

2.4 Optimization by Means of Special Algorithms

Let us focus on CO. The primary obstacle for SIMD optimization is that SIMD operations are performed on integers, thus we should translate floating-point entries of the kernel of interest into fixed-point with acceptable precision loss.

Let us represent the elements of the kernel Γ from (1) in a suitable form:

$$\Gamma_{i,j} = v\gamma_{i,j}, \quad v \in \mathbb{R}, \quad \gamma_{i,j} \in \mathbb{Z} \tag{2}$$

where v is a normalization coefficient. Please note, that Eq. (1) is rather general and perfectly compatible with cv::filter2D(...) function of the OpenCV library [4], but further we will switch gears to square kernels, i.e. $r = c$, and thus from now on we presume kernel to be square-shaped without special mentioning. Now we can perform the most resource-demanding part (additions and multiplications) in a SIMD style and afterwards just normalize the result.

```
1  template <unsigned N>
2  struct func_unroll {
3      template <typename F>
4      static inline void call(F const &f) {
5          f();
6          func_unroll<N - 1>::call(f);
7      }
8  };
9
10 template <> struct func_unroll<0u> {
11     template <typename F> static inline void call(F const &) {}
12 };
13
14 #define unrollDelta 8
15 struct do_unroll {
16     template <typename F>
17     static inline void run(F const &f, int &baseStep, int &restSteps) {
18         for (int j = baseStep; j; --j) {
19             func_unroll<unrollDelta>::call(f);
20         }
21         for (int j = restSteps; j; --j) {
22             f();
23         }
24     }
25 };
```

Fig. 1. Loop unrolling with C++ templates.

Any kernel can be represented in the form (2), but the more precise result we want, the more digits should $\gamma_{i,j}$ have. Thus we should set some constraints on γ to avoid possible overflows due to the target platform limitations.

Suppose, every pixel of the original image is represented as a byte and thus possesses an 8-bit value ranging 0 to 255. The same range is possessed by kernel elements $\gamma_{i,j}$. Intermediate results are stored as 16-bit signed or unsigned values. To warranty that no overflow occurs, we should ensure that it does not happen on any algorithm step. If the kernel has positive elements only, a condition we need looks as follows

$$0 \leq (2^8 - 1) \times \sum_{i=0}^{r} \sum_{j=0}^{r} \gamma_{i,j} \leq 2^{16} - 1. \tag{3}$$

The inequality (3) means that even the largest possible inputs from the image do not lead to overflow.

If the kernel contains negative elements, the condition should be much more complicated and depends on the order of additions when doing CO. Instead, we will use much stronger but more straightforward condition

$$0 \leq (2^8 - 1) \times \sum_{i=0}^{r} \sum_{j=0}^{r} |\gamma_{i,j}| \leq 2^{16-1} - 1, \tag{4}$$

that is independent of the operations' order. Moreover, this condition can be slightly relaxed - we can use it for positive and negative entries of the kernel γ separately. And the last thing to mention: one can easily obtain similar results for signed/unsigned 32-bit intermediate values by substituting $16 \rightarrow 32$ in (3) and (4).

What we propose is selecting for given Γ the largest possible v, such that (3) or (4) is still satisfied (depends on whether the kernel's entries are all positive or not). It can be seen that a plethora of useful kernels can be reduced to a suitable form.

In conclusion, modern hardware provides mechanisms for vectorization, i.e. SIMD technology that programmers can use to enhance the performance of the application. Most cases, this technology is utilized by the compiler to generate binary code without the participation of the programmer. A suitable choice of the library may be handy as well - many libraries contain SIMD-optimized code. But in some cases human intervention is needed to get the most optimal result to make code suitable for the SIMD optimization. However, it is not always possible, but in our case the restrictions (3), (4) are easily satisfied. In the next section we will provide a new method of CO optimization and then compare it with existing results from OpenCV lib.

3   Optimization of Convolution Operation by Means of SIMD

In the current contribution, we propose a new Convolution Operation (CO) optimization method based on the SIMD technique. This section will provide all necessary considerations and an inline assembly code that illustrates the proposed approach. The following section will be devoted to an experimental comparison of this method's performance to known CO implementations of OpenCV.

We presume that the target kernel satisfies condition (3) and provide code listings for type (3) kernels only. Regarding condition (4), provided code should be just slightly modified. We will briefly mention all necessary modifications at the end of the section.

Let's start with the basic implementation of CO (see Fig. 2b). It contains no specific optimizations but still is a good point to begin our considerations. Here v $_n$ are NEON64 vector registers. Regarding syntax and instructions order, we will strictly follow ARM reference manuals. For the sake of simplicity, we avoided normalization by the coefficient v in (see Fig. 2b), but for completeness, let us provide it as a separate (see Fig. 2c). In (Fig. 2c) we suppose the data for normalization to be stored in registers v12 … v15, while v1[0] contains the normalization coefficient v. Presented code is in some sense multipurpose and may be used with different CO implementations.

Now we switch gears to the CO optimization itself by utilizing NEON64. In Fig. 2b we have provided naive implementation of this operation (assembly code). But this version contains one significant drawback - data loading. The data loading/storing process is the slowest operation because it involves sub/inner processes like communication CPU registers with RAM/cache. Even taking CPU cache into account, this operation is still slow.

To avoid the problem, one of the registers was used as a buffer (see Fig. 2a). It is known that simultaneous loading of 16 bytes is quicker than loading them one-by-one. Thus we use one register for preloading extra data and then use this data to perform byte-by-byte shift to exclude redundant load operations [3].

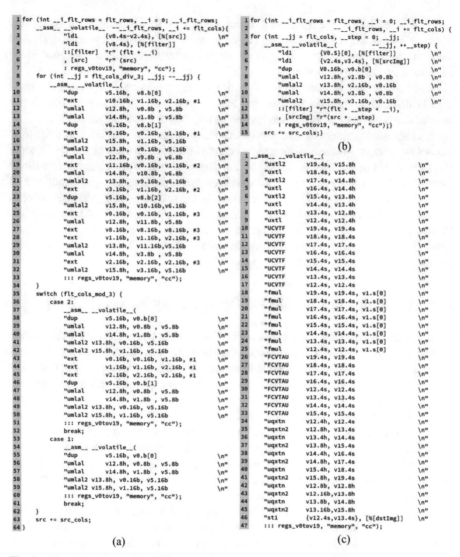

Fig. 2. CO optimization with SIMD NEON64: (a) Optimized with shift approach (b) Naïve approach; (c) Normalization procedure.

The main feature of the presented approach (see Fig. 2a) is the usage of cyclic shift (i.e., ext v10.16b, v0.16b, v1.16b, #1) that allows kernel buffering, and thus, we need fewer operations of loading (for more details, please see comments in (see Fig. 2a)). One more thing that should be mentioned is that the pre-save of the shifted data (see Fig. 2a lines 11, 15, 19, 22) was used for current iteration of CO. Other "ext" operations (lines 25, 28, 31) provide data initialization for the next iteration. The code provided (see Fig. 2a) demands kernel with not more than 16 elements in one row. For more elements one should utilize data reinitialization of the base registers (in lines 3–4).

As we mentioned earlier, this code works for kernels satisfying condition (3). To make it applicable to kernels satisfying (4), we need to change all "umlal" operations to "smlal". These small but crucial changes transform (see Fig. 2a) into code that works with signed integer kernels. Depending on elements in the given kernel, one can choose between these two options.

In conclusion: we found a class of kernels that allow significant optimization of CO with NEON64 and achieved a substantial speedup by exploiting simultaneous 16-byte loading. More detailed results and considerations of the measurement procedure will be presented in the following section.

4 Experimental Setup and Results

Ground Truth. To evaluate our results a certain reference is needed. As a reference we chose functions cv:: filter2d(…) of the OpenCV library. The latter is well-known among AI and DIP researchers due to its high-quality optimized code.

For comparison, we used the latest stable tag for OpenCV available (at the moment we started research) 4.5.2 (2021–04-02 11:23). The compilation was performed with clang-9 - the latest stable clang version. We ensured that libraries utilize vectorization compiling them with flags: -DCMAKE_BUILD_TYPE = RELEASE -DENABLE_NEON = ON…. This leads to setting values to certain critical flags, e.g. "CPU_BASELINE" (NEON F16) and "C++ flags (Release)" (…O3 -DNDEBUG…). Than the OpenCV lib was linked as a dynamic library.

Devices. To make our measurements more relevant, we used Odroid-C4. This helps us understand the influence of the architecture, CPU series, and other parameters on the execution time. The Odroid-C4 CPU is Cortex-A55; OS is Ubuntu 20.04; kernel is Linux 5.7.0-odroid-arm64, and its API is aarch64. CPU of this device is (Amlogic S905X3) more powerful than the latest Raspberry Pi CPU.

Measurement Procedure. The pivoting parameter we need to measure is the execution time for each function. Such measurement might be tricky since it is highly susceptible to transient processes in any GNU OS (Ubuntu, Android, etc.). To avoid this problem, we used the following procedure: each function (cv::filter2d(…) and proposed method - newCO(…)) was successively called seven times and the result was stored into the temporary array. After collecting these 7 data points, we calculated the median time (from temporary array data) and treated it as the function's execution time under consideration. The procedure was performed for various kernel sizes: 2×2, 3×3, …, 15×15.

Results/Experiment Description. We compared the time consumption of the proposed code (see Fig. 2a) and the reference function cv::filter2d(…). All experiments imply convolution on 4500×4500 image size. We collected time consumption over 1000 runs (each datapoint – median of 7 runs as described above) for each CO implementation (cv:filter2D(…) or suggested method newCO(…)) for statistical analysis. The "nice state" of the measuring application was reduced to make it less prone to transient processes in the OS.

Experiment was performed with different kernel sizes. We present results in the form of histograms (time consumption/intervals vs relative frequency), see Fig. 3 for cv:filter2D(…) and Fig. 4 for newCO(…). It can be seen from Fig. 3 and Fig. 4 that the distributions are mostly close to normal and slightly skewed.

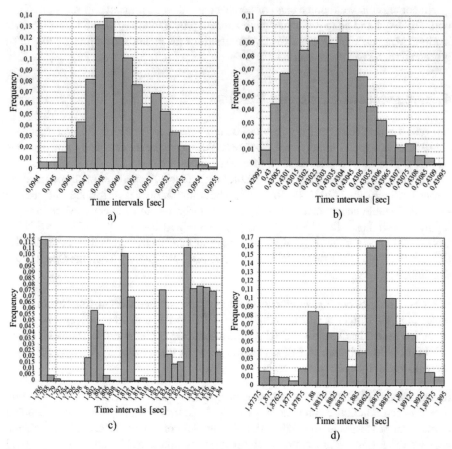

Fig. 3. Time consumption histograms for 1000 runs of the cv::filter2D(…). Kernel sizes: a) 2 × 2 b) 5 × 5; c) 8 × 8; d) 14 × 14;

For large kernels one may observe that the distributions are more like Gaussian mixtures (clearly seen in Fig. 3d and Fig. 4d). We speculate that this is due to the overheat of SOC/CPU. It was observed that running tests leads to temperature increase up to 65 C in about 10 min. We suppose that Linux kernel temperature service decreases the CPU frequency that leads to bimodal distribution.

Now we switch gears to the statistical analysis of the obtained data. We denote $\Omega_N = \{x_l; l = \overline{1, N}\}$, where x_l is the execution time of a single MxM convolution, N is the number of observations. For analysis we compute mean value

$$\bar{x} = \frac{1}{N} \sum_{l=1}^{N} x_l \qquad (5)$$

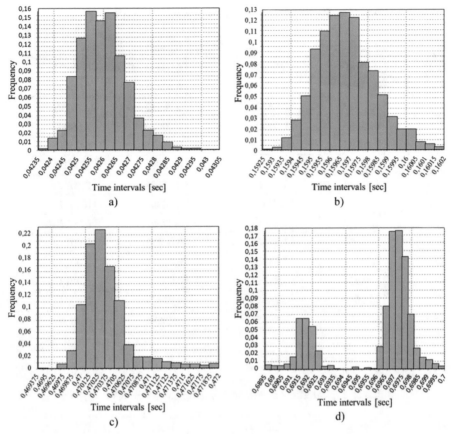

Fig. 4. Time consumption histograms for 1000 runs of the newCO(…). Kernel sizes: a) 2×2 b) 5×5; c) 8×8; d) 14×14;

median (assume that x_l are sorted)

$$MED = \begin{cases} x_{k+1}, & \text{if } N = 2k+1, \\ (x_k + x_{k+1})/2, & \text{if } N = 2k. \end{cases} \tag{6}$$

the unshifted standard deviation

$$S = \sqrt{\frac{1}{N-1} \sum_{l=1}^{N} (x_l - \overline{x})^2} \tag{7}$$

and Pearson's coefficient of variation

$$\overline{W} = \frac{S}{\overline{x}}, \overline{x} \neq 0 \tag{8}$$

The results are summarized in the Tables 1 and 2, first column defines the kernel size (MxM).

Table 1. Typical values and variability of basic (cv::filetr2D(…)) CO.

M	$\bar{x}$ [sec]	MED [sec]	S [sec]	W
2	0,09490	0,09478	0,00019	0,00197
3	0,17566	0,17575	0,00024	0,00139
4	0,28537	0,28521	0,00013	0,00047
5	0,43033	0,43033	0,00020	0,00047
6	0,60653	0,60632	0,00071	0,00117
7	0,81523	0,81541	0,00024	0,00030
8	1,81854	1,82405	0,01654	0,00910
9	1,77349	1,77476	0,00309	0,00174
10	1,87042	1,86961	0,00698	0,00373
11	1,87069	1,87164	0,00335	0,00179
12	1,87757	1,87769	0,00274	0,00146
13	1,88953	1,88958	0,00249	0,00132
14	1,88621	1,88689	0,00397	0,00210
15	1,89801	1,89551	0,00880	0,00464

Table 2. Typical values and variability of optimized (newCO(…)) CO.

M	$\bar{x}$ [sec]	MED [sec]	S [sec]	W
2	0,04259	0,04246	0,00009	0,00213
3	0,10027	0,10032	0,00012	0,00122
4	0,15701	0,15693	0,00023	0,00146
5	0,15969	0,15982	0,00016	0,00097
6	0,19694	0,19686	0,00024	0,00123
7	0,39356	0,39339	0,00017	0,00044
8	0,47034	0,47024	0,00040	0,00085
9	0,34871	0,34861	0,00027	0,00077
10	0,39878	0,39856	0,00046	0,00115
11	0,47659	0,47638	0,00065	0,00136
12	0,53202	0,53206	0,00028	0,00052
13	0,61070	0,61046	0,00060	0,00097
14	0,69725	0,69719	0,00070	0,00100
15	0,79887	0,79904	0,00028	0,00035

One can see that both unshifted standard deviation and coefficient of variation are small, thus computed means and medians are statistically significant. Further we proceed with medians only as they are typically considered due to their robustness to outliers (despite we have none). We use further subscript "b" (base case) for cv::filetr2D and "o" (optimized) for newCO. Table 3 presents comparison between the two.

Table 3. Estimates of CO time execution medians and corresponding differences of values.

M	MED_b [sec]	MED_o [sec]	MED_b - MED_o [sec]	$\frac{MED\_b}{MED\_o}$
2	0,09478	0,04246	0,05232	2,2322
3	0,17575	0,10032	0,07543	1,7519
4	0,28521	0,15693	0,12828	1,8174
5	0,43033	0,15982	0,27051	2,6926
6	0,60632	0,19686	0,40946	3,08
7	0,81541	0,39339	0,42202	2,0728
8	1,82405	0,47024	1,35381	3,879
9	1,77476	0,34861	1,42615	5,091
10	1,86961	0,39856	1,47105	4,6909
11	1,87164	0,47638	1,39526	3,9289
12	1,87769	0,53206	1,34563	3,5291
13	1,88958	0,61046	1,27912	3,0953
14	1,88689	0,69725	1,18964	2,7062
15	1,89551	0,79904	1,09647	2,3722

Plotting relative frequencies of MED_b - MED_o we get histogram as in Fig. 5.

Now let's approximate the growth rates of MED_o and MED_b to get more insights into their efficiency. For MED_o we use linear interpolation

$$MED\_o = 0,0536 \cdot l - 0,071 \qquad (9)$$

where l is the kernels' size ($l = 2..15$), see Fig. 6a.

The base case is much trickier and should be approximated with a more complicated curve (Fig. 6b). The final curve is obtained by sewing together two linear interpolations: one for small kernels

$$MED\_b = 0,144 \cdot l - 0,2467, l = 2..7 \qquad (10)$$

and another for large kernels

$$MED\_b = 0,0134 \cdot l + 1,707, l = 8..15 \qquad (11)$$

Both curves are presented in Fig. 7.

To compare relative methods efficiency it is useful to plot the time consumption of cv::filter2D(...) and newCO(...) against each other (see Fig. 8). It is clearly seen that the observations are quite heterogeneous due to two different methods employed by OpenCV.

For better comparison let's split small and large kernels apart once again. For kernels from 2×2 up to 7×7 (Fig. 9a), the regression line $MED\_o(MED\_b)$ can be written as follows:

$$MED\_o = 0,4134 \cdot MED\_b + 0,0091 \tag{12}$$

which means newCO(...) is 2.4 times faster than cv::filter2D(...).

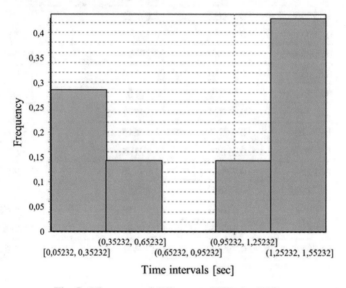

Fig. 5. Histogram of differences MED_b - MED_o

But for the kernel sizes from 8×8 up to 15×15, as (Fig. 9b) shows, the regression line is:

$$MED\_o = 13,437 \cdot MED\_b - 24,701 \tag{13}$$

which means that the efficiency of the newCO(...) drops drastically compared to the cv::filter2D(...). Thus it should not be used for kernels larger than 40×40.

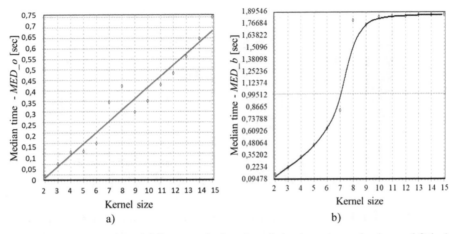

Fig. 6. Time consumption of different methods vs kernel size (experimental points and fitting): a) optimized CO; b) base CO.

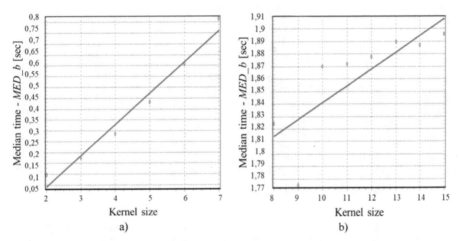

Fig. 7. Time consumption of cv::filter2D vs kernel size (experimental points and fitting): a) kernel sizes from 2 × 2 up to 7 × 7; b) kernel sizes from 8 × 8 up to 15 × 15

It is worth noting, we didn't use parallelism for acceleration. Employing OpenMP or implementing parallelism by any other means may improve presented results twice or even more. As well, additional 10%–20% of acceleration can be achieved by "bottleneck cycles" unrolling. Thus we expect that the upper limit of 40 × 40 kernel's size can be pushed even higher.

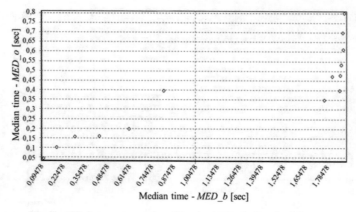

Fig. 8. CO time consumption (X axis: *MED_b*; Y axis: *MED_o*)

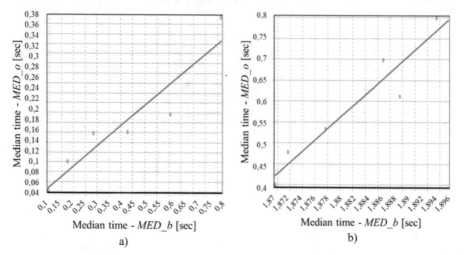

Fig. 9. Time consumption of newCO(…) (MED_o) vs time consumption of cv::filter2D(…) (MED_b): a) kernel sizes 2×2 up to 7×7; b) kernel sizes 8×8 up to 15×15

5   Summary and Conclusion

In conclusion, we propose a method of convolution operation (CO) acceleration. We show that many kernels utilized for practical applications can be reduced to integer form that allows for SIMD optimization. Despite SIMD itself leads to a significant boost of performance, we were able to push the frontiers even further by exploiting simultaneous loading of 48 (32 + 16) image bytes for each kernel row and reusing the data as long as possible compared to naïve one-by-one loading. Register v2 (see Fig. 2b) is used as a buffer/cache for cyclic shifts, i.e. loading operations are partially substituted with cyclic shift (see Fig. 2a).

We performed a series of measurements that show advantage of the presented approach compared to the cv::filter2D(…) function from OpenCV library for kernels smaller than 40×40. Documentation on OpenCV and our measurements suggest that two different methods of convolution are used in OpenCV that compare differently to our method. The most interesting cases of small kernels 2×2 to 7×7 show that the proposed method is almost 2.5 times faster than cv::filter2D(…) for 4500×4500 image and higher (up to 4.5 times) for image sizes lower than 4500×4500. We expect current approach to be useful for real-time image processing and convolutional neural networks training as it significantly reduces processing time.

Since current research was limited to single-threaded application and to using one method for all image sizes, we suppose that more research can be performed in the direction of parallelization and employing different methods for different image and kernel sizes. As well, we expect that cycles enrolling can lead to even higher acceleration and gain advantage for kernels larger than 40×40, but that is left for the future research.

References

1. Приставка, П, Шевченко, А.: Дослідження реалізації лінійного оператора згортки цифрового зображення при 16-бітних обчисленнях. Актуальні проблеми автоматизації та інформаційних технологій **20**, 78–90 (2016). (in Ukrainian)
2. Shevchenko, A., Vitaly, T.: A SIMD-based approach to the enhancement of convolution operation performance. CMiGIN, 447–458 (2019)
3. Shevchenko, A., Pylyp, P.: Enhancement of convolution operation performance using SIMD of AArch64. CMiGIN (2021) (accepted manuscript)
4. Image Filtering: [Online]. https://docs.opencv.org/4.x/d4/d86/group__imgproc__filter.html#ga27c049795ce870216ddfb366086b5a04 (18 Dec 2021). Accessed 20 Jan 2022
5. Flynn, M.J.: Very high-speed computing systems. Proc. IEEE **54**(12), 1901–1909 (1966)
6. Arm Cortex: A53 MPCore Processor Technical Reference Manual [Online]. https://developer.arm.com/documentation/ddi0500/j (18 Dec 2021). Accessed 20 Jan 2022
7. Qualcomm Extends Hexagon DSP [Online]. http://pages.cs.wisc.edu/~danav/pubs/qcom/hexagon_microreport2013_v5.pdf (18 Dec 2021). Accessed 20 Jan 2022
8. Qualcomm Hexagon DSP: An architecture optimized for mobile multimedia and communications [Online]. https://developer.qualcomm.com/download/hexagon/hexagon-dsp-architecture.pdf (18 Dec 2021). Accessed 20 Jan 2022
9. Joseph, P.M., Rajan, J., Kuriakose, K.K., Murty, S.S.: Exploiting SIMD instructions in modern microprocessors to optimize the performance of stream ciphers. Int. J. Comput. Netw. Inform. Secur. **5**(6), 56 (2013)
10. Griffith, A.: GCC: The Complete Reference. McGraw-Hill Inc. (1 Aug 2002)
11. Lopes, B.C., Auler, R.: Getting Started with LLVM Core Libraries. Packt Publishing Ltd. (26 Aug 2014)
12. Suresh, C., Singh, S., Saini, R., Saini, A.K.: A comparative analysis of image scaling algorithms. Int. J. Image Graph. Signal Process. **5**(5), 55 (2013)
13. Mahajan, A., Gill, P.: 2D convolution operation with partial buffering implementation on FPGA. Int. J. Image Graph. Signal Process. (IJIGSP) **8**(12), 55–61 (2016)
14. Arm Compute Library [Online]. https://developer.arm.com/ip-products/processors/machine-learning/compute-library (18 Dec 2021). Accessed 20 Jan 2022
15. Nicolau, A.: Loop Quantization: Unwinding for Fine-grain Parallelism Exploitation. Cornell University, Dept. of Computer Science (1985)
16. Xue, J.: Loop Tiling for Parallelism. Springer Science & Business Media (2000)

Research on the Application of Intelligent Logistics Innovation Technology in Logistics 4.0 Era

MeiE Xie and Hui Ye[✉]

Wuhan Business University, Wuhan 430056, China
yehui1986@126.com

Abstract. With the advent of the era of logistics 4.0, the typical feature is the application of intelligent logistics technology. At present, the representative technology of intelligent logistics includes perception, big data, automation, Internet plus and AI. This paper first describes the development of logistics industry, and then analyzes the role of intelligent logistics and the main intelligent logistics innovation technology, and finally discusses the application of intelligent logistics innovation technology. This paper adopts the research methods of literature research, qualitative analysis and field investigation, which lays a solid foundation for the application of intelligent logistics technology. In order to provide theoretical reference and practical suggestions for the development of China's logistics industry under the current background.

Keywords: Logistics 4.0 · Intelligent logistics · Logistics technology

1 Introduction

Logistics in China has developed rapidly in recent years. Statistics show that the total value of social logistics in China reached 300.1 trillion yuan in 2020, up 3.5% year on year [1]. Total revenue of logistics industry was 10.5 trillion yuan, up 2.2% year on year.

The change trend of Total social logistics in China is showed as Fig. 1:

With the proposal of concepts such as smart earth and smart city, and the rapid development of cloud computing, big data and Internet technologies, the logistics industry is gradually becoming intelligent and automated. With the basic support of technology and the guidance of policy, intelligent logistics has achieved preliminary development. At the micro logistics level, the society's low consumption of natural and social resources has gradually formed a strong desire to develop intelligent logistics. At the macro level of logistics, the deep integration of Internet and logistics industry makes intelligent logistics show a vigorous development trend.

With the continuous progress of science and technology, modern logistics technology has achieved a breakthrough development. At the same time, it has also ushered in a period of rapid development of a new generation of logistics. The fundamental purpose of modern logistics is to reduce cost and increase efficiency, meet customers'

Z. Hu et al. (Eds.): ICCSEEA 2022, LNDECT 134, pp. 76–86, 2022.
https://doi.org/10.1007/978-3-031-04812-8_7

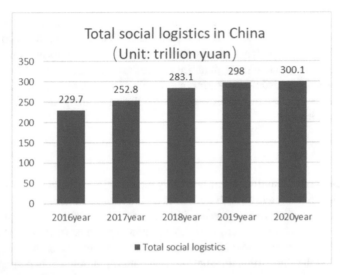

Fig. 1. Chart of total social logistics in China from 2016 to 2020

personalized needs, and show the development trend of informatization, networking and automation. In 2018, the scale of intelligent logistics market exceeded 400 billion yuan, and in 2019, the number of users in China's real-time distribution industry reached 421 million [2]. With the development of big data, cloud computing, artificial intelligence and other technologies, and the continuous improvement of logistics requirements in new retail, intelligent manufacturing and other fields, the scale of China's intelligent logistics market will also continue to expand. This paper will elaborate from the development of logistics industry, the role of intelligent logistics and the types of intelligent logistics technology, and finally study the specific application of intelligent logistics technology. This paper adopts the research methods of literature research, qualitative analysis and field investigation, which not only lays a solid foundation for the application of intelligent logistics technology, but also provides strong theoretical support for the development of intelligent logistics and the construction of logistics system in China.

2 Development of Logistics Industry

2.1 Logistics 1.0 Era

Logistics refers to the flow of raw materials, semi-finished products, finished products or related information through transportation, storage and distribution at the lowest cost in order to meet customer needs. Logistics management is the whole process of planning, implementing and managing goods from origin to consumption place. The traditional logistics industry is the logistics 1.0 era [3].

2.2 Logistics 2.0 Era

With the development of information technology and the progress of science and technology, the information about activities related to raw materials, products in process and finished products from supply to consumption place can be communicated more conveniently through many means. Modern logistics each link can be unified consideration, system operation. On the basis of studying customer demand information, enterprises can efficiently and economically plan, implement and control the activities of various functional links of logistics operation, thus leading to the reform of modern logistics concept. Therefore, modern logistics has entered the era of integrated logistics, namely the era of logistics 2.0.

2.3 Logistics 3.0 Era

Modern logistics and manufacturing are deeply integrated and shared in information. The manufacturing industry can be customer-oriented in procurement, manufacturing support and product sales. The manufacturing industry not only realizes the rapid response of enterprise information system and flexible manufacturing of production line, but also realizes the comprehensive integration of enterprise information flow, logistics and capital flow. At the same time, logistics industry into the supply chain management 3.0 era.

2.4 Logistics 4.0 Era

At present, the rapid development of the Internet has triggered a new industrial revolution. With the Internet at its core and the mobile Internet, big data, cloud computing, the Internet of Things, and automation as the foundation of the technological revolution, deep integration of the Internet and the real industry has been realized. This integration has brought earth-shaking changes to the traditional real industry, which has stepped into the era of "industrial Internet". In this context, the logistics industry ushered in a new round of revolution, so that the logistics industry has entered the era of "logistics Internet", namely, the era of logistics 4.0.

3   Research on Intelligent Logistics

3.1 Research on the Development Status of Intelligent Logistics

He Liming (2017) believes that it should be discussed from seven aspects: connection, data, mode, experience, intelligence, green and supply chain upgrading [4]. Fu Yu (2018), Zhang Chunxia (2018) and Liang Xikun (2016) pointed out that the development prospect of China's intelligent logistics is optimistic, but there are some problems, such as insufficient logistics data sharing, lack of perfect logistics information platform, shortage of intelligent logistics professionals, and the end logistics service capacity needs to be improved [5–7].

3.2 Research on the Development Path of Intelligent Logistics

Zhou Dingbo (2017) and others analyzed the transformation from traditional logistics to intelligent logistics, urgently needed to change the operation thinking of traditional logistics, create a cooperation mode of social logistics resource sharing, and promote the application, research and promotion of new logistics modes, including smart vehicle and goods matching, supply management and so on [8]. Liu Weihua (2019), Wang Xifang (2020) proposed two paths for the development of intelligent logistics: first, driven by e-commerce platforms represented by Ali and JD, it is driven from top to bottom; The second is to form a bottom-up drive based on logistics enterprises, digitize the basic business and realize online operation on the platform, so as to form flexibility and provide customized solutions to customers [9, 10].

3.3   Research on Intelligent Logistics Technology

Tan Hua (2016) and others put forward that informatization, intelligence, intensification and small batch customization are the development trend of logistics in the future from the perspective of the development of logistics technology [11]. Jiang Dali (2018), Zheng Qiuli(2019)believes that the key technologies of intelligent logistics can be classified from three directions: Informatization, intelligence and system integration. Big data and artificial intelligence greatly promote the utilization of logistics operation data. Traditional logistics enterprises are developing towards digitization, intelligence and networking. Intelligent logistics can improve operation efficiency in many ways [12, 13].

4   Role of Intelligent Logistics

4.1   Build a Logistics Ecosystem for Continuous Improvement

The construction of intelligent logistics will accelerate the development of the local logistics industry and become a logistics ecosystem integrating warehousing, transportation, distribution, information services and other functions. Intelligent logistics system can break industry restrictions, coordinate the interests of departments, realize intensive and efficient operation, and optimize the allocation of social logistics resources [14].

4.2   Promote Logistics Cost Reduction and Efficiency

Intelligent logistics can greatly reduce the cost of manufacturing and other industries in the secondary industry. With the development of technology, the application of intelligent logistics key technologies such as wireless positioning, object identification and identification tracking and other new information technology can effectively realize the intelligent scheduling management of logistics, strengthen the scientific management of logistics, so as to reduce logistics costs.

4.3   Promote Integration of Production, Purchase and Sales

The integration of network will certainly promote the integration of intelligent production and intelligent supply chain, so that enterprise logistics can be fully integrated into enterprise management intelligently, and the boundaries of working procedure and process can be broken to create intelligent enterprises [15]. Wisdom than traditional logistics consumes less resources, to improve product competitiveness, promote the integration of supply chain, helps to solve the logistics field information communication, thus improve the competitiveness of the enterprises in the logistics, establish enterprise new economic growth point.

4.4 Promote Local Economic Development

Intelligent logistics can effectively promote the innovation of the logistics industry from the aspects of technology, personnel, organization and standards, provide technical support for the transformation and development of the logistics industry, and make up the shortcomings of the logistics industry. Intelligent logistics integrates a variety of service functions, reflecting the needs of modern economic operation characteristics, so as to improve production efficiency and integrate social resources.

5   Types of Intelligent Logistics Technology

With the in-depth development of the information age, intelligent logistics supported by emerging technologies has become the main mode and basic trend of industry development in the logistics 4.0 era [16].

The Intelligent logistics technology architecture is showed as Fig. 2:

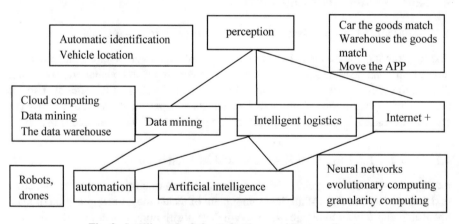

Fig. 2. Intelligent logistics technology architecture diagram

5.1   Internet of Things

The Internet of Things means the Internet connected with all things. It is a network system developed by combining information technology with the Internet and an important representative of modern information technology. The Internet of Things was originally called sensor network in China. The concept of the Internet of Things has been in use since 1999. Since then, the Internet of Things has been widely applied in industry, agriculture, transportation, medical treatment, education, military, environment and other fields, especially in the logistics industry [17]. At present, the Internet of Things technology is mainly composed of radio frequency identification technology, sensor technology, nanotechnology and intelligent embedding technology. It is suitable for pallets, containers, freight and other logistics equipment, and has more and more functions in the field of intelligent storage and intelligent distribution, and has become a key technology to determine the competitive pattern of the logistics industry in the future.

5.2 Big Data

Big data refers to data sets containing huge volumes, and is also a general term for non-traditional strategies and technologies for collecting, organizing and processing large data sets. It can make up for the deficiency of conventional software tools in obtaining, storing, managing and processing information in a short period of time. Big data technology has been widely applied in the fields of national governance, government decision-making, enterprise production and personal life, and the application trend in the logistics industry is increasingly obvious. As we all know, the logistics industry produces a large amount of commercial data and personal information, which need to use data technology to collect, analyze and analyze, which plays an important role in improving the efficiency of logistics and provides a broad space for the transformation of big data from theory to practice. At present, big data technology has become an emerging technology that many logistics enterprises focus on and widely use, playing an important role in logistics demand prediction, equipment maintenance prediction, supply chain risk prediction, network and route planning [18].

5.3 Cloud Computing

Cloud computing can be seen as a more powerful web service technology related to information technology, software and the Internet, which can complete the processing of huge data in an extremely short time. Cloud computing is characterized by technical virtualization, dynamic scalability, high flexibility, strong reliability and high cost performance. Relying on its powerful processing capacity, cloud computing is favored by logistics industry. In fact, from the technical point of view, cloud computing and big data are like two sides of the same coin. Big data gives birth to cloud computing, which means that a single computer cannot process a large amount of data, so cloud computing technology must be used for storage and distributed processing. With the development of the logistics industry and the arrival of the era of big data, cloud computing has become a key technology of intelligent logistics because it can provide powerful storage and processing capabilities for logistics. For example, the cloud computing center in intelligent logistics can sense, collect, store and transfer massive information to provide customized services for users.

5.4 Artificial Intelligence

Artificial intelligence refers to the use of multidisciplinary knowledge, research and development for simulation, extension and expansion of human intelligence a new technology. In essence, artificial intelligence lies in the simulation of human brain, including the simulation of human brain structure and the simulation of human brain function, so as to promote the informatization and intelligence of social development and realize the self-promotion and all-round development of human beings [19]. At present, the application of artificial intelligence in the logistics industry is still in its infancy, mainly used to promote e-commerce platforms, but its advantages of high efficiency and low cost have been explored, and it has become a key technology that the logistics industry urgently needs to develop and apply. Unmanned warehousing, unmanned delivery and unmanned aerial vehicles are all areas of artificial intelligence application. Logistics companies such as Cainiao and JD.com have begun experimenting with new AI technologies such as image recognition.

5.5 Blockchain

Blockchain is a new technology application mode that utilizes distributed data storage, point-to-point transmission, consensus mechanism, encryption algorithm and other computer technologies. In essence, it is a shared data technology. Blockchain is divided into three types: public blockchain, private blockchain and joint blockchain. It has the characteristics of decentralization, traceability, openness, independence, security and anonymity. At present, blockchain technology is widely used in public administration, international trade, equity registration, securities trading, transportation and other fields, and its application in the logistics industry has been rapidly developing. Blockchain technology is also an important technical support for intelligent logistics, playing a huge role in process optimization, goods tracking, main body credit investigation and financial services, helping to build multi-trust, efficient and convenient logistics system, and promoting the deep integration of information technology and the real economy.

6   Application of Intelligent Logistics Innovation Technology

The application scenarios of intelligent logistics are diverse, typical of which are smart distribution, smart storage yard, smart port and smart dock.

The Intelligent logistics core technology application is showed as Fig. 3:

6.1   The Application of Artificial Intelligence

Artificial intelligence can realize intelligent allocation of logistics resources, intelligent optimization of logistics links and intelligent improvement of logistics efficiency by empowering each link and field of logistics [20].

In the storage link, for the problem of warehouse location of logistics enterprises, artificial intelligence can give the optimal site selection scheme according to customer flow and demand, geographical location of suppliers and manufacturers, transportation

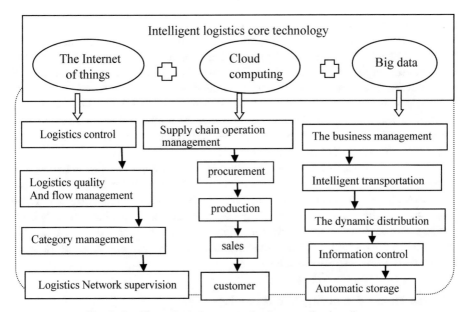

Fig. 3. Intelligent logistics core technology application diagram

economy and other conditions. Artificial intelligence can reduce the interference of human factors, make site selection more accurate, greatly reduce the transportation cost of logistics enterprises, and improve corporate profits.

In inventory management, artificial intelligence can accurately predict users' demand at different times by analyzing users' historical consumption data, and then dynamically adjust the inventory level to maintain the orderly circulation of enterprise inventory and avoid the cost waste of blind production, so that enterprises can always provide high-quality production services.

In the process of cargo sorting, loading and unloading, handling, the application of logistics robot directly improves the efficiency and benefit of logistics system. For example, in the sorting of goods, computer vision technology and intelligent robots can be used for automatic identification, sorting and handling.

In terms of logistics, AI can plan the best transportation route by integrating road traffic information, and the timeliness of logistics will gradually improve. The wide use of intelligent delivery cabinets has also greatly improved the efficiency of logistics terminal distribution and greatly reduced the industry's dependence on human resources.

6.2 The Application of Blockchain Technology

Relying on blockchain technology, capital flow, logistics and information flow of each link can be truly and reliably recorded and transmitted, thus solving the problem of trust in the logistics industry.

To ensure the safety of cargo. Chain blocks, each block contains detailed information, such as the seller, the buyer, price, terms of the contract and any relevant information in detail, through unique cut in verification of the signature of the both and all parties, if the entire network encrypted record is consistent, is the data effectively, and uploaded to the entire network to achieve information sharing and information is absolutely safe.

Optimize cargo transportation route. The blockchain records all steps in the process from delivery to receipt of goods, ensuring traceability of information, so as to optimize the transport route of goods and avoid the occurrence of lost bags and false claims. Enterprises can also master the logistics direction of products through blockchain to prevent channeled goods, which is conducive to cracking down on counterfeits and ensuring the interests of offline dealers at all levels.

To solve the financing of small, medium and micro enterprises in logistics. The assets recorded by block chain technology cannot be changed or forged, which can make all the goods in the logistics chain traceable, falsifiable and untamper with. It can value and capitalize the informationized goods and realize the capitalization of logistics goods. Using the blockchain foundation platform, funds can be effectively and quickly connected to the logistics industry, thus improving the business environment for small and medium-sized enterprises [21].

6.3 The Application of Big Data Technology

Big data has a variety of application scenarios in intelligent logistics, including personalized transport capacity, accurate matching, intelligent fleet management, efficient and collaborative multi-modal transport, supporting regional economic decisions, etc. Big data technology through effective processing and analysis of logistics data, mining valuable information for logistics operation, so as to support scientific and reasonable management decisions, is a universal demand of logistics enterprises.

Data sharing. The application of big data to realize the interconnection of logistics basic data, such as the upstream and downstream parties of the supply chain to share basic data such as goods and vehicles, avoids the repeated collection of logistics information, eliminates the information island of logistics enterprises, and improves the level and efficiency of logistics services.

Sales forecasting. The application of big data to collect massive data of merchants' historical sales and users' consumption characteristics, with the help of big data prediction and analysis model, accurate prediction of sales in various scenarios such as orders, promotions and clearance can provide scientific basis for warehouse goods stocking and operation strategy formulation.

Network planning. By using historical big data and sales forecast, the multi-dimensional operation and research model of cost, timeliness and coverage is constructed to optimize the layout of storage, transportation and distribution networks.

Inventory deployment. Scientific inventory deployment in multi-level logistics network, intelligent forecast replenishment, inventory coordination, inventory turnover, improve spot rate, improve the efficiency of the whole supply chain.

Industry insight. Using big data technology, mining and analyzing the characteristics and rules of logistics operation in different industries such as electronics, household appliances and clothing and different links such as warehouse distribution, express delivery and urban distribution, and summarizing the best solutions to provide valuable solutions for logistics enterprises [22].

6.4 Cloud Computing Applications

Cloud computing is a network model that allows users to easily access, share resources, adapt to changes and be accessed in real time.

Integrate logistics industry resources. Logistics enterprises make statistics of customer information through cloud computing and formulate logistics operation routes. The cloud computing function can also analyze the logistics preferences and needs of current regional users. Logistics companies can design special logistics distribution routes based on the conclusions drawn from cloud computing, which greatly improves the efficiency of logistics distribution.

Provide communication and data sharing platform. Cloud computing companies provide a very convenient cloud communication platform between companies, the same company can realize information sharing through cloud communication, to achieve a win-win situation. Cloud computing platform can realize the whole process of electronic logistics enterprises, customers only need to access the Internet can query the information they want.

Provide storage services for enterprises. Logistics enterprises use cloud storage to ensure data security and obtain data backup and recovery. Cloud computing services can also share data with affiliates, preventing it from being corrupted in transit. Cloud computing can also provide a monitoring system for enterprises, which can control the work of enterprises through remote servers, query logistics information anytime and anywhere, and monitor the running status of enterprises [23].

7 Conclusion

With the advent of the era of logistics 4.0, technological innovation has become the development trend of the logistics industry in the new era. The application of intelligent logistics innovation technology can improve the modernization level of logistics, promote the rapid development of the logistics industry, so as to achieve the goal of leapfrog development of the logistics industry. At the government level, as China is in the period of industrial transformation and development, research on the application of innovative logistics technology is conducive to the government to further promote the integration of intelligent information technology and logistics industry on the basis of vigorously developing smart economy, so as to promote the transformation and development of logistics industry. At the enterprise level, promoting the application of intelligent logistics technology can reduce the total logistics operation cost of enterprises, improve enterprise profits, and promote the motivation and enthusiasm of enterprises to integrate into the construction of intelligent logistics [24].

Under the background of logistics 4.0, studying the impact of innovative logistics technology on the upgrading of logistics industry and exploring the driving role of intelligent logistics construction on regional economy has important practical value for guiding smart economy and logistics industry to accelerate industrial integration and improve circulation efficiency.

Acknowledgment. This paper is supported by project of The National Social Science Fund of China (18BJY138).

References

1. He, L.: Review of the development of logistics industry in China in 2019 and prospect in 2020. China Circul. Econ. **8**(02), 88–92 (2020)
2. Ren, H.X.: Comprehensively promote the high-quality development of intelligent logistics industry. Modern Logist. **7**(1) (2019)
3. Wang, J.X.: "Logistics 4.0" China logistics internet era opens. Logist. Technol. Appl. **12**(10), 77–80 (2018)
4. He, L.: Development status and trend of intelligent logistics in China. China's National Conditions **12**, 9–12 (2017)
5. Fu, Y.: Development trend and promotion strategy of intelligent logistics in China. Foreign Econ. Trade Practice **1**, 90–92 (2018)
6. Zhang, C.X., Peng, D.H.: Development countermeasures of China's intelligent logistics. China's Circul. Econ. **27**(10), 35–39 (2018)
7. Liang, X.K.: Main problems and countermeasures faced by the development of intelligent logistics in China. Modern Econ. Inform. **12**, 376–377 (2016)
8. Zhou, D., Deng, S.: Transformation strategy from traditional logistics to intelligent logistics. Open Guide (6), 105–109 (2017)
9. Liu, W.H.: Development mode, path and trend of China's intelligent logistics. Modern Logist. News (A04) (2019)
10. Wang, X., Zhou, J., Gu, Y.: Research on intelligent logistics allocation mode under the new retail background – taking Ali FRESHIPPO as an example. Logist. Eng. Manag. **42**(1), 22–25 (2020)
11. Tan, H., Lin, K., Yang, S.: Analysis on the development trend of current intelligent logistics technology. Mobile Commun. **40**(21), 45–49, 57 (2016)
12. Jiang, D., Zhang, W., Wang, Q.H.: Research on key technologies and construction countermeasures of intelligent logistics. Packag. Eng. **39**(23), 9–14 (2018)
13. Zheng, Q.: Development model, problems and countermeasures of intelligent logistics in China. Bus. Econ. Res. (18), 108–111 (2019)
14. Anitha, P., Malini, M.P.: A review on data analytics for supply chain management: a case study. Int. J. Inform. Eng. Electron. Bus. (IJIEEB) **10**(5), 30–39 (2018)
15. Zine, B., Ghalem, B., Abdelkader, N.: A cost measurement system of logistics process. Int. J. Inform. Eng. Electron. Bus. (IJIEEB) **10**(5), 23–29 (2018)
16. Si, Y.: Research on the development strategy of intelligent logistics under the background of big data. China Market **03**(33), 121–125 (2019)
17. Kuang, M., Kuang, D.: Analysis on the development and innovation path of China's intelligent logistics industry. Gansu Social Sci. **11**(06), 224–228 (2019)
18. Ma, H.M.: Accelerate the construction of intelligent logistics industry chain. People's Daily **7**(10) (2019)
19. Tiwari, A.: Impact of information and communication technology on logistics industry: an analysi. J. Manag. **02**(12), 201–210 (2021)
20. Jobel Santos, C., Mauro, S.: An exploratory study on emerging technologies applied to logistics 4.0. J. Manag. **10**(35), 68–77 (2020)
21. Khan, S.: Cloud computing: issues and risks of embracing the cloud in a business environment. Int. J. Educ. Manag. Eng. (IJEME) **9**(4), 44–56 (2019)
22. Cheowsuwan, T., Arthan, S., Tongphet, S.: System design of supply chain management and Thai food export to global market via electronic marketing. Int. J. Mod. Educ. Comput. Sci. **9**(8), 1–8 (2017)
23. Zhang, T.: Construction and operation of intelligent logistics business system under the background of big data. Bus. Eco. Res. **21**, 86–89 (2019)
24. Li, J.: Intelligent logistics model reconstruction based on big data cloud computing. China's Circul. Econ. **33**(2), 20–29 (2019)

An Optimal Pricing and Inventory Control Policy for Online Sale

Ming Li[1], Ying Lu[2(✉)], Yueqi Wu[3], and Xuejing Qi[3]

[1] School of Mechanical and Electronic Engineering, Suzhou University, Anhui 234000, China
[2] Suzhou Center for Disease Control and Prevention, Anhui 234000, China
865205856@qq.com
[3] School of Business, Suzhou University, Anhui 234000, China

Abstract. Online sale plays an important role in the sale system and improve rapidly in recent years. A key character of online sale is that the demands are highly sensitive to the prices because the prices of goods are easier to compare between different retailers. In this paper, to deal with the joint pricing and inventory control problem for online sale, first we construct a mathematical model in which the demand is dependent on the price and the aim is to maximize the average profit; The theoretical analysis is conducted to deduce the optimal pricing and inventory policy for the case when the demand is a function with respect to the price. Based on the properties of the objective function, an iterative algorithm is developed to determine the optimal price and order quantity. Finally according to numerical experiments and sensitivity analysis, we examine the performance of the proposed policy and the result reveals that for online sale, it is necessary to consider the relationship between the demand and the price, and the proposed joint policy could improve the sale profit.

Keywords: Pricing · Inventory control · Online sale

1 Introduction

In recent years, with the advent of the era of big data, Internet technology has been further developed, the application of e-commerce become the main trend of the current era, many enterprises in China begin to apply e-commerce to all aspects of production and operation [1]. China's online retail sales reached 11.76 trillion yuan in 2020, up 10.9% year on year, according to the National Bureau of Statistics. The online retail sales of physical goods reached 9.76 trillion yuan, up 14.8% year on year, accounting for 24.9% of the total retail sales of consumer goods, an increase of 4.2 percentage points over the previous year [2]. In the case of urban e-commerce is close to saturation, e-commerce is gradually infiltrating into rural areas. So far, there are more than 30000 agricultural websites in China, in the country's 1000 counties has been built 250000 electricity village service point, initially formed including agricultural futures trading, online electronic trading of agricultural commodities, agricultural B2B e-commerce sites and retail agricultural network platform, such as multi-level agricultural e-commerce

© The Author(s), under exclusive license to Springer Nature Switzerland AG 2022
Z. Hu et al. (Eds.): ICCSEEA 2022, LNDECT 134, pp. 87–96, 2022.
https://doi.org/10.1007/978-3-031-04812-8_8

market system and network system. Online retail sales in rural areas reached 1.37 trillion yuan in 2018, up 30.4% year on year. Online retail sales of agricultural products reached 230.5 billion yuan, up 33.8% year on year. From January to November 2019, online retail sales in rural areas reached 1,522.9 billion yuan, up 19% year on year. Online retail sales of agricultural products reached 355.6 billion yuan, up 26.6% year on year. It can be seen that the market prospect of China's rural e-commerce industry is very broad [3]. There is a key character of the online retail sale that the demand is sensitive to the price. The price reduction on Double Eleven is between 15–30% of the original price, which is very objective. During the period of "double 11" shopping festival, Tmall xtep flagship store of synchronous growth of 40%, the group's online sales rose by about 50%, more than 520 million yuan, business in America and Europe is affected by the epidemic overseas, but under the help of all kinds of online marketing activities, Online sales in the Americas, Europe and the Middle East and Africa (EMEA) grew 45% and 52%, respectively, year on year. Xtep has a great price reduction on Double Eleven. A pair of shoes more than 200 yuan off the price of shoes 60 yuan, but sales increased by 50%. This is a good example to reveal the importance of the joint policy for online retailer. In practice, online retailers could maximize the profit if the reasonable pricing and inventory control policy are determined. This is the main motivation of the research in this paper. We aim to obtain the joint optimal pricing and inventory control policy, which help online retailers to obtain the maximum profit.

The research field is receiving more and more attention [4–6]. In recent decades, an increasing number of firms in many industries have been hit hard by excess inventory as a result of the rapid upgrading and replacement of the products [7, 8]. In the reselling model, the e-tailer purchases promotional product from the firm at a wholesale price and then sells it at a profit to customers [9–11]. The last stream of literature is the effects of price promotions. Previous studies, e.g., [12–14] focus primarily on the effect of the promotional price or storage on promotional demand. PingpingChen [15] develops game theoretic models to explore how the firm and e-tailers, under consideration of retail competition, should strategically use these two business models in promotion. The rest part of this paper is organized as the following. In Sect. 2 we construct the model of the online sale pricing and inventory problem, In Sect. 3, we analyze the joint optimal policy from the view of theory. In Sect. 4, numerical experiments are conducted to examine the performance of the proposed method, and sensitivity analysis is completed to find the managerial insight for online retailers. The last section is the summary and conclusions.

2   Mathematical Model for Online Sale Pricing and Inventory

For the simplification of theoretical analysis we firstly construct a mathematical model in which the demand is the deterministic function of the price, such as constant price elasticity demand function or linear demand function. We introduce the following notations and assumptions:

1) p means the unit selling price
2) c_f means the fixed order cost
3) c means the unit purchase cost

4) h means the unit inventory holding cost
5) $I(t)$ means the inventory level at time t
6) d means the demand rate, which is a known function with respect to the price. Assume that the demand function is $d = D(p)$ and D is a monotone decreasing function with respect to p.
7) The system is operating at an infinite period level, the order quantity of each period is Q.
8) Marginal revenue is a monotonically increasing function of p.

According to above assumptions, the length of each period is $T = Q/d$. the total cost contains the fixed order cost (c_f), the inventory holding cost ($QTh/2$) and the purchase cost (cQ). Then the total cost in one period is $c_f + QTh/2 + cQ$, and total profit in one period is $pQ - c_f - QTh/2 - cQ$, and the average profit is:

$$\frac{pQ - c_f - QTh/2 - cQ}{T} = \frac{pQ - c_f - QTh/2 - cQ}{Q/d} = pd - c_f d/Q - \frac{Qh}{2} - cd$$

It is obvious that the average profit (AP) is the function of p and Q, so the problem is to find the optimal price p the order quantity Q which maximize the average cost in each period. Then we can obtain the mathematical model for this problem (P1):

$$P1 : \max_{p,Q} AP(p, Q) = pd - \frac{c_f d}{Q} - \frac{Qh}{2} - cd \tag{1}$$

$$\text{s.t. } I(t) \in [0, Q] \tag{2}$$

$$d = D(p) \tag{3}$$

3 Theoretical Analysis and Iterative Algorithm

3.1 Theoretical Analysis

Proposition 1. *if* $2D'(p) + \left(p - c - \frac{c_f}{Q}\right)D''(p) < 0$ *holds, the optimal price and optimal order quantity that maximize the average profit exists.*

Proof. When the average profit is maximized, the following condition should be satisfied:

$$\frac{\partial AP(p, Q)}{\partial p} = D(p) + \left(p - c - \frac{c_f}{Q}\right)D'(p) = 0 \tag{4}$$

Considering the second-order partial derivative and assumption 8, we obtain:

$$\frac{\partial^2 AP(p, Q)}{\partial p^2} = D'(p) + \left(p - c - \frac{c_f}{Q}\right)D''(p) + D'(p)$$

$$=2D'(p) + \left(p - c - \frac{c_f}{Q}\right)D''(p) \tag{5}$$

Therefore, if $2D'(p) + \left(p - c - \frac{c_f}{Q}\right)D''(p) < 0$ for any given Q, we can obtain the optimal p that maximize the $AP(p, Q)$ by solving Eq. (4). That is to say, the optimal price is actually the function of the order quantity Q, i.e.,

$$p^* = P^*(Q) \tag{6}$$

Substituting Eq. (6) into Eq. (1) we find that the average profit of the system is a function of one variable Q, so we can denote the objective function as $AP_1(Q)$. If the optimal order quantity Q is determined, then the optimal price p can be obtained by Eq. (4). Up to now, the problem is transformed into finding Q^* to maximize $AP(Q)$. Firstly, the following condition should be satisfied:

$$\frac{dAP_1(Q)}{dQ} = \frac{\partial AP(p, Q)}{\partial Q} + \frac{\partial AP(p, Q)}{\partial p} \cdot \frac{dp}{dQ} = 0 \tag{7}$$

From Eq. (4) we know that when $p = P^*(Q)$, $\frac{\partial AP(p,Q)}{\partial p} = 0$ holds, then we have:

$$\frac{dAP_1(Q)}{dQ} = \frac{\partial AP(p, Q)}{\partial Q} = 0$$

$$Q^* = \sqrt{\frac{2c_f D(P^*(Q^*))}{h}} \tag{8}$$

□.

Proposition 2. If $c_f D'^2 + 2QD(2D'^2 - DD'') = 0$ has the unique solution, then AP1(Q) is a convex-concave function with respect to Q.

Proof. From Eq. (5) yields:

$$p + \frac{D}{D'} - c - \frac{c_f}{Q} = 0 \tag{9}$$

Take the derivative of Eq. (9)

$$\frac{d}{dQ}\left(p + \frac{D}{D'}\right) + \frac{c_f}{Q^2} = 0 \tag{10}$$

$$\frac{dP^*(Q)}{dQ} = \frac{c_f D'^2}{Q^2(DD'' - 2D'^2)} \tag{11}$$

$$\frac{d^2 AP_1(Q)}{dQ^2} = \frac{-2c_f D}{Q^3} + \frac{c_f D'}{Q^2} \cdot \frac{dP^*(Q)}{dQ} = \frac{c_f D}{Q^3}\left(-2 - \frac{c_f D'^2}{QD(2D'^2 - DD'')}\right) \tag{12}$$

Let $\frac{d^2 AP_1(Q)}{dQ^2} = 0$ we can obtain the solution is $\hat{Q} = \frac{-c_f D'^3}{4D - 2D^2 D''} \tag{13}$

If Eq. (13) has the unique solution, then

when $Q < \hat{Q}$, $\frac{d^2 AP_1(Q)}{dQ^2} > 0$; when $Q < \hat{Q}$, $\frac{d^2 AP_1(Q)}{dQ^2} < 0$

Therefore AP1(Q) is a convex-concave function with respect to Q.

□.

3.2 An Iterative Algorithm for Joint Pricing and Inventory Decisions

From Eq. (4) we know that for given order quantity Q, the optimal price p* could be obtained. Take two normal demand functions as examples.

(1) $d = D(p) = \alpha p^{-\beta}$, where α is the scale constant and β is the price elastic coefficient. In this case the average cost of the retailer is:

$$AP(p, Q) = \left(p - c - \frac{c_f}{Q}\right)\alpha p^{-\beta} - \frac{Qh}{2} \tag{14}$$

From Eq. (4) the optimal price for given Q is:

$$p^* = P^*(Q) = \frac{\beta\left(c + \frac{c_f}{Q}\right)}{\beta - 1} \tag{15}$$

(2) $d = D(p) = a - bp, b > 0, p < \frac{a}{b}$. In this case the average cost of the retailer is

$$AP(p, Q) = \left(p - c - \frac{c_f}{Q}\right)(a - bp) - \frac{Qh}{2} \tag{16}$$

From Eq. (4) the optimal price for given Q is:

$$p^* = P^*(Q) = \frac{c + \frac{c_f}{Q} + \frac{a}{b}}{2} \tag{17}$$

So in this section, we propose an iterative algorithm to determine the order quantity Q by Eq. (8).

Step 1: Initialization, let $k = 0$ and $Q_0 = $ infinity.
Step 2: calculate $P^*(Q_k)$ according to Eq. (4)
Step 3: calculate $Q_{k+1} = \sqrt{\frac{2c_f D(P^*(Q_k))}{h}}$
Step 4: terminate if $|Q_{k+1} - Q_k| < \varepsilon$ holds, otherwise let $k = k + 1$ and go to step 2.

4 Numerical Experiments

In this section we conduct numerical experiments to examine the theoretical analysis and the iterative algorithm. It is assumed that the fixed order cost $c_f = 100$, the unit purchase cost is $c = 5$, the unit holding cost is $h = 1$. Two common demand functions in supply chain research are considered, respectively.

In the first experiment we assume the demand function is as Eq. (18) and the price elastic coefficient $\beta = 3$. Table 1 shows the iteration process and the optimal price and the quantity with different initial values of order quantity.

$$d = D(p) = \alpha p^{-\beta} \tag{18}$$

From Table 1 we can see that for different initial values of order quantity, the iterative algorithm converges in 3 or 4 generations, and the optimal order quantity is 1235 and the optimal price is 7.62. Now we conduct another experiment to examine whether above result is right. The average profits when the order quantity is in the interval [100, 2000] and the the price in [5, 10] are calculated. The result is shown in Fig. 1. The average profit function is convex with respect to the price and order quantity. The optimal decision variable are $p = 7.62$ and $Q = 1235$, the same as obtained by the iterative algorithm.

Table 1. The iteration process and experiment result (Eq. (18) case)

Q_0	Q_1	Q_2	Q_3	Q_4	Q^*	p^*
100	962	1226	1235	1235	**1235**	**7.62**
200	1096	1231	1235	1235	**1235**	**7.62**
500	1193	1234	1235	1235	**1235**	**7.62**
3000	1252	1235	1235	1235	**1235**	**7.62**
5000	1257	1235	1235	1235	**1235**	**7.62**

From Eq. (15) and the iterative we can see that the optimal decisions of pricing and order quantity are infected by the price elastic coefficient. So a numerical experiment is conducted to examine the sensitivity. Let β is in the interval [1.2,4] with a increment of 0.2 and record the optimal prices and order quantities in different cases.

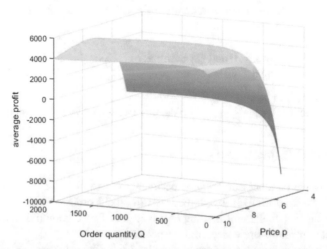

Fig. 1. The average profit with different price and order quantity (Eq. (18) case)

The result of sensitivity analysis is shown in Fig. 2. Along with the increase of price elastic coefficient, the optimal price decreases and the optimal quantity increases. The two optimal decisions are highly sensitive to the coefficient, especially when the coefficient is close to 1. The result of the sensitivity analysis reveals the managerial insight that for online retailers it is of great importance to obtain the accurate demand-price relationship. The incorrect estimation of demand function will reduce a great deal of the profit of online retailers.

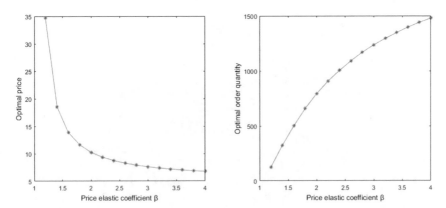

Fig. 2. Sensitivity analysis with respect to price elastic coefficient (Eq. (18) case)

In this section the other demand function is considered to examine whether the conclusion is consistent with the previous experiment. Assume that $a = 30000$, $b = 3000$, other parameters are the same as those in the previous experiment. Table 2 shows the iteration process and the optimal price and the quantity with different initial values of order quantity when the demand function is as Eq. (19).

$$d = D(p) = a - bp, b > 0, p < \frac{a}{b} \tag{19}$$

Table 2. The iteration process and experiment result (Eq. (19) case)

Q_0	Q_1	Q_2	Q_3	Q_4	Q^*	p^*
100	1095	1214	1215	1215	**1215**	**7.54**
200	1162	1214	1215	1215	**1215**	**7.54**
500	1200	1214	1215	1215	**1215**	**7.54**
3000	1221	1215	1215	1215	**1215**	**7.54**
5000	1222	1215	1215	1215	**1215**	**7.54**

From Table 2 we can see that for different initial values of order quantity, the iterative algorithm converges rapidly, and the optimal order quantity is 1215 and the optimal price is 7.54. Now we conduct another experiment to examine whether above result is right. The average profits when the order quantity is in the interval [100, 2000] and the price in [5, 10] are calculated. The result is shown in Fig. 3. The average profit function is convex with respect to the price and order quantity. The optimal decision variable are $p = 7.54$ and $Q = 1215$, the same as obtained by the iterative algorithm.

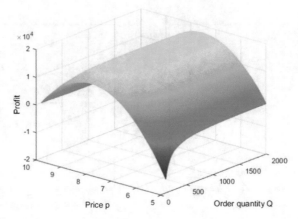

Fig. 3. The average profit with different price and order quantity (Eq. (19) case)

The result of sensitivity analysis is shown in Fig. 4. Along with the increase of parameter b, the optimal price and order quantity decrease. The two optimal decisions are also highly sensitive to the parameter.

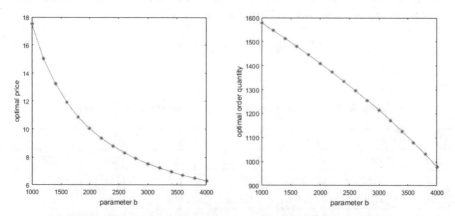

Fig. 4. Sensitivity analysis with respect to parameter b (Eq. (19) case)

5 Summary and Conclusion

In this paper we research the joint pricing and inventory control problem for online sale where the demand is highly sensitive to the price. The mathematical model is constructed for the problem and the theoretical analysis is provided. Based on the analysis, an iterative algorithm is developed to obtain the optimal pricing and order quantity with known demand functions. Numerical experiments show that the iteration process converges in 3 or 4 generations and the result is totally correct. Two common demand functions are considered in the experiments and similar results prove the efficiency of the proposed method. The main conclusions proposed in this paper are:

1) For online retailer sale, to maximize the average profit, it is necessary to conduct the joint pricing and inventory control policy because of the high sensitivity to prices.
2) The optimal price and order quantity exist, and the proposed iterative algorithm converges rapidly to correct optimal values for normal demand functions.
3) Sensitivity analysis reveals the reveals the managerial insight that for online retailers it is of great importance to obtain the accurate demand-price relationship. The incorrect estimation of demand function will reduce a great deal of the profit of online retailers.

Acknowledgments. This research is supported by the Excellent Young Talents support Program of Anhui Universities (gxyq2021219), General project of Natural Science Foundation of Anhui Province (2108085MG235), Quality project of Anhui Province (2020mooc568) and Scientific Research Platform Project of Suzhou University (2019ykf10).

References

1. Hang, T.: Study on problems in the development of forest product e-commerce in Jiangsu Province. Nanjing Forestry University (2019) (in Chinese)
2. Zhang, T.: China's industrial e-commerce transaction scale reaches 27.5 trillion yuan in 2020. Comput. Netw., **47**(10), 2 (2021) (in Chinese)
3. Pan, H., Liu, M., Wu, H.: Current situation, existing problems and development trend of rural e-commerce in China. Rural Science and Technology (2020) (in Chinese)
4. Pakhira, R., Ghosh, U., Sarkar, S.: Application of memory effect in an inventory model with price-dependent demand rate during shortage. Int. J. Educ. Manage. Eng. **9**(3), 51–64 (2019)
5. Eme, O., Uchenna, U.C.A., Uwazuruike, F.O., et al.: Computer-based drug sales and inventory control system and its applications in pharmaceutical stores. Int. J. Educ. Manage. Eng. **8**(1), 30–39 (2018)
6. Pakhira, R., Ghosh, U., Sarkar, S.: Application of memory effects in an inventory model with linear demand and no shortage. Int. J. Math. Sci. Comput. **5**(2), 54–70 (2018)
7. Chen, P., Yan, Y., Geni, X., Zhao, R.: Promotion decisions under asymmetric demand-generation information: self-operated, online-platform and offline-outlet strategies. IEEE Trans. Fuzzy Syst. **27**(5), 928–942 (2019)
8. Elmaghraby, W., Keskinocak, P.: Dynamic pricing in the presence of inventory considerations: research overview, current practices, and future directions. Manage. Sci. **49**(10), 1287–1309 (2003)
9. Abhishek, V., Jerath, K., Zhang, Z.J.: Agency selling or reselling? Channel structures in electronic retailing. Manage. Sci. **62**(8), 2259–2280 (2016)
10. Chun, L., Luo, M., Leng, X., et al.: Pricing the digital version of a book: wholesale vs. agency models. Inform. Syst. Oper. Res. **56**(2), 163–191 (2018)

11. Tian, L., Vakharia, A.J., Tan, Y.R., Yifan, X.: Marketplace, reseller, or hybrid: strategic analysis of an emerging E-commerce model. Prod. Oper. Manage. **27**(8), 1595–1610 (2018)
12. Cohen-Vernik, D., Pazgal, A.: Price adjustment policy with partial refunds. J. Retail. **93**(4), 507–526 (2017)
13. Coughlan, A.T., Soberman, D.A.: Strategic segmentation using outlet malls. Int. J. Res. Mark. **22**(1), 61–86 (2005)
14. Lin, Y.-T., Parlaktuerk, A.K., Swaminathan, J.M.: Are strategic customers bad for a supply chain? Manuf. Serv. Oper. Manage. **20**(3), 481–497 (2018)
15. Chen, P., Zhao, R., Yan, Y., Li, X.: Promotional pricing and online business model choice in the presence of retail competition. Omega **94**, 102085 (2020). https://doi.org/10.1016/j.omega.2019.07.001

Variational Method for Solving the Time-Fractal Heat Conduction Problem in the Claydite-Block Construction

Volodymyr Shymanskyi[1]([✉]), Ivan Sokolovskyy[1], Yaroslav Sokolovskyy[2], and Taras Bubnyak[3]

[1] Department of Artificial Intelligence, Lviv Polytechnic National University, Lviv, Ukraine
vshymanskiy@gmail.com
[2] Department of Computer-Aided Design Systems, Lviv Polytechnic National University, Lviv, Ukraine
[3] Department of Higher Mathematics, Lviv National Agrarian University, Lviv, Ukraine

Abstract. Mathematical models of the heat conduction problem in the claydite-block construction with taking into account the fractal structure of the material is constructed. Integro-differentiation apparatus of fractional order to take into account the fractal structure of the material was used. The variational formulation of the problem was constructed. The variational method for obtaining an approximate solution of the considered problem was proposed. The results of the numerical experiments of studying the thermal conductivity of claydite-block construction depending on the time, wall thickness and materials of different fractions were obtained. Analyzing the founded distributions of temperature fields allows us to more accurately reflect the real speed of the process.

Keywords: Fractal structure · Heat conduction · Variational formulation · Finite element method · Galerkin's method

1 Introduction

The very low thermal conductivity of claydite-block is one of two aspects that claydite-block can be proud of. The first is its thermal conductivity, the second is waterproofing. The thermal conductivity of claydite-block is where it breaks ahead to the level, leaving behind all the same traditional bricks. For those materials that are designed to perform a protective function, the characteristic of thermal conductivity is especially important.

In recent years, among the mathematical models of dynamical systems, models of objects have been positioned in a special way, which are united by one of the most important features – fractality. As a rule, the property of fractality of the simulated systems is considered in relation to space and time [26, 27].

It should be noted that the literal translation of the term "heredity" causes some confusion of terms related to the characteristics of the functioning of systems in time – "lag" and "memory effect". The "memory effect" property determines the dependence

Z. Hu et al. (Eds.): ICCSEEA 2022, LNDECT 134, pp. 97–106, 2022.
https://doi.org/10.1007/978-3-031-04812-8_9

of the state of the system at the current moment of time on the state in which the system was at a certain moment in the past [21, 25].

Mathematically, the transition from a deterministic representation of a dynamic model to its fractal description can be carried out using the apparatus of fractional differentiation and integration. In particular, fractional derivatives by spatial coordinates are used for the mathematical formalization of the characteristics of fractal media and fractional time derivatives are used to represent the hereditary properties of a system or memory effect [6, 13, 14, 17, 19, 21].

There is the practice of using the fractional differential approach in the mathematical modeling of viscoelastic media, filtration in complex heterogeneous porous media, processes of transformation of temperature and humidity fields, processes of anomalous diffusion and diffusion of particles in heterogeneous media, heat conduction processes [5, 7, 10–12, 18, 23].

Since analytical solutions of fractional differential equations with partial derivatives cause special difficulties, in the practice of mathematical modeling, the numerical methods are widely used, including finite-difference ones [4, 15].

Therefore, this paper aims to implement the finite element method to find the numerical solution of the problem of nonstationary thermal conductivity in the claydite-block construction with taking into account the fractal structure of the medium and to analyze the obtained results.

The main contribution of this paper can be summarized as follows:

1. the variational formulation of the problem of nonstationary thermal conductivity in the claydite-block construction with taking into account the fractal structure of the medium is constructed;
2. an algorithm for using the finite element method with a piecewise linear basis to obtain a numerical solution of the problem is proposed;
3. the influence of the fractal structure of the material on the temperature distribution in the claydite-block construction is analyzed.

2 Production of a Problem

One of the effective and widely used approaches to describe the processes of anomalous thermal conductivity is the use of the fractional order integro-differentiation apparatus. In this case, the transport equation is an integro-differential equation containing fractional derivatives by time and/or space variables.

Let us consider the fractional order integro-differentiation operators integral of the function $f(x, y, z)$ over the variable x in Caputo's understanding in more detail [13, 16, 19]:

$$D_x^\alpha f = \frac{1}{\Gamma(1 - \{\alpha\})} \int_a^x \frac{\partial^{[\alpha]+1} f(\xi, y, z)}{\partial \xi^{[\alpha]+1}} \frac{d\xi}{(x - \xi)^{\{\alpha\}}}, \tag{1}$$

$$I_x^\alpha f = \frac{1}{\Gamma(\{\alpha\})} \int_a^x \frac{\partial^{1-[\alpha]} f(\xi, y, z)}{\partial \xi^{1-[\alpha]}} \frac{d\xi}{(x - \xi)^{\{\alpha\}}}, \tag{2}$$

where $\alpha = [\alpha] + \{\alpha\}$, $[\alpha] \in N$, $0 < \alpha < 1$, $\Gamma(\alpha) = \int\limits_0^\infty x^{\alpha-1} e^{-x} dx$ – gamma function.

Mathematical model of the thermal conductivity process in claydite-block construction with taking into account the fractality of the medium in the region $\Omega = \Omega_x \times \Omega_\tau$, $\Omega_x = [a, b]$, $\Omega_\tau = [0, \tau_{end}]$ can be written using a differential equation with partial derivatives of fractional order [22]:

$$c\rho D_\tau^\alpha T = \lambda \frac{\partial^2 T}{\partial x^2} \qquad (3)$$

the corresponding initial conditions are also added:

$$T|_{\tau=0} = T_0(x) \qquad (4)$$

The interaction of the structure with the external environment can be described by the boundary conditions of the third kind:

$$-\lambda \frac{\partial T}{\partial n}\bigg|_{x \in \Gamma} = \beta\left(T_{\text{env}} - T|_{x \in \Gamma}\right) \qquad (5)$$

Thus, the mathematical model of the thermal conductivity process in the claydite-block construction with taking into account the fractality of the medium is described by using the differential Eq. (3), initial condition (4) and the corresponding boundary conditions (5).

3   Variation Formation of the Problem

To date, finite-difference methods are often used to find the numerical solution of differential equations with partial derivatives of fractional order. They are easy to use, but most are relatively consistent and require significant computing resources to ensure the required accuracy. The use of variational methods will allow you to calculate the approximate continuous solution and provide the necessary accuracy [2, 3, 8, 9].

To construct the solution of the initial-boundary value problem (3)–(5) we introduce the space of admissible functions by variational methods [1, 22, 24, 28].

$$U = \{u(x), \ u(x) \in H(\Omega)\} \qquad (6)$$

The scalar product operator was defined as follows:

$$(u, v) = \int_a^b uv \, dx \qquad (7)$$

Multiply Eq. (3) by the function $v \in U$, and integrate over the domain Ω_x:

$$c\rho \int_a^b v D_\tau^\alpha T \, dx = \lambda \int_a^b v \frac{\partial^2 T}{\partial x^2} \, dx \qquad (8)$$

Using the integration in parts formula and taking into account the boundary conditions (5), we obtain the relationship:

$$c\rho \int_a^b vD_\tau^\alpha T dx = v(b)\beta(T_{env} - T|_{x=b}) + v(a)\beta(T_{env} - T|_{x=a}) - \lambda \int_a^b \frac{\partial v}{\partial x}\frac{\partial T}{\partial x}dx$$

$$(9)$$

Multiply the initial condition (4) by the function $v \in U$, and integrate over the domain Ω_x:

$$\int_a^b v(T(x, \tau_0) - T_0(x))dx = 0 \tag{10}$$

We introduce the following bilinear and linear forms:

$$m(T, v) = \int_a^b Tv dx \tag{11}$$

$$a(T, v) = v(b)\beta(T_{env} - T|_{x=b}) + v(a)\beta(T_{env} - T|_{x=a}) - \lambda \int_a^b \frac{\partial v}{\partial x}\frac{\partial T}{\partial x}dx \tag{12}$$

As a result, we come to the variational problem: find a function $T(x, \tau) \in L_2(\Omega_x; \Omega_\tau)$, that satisfies the equation:

$$m(D_\tau^\alpha T, v) = a(T, v) \tag{13}$$

$$m(T(x, \tau_0) - T_0(x), v) = 0 \tag{14}$$

To find an approximate solution of the variational problem (13), (14), we use the Galerkin sampling scheme. We choose a sequence $U_n \in U$ of finite-dimensional subspaces with bases $\phi_1(x)...\phi_n(x)$ in the space of admissible functions U. We give an approximate solution of problem (13), (14) in the form:

$$T(x, \tau) \approx \tilde{T}(x, \tau) = \psi(x, \tau) + \sum_{j=1}^n \tilde{T}_j(\tau)\phi_j(x) \tag{15}$$

where $\psi(x, \tau)$ — some known function, $\tilde{T}_j(\tau)$ — unknown functions, $\phi_j(x)$ — piecewise-linear basic functions.

According to the method, we substitute in Eqs. (13), (14) instead of T developing an approximate solution (15) and instead of a function v consistently basic functions $\phi_j(x)$; $j = \overline{1, n}$. As a result, we obtain the following problem for a system of ordinary differential equations with a derivative of fractional order for an unknown function:

$$MD_\tau^\alpha \tilde{T}_j(\tau) = A\tilde{T}_j(\tau) \tag{16}$$

$$M\tilde{T}_j(\tau_0) = 0 \tag{17}$$

The coefficients of the matrices M, A are calculated according to the rules:

$$M = \left[m(\phi_i, \phi_j)\right]_{i,j=1}^n \tag{18}$$

$$A = \left[a(\phi_i, \phi_j)\right]_{i,j=1}^n \tag{19}$$

Using relation of the approximation of derivative of fractional order α on the interval $\left[\tau^n; \tau^{n+1}\right]$ [22, 25] we can rewrite (16) in the form

$$M \frac{\tilde{T}_j^{n+1} - \alpha \tilde{T}_j^n}{\Gamma(1-\alpha)(1-\alpha)\Delta\tau^\alpha} = A\tilde{T}_j(\tau) \tag{20}$$

Thus, the problem is reduced to finding unknown functions $\tilde{T}_j(\tau)$ that satisfy the differential Eq. (16) and the initial condition (17). We obtain a numerical solution of the problem using relation (20) and apply the iterative search procedure.

4   Obtained Results

Numerical experiments of studying the thermal conductivity of claydite-block construction were conducted depending on the time and wall thickness of the structure. Also, materials of different fractions were considered, which was described by the degree of material fractality. The obtained results with taking into account the fractal structure of the material were compared with the results obtained using the traditional approach, ie $\alpha = 1$.

Numerical experiments were performed for the 25 cm. thick claydite-block construction $\Omega_x = [0; 0.25]$ during the first 12 h $\Omega_\tau = [0; 12]$. The temperature inside the building was set 18 °C, statistically – the most common room temperature.

Consider a construction that interacts with the external environment with a temperature 6 °C. Accordingly, the initial temperature will vary linearly from the ambient temperature to the temperature inside the building. Let there be a decrease in ambient temperature too -8 °C. Let's calculate the dynamics of temperature change in the structure at given parameters.

Figure 1 shows the dynamics of temperature change in the claydite-block construction depending on the time and spatial variable.

Analyzing the obtained results we see that at the boundary adjacent to the external environment the temperature of the structure decreases to ambient temperature. The temperature in the middle of the structure also gradually decreases.

To verify the proposed numerical method, let's calculate the maximum in a modular relative error of the method at different numbers of nodes. The obtained values of relative errors are shown in Table 1.

Analyzing the data in Table 1 we conclude that the method is convergent because with an increasing number of nodes the maximum relative error decreases by an order of magnitude.

Figure 2 shows the dependence of temperature on the spatial coordinate at the final modeling time and depending on the degree of fractality of the material. Figure 3 shows a similar relationship only at the initial temperature of the construction is equal to 18 °C.

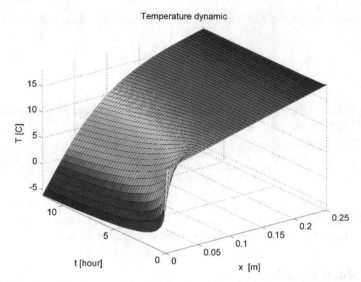

Fig. 1. Changing of the structure temperature depending on the time and spatial coordinates.

Table 1. The relative error of the method at different numbers of nodes

Number of nodes	Error in geometric points of the sample		
	$x = 0$	$x = 0.125$	$x = 0.25$
1000	0.0353	0.0368	0.0374
2000	0.0297	0.0301	0.0299
10000	0.0021	0.0028	0.0024

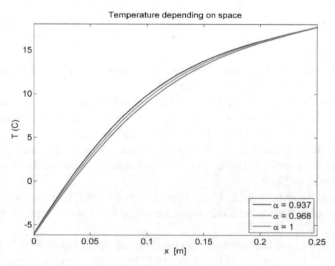

Fig. 2. The temperature depends on the space coordinates on $T_{env} = -8\ °C$.

After analyzing the graphical dependences in Fig. 2 and Fig. 3 it can be concluded that the structure with a wall thickness of 25 cm provides a good level of thermal insulation of the room under given weather conditions, as changes in the temperature of the structure in layers close to the boundary with the room are almost absent. It should also be noted the influence of the fractal structure of the medium on the temperature distribution. In particular, with an increasing degree of medium fractality the thermal insulation characteristics increase which can be seen from the graphical dependencies. The maximum by the modulus difference between the temperature at a certain spatial point of the structure at $\alpha = 1$ and $\alpha = 0.937$ is -1.47 °C, and between $\alpha = 1$ and $\alpha = 0.968$ is -0.53 °C.

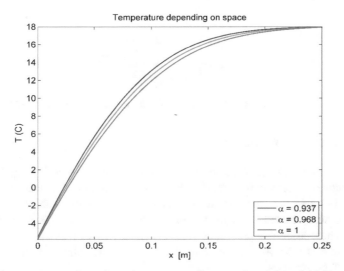

Fig. 3. The temperature depends on the space coordinate on the constant initial temperature.

Figure 4 shows the dependence of temperature on the spatial coordinate at the final modeling time and depending on the degree of the material fractality when interacting with the external environment with a temperature of 33 °C.

As you can see from Fig. 4 the given design of the construction creates good thermal insulation characteristics of the interior. It is seen that significant changes in the temperature of the structure occur in the range of space coordinates from [0; 0.2]. Also analyzing the influence of the fractal structure of the medium, it can be noted that with increasing the degree of medium fractality increases the thermal insulation characteristics. In particular, the maximum by the modulus difference between the temperature at a certain spatial point of the structure at $\alpha = 1$ and $\alpha = 0.937$ is -1.68 °C, and between $\alpha = 1$ and $\alpha = 0.968$ is -0.59 °C.

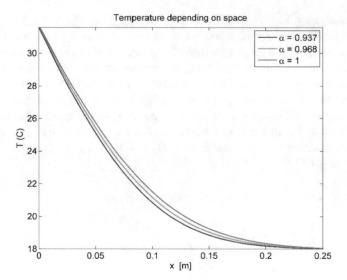

Fig. 4. The temperature depends on the space coordinate on $T_{\text{env}} = 33\,^\circ\text{C}$

5   Summary and Conclusion

In this article, the actual problem of construction of the mathematical model of time fractal thermal conductivity in expanded clay-block construction is solved. A variational method for obtaining an approximate continuous solution of the problem is developed. The method was verified by analyzing the relative errors. The influence of material fractions on heat-insulating properties of expanded clay-block construction is analyzed. It was found that the material with a higher degree of fractality has the best thermal insulation characteristics.

The further development of the study can be implemented in such direction. This model can form the basis of systems for calculating the optimal thickness of claydite-block construction which will provide good thermal insulation of the building. However, such studies must take into account the stability of the structure, as increasing the degree of fractality of the material increases the thermal insulation properties but reduces the strength of the material.

References

1. Acosta, G., Borthagaray, J.P., Bruno, O., Maas, M.: Regularity theory and high order numerical methods for the (1D)-fractional Laplacian. Math. Comput. **87**, 1821–1857 (2018). https://doi.org/10.1090/mcom/3276
2. Boffi, D.: Finite Element Methods and Applications. Springer Series in Computational Mathematics, p. 575 (2013)
3. Cai, M., Li, C.: Numerical approaches to fractional integrals and derivatives: a review. Mathematics **8**, 43 (2020). https://doi.org/10.3390/math8010043
4. Diethelm, K., Garrappa, R., Stynes, M.: Good (and not so good) practices in computational methods for fractional calculus. Mathematics **8**, 324 (2020). https://doi.org/10.3390/math8030324

5. Edelman, M.: Dynamics of nonlinear systems with power-law memory. Handbk. Fraction. Calculus Appl.: Appl. Phys. A, 103–132 (2019). https://doi.org/10.1515/9783110571707-005
6. Falade, K.I., Tiamiyu, A.T.: Numerical solution of partial differential equations with fractional variable coefficients using new iterative method (NIM). IJMSC **6**(3), 12–21 (2020). https://doi.org/10.5815/ijmsc.2020.03.02
7. Ford, N.J., Morgado, M.L., Rebelo, M.: A nonpolynomial collocation method for fractional terminal value problems. J. Comput. Appl. Math. **275**, 392–402 (2015). https://doi.org/10.1016/j.cam.2014.06.013
8. Garrappa, R.: Numerical solution of fractional differential equations: a survey and a software tutorial. Mathematics **6**, 16 (2018). https://doi.org/10.3390/math6020016
9. Hilfer, R., Luchko, Y., Tomovski, Z.: Operational method for the solution of fractional differential equations with generalized Riemann-Liouville fractional derivatives. Fract. Calc. Appl. Anal (12), 299–318 (2009)
10. Hinze, M., Schmidt, A., Leine, R.I.: Numerical solution of fractional order ordinary differential equations using the reformulated infinite state representation. Fract. Calc. Appl. Anal. **22**, 1321–1350 (2019). https://doi.org/10.1515/fca-2019-0070
11. Ismail, M., Saeed, U., Alzabut, J., Rehman, M.: Approximate solutions for fractional boundary value problems via green-CAS wavelet method. Mathematics **7**, 1164 (2019). https://doi.org/10.3390/math7121164
12. Kelly, J.F., Sankaranarayanan, H., Meerschaert, M.M.: Boundary conditions for two-sided fractional diffusion. J. Comput. Phys. **376**, 1089–1107 (2019). https://doi.org/10.1016/j.jcp.2018.10.010
13. Kilbas, A., Srivastava, H.M., Trujillo, J.J.: Theory and Applications of Fractional Differential Equations. Elsevier (2006)
14. Kochubei, A.N.: Equations with general fractional time derivatives. Cauchy problem. In: Handbook of Fractional Calculus with Applications, vol. 2: Fractional Differential Equations, pp. 223–234 (2019). https://doi.org/10.1515/97831105716620-011
15. Lischke, A., Zayernouri, M., Zhang, Z.: Spectral and spectral element methods for fractional advection-diffusion-reaction equations. In: Karniadakis, G.E. (ed.) Handbook of Fractional Calculus with Applications, vol. 3: Numerical Methods, pp. 157–183 (2019). https://doi.org/10.1515/9783110571684-006
16. Luchko, Y., Yamamoto, M.: The general fractional derivative and related fractional differential equations. Mathematics **8**(12), 2115 (2020). https://doi.org/10.3390/math8122115
17. Madhu, J., Maneesha, G.: Design of fractional order recursive digital differintegrators using different approximation techniques. IJISA **12**(1), 33–42 (2020). https://doi.org/10.5815/ijisa.2020.01.04
18. Pezza, L., Pitolli, F.: A multiscale collocation method for fractional differential problems. Math. Comput. Simul. **147**, 210–219 (2018). https://doi.org/10.1016/j.matcom.2017.07.005
19. Podlubny, I.: Fractional Differential Equations. Academic Press (1999)
20. Povstenko, Y.: Fractional Thermoelasticity. Springer International Publishing, Cham, Heidelberg, New York, Dordrecht, London (2015). https://doi.org/10.1007/978-3-319-15335-3
21. Rituparna, P., Uttam, G.h., Susmita, S.: Application of memory effect in an inventory model with price dependent demand rate during shortage. IJEME **9**(3), pp. 51–64 (2019). https://doi.org/10.5815/ijeme.2019.03.05
22. Shymanskyi, V., Protsyk, Y.: Simulation of the heat conduction process in the claydite-block construction with taking into account the fractal structure of the material. In: XIII-th International Scientific and Technical Conference; Computer Science and Information Technologies, CSIT-2018, pp. 151–154. https://doi.org/10.1109/STC-CSIT.2018.8526747
23. Shymanskyi, V., Sokolovskyy, Ya.: Finite element calculation of the linear elasticity problem for biomaterials with fractal structure. Open Bioinform. J. **14**(1), 114–122. https://doi.org/10.2174/18750362021140100114

24. Shymanskyi, V., Sokolovskyy, Ya.: Variational formulation of viscoelastic problem in biomaterials with fractal structure. CEUR Workshop Proc. **2753**, 360–369 (2020)
25. Sokolovskyy, Y., Levkovych, M., Sokolovskyy, I.: The study of heat transfer and stress-strain state of a material, taking into account its fractal structure. Math. Model. Comput. **7**(2), 400–409 (2020). https://doi.org/10.23939/mmc2020.02.400
26. Tarasov, V.E.: General fractional dynamics. Mathematics **9**(13), 1464 (2021). https://doi.org/10.3390/math9131464
27. Tarasov, V.E.: Self-organization with memory. Commun. Nonlinear Sci. Num. Simul. **72**, 240–271 (2019). https://doi.org/10.1016/j.cnsns.2018.12.018
28. Washizu, K.: Variational Methods in Elasticity and Plasticity, 3rd edn. Pergamon Press, New York (1982)

Programming Language ASAMPL 2.0 for Mulsemedia Applications Development

Ivan Dychka, Yevgeniya Sulema[(✉)], Dmytro Rvach, and Liubov Drozdenko

Igor Sikorsky Kyiv Polytechnic Institute, 37 Peremohy pr., Kyiv 03056, Ukraine
sulema@pzks.fpm.kpi.ua

Abstract. ASAMPL is a domain-specific programming language designed for the easy development of mulsemedia applications. This paper presents the updated version of ASAMPL 2.0 that enables simplification of a program code development and allows achieving better metrics of the developed code in comparison with the original version of ASAMPL. The paper provides information on the changes in ASAMPL according to the updated version of the semantics and syntax of this programming language. The comparison of ASAMPL versions is presented and discussed, as well.

Keywords: Domain-specific programming language · Mulsemedia applications · Immersive technologies

1 Introduction

Mulsemedia (multiple sensorial media) [1–3] and immersive technologies [4] implement a human-centred approach that involves the registration, presentation, processing, and reproduction of multimodal information about physical objects that is perceived by humans through sense organs.

ASAMPL programming language [5, 6] is a domain-specific programming language designed for processing mulsemedia data. This programming language got its name from "Algebraic System of Aggregates" and "Mulsemedia data Processing Language". The purpose of developing the ASAMPL language is to simplify the processing of multimodal, in particular, mulsemedia data, the main feature of which is that this data is timewise. Besides, the processing of such data should consider the modality, data compatibility for different modalities and the possible interaction between the data of different modalities. A key concept in ASAMPL is the concept of a multi-image of an object. To present a multi-image of the object in this programming language, special data structures are used; they are tuples and aggregates. The processing of aggregates is performed in accordance with the rules specified in the Algebraic System of Aggregates (ASA) [7, 8].

Z. Hu et al. (Eds.): ICCSEEA 2022, LNDECT 134, pp. 107–116, 2022.
https://doi.org/10.1007/978-3-031-04812-8_10

The main features of the ASAMPL language, which specifies the principles for processing multimodal data structures:

1. Relationship between the data processing process and the timeline.
2. Presentation of multimodal data using aggregates and tuples.
3. Simultaneous use of different sources of multimodal data and different formats.
4. Synchronization of data of different modalities.
5. Use of the apparatus of ASA for processing of multi-images and their components.

Therefore, the main principle of organizing the program in ASAMPL is that the execution of the program is based on two entities: time and modality of data that corresponds to the multi-images programming paradigm [9]. This allows simplifying mulsemedia applications development because mulsemedia data is temporal and multimodal and ASAMPL processes this data according to their features.

This approach can simplify the development of not only mulsemedia applications but also multimedia software [10–12].

2 Related Research

The review of recent papers published in the field of mulsemedia allows us to conclude that there is a lack of research on programming approaches for mulsemedia applications development. Mostly, recent papers deal with using authoring tools, but not with program code development.

Thus, in [13], the authors propose an interoperable mulsemedia framework for delivering sensory effects to heterogeneous systems. They propose the evolution of an open distributed mulsemedia system by changing its core following architectural and design patterns to meet the modern requirements to mulsemedia applications.

The paper [14] presents MulSeMaker which is a tool for the development of mulsemedia applications in the Web domain. This tool is based on the application family concept from generative software development.

The authors of papers [2, 15] propose ways to handle sensory effects using annotation options for multimedia components.

The survey paper [16] reviews existing mulsemedia authoring tools and proposals for representing sensory effects and their characteristics. The authors discuss future directions for the mulsemedia authoring and challenges in the field of mulsemedia.

The most promising tool for the creation of mulsemedia applications is Play-SEM SER 2 [17]. This framework allows working with heterogeneous applications and devices and supports timeline-based multimedia applications [18].

Although several interesting tools described in the recent research papers can be used to create mulsemedia applications, none of them proposes the programming way of mulsemedia applications development.

3   Programming Language ASAMPL 2.0

The programming approach provides the developers of mulsemedia applications with flexibility in the mulsemedia data processing. In particular, the ASAMPL programming language offers to programmers many advantages such as the possibility of working with arbitrary data sources and devices, temporal multimodal data handling and many others. However, its first version [5, 6] has certain limitations which can cause worsening program code metrics. To improve the ASAMPL programming language, several new features are added in ASAMPL 2.0.

3.1   User-defined Functions

ASAMPL 2.0 programming language enables creating user-defined functions. This new feature allows a programmer to describe parts of the code with individual functions that can be called anywhere in the program code and more than once.

Declaring a user-defined function is expressed by the syntactic rule (1).

$$
\begin{aligned}
&\textit{user-defined\_function\_declaration} = \text{function}, \textit{identifier}, \text{"("}, \textit{identifier} \mid \\
&\quad \textit{tuple} \mid \textit{identifiers\_set} \mid \textit{tuples\_set} \mid \textit{identifiers\_and\_tuples\_set}, \text{")"}, \text{"\{"}, \\
&\quad\quad \textit{operator} \left[\text{return}, \left[\textit{identifier} \mid \textit{value} \right], \text{";"} \right], \text{"\}"}
\end{aligned} \tag{1}
$$

Calling a user-defined function is defined by the syntactic rule (2).

$$
\begin{aligned}
&\textit{user-defined\_function\_calling} = \left[\textit{identifier}, \text{"="} \right] \textit{identifier}, \text{"("}, \textit{identifier} \mid \\
&\quad \textit{tuple} \mid \textit{identifiers\_set} \mid \textit{tuples\_set} \mid \textit{identifiers\_and\_tuples\_set}, \text{")"}, \text{";"}
\end{aligned} \tag{2}
$$

The state diagram of a user-defined function is shown in Fig. 1.

An optional element in a user-defined function is the *return* operator, which allows returning a value to the location in the code where the user-defined function was called. The function which can return values must be linked with a variable to avoid losing the data obtained in this function. After using the return operator, the execution of further program instructions is interrupted. Besides, a programmer can use return without any values to abort the function when necessary.

ASAMPL 2.0 also defines a *break* operator that allows aborting a loop at any time, regardless of the number of iterations remaining.

So, user-defined functions simplify the development of code because they allow a programmer to avoid program code duplication as well as they enable using recursion.

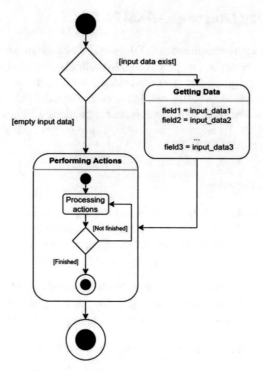

Fig. 1. User-defined function state diagram

3.2 Simplification of ASAMPL Program Code

ASAMPL 2.0 improves the import procedure for objects such as libraries, sources, handlers, renders, and more. In ASAMPL 2.0, it is possible to import these objects in one line and indicate to which type this object belongs and its name. Object types are declared by the following keywords: *handlers, renders, libraries.* Importing objects has a syntax that is expressed by the syntactic rule (3).

$$import = \text{import} \, , \, object\_type \, , \, \text{"."} \, , \, object\_name \\ \left[\, \text{as} \, , \, identifier \, \right] \, , \, \text{"."}, \tag{3}$$

If the optional part (with the *as* operator) is not specified, the object (handler, render, etc.) will have a default name that corresponds to the name of the object. These simplifications allow reducing the program code size.

In ASAMPL 2.0, the *timeline* operator has only one variation (the original version of ASAMPL syntax included three types of this statement), and its execution options depend on the parameters that are passed to it. These options are:

1. Three parameters in the following order: a start time value, an end time value, and a time step.
2. One parameter: a tuple of time values.
3. One parameter of a logical data type or a logical expression.

The timeline operator is defined by the syntactic rule (4).

$$timeline\_operator = \text{timeline} , "(" , identifier \mid time\_value, "," , identifier \mid$$
$$time\_value , "," , identifier \mid time\_value , ")" \mid "(" , time\_values\_tuple , ")" \mid \quad (4)$$
$$"(" , identifier \mid logical\_expression , ")" , "\{" , operator , "\}"$$

Declaring data sources has been also simplified in ASAMPL 2.0. In order not to separate the list of sources to be used in the program, a programmer can use the keyword *source* with the identifier and assign the location of the source as a local path (relative or absolute) or a network address with a protocol, by which this data can be obtained. The syntax of source declaration is defined by the syntactic rule (5).

$$data\_source = \text{source} , identifier , "=" , data\_location , ";" \quad (5)$$

Declaring data sets programming language has also been changed in ASAMPL 2.0. Now, it is declared with *set* keyword. The syntactic rule (6) defines this.

$$data\_set = \text{set} , identifier , "=" , type , ";" \quad (6)$$

The declaration of tuples works in the same way, but with the difference that now a programmer does not need to use the keyword "tuple", but just define the tuple by using square brackets "[]" next to the identifier of the data set as shown in the syntactic rule (7). Arrays are represented in a similar way described by the syntactic rule (8).

$$tuple = set\_type , "[]" , identifier , ";" \quad (7)$$

$$array = identifier , "[]" , "=" , "[" , value\_1 , "," , value\_2 , "]" , ";" \quad (8)$$

The aggregate declaration is defined according to the syntactic rule (9).

$$aggregate = \text{aggregate} , identifier , "=" , "[" , tuple , "]" , ";" \quad (9)$$

The *download, unload, substitute,* and *render* operators have also been simplified. The changes made reduce the program code size and make it easier to write this code.

In download and upload operators, the number of keywords has been reduced. In the download operator, the result of the execution must be assigned to a certain variable. In turn, in the upload operator, the first parameter is the identifier of data to be uploaded. The second identifier is the path to/from the location where data should be uploaded/downloaded. The third parameter is optional and specifies which library to use for the defined operation. If a programmer does not specify this option, a default library will be used. The syntactic rules for downloading (10) and unloading (11) in ASAMPL 2.0 programming language are as follows.

$$downloading\_operator = identifier , "=" , \text{download} , "(" , identifier , "," , \quad (10)$$
$$identifier , ["," , identifier ,] , ";"$$

$$uploading\_operator = \text{upload} , "(" , identifier , "," , identifier , ["," , \quad (11)$$
$$identifier ,] , ";"$$

The substitute operator also has parameters. The first parameter is the object to be replaced, and the second parameter is another object to replace the first object. This operator is defined by the syntactic rule (12).

$$substitute\_\ operator\ =\ substitute\ ,\ "("\ ,\ identifier\ ,\ identifier\ ,\ ")"\ ,\ ";" \qquad (12)$$

The render operator has two parameters, where the first parameter is the identifier for the set of data to be reproduced using the renders listed in the tuple, which is the second parameter of this operator. This is determined by the syntactic rule (13).

$$render\_operator\ =\ render\ ,\ "("\ ,\ identifier\ ,\ "["\ ,\ identifier\ ,\ "]"\ ,\ ")"\ ,\ ";" \qquad (13)$$

Thus, the proposed changes allow simplifying program code development due to more evident definitions of the main operators of the original ASAMPL. The major objective of these changes is to improve program code metrics and, therefore, the efficiency of application execution.

4   Results and Discussion

In order to evaluate the usefulness of the proposed changes that distinguish the ASAMPL 2.0 programming language from its original version, it is necessary to perform a comparative analysis of two versions of the programming language on the same example. Let us consider the task to form a complex data set obtained from remote data sources, namely, video, audio, and odour data, recorded in the same location. Program code in the ASAMPL 2.0 programming language which realises this data processing is given in Listing 1.

```
import libraries.video.VisualLib;
import libraries.audio.AudioLib;
import libraries.scent.OlfactLib;
import handlers.mpeg2tup as MPEG2tuple;
import handlers.tup2mpeg as Tuple2MPEG;
import renders.video.VisualRen;
import renders.audio.AudioRen;
import renders.scent.OlfactRen;
Source VideoFile = 'http://edu.net/Forest&WildLife.mp4';
Source VisualDataStream = 'http://webcam.edu.net/005201';
Source OlfactoryFile = 'D:\Lesson05\forest.dat';
Source SceneFile = 'D:\Lesson05\forest.agg';
Set Frame = int[1920, 1080];
Set Audio = float();
Set Scent = str();
Time time1 = 00:00:01;
Time time2 = 00:00:01;
int step = 1;
Frame[] VisualDat;
Audio[] AudioDat;
Scent[] OlfactoryDat;
Agregate ForestScene = [VisualDat,AudioDat,OlfactoryDat];
AudioDat = download(VideoFile.audio, MPEG2tuple);
VisualDat2 = download(VideoFile.visual, MPEG2tuple);
OlfactoryDat = download(OlfactoryFile, default.OlfactLib);
timeline(time1, time2, step) {
VisualDat = download(VisualDataStream, default.VisualLib); }
if (VisualDataStream == Null) {
substitute(VisualDat2, VisualDat); }
upload(ForestScene, SceneFile, default.all);
render(ForestScene, [VisualRen, AudioRen, OlfactRen]);
```

Listing 1. ASAMPL 2.0 program code example

As a result, we get 1123 characters of program code. In the previous version of the ASAMPL programming language, the code which solves the same problem would be 1453 characters long. Let us define the basic metrics of the program code.

Cyclomatic complexity [19] can be determined by the formula (14).

$$\Psi_c = \upsilon_{edge} - \upsilon_{node} + 2 \cdot \upsilon_{comp}, \tag{14}$$

where υ_{edge} is a number of edges in a cyclomatic graph (Fig. 2); υ_{node} is a number of nodes; υ_{comp} is a number of connectivity components.

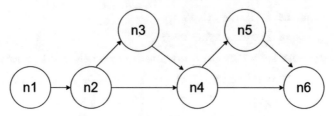

Fig. 2. An example of the cyclomatic graph

The program code size is one of the indicators of Halstead's metrics [20] which are based on the measured properties of the program code and allow to establish a connection between them.

The program code size can be calculated by the formula (15).

$$V = (\eta_o + \eta_d) \cdot \log_2(\eta_{uo} + \eta_{ud}), \tag{15}$$

where η_o is a total number of operators, η_d is a total number of operands, η_{uo} is a number of unique operators, η_{ud} is a number of unique operands.

The maintainability index [21] allows assessing how easy it is to maintain and change the program code. There are several approaches to calculating this indicator. One of them is determined by the formula (16).

$$\Psi_m = \max\left(0, \ 100 \cdot \frac{171 - 5.2 \cdot \ln V - 0.23 \cdot \Psi_c - 16.2 \cdot \ln L}{171}\right), \tag{16}$$

where V is a volume of program code, L is a number of program code lines.

Table 1 shows program code metrics for both ASAMPL 2.0 and the original version of this programming language. As the table demonstrates, the program code metrics for ASAMPL 2.0 are improved compared to the original ASAMPL: program code size is decreased by 20%, and the maintainability index is increased by 13% on average.

This research was conducted on many program code samples and it allows us to conclude that program code in ASAMPL 2.0 has better performance than code developed by using the original version of the ASAMPL programming language.

Table 1. Program code metrics comparison

A version of programming language	ASAMPL 2.0	ASAMPL 1.0
Cyclomatic complexity	1	1
Program code size	437.52	549.23
Maintainability index	48.248	42.052

5   Conclusions

The main contribution of this research is the advancement of the ASAMPL programming language specially designed for the development of mulsemedia applications. Since some rules in the original ASAMPL were not useful enough, the syntax of the ASAMPL 2.0 programming language has been redesigned and updated to meet modern programming language standards, as well as to simplify software development. ASAMPL 2.0 provides additional features that were missing in the original version of the ASAMPL programming language. These include the possibility to create user-defined functions and to forcibly terminate cyclic functions, regardless of the number of iterations remaining.

The results obtained in this study were analysed using metrics such as cyclomatic complexity, program code size, and maintainability index. The comparison has shown that the second version of the programming language ASAMPL allows a programmer to reduce the program code size and increase the maintainability index.

Further research will be focused on improving the ASAMPL 2.0 language interpreter to allow using it as a microservice in a web application. Also, the updated syntax will simplify the creation of an integrated environment for visual programming in ASAMPL 2.0 that will contribute to raising the popularity of creating mulsemedia applications in this programming language.

References

1. Ghinea, G., Timmerer, C., Lin, W., Gulliver, S.R.: Mulsemedia: state of the art, perspectives, and challenges. ACM Trans. MCCA **11**, 17:1–17:23 (2014)
2. Mattos, D.P., Muchaluat-Saade, D.C., Ghinea, G.: An approach for authoring mulsemedia documents based on events. In: Proceedings of the IEEE International Conference on Computing, Networking and Communications (ICNC), pp. 273–277 (2020)
3. Abreu, R., Mattos, D., Santos, J.A.F., Muchaluat-Saade, D.C.: Semi-automatic synchronization of sensory effects in mulsemedia authoring tools. In: Proceedings of the 25th Brazilian Symp. on Multimedia and the Web WebMedia'2019, pp. 201–208 (2019)
4. Handa, M., Aul, G., Bajaj, S.: Immersive technology – uses, challenges and opportunities. Int. J. Comput. Bus. Res., pp. 1–11 (2012)
5. Sulema, Y.: ASAMPL: programming language for mulsemedia data processing based on algebraic system of aggregates. AISC **725**, 431–442 (2018)
6. ASAMPL Project. https://github.com/orgs/Asampl-development-team/repositories
7. Sulema, Ye., et al.: Chakraverty S. (ed.) Mathematical Methods in Interdisciplinary Sciences, 464 p. Wiley, USA (2020)

8. Sulema, Y., Kerre, E.: On fuzziness in algebraic system of aggregates. New Math. Nat. Comput. **17**(1), 145–152 (2021)

9. Sulema, Ye.S., Rvach, D.V.: Models of computation for Digital Twins data processing. KPI Science News **2**, 74–81 (2020)

10. Marsono, Wu, M.: Designing a digital multimedia interactive book for industrial metrology measurement learning. I. J. Modern Educ. Comput. Sci. **5**, 39–46 (2016)

11. Chun-ko Hsien, et al.: Easy and deep media in cultural heritage field—the development of Mau-kung Ting educational media for the national palace museum. I.J. Educ. Manage. Eng. **2**, 26–34 (2013)

12. Jiang, L., Sun, K.: Application of task-based approach in college English teaching based on internet-assisted multimedia. I.J. Educ. Manage. Eng. **8**, 58–64 (2012)

13. Saleme, E.B., Santos, C.A.S., Ghinea, G.: A mulsemedia framework for delivering sensory effects to heterogeneous systems. Multimedia Syst. **25**(4), 421–447 (2019). https://doi.org/10.1007/s00530-019-00618-8

14. de Sousa, M.F., et al.: MulSeMaker: an MDD Tool for MulSeMedia web application development. In: Proc. of the 23rd Braz. Symp. WebMedia'2017, pp. 317–324

15. Abreu, R., Mattos, D., Santos, J.A.F., Muchaluat-Saade, D.C.: Semi-automatic synchronization of sensory effects in mulsemedia authoring tools. In: Proceedings of the 25th Brazilian Symp. on Multimedia and the Web WebMedia'2019, pp. 201–208 (2019)

16. De Mattos, D.P., et al.: Beyond multimedia authoring: on the need for mulsemedia authoring tools. ACM Comput. Surv. **54**(7), 150:1–150:31 (2021)

17. PlaySEM SER 2.0.0.: https://github.com/estevaosaleme/PlaySEM_SERenderer/releases/tag/2.0.0

18. Saleme, E.B., Santos, C.A.S., Ghinea, G.: A mulsemedia framework for delivering sensory effects to heterogeneous systems. Multimedia Syst. **25**, 421–447 (2019)

19. Code Metrics Values. Microsoft (2018). https://docs.microsoft.com/en-us/visualstudio/code-quality/code-metrics-values?view=vs-2019

20. Halstead, M.H., McCabe, T.A.: Software complexity measure. IEEE Trans. Software Eng. **2**(12), 308–320 (1976)

21. Maintainability Index Range and Meaning. Microsoft (2007). https://docs.microsoft.com/en-us/archive/blogs/codeanalysis/maintainability-index-range-and-meaning

Location-Based Threats for User Equipment in 5G Network

Giorgi Akhalaia[1]([✉]), Maksim Iavich[2], and Sergiy Gnatyuk[3]

[1] Georgian Technical University, 0171 Tbilisi, Georgia
Akhalaia.g@gtu.ge
[2] Caucasus University, CST, 0102 Tbilisi, Georgia
miavich@cu.edu.ge
[3] National Aviation University, Kyiv 03058, Ukraine
s.gnatyuk@nau.edu.ua

Abstract. Over the last decade rate of mobile device development has extremely increased. Mobility and flexibility of new products takes one of the most important objectives for manufacturers. Microcomputers, smartphones, IoT devices can provide majority of everyday service, including emergency, security, healthcare, and education. Development of mobile devices itself triggered the 5G network deployment. The new telecom standard will create new ecosystem with variety of industries and will exceed the limit of telecom communication. New standards, functionality, services, products always arise new cyber threats. When we are talking about mobile devices, generally they are caried with people during the day or sometimes at night too. Our research idea was to study location-based vulnerabilities for user equipment in 5G network. Study objectives were to assess if new standard increased the risk of locating devices without their prior permissions. Determine the scale of the problem and to find from which part of the system comes the threats. During the experimental work, we have simulated different methods of tracking devices and found that some of them are less noisy and can be easily done without target's attention. According to our research, there is chance to track, locate the device in 5G network only with one cell-tower. Study shows that, using MITM, when there are fake base stations in 5G network, users might suffer from relocation. As the locating devices sometimes are used for emergency services, this might cause the serious problems. We have compared existing location-based threats with newly arisen and assessed which one is more vulnerable. Research was oriented to study how technical changes affect and can be used to locate devices without prior permission. Hence the study was done from different point of view and not only using already known methods. According to our results, new architecture might be more vulnerable for locating devices then already known methods. The ideas developed in this article and results from experimental works, triggered new problems and targets for continuing research. As the scale of new network is so huge, mathematical model should be implemented and deployed to find the solution for exiting vulnerabilities.

Keywords: Triangulation · Device tracking · UE security · MITM in 5G design · Fake base station

1 Introduction

World leading companies, operating in technical market, are working to make gadgets, robots more mobile, portable to increase their usage in every situation. Extensive development of mobile devices such a mobile phone, IoT, microcomputers like raspberry pi, arduino approves their importance and future trend. Developing mobile networks acts as catalyst for usage of mobile devices in every field of our life.

New arrival standard of telecom communication, 5G Network will incorporate multiple systems into one huge network. Which itself arises new cyber threats. Idea of this research is to reassess vulnerabilities related to the location of UE in 5G ecosystem. 3GPP has announced 3 KPIs for 5G standard: Enhanced mobile broadband - more than 10 Gbps, ultra-reliable low latency communication – up to 1 ms latency and massive machine type communications – more than one million connected UE per square km [1]. Hence, engineers have to make some technical and software changes to achieve these requirements. Study was done on these changes to check how technical/software improvement affects on security of UE in terms of location. Our research was regarding 3 objectives:

- Does 5G architecture affect on UE location privacy?
- Which Band is more vulnerable?
- What is a scale of new vulnerability?

During the study different scientific and technical articles about the device locating methods were analyzed. We have simulated major techniques to compare which one is more accurate and represents higher risk by violating user privacy. Location privacy issue is not new vulnerability. The problem exists in previous generation networks too. Our goal was to check these vulnerabilities in 5G architecture and compare it's scale with the lack of security caused by new architecture – operating spectrum diversity and it's limitation. According to our study and experimental work, technical changes of 5G network might be more vulnerable to locating devices without their prior permission than vulnerabilities.

2 Literature Overview

Ideas overview and developed in our research are fulfilled with the latest scientific articles. International research papers, technical documents and overviews from leading organizations, network operators, experienced in mobile network deployment were analyzed. Authors have discussed general improvement of 5G network, telecom communication, their research results and the various threats. For our experimental work, we have used open-source projects, tools available on github. Our study was done on theoretical experience and was approved with practical experiments.

Articles processed during working on the research covers security aspects of mobile communication, cyber attack vectors against 5G network and in most cases provide ideas about solutions. Germany scientist, A. Shaik explained their experimental works, which approves MITM type attack existence in 5G design. One of our research ideas – UE location tracking and its accuracy, was developed with relevance of MITM. How the vulnerability explained by A. Shaik can be used in location-based attacks. Generally, most of researchers discuss location-based problems during the UE attach process with cell-tower or by the malware. We tried to simulate major techniques and compare them with new, 5G network design related problem (location-based).

3   Methods of Locating Device

There are various methods for tracking mobile devices. Tracking means to find their precise location and it's changes in a time. Concept of determining device location for different techniques is the same: reference system should be chosen and after that UE calculates its coordinates related to it (reference system). Usually reference systems are GPS satellites or cell-towers of telecom operators. By measuring and processing signals tracked from satellites or cell-towers UE determines it's location. Usually, frequencies, arrival time, angle and signal strength are used to locate device. When cell-towers are used, degree of accuracy depends on telecon operators, how properly they configure cell-towers. In some countries, like US there are some regulations by the government to achieve certain quality of accuracy. This is for emergency services like 911 or 112 [2].

Mobile positioning are used for positive purposes, like a navigation, emergency cases, location-based services such a marketing services, checking in for different places (for social networks) and so on. However, intruders use this technique to track humans/devices without their prior permissions. Our study was oriented to assess how 5G standards will affect on location-based threats. During the research three main methods were analyzed: GNSS (GPS, A-GPS); Trilateration; Triangulation.

Device operating on telecom networks, periodically scans for cell-towers and roams to the tower with the strongest signal. So, at UE level can be found information about cell-towers, which itself means to locate towers. Hence, catching that information, attacker is able to calculate approximate location of device. Tracking cell-towers process run in background so, attacker does not need to run additional software and make some noise at this point. Next sections cover main techniques for tracking device and experimental work.

3.1   GNSS (Global Navigation Satellite System)

The most precise method, technique for determining device location on the earth is GNSS – Global Navigation Satellite System. Systems consist with Satellites sending signals, and the clients with antenna and receiver. GNSS satellites provides signal from the space, which transmits timing and positioning data to clients (Fig. 1). Client devices process this data to determine their real location. Accuracy of this technique is about mm. But this accuracy and high precision can be achieved only by scientific GNSS systems, which support multiple signals tracking to overcome ionospheric affect and different type noises. User friendly devices like mobile phones and other smart gadgets is compatible (generally) only with L1 band which cannot overcome effects of different layers around the earth, so their accuracy is about 3–5 m. However, it is enough to locate and find address of human or any device. This technique is by devices to determine their location and for location-based services like timing and so on. As any other methods, GNSS also have limitations. In our case, most interesting is that it cannot be used indoor. Our device has to has open sky, good satellite view for tracking GNSS signals. This method is also called - GPS (Global Positioning System). Actually, GPS represents satellites managed by US. As they were first providers, usually GPS is used to talk about GNSS.

There is also method A-GPS (Assisted GPS). This method uses cell-towers for locating device. It is a good solution for in-door use, but has a lower accuracy then previous one (Fig. 2).

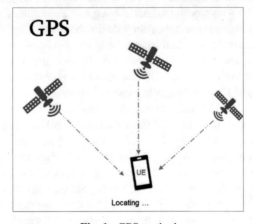

Fig. 1. GPS method

Fig. 2. A-GPS method

3.2 Trilateration/Triangulation

GSM operators cover cities with cell-towers for better network coverage and service delivery. By knowing their coordinates (x, y) we can determine UE's location. Actually there are 2 common methods to use: trilateration and triangulation. There are some technical differences during the calculation. In trilateration distances are calculated from each station and common area is a estimated location of UE. While in case of triangulation 2 lines from the BS to UE and third line between BSs are used. This will create the triangle (Fig. 4). Sides of BSs' and Alfa/Beta angles are known. So, the location is determined by processing the angles related to reference points. In some article trilateration is called as distance measuring techniques, as it measures distances from different cell-towers [2].

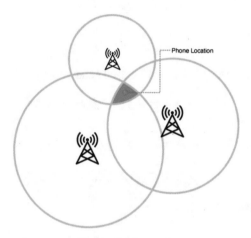

Fig. 3. Trilateration

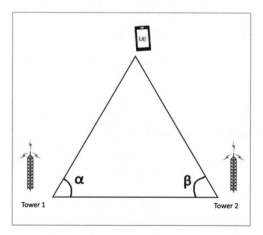

Fig. 4. Triangulation

Latitude and longitude of nearby cell-towers are used to compute estimate coordinates of UE. In the Experimental Work section, we have explained how we can get the coordinates of telecom towers. Measurements are done based on time delay of signal travel or based on signal quality, strength.

Figure 3 demonstrates trilateration method. Green circles indicate possible locations of UE for a specific radio of BS. Our goal is to compute the intersection coordinates of these circles. For simplifying the calculation process, 2D model is used (x, y):

First run, equations per circle:

$$(x - x_1)^2 + (y - y_1)^2 = r_1^2$$

$$(x - x_2)^2 + (y - y_2)^2 = r_2^2$$

$$(x - x_3)^2 + (y - y_3)^2 = r_3^2$$

Second stage, open the parentheses for each equation:

$$x^2 - 2x_1x + x_1^2 + y^2 - 2y_1y + y_1^2 = r_1^2$$

$$x^2 - 2x_2x + x_2^2 + y^2 - 2y_2y + y_2^2 = r_2^2$$

$$x^2 - 2x_3x + x_3^2 + y^2 - 2y_3y + y_3^2 = r_3^2$$

Third stage, subtract the equation:

$$(-2x_1 + 2x_2)x + (-2y_1 + 2y_2)y = r_1^2 - r_2^2 - x_1^2 + x_2^2 - y_1^2 + y_2^2$$

$$(-2x_2 + 2x_3)x + (-2y_2 + 2y_3)y = r_2^2 - r_3^2 - x_2^2 + x_3^2 - y_2^2 + y_3^2$$

Fourth stage, rewrite the system with A, B, C, D, E, F

$$Ax + By = C$$

$$Dx + Ey = F$$

Last stage, solution for this system is:

$$x = \frac{CE - FB}{EA - BD}$$

$$y = \frac{CD - AF}{BD - AE}$$

This is a simplified 2D version of the trilateration [3].

4 Experimental Work

4.1 Get GPS Data from Smartphones

During the search we used different cases to determine, which solution is the easiest way to track the device (Table 1). There is various software for stealing coordinates from UE. For experimental work, we have used storm-braker (Fig. 5, 6) [4].

This tool runs on Linux-based systems and work very well. There are some limitations: it is too noise and requires enabled GPS module. As it tries to steal GPS information, users were alerted several times, that someone was trying to access on location data (Fig. 7, 8). GPS service is not mandatory for devices, so part of users, disable GPS module to save the battery and increase the security, to avoid extra trackers.

By default, devices do not track GPS signals simultaneously, so if we need to locate them using GPS satellites, we have to start GPS measurements on mobile device. Modern software automatically throws the sign of "location", which means that GPS measurements have been started. This is also very noisy. Another limitation comes with GPS

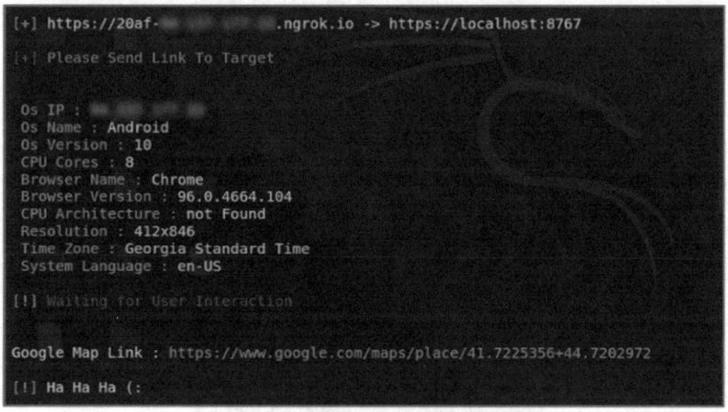

Fig. 5. .

Fig. 6. .

Fig. 7. .

Fig. 8. .

techniques. As it needs open sky, for good satellite view, if victim is in the building it will not be easy to compute coordinates only via GPS. Hence, according to the research this method does not seem to be the best solution for tracking devices without their prior permissions.

4.2 Collect Cell-Towers Using Smartphones

There are several reliable products, software available on mobile markets to collect information about active cell-towers. During the research we used "Tower Collector" from "Play Store" market.

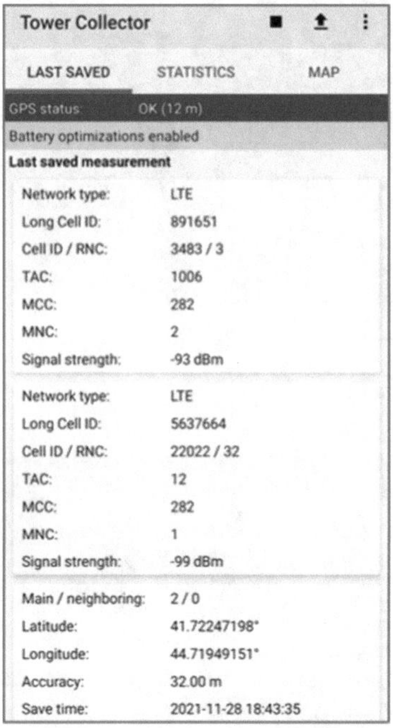

Tower Collector	■ ↥ ⋮
LAST SAVED	STATISTICS	MAP

GPS status:	OK (12 m)
Battery optimizations enabled	
Last saved measurement	
Network type:	LTE
Long Cell ID:	891651
Cell ID / RNC:	3483 / 3
TAC:	1006
MCC:	282
MNC:	2
Signal strength:	-93 dBm
Network type:	LTE
Long Cell ID:	5637664
Cell ID / RNC:	22022 / 32
TAC:	12
MCC:	282
MNC:	1
Signal strength:	-99 dBm
Main / neighboring:	2 / 0
Latitude:	41.72247198°
Longitude:	44.71949151°
Accuracy:	32.00 m
Save time:	2021-11-28 18:43:35

Fig. 9. .

Figure 9 shows detailed information about the specific tower. Network type describes that it is LTE network tower. RNC stands for the Radio Network Controller, which is responsible for managing, controlling connected NODE BS. Encryption is also done at this level, before the data is exchanged between mobile and to it. LAC for UMTS networks and TAC for LTE network is a unique id for current location area. MCC stands for a Mobile Country Code, which identifies the country. MNC – Mobile Network Code, which identifies the telecom operator. Cell-ID represents unique ID for identification each BTS or sector for specific LAC (Base Transceiver Station) [5]. The signal strength parameter is very interesting. There are lots of argues, can it be used to compute precise location. However, it is too complicated, because there are lots of things that can affect on signal strength. So, weak signal does now always mean long distance from

Fig. 10. .

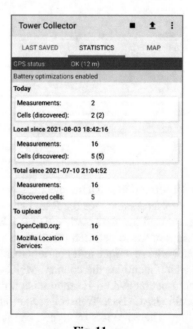

Fig. 11. .

cell-towers, it may be caused by buildings or any influencer that damage radio waves. Software prints latitude and longitude of the tower. That information is used during the calculation distances. Figure 10 prints the results (towers and UE location) on map for better visualization. Figure 11 summarizes results and gives statistical information.

As detailed information is available and sent by cell-towers, we can map whole city and use this data to track the devices. The only thing to do is to collect information received by mobile phones. There are two most important things: 1. this method does not require to enable GPS module on mobile, as it uses locations of towers and 2. Process of scanning cell-towers is always activated and running in background. So, this will be less noise than catching GPS data from UE.

Table 1. Equipment and methods used for experimental work.

Device	Quantity	Usage
Raspberry Pi with LTE and GPS Modules	30	10 - For Base Station, 15 - For Fake Base Station, 5 - For User Equipment
Mobile Phones with GPS support	5	For User Equipment
Laptop with Kali OS	2	Manage and Monitor Experimental Work
Results		
Algorithm type	Success/Fail	Comment
GPS (Catch data from UE)	Success	Success with noise if GPS module was enabled. User interaction was needed. As they were alerted by the system
A-GPS (Catch data from UE)	Success	10/10
MITM by Fake BS	Success	10/10
Catching Frequency Info	Success	8/10

5 5G Network Design

New standard of telecom communication will overcome the limitation of existing mobile networks and create new ecosystem with incorporated various sectors. Involvement of different industries in 5G network, itself arises new threats. 3 KPIs of 5G network requires some major technical and software changes. The core components are network virtualization, which arises software-based threats, operating-spectrum division, which by our research influence on tracking devices, massive multiple input multiple output technology and flexible beamforming [7].

Operating spectrum is divided into three categories:

1. Below 1 GHz (Low-Band). These frequencies are less affected by buildings. Hence, they can be used in urban areas. But, for bandwidth for this range is limited to 100 Mbps.

2. 1 GHz to 6 GHz (Mid-Band). This range has better bandwidth (up to 1 Gbps), but the frequencies from this range are more affected by buildings, than from first category - Low-band.
3. 6 GHz to 100 GHz (High-Band / mmWave). Provides the best bandwidth in categories - 10 Gbps. But, is critically affected by buildings [1].

The limitation of third category, High-Band can be used for tracking devices. UE has to be near to the cell-tower of this category to achieve full capabilities. So, if we catch the information that UE is in a range of mmWave tower, we will be able to determine estimated location. Hence, we do not need to know information about three nearby cell-towers or the data from GPS satellites.

5G Network is not protected from MITM type of attack [6]. Hence there are chance of fake base stations. As we discussed in previous section, cell-towers are used to estimate the location of UE. Coordinates of cell-towers are used in A-GPS method and/or during the trilateration. Every result depends on the data quality. The most important input in our equation is the coordinates of cell-towers. So, if the network suffers from fake-base stations (MITM), which sends forged x, y coordinates, device estimated location will not be accurate.

6   Results

We have reviewed various scientific articles. From theoretical aspects, according to our study and analyzing results of other researchers, technical changes in 5G architecture can cause more significant cyber threats related to location privacy than it was transferred from previous generation networks. Our experimental work approved theoretical assumptions. Stealing GPS data from devices is very noisy as the system alerts to UE for permission. Also, UE has to equipped with GPS module and enabled it. The last limitation does not exist if we use A-GPS method for locating UE, but it requires permission from user. The most important objective of our study was to determine how spectrum diversity can be used for locating devices. According to the results of experimental work, 80% of cases UE can be tracked by knowing on which band range it is using. Hence, in case of 5G network if UE is attached to 3rd band, known as mmWave band, can be located by 1 cell-tower (because of band limitation).

7   Conclusion

Deployment of new telecom standard will play a significant role for existing services and for future development. Successfully and securely implemented network will solve the existing limitations and create new possibilities. Because of the scale of 5G target group, interest of illegally motivated persons will totally arise. Hence, working on security functions, protocols, policies of 5G network should be on high priority. As every new technology, functionality, software, hardware, solution, 5G also has vulnerabilities. Our goal was to assess vulnerabilities related to device location privacy. Does the changes in 5G standard affect on location privacy? What is the scale and which band is more vulnerable? Theoretical and experimental work done by international experts and by our group has approved that there are threats against device tracking. We have overview different type of methods for device tracking. According to our research:

- MITM in 5G network can cause to relocate UE location and decrease the accuracy. As the Fake base stations can provide false coordinates, result from computation will not be real. This might be very problematic during emergency situation, when there is a need to rapidly locate the device.
- The second experiment was to collect the GPS data from smartphones. However, this method gives the most accurate coordinates of UE, but experiment shows, that it is too noisy. As user(victim) was alerted several times before granting access to GPS data. Also, because of method limitation, sometimes it is impossible to determine the device location, when the device is in the building.
- The third experiment – locating UE using A-GPS, by the knowing details about nearby cell-towers was more effective as it does not require user interaction.
- The fourth experiment was to determine how third band gives ability to locate device. From theoretical and practical experience, if UE want to use the towers operating on these frequencies (known as mmWave), it must be near the tower. Hence, only one tower will be enough to locate the device. This will accelerate process of device tracking for emergency situations but represents the huge problem for security.

Our future plan is to create the mathematical model of the solution, how to hide the details about cell-towers to minimize the risk of device tracking for third parties – without permissions.

Acknowledgment. The work was conducted as a part of PHDF-21-088 financed by Shota Rustaveli National Science Foundation of Georgia.

References

1. Huawei Technologies Co., Ltd. in 5G Network Architecture – A high Level Perspective (2016)
2. Asad Hussain, S., Ahmed, S., Emran, M.: Positioning a mobile subscriber in a cellular network system based on signal strength. IAENG Int. J. Comput. Sci. **34**(2), IJCS_34_2_13 (2007). https://www.researchgate.net/publication/26492533
3. Cell Phone Trilateration Algorithm: Online Journal "Computer Science" (2019). https://www.101computing.net/cell-phone-trilateration-algorithm/. Accessed 10 Dec 2021
4. Ultrasecurity: "Strom-Breaked" (Software Package). https://github.com/ultrasecurity/Storm-Breaker. Accessed 8 Dec 2021
5. Johhny: How to find the Cell Id location with MCC, MNC, LAC and CellID (CID) (2015). https://cellidfinder.com/articles/how-to-find-cellid-location-with-mcc-mnc-lac-i-cellid-cid
6. Iavich, M., Akhalaia, G., Gnatyuk, S.: Method of improving the security of 5G network architecture concept for energy and other sectors of the critical infrastructure. In: Zaporozhets, A. (ed.) Systems, Decision and Control in Energy III. SSDC, vol. 399, pp. 237–246. Springer, Cham (2022). https://doi.org/10.1007/978-3-030-87675-3_14
7. Maheshwari, M.K., Agiwal, M., Saxena, N., Abhishek, R.: Flexible beamforming in 5G wireless for internet of things. IETE Tech. Rev. **36**(1), 3–16 (2017). https://doi.org/10.1080/02564602.2017.1381048

8. Ivezic, M., Ivezic, L.: "5G Security & Privacy Challenges" in 5G. Security Personal Blog (2019). https://5g.security/cyber-kinetic/5g-security-privacy-challenges/

9. Shaik, A., Borgaonkar, R., Park, S., Selfert, J.P.: New vulnerabilities in 4G and 5G cellular access network protocols: exposing device capabilities. In: Proceedings of the 12th Conference on Security and Privacy in Wireless and Mobile Networks, WiSec 2019 (2019). https://doi.org/10.1145/3317549. ISBN 9781450367264

10. Purdy, A.: "Why 5G Can Be More Secure Than 4G" in Forbes Online Journal (2019). https://www.forbes.com/sites/forbestechcouncil/2019/09/23/why-5g-can-be-more-secure-than-4g/?sh=2ffcdf1657b2

11. Qualcomm Technologies Inc. "What is 5G", in Online Article. https://www.qualcomm.com/5g/what-is-5g

12. SK Telecom: 5G architecture design and implementation guideline (2015)

13. Hanif, M.: "5G Phones Will Drain Your Battery Faster Than You Think", in Online Journal (2020). https://www.rumblerum.com/5g-phones-drain-battery-life/

14. Samsung in online report "Samsung Phone Battery Drains Quickly on 5G Service". https://www.samsung.com/us/support/troubleshooting/TSG01201462/

15. Yusof, R., Khairuddin, U., Khalid, M.: A new mutation operation for faster convergence in genetic algorithm feature selection. Int. J. Innov. Comput. Inf. Control **18**(10), 7363–7380 (2012)

16. Shehu, I.S., Adewale, O.S., Abdullahi, M.B.: Vehicle theft alert and location identification using GSM, GPS and web technologies. Int. J. Inf. Technol. Comput. Sci. **7**, 1–7 (2016)

17. The EU Space Programme. https://www.euspa.europa.eu/european-space/eu-space-programme. Accessed 10 Dec 2021

18. Hu, Z., Odarchenko, R., Gnatyuk, S.: Statistical techniques for detecting cyberattacks on computer networks based on an analysis of abnormal traffic behaviour. Int. J. Comput. Netw. Inf. Secur. **6**, 1–13 (2020)

19. Iavich, M., Kuchukhidze, T., Gnatyuk, S.: Novel certification method for quantum random number generators. Int. J. Comput. Netw. Inf. Secur. **3**, 28–38 (2021)

Design and Implementation of Virtual Laboratory Construction Scheme for Transportation Logistics Specialty Group

Yanzhi Pang and Jianqiu Chen[✉]

School of Transportation, Nanning University, Nanning 530200, Guangxi, China
153862839@qq.com

Abstract. It is the development of information technology that promotes the reform of the education industry, and the teaching mode and method keep pace with the times. This paper analyzes the impact of virtual laboratory on traditional education, makes a qualitative and quantitative analysis on the necessity and advantages and disadvantages of virtual laboratory construction, and puts forward the construction scheme of virtual laboratory. Taking the author's college as the research object, on the premise of meeting the talent training scheme, this paper puts forward a scheme for the construction of Virtual Laboratory of transportation logistics specialty group, follows the principles of safety and openness, and forms an open experiment mode. The construction of virtual laboratory alleviates the problem of insufficient equipment in practical teaching, breaks through the traditional path dependence, and provides some research ideas for talent training in colleges and universities.

Keywords: Information technology · Virtual laboratory · Transportation logistics professional group

1 Introduction

Virtual experiment has been widely used in the construction of transportation and logistics specialty [1]. Virtual experiment refers to the creation of relevant software and hardware operating environment on the computer that can assist, partially replace or even completely replace each operation link of traditional experiment with the help of multimedia, simulation and virtual reality (also known as VR) and other technologies. Experimenters can complete various experimental projects as in the real environment [2, 3].

The concept of virtual simulation experiment was first proposed by Professor William Wolff of the University of Virginia it in 1989, which was called "a research center without fences". With the development of information technology, there is a clearer definition of virtual simulation laboratory, which refers to the interactive environment used to create and simulate experiments [4]. China's virtual experiment research started late, and 2016 is called the first year of VR. As of March 2020, a total of 157 colleges and universities in China have opened the specialty of "virtual reality application technology" [5]. The general office of the Ministry of Education decided to carry out the construction of demonstration virtual simulation experiment teaching project in ordinary undergraduate colleges and universities from 2017 to 2020 [6]. In this sense, it is imperative to increase the construction of virtual laboratory in Colleges and universities.

2   Necessity and Advantages and Disadvantages of Virtual Laboratory Construction

The realization of virtual experiment will effectively alleviate the difficulties and pressures faced by many colleges and universities in terms of funds, venues and equipment. Moreover, the development of online virtual experiment teaching can break through the limitations of traditional experiments on "time and space". Both students and teachers can freely and without worry access to the virtual laboratory anytime, anywhere, operate instruments and carry out various experiments. It is helpful to improve the quality of experimental teaching [7–9].

2.1   Necessity Analysis of Virtual Laboratory Construction

2.1.1   Traditional Teaching Methods are Difficult to Meet the Needs of the Information Age

It is difficult for students to understand the knowledge in books. The reason is that students have no or little access to equipment in life, and there is little reserve of relevant experience and knowledge. Classroom heuristic teaching is difficult for students, and the teaching materials students learn in class are only book knowledge integrated by Chinese characters and grammar, which requires students to think clearly and even "live" the actual contents they represent through language, symbols and charts. After COVID-19 broke out, this virtual laboratory has shown hitherto unknown superiority [10, 11]. It can provide remote laboratory support for it projects in distance education, realize the automatic adjustment of job scheduling parameters in grid environment, construct the case teaching mode of management specialty under the support of information technology, and establish a new computer network practice teaching system [12–15].

The traditional classroom teaching method pays more attention to rational knowledge than perceptual knowledge; The idea of emphasizing rationality and neglecting sensibility will affect the pursuit of theorization and abstraction in teaching, which is not conducive to students' mastery of knowledge.

There are many knowledge points, which are difficult for students to digest and remember in a short time, and the information queried on the network is limited.

Due to the huge knowledge system, lack of network resources and limited teaching laboratories, students can't start through self-study or after-school review;

The traditional classroom teaching method pays attention to the conclusion and despises the process. In the learning process of many subjects, the relationship between conclusion and process is a very important relationship in the process of students' seeking knowledge. Emphasizing conclusion and neglecting process is the reflection and embodiment of the traditional view of static knowledge.

2.1.2 Social Development Needs

Modern teaching methods have made great progress and development, which are better than traditional teaching in both form and content. With the in-depth development of computer technology, multimedia technology and network technology, virtual simulation technology has gradually moved from business environment to modern education and teaching field.

With the characteristics of short cycle, high security and strong sense of reality, virtual simulation technology has gradually become an indispensable part in the field of modern education. The time when teachers write on the blackboard and students take notes is over. Various emerging virtual simulation training platforms, virtual simulation animation and network courses have improved the strength of modern education, changed the concept of modern education and sublimated the level of modern education. A series of computer virtual simulation technologies play an irreplaceable role in the application of modern education.

2.1.3 Achievement of Educational Objectives

The research and development of modern educational digitization cater to the development of science and technology, apply simulation technology to education and teaching, improve the process of modern educational digitization and informatization, and improve the structure of modern education. The development of virtual simulation technology affects modern education. The upload and release mode is changed to independent mode, which is transferred from classroom to extracurricular, so as to lay the foundation for the cultivation of five talents.

Table 1 and Fig. 1 show the conclusions drawn from the data of students' teaching evaluation and the comparison of the results between the traditional laboratory and the virtual laboratory.

2.2 Advantages and Disadvantages of Virtual Laboratory

2.2.1 Advantage Analysis

The advantages of virtual experiment are mainly reflected in three aspects.

(1) High Degree of Informatization

University Virtual Laboratory is to complete or improve the experiment by using computer system, combined with communication and information technology processing, integration and innovation [16]. Through the virtual laboratory, we can reproduce the environment, have a more intuitive understanding of knowledge principles, and have a deeper understanding of the engineering background. At the same time, it is of great help to students' innovative learning.

(2) Meet the Needs of Entrepreneurship and Innovation Education

In response to the call of "mass entrepreneurship and innovation", we are committed to building a "multi-level, interdisciplinary, realistic and complementary" entrepreneurship and innovation talent training ecosystem. According to the needs of regional economic development for logistics engineering professional post

Table 1. Teaching evaluation data of students in traditional and virtual laboratory

Evaluation index	Full score	Traditional laboratory evaluation	Scoring rate-T	Virtual laboratory evaluation	Scoring rate-V
Respect and care for students. Enthusiastic teaching; Strictly manage the classroom and keep it in good order	10	9.3	93.00%	9.4	94.00%
Proficient in lectures and strong logic. (experimental class: be familiar with the experimental contents and instruments and equipment, be able to eliminate faults in time, explain concisely and concisely, and demonstrate the operation specifications)	10	9.4	94.00%	9.74	97.40%
The experimental instruction or comprehensive design experimental instruction materials are comprehensive	5	5	100.00%	5	100.00%
The experimental course is rich in information and depth, which can be accepted by students through their efforts; Highlight the key points and clarify the difficulties	5	4.5	90.00%	4.7	94.00%

(*continued*)

Table 1. (*continued*)

Evaluation index	Full score	Traditional laboratory evaluation	Scoring rate-T	Virtual laboratory evaluation	Scoring rate-V
Give comprehensive guidance before the experimental class, and explain the experimental process and precautions	10	9.5	95.00%	9.66	96.60%
Pay attention to the whole process of the experiment, and give guidance and demonstration to key links	10	8	80.00%	9	90.00%
Assign appropriate homework, tasks or group activities after class; Carefully tutor, answer questions and correct homework in time	10	9.5	95.00%	9.75	97.50%
Collect and record the experimental data comprehensively to facilitate students to understand and master the experimental content	15	10	66.67%	12	80.00%
Inspire students' innovative thinking and improve students' ability. (experimental class: enable students to master the operation essentials, and their ability to analyze, solve problems and innovate has been improved)	15	8	53.33%	13	86.67%

(*continued*)

Table 1. (*continued*)

Evaluation index	Full score	Traditional laboratory evaluation	Scoring rate-T	Virtual laboratory evaluation	Scoring rate-V
Satisfaction with the environment of the classroom where the experimental training course is located (including whether the seats, lighting, ventilation, temperature, etc. are sufficient and comfortable)	10	5	50.00%	8	80.00%

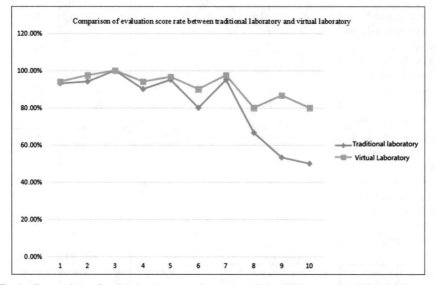

Fig. 1. Comparison of evaluation score rate between traditional laboratory and virtual laboratory

groups, we closely follow the training objectives of technology applied talents, and adopt the school running idea of "school enterprise cooperation, school bank cooperation and school school cooperation", Strive to build a diversified talent training mode of "combination of production and learning and alternation of learning and application".

(3) Make up for the Lack of Practical Teaching in Colleges and Universities

Students' demand for the laboratory is no longer only to complete learning tasks and experimental verification. More and more students participate in Teachers' topics or discipline competitions and need to carry out scientific research and Technological Development in the laboratory. Therefore, they have higher requirements for laboratory resources, the existing equipment can no longer meet the requirements, and the real equipment is limited in terms of cost and site, Therefore, it is urgent to broaden the vision through virtual laboratory to meet the needs of practice and training [17, 18]. At the same time, as long as students have computers or smart phones, they can log in and operate training at any time, regardless of time and place, so as to effectively improve students' learning autonomy, stimulate their initiative to rest and greatly improve learning efficiency.

2.2.2 Disadvantage Analysis

The deficiency of virtual experiment mainly exists in two aspects discussed below.

(1) Incomplete scene simulation
    In reality, faults occur under different conditions, and there are many emergencies, so the laboratory cannot simulate them all, so the training for students to deal with emergencies is not enough.
(2) Over reliance on computer hypothetical data
    During the operation of the virtual laboratory, students conduct simulation calculation by assuming specific data. Therefore, the data lacks the randomness of natural variation in the corresponding experimental process, which will lead to students' unfamiliar and understanding of bad or atypical data. At the same time, it is easy to ignore the importance of some experimental variables, which may lead to fixed and rigid experimental results, and the experimental process can not reflect the complexity.

3 Construction Scheme of Virtual Laboratory

The discussion on the construction scheme of virtual laboratory here is carried out from three aspects: theoretical basis, design concept and structural framework.

3.1 Theoretical Basis

The construction of virtual laboratory is based on the relevant theories of situational learning and embodied cognitive teaching environment [19].

3.1.1 Situational Learning

The construction of virtual laboratory promotes students to participate in practical training through high-precision three-dimensional experimental resources and real experimental scenes, so as to obtain better learning results in this scene. Based on the situational theory that "knowledge is situational", "learning is a situational activity", "learning is the social coordination of knowledge" and "students are the legitimate marginal participation in the community of practice", the above theory is the basis of the feasibility and necessity of the construction of virtual laboratory.

3.1.2 Embodied Cognition

Take students as the center, highlight the central position of students and emphasize the participation of the process. With the development of information technology, Internet-based education technology perfectly integrates the virtual environment with the real scene to form a mixed reality environment. In this environment, teachers can make full use of information technology and teaching media resources to mobilize students' enthusiasm, so as to promote students to learn actively, rather than forced and cramming learning.

3.2 Design Concept

3.2.1 Advanced Nature

The virtual reality system structure should be able to reflect the current mainstream of technology development, flexible configuration, scalability and other technical research needs, and provide a virtual reality environment that can track the leading edge of technology [20].

3.2.2 Economic and Practical

Virtual reality system is different from general system integration, not simply device stacking. The system requires a complete demonstration and development platform. All equipment and software of the system are configured for system research, development and demonstration.

3.2.3 Openness

Virtual reality system should be scalable and can be upgraded with the development of technology. Scalability includes equipment replacement, software update and replacement.

3.2.4 Scalability and Maintainability

Adapt to the requirements of system changes, and consider the simplest way and the lowest investment to realize the expansion and maintenance of the system.

3.2.5 Safety

In the system design, we not only consider the full sharing of information resources, but also pay more attention to the protection and isolation of information. The system adopts the same security measures for different applications and different network environments, including system security mechanism, data storage security and authority control [21–23].

3.3 Structural Framework

Taking the construction of Virtual Laboratory of transportation logistics specialty group of School of transportation of Nanning University as an example, considering the needs of teachers and students, the construction framework of virtual laboratory is shown in Fig. 2.

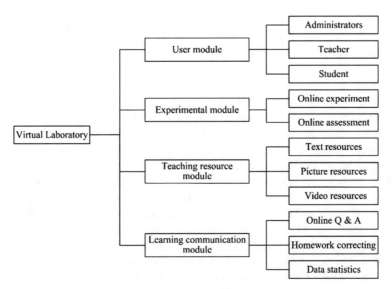

Fig. 2. Framework of virtual laboratory

4   Construction Example of Virtual Laboratory of Professional Group

This paper takes the construction of Virtual Laboratory of transportation logistics specialty group of Transportation College of Nanning University as an example.

4.1 Urban Rail Transit Integrated Virtual Simulation Laboratory

In recent years, China's rail transit has developed rapidly, the scale of construction and operation has been growing, and the demand for talents in the industry has been expanding. At the same time, problems such as insufficient supply of talents and unreasonable structure in the industry have gradually become prominent, which has become a short board restricting the healthy, safe and sustainable development of rail transit.

Today, with the continuous upgrading of information technology and high-tech products, the traditional mode still adopts the mode of "master with apprentice", which leads to a series of problems, such as low efficiency, few training opportunities, difficult learning, slow data update, inability to objectively assess the real level, inability to carry out continuous training, gap between training effect and reality, and inability of technical equipment to keep up with the times, It cannot keep up with the speed and scale of urban rail transit development in China. In order to promote the smooth development of rail transit industry, more efficient, high-quality and low-cost virtual laboratory technology is put on the agenda [24, 25].

(1) Construction Scheme

The Simu Matri industry simulation engine development platform, virtual reality technology, augmented reality technology and system simulation technology are adopted to build a comprehensive and multi type cooperative training base for rail transit through the combination of physical equipment and virtual simulation system. The research and development of VR/AR technology provides a more efficient and high-quality possibility for urban rail transit talent training. Its application in urban rail transit talent training opens a new mode of training + 3D virtual simulation. The product design is based on the basic idea of "career guidance and progressive ability", and the training course system is built based on the working process with "teaching, practice and examination" as the main line. At the same time, the enterprise technical specifications and standards are taken as the training evaluation standards, and the training site is built according to the real work scene, real equipment and real operation process to meet the training needs of students.

Combining 3D virtual technology with real environment, the strong sense of immersion and animation demonstration effect are very attractive, which is easy to arouse students' interest in learning and produce good learning effect; Virtual plus practical training, practice plus assessment, the training quality is higher than the traditional theoretical teaching; Computer three-dimensional modeling to build a virtual scene, high degree of repeated training and cost saving. A series of advantages, such as high efficiency, high quality, low cost, strong interest, easy learning and high level, are the trend of urban rail transit talent training in the future. Taking train operation control as the core, it includes train operation, passenger transport, vehicle driving and maintenance, communication, signal, power supply, AFC and other systems as well as simulation interfaces of relevant systems. By building a real working environment for OCC dispatching center, depot, station control room, station hall and platform public area of urban rail operation, and coordinating and cooperating among dispatching center, station, station affairs and trains, the whole process of normal operation organization, abnormal operation organization and emergency treatment is completed, so that students can master the basic operation of relevant terminal software at the same time, Improve practical ability and emergency handling ability. Through the cooperation and cooperation between different posts, enhance the students' cooperation ability and safety awareness, and be proficient in the work content of each link, so as to improve their professional quality (Fig. 3).

(2) Construction Tasks

According to the framework diagram, the urban rail transit comprehensive virtual simulation laboratory mainly completes four parts: user module, experiment module, teaching resource module and learning communication module.

Fig. 3. Screenshot of urban rail transit simulation

For students, it mainly uses two parts: experiment module and teaching resource module.

- Experimental module: including mature online simulation experiment and innovative open experiment. Mature online simulation experiments are used for learning and simulation operation, which can be repeated for many times, and meet the needs of experimental training. Innovative and open experiment is designed by students with their own topics. Its purpose is to cultivate students' practical ability and innovative spirit.
- Teaching resource module: it is a gradually enriched and improved link. At present, it mainly includes text, pictures and video resources. Theoretical, practical and interesting coexist, which can not only learn knowledge, but also not boring and stimulate learning interest. This module is added by students in the process of innovation experiment.

4.2   Intelligent Logistics Virtual Laboratory and its Construction Scheme

Logistics industry is a compound or aggregate industry formed by the industrialization of logistics resources. With the development of economic globalization and the rise of network economy, the global logistics service industry has accelerated its development. The shortage of logistics talents is becoming more and more prominent. At present, more than 400 colleges and universities in China have logistics majors, but the teaching of Logistics Majors in most schools stays in traditional teaching content modules such as logistics management, warehousing and transportation. In recent years, more and more schools have begun to pay attention to combining the theoretical basis of professional students with comprehensive practical training, but more emphasis is on the business process processing and simulation practice of the whole process of logistics operation, and there is a lack of platforms and opportunities for students' independent design, independent research and development and independent innovation.

Based on the integration of logistics industry and Internet and Internet of things, a powerful platform based on mobile Internet, cloud computing and big data technology has been formed to build a new mode of Internet plus logistics. Intelligent logistics virtual laboratory strengthens the advanced technology represented by the Internet of things and mobile terminals as the breakthrough point, breaks through the traditional training practice concept in the logistics professional training, changes the management mode and service mode of the traditional training room, and collocation the logistics training room with the idea of the national college skills competition as the building idea, so that it has more modernity, advanced nature and standardization. The comprehensive management mode of service makes students better adapt to the changes of talent ability demand in the industry and market.

Taking the application event of intelligent logistics warehousing system as an example, the event takes intelligent warehousing as the core, including various soft and hard technology links such as logistics planning, logistics information technology and automatic logistics system. Through the standard maker kit, students can fully understand the logistics automation system and the necessary links of logistics system planning, and enable students to more fully understand and understand logistics automation through assembly, programming, system planning and other operation contents. The creative model system can be realized through software programming, hardware debugging, system debugging and other contents, It puts forward more substantive requirements for logistics engineering and technical talents (Figs. 4 and 5).

Fig. 4. Screenshot of port logistics simulation

Fig. 5. Screenshot of logistics warehousing simulation

The application results show that the virtual laboratory project has played a prominent role in the construction of transportation logistics professional group. By providing a new teaching environment and using the equipment in the virtual equipment library to freely build any reasonable typical experiments or experimental cases, students can configure, connect, adjust and use experimental instruments and equipment by themselves, so that they have a deeper understanding of professional knowledge and its application.

5   Conclusion

In this paper, the necessity, advantages and disadvantages of virtual laboratory construction are analyzed in detail. Combined with the theoretical basis and design concept of virtual laboratory construction, the construction scheme is put forward. Taking the construction of Virtual Laboratory of transportation and logistics specialty group as an example, an open network virtual experiment teaching system is constructed by using advanced information technology and network technology. It realizes the digitization and virtualization of various existing teaching laboratories. Under this new teaching mode, the content of the experiment is richer and the method is more flexible, which complements the shortcomings of the traditional laboratory, greatly alleviates the problem of insufficient equipment in practical teaching, breaks through the traditional path dependence and improves the talent training mode.

The necessity of virtual laboratory construction is analyzed in detail from the qualitative and quantitative perspectives, which can more objectively reflect the important role of virtual laboratory in talent training. Virtual laboratory is an important development direction of internet teaching technology in the future. It is changing the cognitive concept of experimental personnel in terms of personalization and innovation, providing some research ideas for talent training in Colleges and universities.

Acknowledgment. This project is supported by:

(1) Sub-project of Construction of China-ASEAN International Joint Laboratory for Comprehensive Transportation (Phase I), No. GuiKeAA21077011-7;

(2) 2021 District Level Undergraduate Teaching Reform Project: Exploration of Innovative Teaching Methods for Seamless Integration of Transportation Courses and Ideological and Political Education Based on Semantic Analysis (2021JGB428);

(3) "Curriculum Ideological and Political Education" of Nanning University Demonstration Courses: Train Operation Control System (2020SZSFK02);

(4) Core Construction Courses of Undergraduate Major of Nanning University: Train Operation Control System (2020BKHXK16);

(5) School Level Teaching Team Cultivation Project of Nanning University Rail Transit Teaching Team Project (2021JXTDPY04).

References

1. Yan, M., Zhu, L., Lou, J., Wang, G.: Construction of traffic information and control virtual simulation experiment teaching center. J. Electr. Electron. Teach. **39**(02), 131–135 (2017). (in Chinese)
2. Sun, S., Xu, T., Zhou, J.: The design and implementation of computer hardware assembling virtual laboratory in the VR environment. In: Proceedings of 2018 2nd International Conference on Electronic Information Technology and Computer Engineering (EITCE 2018), pp. 238–243 (2018)

3. Widodo, A., Maria, R.A., Fitriani, A.: Constructivist learning environment during virtual and real laboratory activities. Biosaintifika: J. Biol. Biol. Educ. **9**(1), 11–18 (2017)
4. Hu, Z.: Exploration and research on virtual laboratory. Chin. Character Cult. **03**, 137–139 (2021). (in Chinese)
5. Wang, F., Li, J., Wang, W.: Application and practice of VR/AR technology in college teaching. Comput. Knowl. Technol. **17**(25), 31–35 (2021). (in Chinese)
6. Fu, F.: Research and practice of open computer network virtual laboratory based on VR. Digit. World **01**, 131 (2020). (in Chinese)
7. Ghazala, R., Muzafar, K., Noman, M., Adnan, A.: Measuring learnability through virtual reality laboratory application: a user study. Sustainability **13**(19), 1–16 (2021). 10812
8. Nataro, C., Johnson, A.R.: A community springs to action to enable virtual laboratory instruction. J. Chem. Educ. **97**(9), 3033–3037 (2020)
9. Rivera, L.F.Z., Suescun, C.A.: Enhanced virtual laboratory experience for wireless networks planning learning. IEEE Revista Iberoamericana de Tecnologias del Aprendizaje **15**(2), 105–112 (2020)
10. Rakhi, R., Dhanush, K., Nijin, N., et al.: What virtual laboratory usage tells us about laboratory skill education pre- and post-COVID-19: focus on usage, behavior, intention and adoption. Educ. Inf. Technol. **26**(6), 11–19 (2021)
11. Amayri, M., Ploix, S., Bouguila, N., Wurtz, F.: Estimating occupancy using interactive learning with a sensor environment: real-time experiments. IEEE Access **7**, 53932–53944 (2019)
12. Senthilkumar, L.: Provisioning remote lab support for IT programs in distance education. Int. J. Mod. Educ. Comput. Sci. (IJMECS) **4**(4), 1–7 (2012)
13. Sharma, A., Sahana, S.K.: An automated parameter tuning method for ant colony optimization for scheduling jobs in grid environment. Int. J. Intell. Syst. Appl. **11**(3), 11–21 (2019). https://doi.org/10.5815/ijisa.2019.03.02
14. Li, W., Zhou, R., Deng, P., Fang, Q., Zhang, P.: Construction of case teaching model for management specialty supported by information technology. Int. J. Educ. Manag. Eng. (IJEME) **2**(9), 44–48 (2012)
15. Guo, N.: Construction and implementation of innovation computer network practical teaching system. Int. J. Educ. Manag. Eng. (IJEME) **1**(2), 30–35 (2011)
16. Xia, X., Zou, G., Qu, J.: Discussion on the construction of virtual laboratory in Colleges and universities. Heilongjiang Educ. (Theory Pract.) **05**, 57–58 (2020). (in Chinese)
17. Cai, J., Zhu, Y.: Reform of practical teaching mode of open courses based on virtual simulation technology. China Educ. Technol. Equipment **02**, 105–107 (2021). (in Chinese)
18. Hao, C., Zheng, A., Jiang, B., et al.: Discussion on mixed learning scheme of experimental course. Educ. Teach. Forum **33**, 326–328 (2020). (in Chinese)
19. Jiao, Y., Gao, X., Wei, B., Hu, Y.: Research on the design of immersive virtual laboratory based on VR. Digit. Educ. **6**(04), 38–42 (2020). (in Chinese)
20. Liu, Y., Zhang, Q., Wang, D.: Thoughts on strengthening laboratory management in colleges and universities. Lab. Res. Explor. **39**(05), 244–246+251 (2020). (in Chinese)
21. Deng, H.: Design and implementation of computer virtual laboratory based on cloud computing. Ind. Sci. Technol. Innov. **2**(09), 21–22 (2020). (in Chinese)
22. Jin, H., Cui, J., Sun, Y.: Discussion on the integration of practice teaching management and Internet plus. J. Electr. Electron. Teach. **42**(02), 84–87+115 (2020). (in Chinese)
23. Deng, Z.: Research on the construction and management of computer virtual laboratory. Comput. Prod. Circ. **01**, 123 (2020). (in Chinese)
24. Li, H., Wang, J., Yi, Y., Gu, C.: Implementation of Internet-based teaching inquiry virtual laboratory. Intelligence **36**, 175 (2019). (in Chinese)
25. Wang, K., Wu, X., Guo, T., Kong, M.: Research status and development trend of virtual laboratory based on VR. Sci. Technol. Innov. Prod. **06**, 7–9 (2021). (in Chinese)

Application Research on Construction and Evaluation of Logistics Enterprises Performance

Qian Lu[1,3], Qing Liu[1], Yong Wang[2,4(✉)], and Tian Liu[4]

[1] School of Transportation and Logistics Engineering,
Wuhan University of Technology, Wuhan 430063, China
[2] Evergrande School of Management,
Wuhan University of Science and Technology, Wuhan 430080, China
wang-yong@whut.edu.cn
[3] Department of Planning, Puren Hospital, Wuhan 430080, China
[4] School of Logistics, Wuhan Technology and Business University, Wuhan 430065, China

Abstract. Over the past decade, the logistics industry in China has experienced extremely rapid development. China's total social logistics value continues to rise, significantly higher than the growth rate of GDP over the same period. But in most business operations, logistics costs have remained high. Entrepreneurs still have major problems in the introduction and management of the internal logistics system. The internal management level is weak and the ability of performance evaluation is poor. However, logistics performance evaluation is not a static process, it is a dynamic development. The formulation of performance evaluation index needs to be based on the company's own situation. Clear performance evaluation indexes can help the company to clarify its development direction, stimulate the enthusiasm of employees, and then obtain greater economic benefits. This paper uses AHP to analyze the problems existing in the establishment of the performance evaluation index of company D's logistics, and proposes corresponding improvement measures. It also helps business managers to clarify the direction of the company, so as to enhance market share and increase profit levels.

Keywords: Construction and evaluation · Logistics performance evaluation · Improvement measures · TPL

1 Introduction

1.1 Research Background

With the development of information and economy, a large number of commodities and information flow speed is accelerating day by day and the demand for logistics is therefore increasing exponentially. In recent years, China's total value of social logistics has been growing rapidly. Comparing with the comparable price, it has increased by 15%, which is obviously higher than the growth rate of GDP in the same period.

© The Author(s), under exclusive license to Springer Nature Switzerland AG 2022
Z. Hu et al. (Eds.): ICCSEEA 2022, LNDECT 134, pp. 145–159, 2022.
https://doi.org/10.1007/978-3-031-04812-8_13

The efficiency of logistics operation in China has been improved, but the competition within the logistics industry has also followed. Enterprises are facing the double competitive pressures brought by domestic and foreign related enterprises. For their own survival and development, many enterprises are exploring effective ways to improve productivity and logistics performance [1]. Improving the existing logistics performance appraisal can be an important measure for enterprises to reduce costs and clear the way forward. Whether large enterprises, medium-sized enterprises or small enterprises can benefit from an excellent logistics performance system. It can help managers to clarify the direction of the enterprise, provide decision-making information for the strategic management of enterprise development, establish a suitable environment for enterprise development, strengthen team cooperation, formulate clear performance plans for each employee and improve the enterprise. The work flow of the industry, the establishment of relatively perfect incentive and incentive system, to promote better development of enterprises [2]. This paper mainly analyses the existing performance evaluation index formulation methods of D company's logistics, finds out the problems existing in the formulation of D company's logistics performance evaluation index, puts forward corresponding optimization strategies for the rational formulation of performance evaluation index, helps enterprise managers to clarify the direction of the enterprise's progress, and achieves the increase of market share and the increase of profit level.

1.2 Research Status and Significance

1.2.1 Research Status

At present, a common problem in the logistics performance appraisal of enterprises is that the formulation of strategic index is not clear. The formulation of strategic index is to clarify the direction of progress, standardize the process of enterprises, determine the powers and responsibilities of departments, so that all departments within enterprises can participate in the work according to the standardized process. The ultimate goal is to reduce the cost of enterprises, increase profits and market share. However, in recent years, China's logistics enterprises have encountered some problems in the process of exploring the performance appraisal system to improve productivity and reduce logistics costs, and to find a suitable performance appraisal system for their own development. The formulation of strategic index is not clear, while the key performance index (KPI) of enterprises can be directly quantified and closely linked with the overall strategic objectives of the company and the key index of concern of the department. Objectives: By setting up projects, sampling, calculation and analysis, a goal of measuring performance process with standardized processes is to quantify management objectives, and to decompose the strategic objectives of enterprises into practical tools. PI can clarify the main responsibilities of each department, and then define the performance measurement index of personnel. Therefore, the establishment of a practicable key performance index assessment system is the key to do a good job in performance management and monitoring. However, at present, enterprises are facing some difficulties in implementing KPI system in all aspects, such as: if the assessment index lack accurate historical data as a basis, especially for the performance appraisal of business departments will bring some difficulties [3].

1.2.2 Research Significance

This paper analyses the problems existing in the process of formulating performance appraisal index of modern logistics enterprises. Taking Wuhan D Company Logistics Co., Ltd. as an example, by introducing the standards and actual situation of formulating performance appraisal index of D Company's logistics, this paper uses AHP method to analyze the problems existing in formulating logistics performance appraisal index of D Company and the causes of these problems, and to formulate performance appraisal reasonably. The core index put forward corresponding optimization strategies to help business managers improve the level of performance management, better play the role of performance appraisal in the selection and training of enterprise talents, and increase the speed and efficiency for enterprise development [4].

2 Overview of AHP

2.1 Definition of AHP

Analytic Hierarchy Process is abbreviated as AHP, or Analytic Hierarchy Process (AHP), which is a quantitative analysis method for qualitative problems. It is a simple, flexible and practical method to divide complex problems into multi-level, clear and simple problems through multiple related factors. AHP generally combines subjective judgment with objective analysis results directly and effectively, compares the importance of one factor in the problem in two ways, and then describes it quantitatively. Then, the weight of the order of relative importance of each level factor is calculated by mathematical method, and the relative weight of each level factor is calculated by the total ranking of all levels. This method was introduced into China in the 1990s, because it can analyze and deal with various decision-making factors qualitatively and quantitatively, and the tomographic analysis system is relatively simple and flexible, which has been rapidly applied in various fields of our social experience, such as: urban planning, economic management, energy analysis, scientific testing, logistics performance appraisal, etc.

2.2 Advantages, Disadvantages and Functions of AHP

1) *Advantages of Analytic Hierarchy Process*
    Establish all elements hierarchy, including quantifiable and non-quantifiable elements, and clarify the relationship of each element at different levels;
    Make the analysis program concise and clear, and optimize the calculation process;
    In the case of missing data in the analysis, the weight of each element can be calculated by AHP analysis method.
2) *Disadvantages of Analytic Hierarchy Process*
    It is difficult to make one-to-one comparisons if elements have similar attributes.
    Whether or not the elements have correlations can't be considered in the analysis;
    Within seven elements, they can pass the consistency test, but they may not pass when there are more elements.

3) *The Role of Analytic Hierarchy Process*

Chromatographic analysis can clarify and simplify complex problems. Both quantifiable and non-quantifiable problems can be quantitatively analyzed by chromatography. Moreover, the system is flexible and concise, which can clarify the scope of planning and decision-making, the measures and policies to be taken, the criteria, strategies and various constraints to achieve the goals, etc.

2.3   Overview of Qualitative and Quantitative Analysis

1) *Qualitative analysis*

Qualitative analysis method, namely "non-quantifiable analysis method", is an analysis method to infer the nature and development trend of events based on the practical experience and analysis ability of judges. It is a basic method of analysis. It is mainly applicable to the analysis of some things that do not need data consideration and some things that do not have complete data and data.

2) *Quantitative analysis*

Quantitative analysis method, namely "quantifiable analysis method", refers to the use of specific data (data characteristics, data relations, data changes) to analyze a method of matters. In the management of enterprises, quantitative analysis is based on financial statements as the main data source, sorted out according to mathematical formulas and principles, and then used mathematical theory to analyze the quantifiable data of enterprises, calculate the operation of the company and make investment judgments.

3   Analysis on Current Situation of Logistics Performance Assessment in Company D

3.1   Brief Introduction of Logistics Company D

Hubei D Company Logistics Group Co., Ltd. was established in 1999, is a comprehensive logistics service enterprise, including warehousing and distribution, vehicle transportation, LTL express transportation, supply chain management, information technology research and development, etc.

In 2015, it joined forces with five other enterprises to create one meter tick, and now ranks among the top five in China.

Wuhan D Company Logistics Co., Ltd. is the largest subsidiary of D Company Group. It is mainly engaged in LTL express, payment collection and city distribution in Hubei Province.

Company D pioneered the "five-oriented" management method of logistics, namely, trunk transportation shuttle bus, payment collection electronic, guaranteed price transportation normalization, goods delivery express delivery, and information management platform.In order to ensure the standardization and scientization of D Company's logistics performance management, an efficient performance management system is specially established to clarify the responsibilities and operation standards of all departments of the Company and to stimulate the enthusiasm of employees.

The company's performance evaluation index is decomposed through the system hierarchy and implemented to everyone to realize the ultimate goal of the company's strategic development, so that the company and individuals can develop together. Through the formulation of performance evaluation index, employees can find their own advantages and disadvantages, and find ways to improve organizational performance as soon as possible.

According to the nature of different posts, index should be set, both qualitative and quantitative index should be designed. In order to give full play to the role of assessment, set up a performance team to assist and supervise the formulation and implementation of assessment plans of various departments, and always adhere to the principle of "fairness, impartiality and openness" throughout the entire performance management system. Handle the abnormal and problems in performance work in time to improve the overall performance of the company [6].

3.2 Qualitative and Quantitative Analysis on the Selection of Logistics Performance Evaluation Index of D Company

1) *Job responsibilities*

Job responsibilities are the basis of performance evaluation index. The job responsibilities of personnel are analyzed, and the daily work priorities are determined in the performance evaluation process, i.e. the index with the strongest work relevance are selected from the five dimensions of timeliness, quality, cost, business and profit for evaluation.

2) *Organizational performance*

Organizational performance is an "optional item" of performance appraisal index. It determines which personnel need to be linked to organizational performance appraisal according to job level.

For example, the department heads of all first-level units need to assess the company's income, profits, business operations and other index.

3) *Talent contribution*

Talent contribution is the "bonus item" of the performance appraisal index, mainly assessing the contribution of all levels of units to the establishment of the company's talent system.

For example, the output rate of reserve cadres, the turnover rate of department employees, the personnel training rate, etc.

4) *Main Performance*

The main performance is the "bright spot" of the performance appraisal index, that is, the most valuable thing that the post outputs. For example, the main assessment income of business outlets [7].

3.3 Selection of Quantitative Performance Evaluation Index

1) *Peer analysis*

Peer analysis refers to comparing and analyzing the actual value of enterprise index with the average value of the same industry.

General enterprises usually combine the average standard of their industry when setting goals.

For example, compare and analyze various assessment indexes (historical value, index value growth rate, etc.) of D Company's logistics with Debang Logistics.

The revenue of Debang Logistics will increase by 40% in 2017, while that of Company D will be set at 40%. Debang Logistics will be taken as the average standard value to measure its own profit.

2) *Year-on-year analysis*

The year-on-year analysis refers to the comparison between the current period and the previous period. Compared with the same period of history.

For example, Company D will set targets by comparing the completion values in July 2017 and July 2016.

Year-on-year growth rate = (current period-current period)/current period × 100%.

For example, the goods difference index, the third quarter of 2017 D company logistics value index is the highest.

Due to the influence of some objective factors (hot weather and high volume of goods), the index can only be determined by comparing the difference index with the completion value in June 2016.

3) *Ring comparison analysis*

Ring comparison analysis is to compare the data of a certain period with the data of the previous period, and calculate the trend percentage to observe the annual increase or decrease. For example, Company D logistics in June 2017 and July 2017 are compared to establish index.

The formula is as follows: 1) Ring-on-Ring Growth Rate = (Current Period Number-Previous Period Number)/Previous Period Number × 100%.

3.4 Problems in the Formulation of Logistics Performance Evaluation Index of Company D

1) *The formulation of strategic index for various departments is unreasonable.*

There is no historical basis for the formulation of strategic performance index.

Due to the continuous change of D Company's logistics operation system, historical data are missing 60% of the index lack accurate historical values as the basis for the strategic index for the next quarter.

Strategic index are drawn up entirely on the basis of the opinions of superior leaders, without adequate communication with the appraisers. The appraisers and the appraisees do not have the same opinions on the objectives. Performance serves the management of salary more, and does not promote the achievement of employee performance and the achievement of strategic index of the whole company.

However, insufficient communication with the appraisee when setting performance appraisal targets leads to passive execution of the appraisee during the execution process. At the same time, the lack of peer and market research, strategic index focus on meeting high-level wishes.

As a result, the formulated strategic value is often on the high side, which violates one of the five principles of performance appraisal, but realization principle, and to some extent kills the enthusiasm of employees.

In 2017, the average achievement rate of department heads at all levels was less than 60%, and that of branch heads was less than 58%. The performance appraisal did not rectify the deviation in time during the implementation process, and did not play an encouraging role in the performance itself.

Taking 2016 and 2017 as examples, the company's revenue index dropped to 90% in the fourth quarter of 2016 based on the annual planning index value, and even dropped to 80% in the fourth quarter of 2017.

2) *The application of performance appraisal results is less*

The implementation of performance assistance is poor.

When employees encounter difficulties in their work and performance index cannot be reached, direct leaders seldom take the initiative to communicate with subordinates about problems encountered in the performance implementation process, and employees' performance is entirely their own responsibility.

After the performance appraisal cycle is over, the performance interview process tends to be formalized, the difficulties and problems of employees are not fully understood, and there is no effective improvement plan after the start of the next appraisal cycle. As shown in Fig. 1:

The leader's role of "leader" was not well implemented, resulting in the following members being ambiguous about their performance index, mistaking flowers for mists and inefficiency.

3.5 The Performance Results Have not yet Provided an Important Basis for Enterprise Training

In 2017, the company vigorously promoted full-process scanning. Performance appraisal increased the weight of scanning rate and the completion of scanning rate was poor.

Based on the age of workers, combined with the results of performance evaluation, according to the process, performance needs to take stock of the weaknesses of the evaluation, which is used as the basis for training courses, and organize workers to carry out special training for scanning, but it has not been implemented.

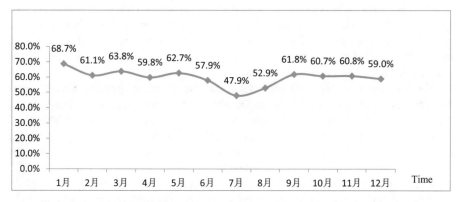

Fig. 1. Percentage of D company logistics network manager performance scores rate

1) *Performance results are not used as the basis for promotion and elimination of employees.*

The promotion of employees is mostly based on the demand of job gap, not on performance and ability, and the enterprise does not eliminate employees based on performance results in the development process.

Looking at 2017, there were 140 branch managers, less than 5 of whom were actively eliminated by enterprises, and 15% of whom performed poorly, as shown in Fig. 2.

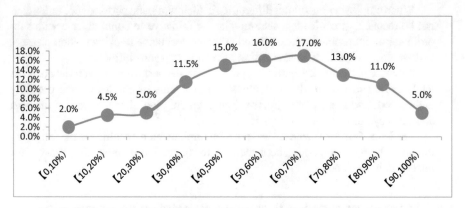

Fig. 2. Performance achievement rate of logistics network managers of D company

4   Application Research in Logistics Performance Evaluation Index

4.1   Analysis of AHP in D Company's Logistics Performance Evaluation Index

Company D logistics employee performance appraisal mode is A, B, C, D appraisal mode (A is excellent, B is good, C is qualified and D is unqualified). It consists of two appraisal parts. One is ability score, which is an appraisal item designed from the index or event dimension based on the key work of the department. The second is the quality score, which includes self-evaluation and comprehensive evaluation of leaders, and is designed from the team cooperation dimension.

The principle of A, B, C, D appraisal is that the appraisers (usually directly superior) must conduct appraisal in strict accordance with the completion of the appraisee's appraisal scheme. The number of appraisers A shall not exceed 30% (rounded) of the total number of departments participating in A, B, C, D appraisal (excluding department heads and appraisers). The proportion of appraisers C or D is not mandatory and can be determined according to the actual situation [8].

Case: AHP is used to analyze the monthly performance evaluation index of an employee of D Company's Logistics Human Resources Department in November 2017 (Tables 1, 2, 3 and 4).

Table 1. Judgment matrix B

B	Difficulty	Time	Efficiency
	V_1	V_2	V_3
V_1	1	4	3
V_2	1/4	1	1/2
V_3	1/3	2	1

Table 2. Degree of difficulty

B_1	φ	σ	λ	β
φ	1	2	3	4
σ	1/2	1	1/2	2
λ	1/3	2	1	1/2
β	1/4	1/2	2	1

Table 3. Completion time

B_2	φ	σ	λ	β
φ	1	1/4	1/3	1/2
σ	4	3	2	1
λ	3	1/2	1	2
β	2	1/3	1/2	1

Table 4. Completion rate

B_3	φ	σ	λ	β
φ	1	1/3	1/3	1/3
σ	3	1	1/2	2
λ	3	2	1	3
β	3	1/2	1/3	1

The assessment index for this month are "Competition for the manager of the quality control department, Rectification of the training room, Inventory of the deputy in the network, Training of new employees", which are indicated by A, B, C and D.

The three related attributes used are: difficulty coefficient, completion time and completion rate are represented by V_1, V_2 and V_3 [9].

Score the attributes of the individual objectives of the four assessment index (expressed as "quality control department manager competition, training room rectification, branch deputy inventory, and new employee training").

Solution

1) *Step 1: Draw a hierarchical analysis diagram:*
2) *Step 2: Calculate the weight estimation of the target layer and calculate its maximum eigenvector judgment matrix B by "sum and product method":*
   Normalization of Vector $W = (W1, W2, W3, W4, W5)$ τ

$$Wi = Wi/(\sum Wi)(i - 1, 2, \ldots \ldots n) \tag{1}$$

   Calculate that maximum characteristic of the judgment matrix λ_{max}

$$\lambda_{max} = 1.58/3 * 0.62 + 7/3 * 0.14 + 4.5/3 * 0.24 = 2.9$$
$$C.I. = (\lambda_{max} - N)/(N - 10 = (2.9 - 3)/(3 - 1) = -0.05$$
$$C.R. = -0.05/1.12 = -0.08 < 0.10$$

3) *Step 3: Find out the maximum feature vector of the scheme layer to the target layer, and find out:*

$$(W_1, W_2, W_3)^1 = (0.1, 0.25, 0.30, 0.35)$$
$$(W_1, W_2, W_3)^2 = (0.43, 0.09, 0.16, 0.32)$$
$$(W_1, W_2, W_3)^3 = (0.45, 0.17, 0.10, 0.28)$$

4) *Step 4: Get the total score of the four performance evaluation index:*

The score of φ: $= \Sigma Wi * wi^1 = 0.62 * 0.10 + 0.14 * 0.43 + 0.24 * 0.45 = 0.2302$.
The score of σ: $= \Sigma Wi * wi^2 = 0.62 * 0.25 + 0.14 * 0.09 + 0.24 * 0.17 = 0.2084$.
Score of λ: $= \Sigma Wi * wi^3 = 0.62 * 0.30 + 0.14 * 0.16 + 0.24 * 0.10 = 0.2324$.
Score of $\beta = \Sigma Wi * wi^4 = 0.62 * 0.35 + 0.14 * 0.32 + 0.24 * 0.28 = 0.3290$
To sum up, $\beta > \lambda > \varphi > \sigma$.

Conclusion: Therefore, the reasonable sequence and weight of performance appraisal objectives should be the training of new employees, the inventory of network deputies, the competition of quality control department managers, and the rectification of training rooms.

However, the actual situation is that the department lacks historical basis when formulating strategic performance index and accurate historical values as the basis for the strategic index for the next quarter.

The strategic index are drawn up entirely according to the opinions of the superior leaders, without full communication with the appraisers, and the appraisers and the appraisees do not have a consistent opinion on the objectives, which does not promote the achievement of employee performance and the realization of the strategic index of the whole company.

Due to insufficient communication in formulating the performance plan, the examinee prefers passive execution in the execution process. Finally, this month's performance was poor and the performance evaluation index did not play an encouraging role.

4.2   Application of AHP in Logistics Performance Evaluation Index of D Company

AHP is relatively simple to apply in D company's logistics at present. Although D company's logistics will pass qualitative and quantitative analysis when making logistics performance evaluation indexes.

Qualitative analysis and formulation are carried out on the basis of job responsibilities, organizational performance, contribution of talents, main performance, etc. in a quantitative manner, including peer analysis, year-on-year analysis, month-on-month analysis, etc., but the durability is not high.

In the third and fourth quarters, performance appraisal often slackened off and there was no quantifiable data support with specific data.

Moreover, when qualitative index cannot be quantified, AHP method has not been applied well, and relevant factors have not been given weight after chromatographic analysis. Some assessment objectives are formulated through subjective evaluation by decision makers, so there will be deviation of assessment index leading to poor final performance [10].

4.3   Measures to Improve Logistics Performance Evaluation Index Formulation

1) *Reasonably Formulate Logistics Performance Evaluation Index Based on AHP Analysis*

The performance of the previous quarter can be taken as the historical basis for the formulation of the next quarter's performance appraisal, and can be formulated after qualitative and quantitative analysis of the realizability according to the analysis results of AHP method and the actual situation.

In addition, when formulating employee performance appraisal index, the employee should fully communicate with the appraisee to make the appraisee clear and clear about the appraisal objectives, so that the appraisal index can also play a corresponding incentive role.

For example, the loading and unloading capacity of the stevedores in the logistics operation department of company d should be reasonably determined according to the actual situation, and corresponding adjustments should be made in case of peak load [11].

2) *Organize regular performance interviews*

After the performance appraisal cycle is over, the department shall organize regular performance interviews, confirm the good parts of the employees according to the results of this performance appraisal, guide and encourage the insufficient parts, and jointly formulate performance improvement plans to avoid the same problems next time.

In addition, regular performance interviews can be incorporated into managers' performance appraisal, which can not only help employees but also play a supervisory role in managers' work.

4.4 Apply the Quantitative Results of Performance Appraisal to Maximization

1) *Applying Performance Results to Talent Cultivation*

The establishment of performance appraisal index is not only to help employees to clarify the work content and improve work efficiency, but also to make contributions to personnel training.

Enterprises can sort out and make corresponding training courseware according to the problems in performance appraisal to organize regular training for employees.

To solve the weaknesses in the work of the staff, the staff's work passion has also been improved accordingly, to avoid the loss of personnel who are not suitable for the work, and to promote the cultivation of talents.

For example: sort out the SOP process for the invoicing process of the drawer and the tally process of the tallyman, make special courseware, and organize the staff to carry out training.

2) *The results of the performance appraisal will be included in the year-end debriefing report*

Incorporating employee's performance appraisal results into debriefing reports has aroused managers' high attention.

360-degree evaluation refers to all-round evaluation of employees through several dimensions and different levels, namely, superior evaluation of subordinates, peer evaluation of each other, and subordinate evaluation of superiors. At the same time, 360-degree evaluation is also an integral part of employee debriefing report.

Debriefing report is a promotion or elimination method for managers above the manager level in the year-end assessment of enterprises. Including the results of performance assessment can arouse the attention of employees and managers.

3) *Organize regular performance assistance work*

When employees encounter difficulties in their work and the performance index cannot be reached, the direct leader should actively communicate with the subordinates about the problems encountered in the performance implementation process to guide the employees in their work.

After the performance appraisal cycle is over, if the achievement rate of the performance appraisal index in the previous quarter is relatively low, the organization has conducted performance interviews, fully understood the difficulties and problems of the employees, and implemented effective improvement plans after the start of the next appraisal cycle. To avoid the recurrence of problems in the last quarter, and to make employees work passively to actively [11].

4) *Strictly abide by the five principles of performance evaluation index*

In the formulation of logistics performance evaluation index must strictly abide by the five principles, namely SMART principle, are:

- S: (Specific)-Principle of Clarity

The establishment of assessment index must be clear and definite so that the assessed can understand the assessment objectives.

Take the assessment index of an employee in the customer service department as an example: if its assessment index is positioned as "paying attention to customer relationship", this index is very vague and everyone has different understanding, so there are many ways to pay attention to customer relationship.

For example, if the index is set to "reduce the customer complaint rate", the complaint rate above 5% is unqualified, 3%–5% is qualified, 3%–1% is excellent, and 1% is excellent, then it is very clear, the examinee clearly knows how to implement it, and the examinee also knows what criteria to evaluate according to. Therefore, it is necessary to set clear project definitions and standards when formulating assessment objectives.

- M:(measurable)-principle of measurability

The assessment index must be measurable and digital and need a clear set of data to support them.

Avoid using the evaluation word "Nice, OK" to measure horizontally, which will lead to fuzzy standards and assessment errors.

In order to further train the stevedores of the operation department, "further" is a concept that is neither clear nor easy to measure. is it only necessary to arrange the training, regardless of the content and effect of the training is further training? However, if such assessment is made, the stevedores shall be given safety training at a certain time and place, and the actual working conditions of the employees shall be assessed after the training is completed.

If the error rate of safety accidents is higher than 10%, the training result is not ideal; if the error rate is lower than 10%, the result is ideal. Such assessment criteria are measurable.

- A: (Attainable)-But realization principle

The assessment index formulated must be achievable and acceptable to the executor. If it is formulated completely according to the wishes of the manager, then the typical psychology and behavior of subordinate employees are resisted and may be considered acceptable at that time. However, there is no final assurance to complete this index, which leads to low working passion of employees and the assessment index does not play an incentive role.

Therefore, when setting goals, employees should be insisted to participate, and the set assessment goals should be agreed between the organization and employees in combination with personal situation and post situation to ensure that the index can be realized.

- R: (relevance)-principle of relevance

When making assessment index, attention should be paid to the relevance of assessment items. If the achieved indicator is completely unrelated to other index or the relevance is very low, then the achievement of this indicator is not of great significance.

Because the setting of assessment index is related to job responsibilities.

For example, a tallyman can learn the knowledge of standard loading and unloading, and can be used as a support during the peak load period. Both tallyman and stevedore work in the operation department, and there are many overlapping areas of work.

If the tallyman is allowed to learn financial knowledge, the focus of his work will be deviated, because the goal of learning financial knowledge has very low correlation with the goal of improving the working standard of the tallyman.

- T: (timebound)-principle of timeliness

There must be a time limit when making the assessment. The completion of the assessment project must be within the specified time. The deadline is the completion result of the assessment project.

For example, an employee of company d logistics will complete the research work for the deputy of the branch before November 17, 2017. November 17 is a definite time limit.

There is no way to assess if there is no definite time limit, because superiors and subordinates have different perceptions of the priority of index. Some indexes are anxious but subordinates do not know. If there is no clear assessment time, the final thing is to complete, the superiors are very angry and the subordinates are very aggrieved. This will also lead to unfair assessment and hurt the relationship between superiors and subordinates and their enthusiasm for work [12].

Therefore, specific completion time should be set, the completion time of index items should be determined according to the priorities and weights of index events, and the completion and changes of indexes should be checked regularly, so as to facilitate managers to guide and assist employees in their work.

In a word, five principles are indispensable in the formulation of logistics performance evaluation objectives, and the formulation process is to enhance the overall planning ability of the enterprise's preliminary work, and the completion process is also the practice and experience process of the enterprise's management ability.

5   Conclusion

At present, China's logistics industry is developing in the direction of economization and informatization and has become an indispensable part.

However, in the process of development, it is necessary for enterprises to establish a set of perfect performance appraisal system, formulate clear strategic objectives of performance appraisal, and improve the role of performance appraisal in logistics enterprises.

Taking D company logistics as an example, this paper first analyzes the unreasonable part of the current performance evaluation index of D company logistics, then uses AHP analysis method and discusses how to select strategic index qualitatively and quantitatively, and finally puts forward improvement measures. It is necessary to formulate reasonable performance plan and strictly follow the five principles of performance evaluation. In addition, it is necessary to introduce competition mechanism to comprehensively consider from inside and outside to avoid "rigidity" of enterprise organization.

In order to promote the continuous development of logistics enterprises, every employee needs to formulate performance appraisal plans and objectives according to his own work, and the managers of enterprises should also standardize themselves and promote the full implementation of performance appraisal, so as to promote the development of logistics enterprises in a better and clearer direction [13].

Acknowledgment. The research was supported by the Hainan Provincial Joint Project of Sanya Yazhou Bay Science and Technology City; Hainan Special PhD Scientific Research Foundation of Sanya Yazhou Bay Science and Technology City (HSPHDSRF-2022-03-032).

References

1. Anna, W.: Overview of logistics enterprise performance appraisal. SME Manag. Technol. **8**, 6–9 (2014)
2. Anitha, P., Patil, M.M.: A review on data analytics for supply chain management: a case study. Int. J. Inf. Eng. Electron. Bus. **10**(5), 30–39 (2018)
3. Zhang, X.: Research on the performance evaluation system of logistics enterprises. Electron. Commer. **11**, 5–6 (2017)
4. Zhou, T., Wang, L., Mo, Y., et al.: Performance evaluation of a logistics distribution center of an automobile factory based on KPI and AHP. Logist. Technol. **35**(12), 151–154 (2016)
5. Benotmane, Z., Belalem, G., Neki, A.: A cost measurement system of logistics process. Int. J. Inf. Eng. Electron. Bus. **10**(5), 23–29 (2018)
6. Yang, W.: Performance evaluation of Huizhou port logistics enterprises based on super-efficiency DEA model. Market Modernization (9), 52–53 (2016)
7. Ting, P.-H.: An efficient and guaranteed cold-chain logistics for temperature-sensitive foods: applications of RFID and sensor networks. Int. J. Inf. Eng. Electron. Bus. **5**(6), 1–5 (2013)
8. Wang, Y., Zhang, P., Semere, D.T., et al.: Application on performance assessment of international food supply chain. J. Coastal Res. **103**(sp1), 52 (2020)
9. Jia, X., Shu, Z.: Research on AHP-based SME logistics service performance evaluation. Urban Constr. Theor. Res. (Electron. Ed.) **18**, 33–35 (2017)
10. Yi, X.: Performance evaluation analysis and countermeasure research of logistics enterprises. Econ. Trade Pract. **1**, 11–13 (2017)
11. Bakar, M.A.: Malaysian logistics performance: a manufacturer's perspective. Procedia-Soc. Behav. Sci. **224**, 571–578 (2016)
12. Shukla, S.K.: Integrated logistics system for indigenous fighter aircraft development program. Procedia Eng. **97**(1–2), 2238–2247 (2014)

Human Resource Structure and Scientific Research Productivity of Chinese Universities Based on Entropy Weight Method and the Relationship Model

Qiuyan Zhang and Huanhuan Mao[✉]

Personnel Department, Wuhan University of Technology, Wuhan 430070, China
mhhyue@163.com

Abstract. Scientific research performance is one of the significant contents of personnel management in colleges and universities. Effectively introducing talents and encouraging scientific researchers to innovate requires not only the improvement of personnel management in colleges and universities, but also the reform of personnel assessment system. Focusing on the scientific research productivity of colleges and universities which fits an important proposition that determines the future development of colleges and universities, an index system of scientific research productivity is constructed in the paper, which includes two primary dimensions of academic and practical productivity, four secondary dimensions: paper publication, monograph publication, valid invention patent and the formation of national or industrial standards. The entropy weight method is used to comprehensively evaluate the multi index panel data. The paper empirically tests the impact of human resource structure on scientific research productivity in colleges and universities in China. The research conclusion can provide a scientific basis for the practice of personnel management in colleges and universities.

Keywords: Human resource · Research performance · Scientific research productivity · Entropy method

1 Introduction

Scientific research productivity in colleges and universities is an important cornerstone of national scientific and technological innovation and industrial development, which determines a country's global competitiveness. In the personnel management of colleges and universities, it is of great significance to give full play to the cornerstone role of universities in serving the national development through the reform of personnel system and the innovative practice of personnel management. Scientific research performance appraisal is the key point, which not only affects the personal development of university scientific researchers and the reputation of departments and institutions, but also an important basis for promotion and salary decision-making in personnel management. More objective evaluation system and influencing factor analysis are needed to provide reference for relevant personnel management decisions. In this sense, through the reconstruction of the evaluation system of scientific research productivity and the empirical analysis of the impact of human resources (HR) factors on scientific research productivity (SRP), this paper discusses how to introduce talents with innovative ability and stimulate the scientific research productivity of talent team.

2 Study Design

2.1 Construction of Index System for Scientific Research Productivity Evaluation

Existing studies mainly use the publication of academic papers to measure scientific research output. Some scholars also define it as academic productivity or scientific research performance [1, 2]. In recent years, relevant studies also use "H index" and "G index" to measure scientific research performance. On the basis of the traditional quantitative indicators of paper publication, quality indicators such as citation times and journal impact factors are added [3, 4], However, these studies are still academic paper oriented analysis, the consideration of other factors is not comprehensive enough, and there is a certain gap with the actual situation [5–7].

On February 23, 2020, the Ministry of Science and Technology issued "several measures on breaking the bad orientation of "paper only" in science and technology evaluation (Trial)", which pointed out that the representative work system should be implemented for the paper evaluation of basic research scientific research activities, and the paper should not be used as the main evaluation index for applied research and technology development scientific research activities, but should pay attention to the transformation of evaluation results support the performance of industrial development. In the relevant estimate system, the evaluation weight should be increased for the achievements with certain academic influence, practical application effect and driving effect of scientific and technological innovation.

Based on the above considerations, the evaluation of scientific research productivity in this paper breaks the inherent principle of "paper only". Referring to the suggestions of the document, the evaluation system is divided into two primary indicators (as shown in Table 1), namely, academic productivity based on basic research and practical productivity based on Application of R & D.

Table 1. Index system framework of scientific research productivity evaluation

Target layer	First index layer	Second index layer
Scientific research productivity	Academic productivity	Publication of paper (X_1)
		Publication of monograph (X_2)
	Practical productivity	Valid invention patent (X_3)
		Formation of national or industrial standards (X_4)

Among them, the secondary index layer of academic productivity adds the mono-graph publishing index (X_2) on the basis of the traditional paper publishing index (X_1), and no longer emphasizes the importance of foreign paper publishing index alone. The secondary index layer of practical productivity is guided by the application of scientific and technological innovation and industrial driving effect, adopting valid invention patent index (X_3) and forming national or industrial standard index (X_4). All of the above indicators are positive ones.

2.2 Comprehensive Evaluation of SRP Based on Entropy Weight Method

Entropy weight method is an objective weighting method in comprehensive evaluation. Its basic principle is to determine the importance of the index according to the dispersion degree between the observed values of the same index. The bigger the dispersion degree of the index, the greater the amount of information provided, and the greater the impact of the index on the evaluation system [8]. This method measures the amount of information by calculating the information entropy of each index, which can effectively avoid the interference of human factors in the subjective evaluation method, and ensure that the constructed index system can reflect most of the original information, so as to make the evaluation results more in line with the actual situation.

Common comprehensive evaluation methods are mostly applied to cross-sectional data or time series data, while the comprehensive evaluation method of multi index panel data is still mainly discussed at the level of statistical theory because it involves the comprehensive extraction of spatial dimension and time dimension information. Euclidean distance is usually used to measure the index distance without constraints, Ignoring the influence of index correlation emphasized in panel data analysis, it is very easy to produce analysis deviation [9, 10]. This paper mainly refers to Dong Feng's method of re synthesis after evaluation in a single dimension [11]. Although it cannot meet the requirements of statistical system analysis for temporal and spatial unity, in practical application, this method basically meets the characteristics of short panel samples in this paper and the needs of individual effect model used in empirical analysis. Specifically, the calculation method of entropy weight method is improved to the following calculation process:

It is defined that the sample matrix contains m evaluation objects, n evaluation indicators, and the time length is T. The sample matrix is expressed as,

$$X = \begin{pmatrix} X_1 & \cdots & X_i & \cdots & X_m \end{pmatrix} \tag{1}$$

$$X_i = \begin{bmatrix} x_{i1}^1 & \cdots & x_{i1}^j & \cdots & x_{i1}^n \\ \vdots & \ddots & \vdots & \ddots & \vdots \\ x_{it}^1 & \cdots & x_{it}^j & \cdots & x_{it}^n \\ \vdots & \ddots & \vdots & \ddots & \vdots \\ x_{iT}^1 & \cdots & x_{iT}^j & \cdots & x_{iT}^n \end{bmatrix} \tag{2}$$

where, x_{it}^j ($i = 1, 2, \ldots, m; j = 1, 2, \ldots, n; t = 1, 2, \ldots, T$) is the value of the j-th index of the i-th evaluation object at time t.

Since all indexes in this paper are positive ones, the range transformation method is firstly used to dimensionless process each index, and the formula is,

$$v_{it}^j = \frac{x_{it}^j - \min\{x_{it}^j\}}{\max\{x_{it}^j\} - \min\{x_{it}^j\}} \tag{3}$$

where, v_{it}^j is the new value of x_{it}^j after dimensionless treatment.

Secondly, calculate the proportion y_{it}^j of the j-th index of the i-th evaluation object in time t, and obtain the proportion matrix $Y_i = (y_{it}^j)_{T \times n}$. The formula is,

$$y_{it}^j = \frac{v_{it}^j}{\sum_{t=1}^{T} v_{it}^j} \tag{4}$$

Then, calculate the information entropy e_i^j and difference coefficient d_i^j of the j-th index of the i-th evaluation object and the formula is,

$$e_i^j = -K \sum_{t=1}^{T} y_{it}^j \ln y_{it}^j \tag{5}$$

$$d_i^j = 1 - e_i^j \tag{6}$$

where, $K = \frac{1}{\ln T}$ is a nonnegative constant, and $0 \leq e_i^j \leq 1$. And stipulate that $y_{it}^j \ln y_{it}^j = 0$ when $y_{it}^j = 0$.

Finally, calculate the weight w_i^j and evaluation value U_{it} of the j-th index of the i-th evaluation object. The formula is,

$$w_i^j = \frac{d_i^j}{\sum_{j=1}^{n} d_i^j} \tag{7}$$

$$U_{it} = \sum_{j=1}^{n} y_{it}^j w_i^j \tag{8}$$

Wherein, it is the evaluation value U_{it} of the i-th evaluation object at time t, which is a set of panel data. For the short panel, the time length T is small and the evaluation object m is large. The above method of calculating the evaluation value separately in each evaluation object space can intuitively reflect the trend of each evaluation object on the time axis. At the same time, the evaluation value constructed based on the same index system can also reflect the common characteristics between different evaluation objects, which conforms to the characteristics of compound disturbance of individual effect model adopted.

2.3   Relationship Model Between HR and SRP

The research on the influencing factors of scientific research productivity is mainly focused on personal characteristics, family relations, professional background, working environment and so on. In terms of personal characteristics, relevant studies is usually carried out on the impact of gender, age and race on scientific research productivity [12–14]. As for family relations, the researches mainly combine gender characteristics to study whether female scholars are different from male scholars in scientific research output due to housework, childbirth and parent-child relationship [2, 15, 16]. About professional background, the studies is made around the influence of factors such as doctoral degree, professional title, research experience and academic network [6, 17, 18]. Considering working environment, it is to investigate the impact of factors such as the ranking of colleges and departments, research funds and infrastructure guarantee [5, 18, 19].

However, most of these studies are based on the investigation of paper publication indicators, which has certain limitations, is not comprehensive, and is inconsistent with the actual situation. Based on the expanded and reconstructed evaluation system of scientific research productivity, this paper summarizes the influencing factors of the above scientific research productivity into four categories: human resource investment, capital investment, platform strength and external economic environment, and focuses on the impact of human resources on scientific research productivity. The input of human resources mainly includes the personal ability of scientific researchers and the workload of scientific research. Among them, the ability factor mainly focuses on educational background, professional title, full-time and part-time status of teaching and scientific research and the workload factor mainly focuses on the impact of gender and the opportunity cost of time-consuming teaching and other work. The research framework is shown in Fig. 1.

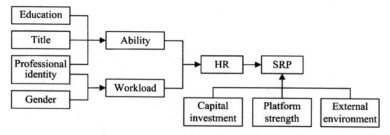

Fig. 1. Relationship framework between HR and SRP

Due to the differences of provincial conditions in the areas where colleges and universities are located, there may be individual effects that do not change with time. Therefore, this paper constructs a fixed effect model with reference to the research method of Rodgers & Neri [17], as follows:

$$productivity_{it} = hrability'_{it} \beta_1 + hrtime'_{it} \beta_2 + fund'_{it} \beta_3 + univ'_{it} \beta_4 + econ'_{it} \beta_5 + z'_i \delta + u_i + \varepsilon_{it}$$
$$(i = 1, 2, \ldots, m; \quad t = 1, 2, \ldots, T) \tag{9}$$

Among them, the dependent variable *productivity*$_{it}$ is the calculated evaluation value of scientific research productivity U_{it}, the core independent variable is the personal ability *hrability*$_{it}$ and scientific research workload *hrtime*$_{it}$ of scientific researchers; the control variables are capital investment *fund*$_{it}$, platform strength *univ*$_{it}$ and external economic environment *econ*$_{it}$, while controlling the provincial situation characteristics z_i that do not change with time. In addition, $(u_i + \varepsilon_{it})$ is the compound disturbance term of the model.

3 Evaluation of Scientific Research Productivity

According to the index system of scientific research productivity constructed in this paper, the data of paper publication (X_1), monograph publication (X_2), effective invention patent (X_3) and formation of national or industrial standard (X_4) of 31 provinces, autonomous regions and municipalities directly under the central government from 2009 to 2019 are selected for scientific research productivity evaluation. The data are from China Science and technology statistical yearbook.

The weight of each index is calculated according to the entropy weight method of comprehensive evaluation of multi index panel data, and the results are shown in Table 2. The calculation results of index weight show that, on the whole, the national or industrial standard (X_4) has the highest weight, with an average level of 0.4314, reflecting the importance of the index in leading industrial development. Second only to it is the effective invention patent (X_3), with an average weight of 0.2442, reflecting the importance of this index in scientific research and innovation application. The weighted mean values of academic paper publication and monograph publication indicators are only 0.1591 and 0.1652.

The objective evaluation based on entropy weight method in this paper shows that the importance of improving scientific research productivity in practice has an objective factual basis, and is in line with the policy guidance of the Ministry of science and technology on some measures to get rid of the bad orientation of "paper only" in scientific and technological evaluation (trial).

On the basis of determining the weight, the evaluation value of local scientific research productivity is further calculated, and the results are shown in Fig. 2.

The evaluation results of scientific research productivity show that the overall scientific research productivity has increased significantly from 2009 to 2019, but there are great differences in the time trend between regions. For example, Jilin (*code 7*) and Heilongjiang (*code 8*) are relatively stable, Hubei (*code 17*), Hunan (*code 18*) and Guangdong (*code 19*) have an obvious upward trend, and Liaoning (*code 6*), Ningxia (*code 30*) and Xinjiang (*code*) shows a certain downward trend in recent years. The overall level of Tianjin (*code 2*), Hainan (*code 21*) and Tibet (*code 26*) is low, but there are abnormally high values in a few years. Therefore, further empirical tests are needed to judge the possible causes of these differences.

4 Empirical Analysis of the Impact of HR Structure on SRP

Adopting the method of empirical analysis, the paper selects the relevant data of China Education Statistical Yearbook, China Science and Technology Statistical Yearbook and

Table 2. Index weight based on entropy weight method

Code	Region	W_1	W_2	W_3	W_4
1	Beijing	0.1339	0.2183	0.2532	0.3945
2	Tianjin	0.0936	0.1024	0.1360	0.6680
3	Hebei	0.2402	0.2820	0.2611	0.2168
4	Shanxi	0.2239	0.1780	0.2234	0.3746
5	Inner Mongolia	0.1035	0.2191	0.3286	0.3487
6	Liaoning	0.1397	0.1091	0.3642	0.3869
7	Jilin	0.1506	0.1145	0.2695	0.4654
8	Heilongjiang	0.1487	0.1184	0.2847	0.4482
9	Shanghai	0.1936	0.1101	0.1064	0.5899
10	Jiangsu	0.2119	0.1866	0.3086	0.2928
11	Zhejiang	0.2179	0.2215	0.2627	0.2979
12	Anhui	0.1341	0.1905	0.3862	0.2892
13	Fujian	0.3461	0.1146	0.2042	0.3350
14	Jiangxi	0.1000	0.2215	0.2036	0.4750
15	Shandong	0.2396	0.1361	0.2720	0.3524
16	Henan	0.1840	0.2122	0.3540	0.2498
17	Hubei	0.1216	0.0876	0.2052	0.5856
18	Hunan	0.2655	0.1513	0.2696	0.3136
19	Guangdong	0.2786	0.1546	0.1942	0.3727
20	Guangxi	0.1230	0.1470	0.2883	0.4417
21	Hainan	0.0452	0.1021	0.1326	0.7201
22	Chongqing	0.1217	0.1631	0.2722	0.4429
23	Sichuan	0.1783	0.1856	0.2338	0.4023
24	Guizhou	0.1243	0.1808	0.2550	0.4399
25	Yunnan	0.1315	0.1397	0.2304	0.4984
26	Xizang	0.0539	0.1290	0.1629	0.6542
27	Shanxi	0.1915	0.2054	0.2106	0.3926
28	Gansu	0.1085	0.2502	0.2752	0.3662
29	Qinghai	0.1694	0.1517	0.2503	0.4286
30	Ningxia	0.0793	0.2036	0.1742	0.5428
31	Xinjiang	0.0800	0.1359	0.1961	0.5879
Average weight		0.1591	0.1652	0.2442	0.4314

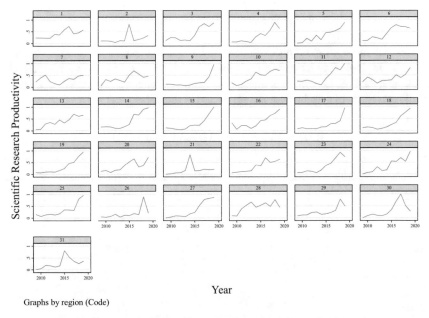

Scientific Research Productivity

Year

Graphs by region (Code)

Fig. 2. Time trend of scientific research productivity evaluation value

China Statistical Yearbook to calculate the proxy variables of scientific research capacity, scientific research workload, capital investment, platform strength and external economic environment. Among them, the ability variable ($hrability_{it}$) in the core independent variable selects and calculates the number and proportion of researchers, the number and proportion of female researchers, the number and proportion of doctoral researchers (*PhD*), the number and proportion of personnel with senior professional titles (*title*) as proxy variables. Workload ($hrtime_{it}$) select and calculate the full-time equivalent and proportion (time), basic research workload and proportion (*fdmt*), applied research workload and proportion (*applic*), experimental development workload and proportion (*experi*). The control variable capital input ($fund_{it}$) selects and calculates the natural logarithm (*fund*) of the total internal and external R & D expenditure, the platform strength ($univ_{it}$) selects and calculates the proportion of central subordinate universities (*cuniv*), and the external economic environment ($econ_{it}$) selects and calculates the natural logarithm (*econ*) of local GDP growth rate.

In addition, due to the high coincidence of scientific researchers and teaching personnel in the personnel structure of colleges and universities in China, there is a serious multicollinearity problem in the relevant indicators. Therefore, among the scientific researchers indicators and full-time teachers indicators, this paper chooses the scientific researchers related indicators with higher correlation with SRP-scientific research productivity. At the same time, owing to the alternative collinearity of basic research, applied research and experimental development indicators, this paper only retains the relevant indicators of applied research and experimental development. The above data selection and processing do not affect the empirical results. The descriptive statistical results of the main variables are shown in Table 3, in which the evaluation object $m =$

31 and the time length $t = 11$. The sample belongs to short panel data and is a balanced panel without missing data.

The data are used to empirically test the fixed effect model (9), and the results are shown in Table 4. Empirical results (1) and (2) show that all scale variables such as the number of R & D personnel and the full-time equivalent of research work are not significant (space constraints are not presented in the paper).

Table 3. Descriptive statistics of main variables

Variable	Obs	Mean	Std. dev.	Min	Max
productivity	341	0.3319	0.2621	0.0000	1.0000
researcher	341	0.2813	0.1086	0.0987	0.7430
time	341	0.8428	0.0391	0.7466	0.9516
fund	341	11.9840	1.4380	7.5088	14.9385
cuniv	341	0.0415	0.0728	0.0000	0.4194
econ	341	2.1591	0.3483	−0.6931	2.8449
female	341	0.3935	0.0592	0.2202	0.5489
phd	341	0.2348	0.0811	0.0794	0.5098
title	341	0.2688	0.0316	0.1807	0.3495
applic	341	0.4776	0.0804	0.2204	0.7424
experi	341	0.0582	0.0338	0.0000	0.2012

However, the structural variables of HR concerned in this paper, such as the proportion of researchers and the proportion of scientific research workload (*time*), can significantly promote the improvement of scientific research productivity. The control variables, such as R & D expenditure (*Fund*), the proportion of central subordinate universities (*cuniv*) and the external economic environment (*econ*) are not significant, and do not affect the robustness of the core independent variables. On this basis, the benchmark model of the relationship between the human resource structure of colleges and universities and scientific research productivity can be constructed as follows:

$$productivity = 1.1315 * researcher + 3.4673 * time - 3.0312 \qquad (10)$$

Further, the ability variables of human resource structure are subdivided according to educational background, gender and professional title, and the workload of human resource structure is subdivided into two categories: applied research and experimental development. Results (3) and (4) are obtained. The empirical results show that the proportion of doctoral researchers (*PhD*) and female researchers (*female*) have a significant impact on scientific research productivity, while the proportion of senior title researchers (*title*) and the proportion of different types of research workload (*apple, experi*) have no significant impact, and do not affect the significance of the proportion of researchers (*PhD*) and female researchers (*female*). On this basis, the relationship model further considering the educational structure and gender structure of human resources in colleges

Table 4. Empirical test results

Dependent variable: productivity

Independent variable	Result (1)	Result (2)	Independent variable	Result (3)	Result (4)
researcher	1.1362***	1.1315***	phd	3.6487***	3.5553***
	(0.2603)	(0.2631)		(0.4161)	(0.4199)
time	3.4356***	3.4673***	female	1.2059***	1.1034***
	(0.7532)	(0.7534)		(0.4033)	(0.3934)
fund	0.0112		title	0.2525	
	(0.0159)			(0.4113)	
cuniv	0.0398		applic	0.2206	
	(0.2980)			(0.2414)	
econ	0.0802		experi	0.7328	
	(0.0618)			(0.6589)	
constant	−3.3299***	−3.0312***	constant	−1.7593***	−1.4682***
	(0.5870)	(0.5936)		(0.2733)	(0.1435)
R2	0.3791	0.3720	R2	0.5980	0.5945
N	341	341	N	341	341

Note: Robust standard errors in parentheses, $*p < 0.1$, $**p < 0.05$, $***p < 0.01$.

and universities can be constructed as follows:

$$productivity = 3.5553 * phd + 1.1034 * female - 1.4682 \qquad (11)$$

5 Conclusion

Colleges and universities undertake the sacred responsibility of teaching and educating people, scientific research and serving the society. The effective development of human resources and the growth of scientific research productivity in colleges and universities are directly related to the future development and hope of colleges and universities.

To realize the reform of personnel management system in colleges and universities, we should get rid of the "paper only" scientific research performance evaluation orientation, appropriately increase the weight of evaluation indicators with practical application effect and industrial driving effect, and give full play to the innovation and leading role of colleges and universities in serving social development, serving the real economy and industrial transformation.

In the specific practice of personnel management in colleges and universities, we should fully optimize the human resource structure of colleges and universities and promote the improvement of scientific research productivity in colleges and universities, including efforts to improve the proportion of scientific researchers, reduce the proportion of non scientific research teaching workload of scientific researchers, and encourage female scholars to participate in scientific research.

References

1. Diamond, A.: An economic model of the life-cycle research productivity of scientists. Scientometrics **6**(3), 189–196 (1984)
2. Zhu, Y., Ma, Y.: Gender, time allocation and scholarly productivity among Chinese university faculty. Collection Women's Stud. **130**(4), 24–30+49 (2015) (in Chinese)
3. Hirsch, J.E.: An index to quantify an individual's scientific research output. Natl. Acad. Sci. USA **102**(46), 16569–16572 (2005)
4. Egghe, L.: Theory and practice of the g-index. Scientometrics **69**, 131–135 (2006)
5. Qin, Z., Ma, W., Xu, Z., Sai, F.: Impact path of human capital and social capital on scholars' research performance: a fuzzy-set qualitative comparative analysis. Sci. Technol. Manag. Res. **466**(24), 159–167 (2020). (in Chinese)
6. Lei, Y., Chen, X.: How does scientific research papers network capital affect scientific research performance of scientists: microscopic evidence from Chinese scientists in universities. Sci. Technol. Manag. Res. **484**(18), 121–130 (2021). (in Chinese)
7. Ting, W.: Study on the scientific research group management of Chinese high technology enterprises from the perspective of knowledge transfer. Int. J. Educ. Manag. Eng. (IJEME) **2**(3), 57–63 (2012)
8. Abdullah, L., Otheman, A.: A new entropy weight for sub-criteria in interval type-2 fuzzy TOPSIS and its application. Int. J. Intell. Syst. Appl. (IJISA) **5**(2), 25–33 (2013)
9. Bonzo, D.C., Hermosilla, A.Y.: Clustering panel data via perturbed adaptive simulated annealing and genetic algorithms. Adv. Complex Syst. **5**(4), 339–360 (2002)
10. Zheng, B.: The clustering analysis of multivariable panel data and its application. Appl. Stat. Manag. **154**(2), 265–270 (2008). (in Chinese)
11. Dong, F., Tan, Q., Zhou, D.: Factor analysis on enterprises' R&D ability under multi-index panel data. R&D Manag. **21**(3), 50–56 (2009). (in Chinese)
12. Leahey, E.: Gender differences in productivity: research specialization as a missing link. Gend. Soc. **20**(6), 754–780 (2006)
13. Oster, S.M., Hamermesh, D.S.: Aging and productivity among economists. Rev. Econ. Stat. **80**(1), 154–156 (1998)
14. Sreeramana Aithal, P., Suresh Kumar, P.M.: ABC model of research productivity and higher educational institutional ranking. Int. J. Educ. Manag. Eng. (IJEME) **6**(6), 74–84 (2016)
15. Sax, L.J., Hagedorn, L.S., Arredondo, M., Dicrisi, F.A.: Faculty research productivity: exploring the role of gender and family-related factors. Res. High. Educ. **43**(4), 423–445 (2002)
16. Sunish Kumar, O.S.: A fuzzy based comprehensive study of factors affecting teacher's performance in higher technical education. Int. J. Educ. Manag. Eng. (IJEME), **5**(3), 26–32 (2013)
17. Rodgers, J.R., Neri, F.: Research productivity of Australian academic economists: human-capital and fixed effect. Aust. Econ. Pap. **46**(1), 67–87 (2007)
18. Taylor, S.W., Fender, B.F., Burke, K.G.: Unraveling the academic productivity of economists: the opportunity costs of teaching and service. South. Econ. J. **72**(4), 846–859 (2006)
19. Zong, X., Fu, C.: The research performance of chinese research universities and its influencing factors: an empirical analysis based on the relevant data of universities directly under the ministry of education. J. High. Educ. Manag. **77**(5), 26–35 (2019). (in Chinese)

Dynamic Pricing Strategy and Simulation of Electricity Enterprises Based on Supply Chain Revenue Management

Huaye Huang[1(✉)] and Lingchunzi Li[2]

[1] School of Transportation and Logistics Engineering,
Wuhan University of Technology, Wuhan 430063, China
yezyhuang@163.com

[2] School of Management, Huazhong University of Science and Technology, Wuhan 430074,
China

Abstract. The electricity system reform has put forward new requirements on the traditional electricity trading mode and pricing mechanism. The deregulation of both sides of electricity generation and electricity sale and the control of transmission has become an important guide for the reform. In this paper, reference the centralized transaction mode of foreign electricity market, the idea of game theory and dynamic pricing method are used to introduce the competition mechanism into the electricity supply chain. The dynamic Stackelberg game model considering the strategic choice behavior of users is established under the centralized transaction mode. The time-sharing dynamic pricing scheme for maximizing the revenue of electricity producers is obtained. The effectiveness of the pricing strategy is verified by introducing the equilibrium validity factor through the example simulation. The results show that the introduction of competition in the electricity market can effectively reduce the electricity price of electricity producers and improve the electricity supply. At the same time, the dynamic pricing strategy has higher equilibrium effectiveness in the electricity market where the price demand elasticity of electricity users, the leader's initial market share and the proportion of active users are all higher.

Keywords: Dynamic pricing · Revenue management · Electricity market · Game theory

1 Introduction

On March 5, 2017, Premier Li Keqiang pointed out in the 2017 Government Work report that the reform of mixed ownership should be deepened and take substantive steps in electricity and other sectors. China will reform the electricity, oil and gas systems and open up competitive businesses.

For a long time, the inadaptability of China's electricity monopoly management system and socialist market-oriented economic system has gradually revealed. It is difficult to realize the optimal allocation of electric resource so that the electricity industry

© The Author(s), under exclusive license to Springer Nature Switzerland AG 2022
Z. Hu et al. (Eds.): ICCSEEA 2022, LNDECT 134, pp. 171–185, 2022.
https://doi.org/10.1007/978-3-031-04812-8_15

is struggling. In order to promote the development of electricity industry and improve the effective use of electricity resource, electricity market reform is imperative. Since 2002, China has been implementing the electricity market reform policy of controlling the middle and liberalizing the two ends. The competition mechanism is introduced on the generating side and the selling side. The nodes in the electricity supply chain are gradually to be separated from the initial vertical integration to independent operating subjects. By the end of 2016, a total of 31 provinces, municipalities and autonomous regions had carried out comprehensive trials for the reform of the electricity system, covering all provincial-level electricity transmission and distribution price reforms. In August 2017, 8 provinces (regions) carried out electric-selling side reform pilot. By the end of 2019, the first batch of 8 spot electricity pilot projects had carried out settlement trial operation. On April 26, 2021, 6 provinces and cities were selected as the second batch of spot electric power pilot.

China's electricity system reform has made many achievements. However, in the reform process, especially in the process of electricity trading, there are some problems such as uneven distribution of interests, single trading mechanism, and unreasonable pricing strategy and so on. Reasonable electricity pricing is an important guarantee for economic development and daily life of people as well as the focus of research and pursuit of all countries. In order to solve the above problems, it is necessary to study the electricity pricing model and strategy in line with China's national conditions.

2 Literature Review

2.1 Electricity Industry Pricing

In the literature, many studies show that the electricity trade markets are open in most developed countries. Foreign scholars mainly study their pricing strategies from three aspects: electricity generators, sellers and users. For generation companies, Philpott [1] established a two-stage transaction model of day-ahead market and regulated market between a single power generation company and the electricity buyer. For electricity sellers, Bu [2] proposed dynamic real-time pricing strategy for sellers in the context of complete competition and cooperation considering traditional and strategic electricity demand in order to achieve the highest income for the enterprise itself or the whole. Kanellou [3] assumed that customers had the right to choose any company that they desired in an hourly basis. He introduced a real-time pricing algorithm which attempted to model the interaction between companies and customers as a Stackelberg game.

Since China's electricity market has not been fully opened, the power transmission link led by the grid is still an indispensable way of power supply. Domestic scholars mainly study the electricity price of power generation providers, transmission providers and electricity sellers. On the power generation side, Dai [4] established a dynamic Stackelberg game model and proposed a dynamic real-time pricing strategy to maximize the revenue of each distributed power supply. Zhu [5] combined the clearing mechanism of electric power spot market with the operating characteristics of thermal power enterprises to construct a short-term pricing game decision model considering marginal cost.

On the transmission side, Chen [6] proposed the zonal transmission pricing method, that is, in a region, the British method is used to calculate the cost related to location, the revised stamp method is used to calculate the cost unrelated to location, and the cost after the regional connection line is equivalent is recovered through the trend tracking method. Zhang [7] discussed the influence of transmission price mechanism on the competitive efficiency of electricity spot market, and proposed a quantitative evaluation index of transmission pricing mechanism taking into account the competitive efficiency of electricity spot market.

Most of the research on pricing in China's electric power industry focuses on the selling side. Considering the competitive environment and discriminatory pricing strategies, Lv [8] set price discrimination strategies for new and old users in the duopoly electricity retail market. Considering multi-objective and single pricing strategies, Zhang [9] studied the selling price and market competitiveness of different electricity selling companies under the new electric power system reform, and proposed the pricing model of electricity selling companies based on Bertrand model.

2.2 Revenue Management and Dynamic Pricing

Revenue management originated from the American aviation industry in the 1970s, and has become an important branch of management science after nearly 50 years of development. Kimes [10] proposed the 4R theory to define revenue management, that is, to provide the right products or services to the right customers at the right time and place the right price, so as to achieve the goal of maximizing corporate revenue under resource constraints. Chopra [11] believed that revenue management was to use pricing to improve profits under the constraints of limited supply chain assets, the core of which was differential pricing.

As one of the important means of revenue management, dynamic pricing has attracted more and more scholars' attention. Kincaid [12] were the first to study the continuous time dynamic pricing of perishable products. At present, the research of dynamic pricing focus on multi-product joint pricing, uncertain demand, market competitiveness, user's strategic behavior, decision maker's risk preference and so on. For example, Bi [13] studied the two-stage dynamic pricing of multiple products. Hou [14] studied the influence of customer inertia on dynamic pricing of enterprises in a competitive environment.

Dynamic pricing is becoming more widely used as revenue management moves from traditional applications (aviation, hotels, car rental, etc.) to other industries such as broadcast advertising, medical services, real estate, transportation, manufacturing, sports and entertainment event management.

To sum up, the existing income management research focuses on the products with fixed output and time-limited, and there are few research literature on intangible products such as electricity, which cannot be stored and whose demand has great volatility and dynamics. A few researches on power pricing based on revenue management seldom consider the competitive factors under market conditions and do not involve dynamic pricing strategies.

Compared with the traditional pricing methods such as discrimination pricing and ladder pricing, dynamic pricing is more suitable for the increasingly fierce competition in the current electricity market. Based on the existing research results of dynamic pricing, this paper builds a two-level supply chain model of duopoly competitive electricity sellers and power users to study the dynamic pricing strategy of electricity generation side of supply chain.

3 Dynamic Pricing Strategy of Single Electricity Producer in Competitive Environment

3.1 Electricity Supply Chain and Pricing of China

China's electricity system reform aims to reduce the cost of electricity sales, optimize resource allocation and promote electricity development by breaking the natural monopoly of the grid and introducing competition in the wholesale and retail market [15]. Under this reform policy, the electricity supply chain has gradually changed from vertical integration into a number of new supply chains that are mutually independent (economic) and coordinated (supply and demand). Its structure and electricity supply process are shown in Fig. 1.

The new electricity pricing system consists of the on-grid price, transmission and distribution price, wholesale price and retail price. The on-grid price is influenced by government monitoring and market competition. The electricity price of transmission and distribution will be gradually changed from controlled by electricity grid enterprises to unified pricing by the government. Wholesale electricity price and retail electricity price will gradually change from government monitoring to market competition through bilateral transactions and centralized transactions at present.

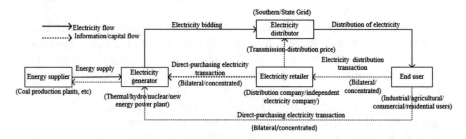

Fig. 1. Electricity supply chain of china

In this paper, the part of electricity supply chain is considered. It only includes the electricity generating enterprises and large electricity users. The former directly sells electricity to the latter. This paper studies the dynamic pricing strategy of electricity producer.

3.2 Dynamic Pricing Analysis of Single Electricity Producer

3.2.1 Analysis of the Electricity Market in Which Single Electricity Producer Participates

1) **Market participants and trading patterns**

Electricity seller is the electricity producer. Electricity buyer is the large electricity users who can only choose one electricity supplier to purchase electricity during the same period. Centralized trading platform is a non-profit trading institution that organizes and supervises all kinds of transactions and discloses information to trading entities. Let $N = \{1, 2, \ldots, n\}$, $S = \{1, 2, \ldots, m\}$, which indicate respectively the set of large electricity users and the set of electricity producer. Divide the day into T periods. At the beginning of the period $[T, T + 1]$, electricity producers provide the centralized trading platform with the planned electricity production of q_{it} and the unit price of p_{it} to maximize their own interests. Through the p_{it} and the historical data, the trading platform calculates D_t of the total electricity consumption of users, finds out the price adjustment function according to the relationship between supply and demand, and determines the final price P_{it} based on the unit quotation of each electricity producer. Large electricity users obtain P_{it} through the trading platform, at which time there is a balance between supply and demand of $D'_t = Q_t$. The trading platform realizes market clearing through the price mechanism.

2) **Demand analysis**

For the j node user, its demand function reference [16] is as Formula (1).

$$d_{jt} = N_j(t) - e_j p_{it} \tag{1}$$

where, $N_j(t)$ is the demand of j user at time t when the electricity price is zero. e_j is the elasticity of j user's demand to price.

3) **Balance analysis**

The characteristics of electric power products are that they are produced and consumed immediately and cannot be stored. Supply and demand should always be the same in each period. Therefore, price adjustment function is introduced here to change users' demand for electricity purchase through pricing mechanism and directly adjust the imbalance between supply and demand that may occur in the current electricity trading market. Assuming that the trading platform can obtain more accurate $N_j(t)$ and e_j through historical data. When the electricity producer's quotation is p_{it}, the demand is $D_t = \sum\limits_{j=1}^{n} (Nj\ (t) - ej\ pit)$.

Let $P_{it} = p_{it} + k(Q_t)$ and make $D'_t = Q_t$, it can be derived as Formula (2).

$$\sum_{j=1}^{n} \left(N_j\ (t) - e_j \left(p_{it} + k(Q_t) \right) \right) = Q_t \tag{2}$$

Formula (3) is obtained by simplifying Formula (2).

$$k(Q_t) = \frac{D_t - Q_t}{ne_j} \tag{3}$$

If $Q_t > D_t$, then k $(Q_t) < 0$. The result is that the electricity producer sells electricity at a price lower than its own and the demand increases. Subsequently, the electricity producer will reduce the electricity supply in the next stage. If $Q_t < D_t$, then k $(Q_t) > 0$. The generator sells electricity at a higher price than its own and the demand decreases. The result is that the generator increases the supply at the next stage.

4) **Cost analysis**

For the generator at the i node, its marginal cost function reference [16] is as Formula (4).

$$C'_{it}(q_{it}) = a_i q_{it} + b_i \quad i \in S \tag{4}$$

The real cost function is as Formula (5).

$$C_{it}(q_{it}) = \int_0^{q_{it}} (a_i q + b_i) dq \quad i \in S \tag{5}$$

3.2.2 Dynamic Pricing Basic Model of Single Electricity Producer

In the early stage of $[T, T+1]$, the electricity producer provides the trading platform with planned production q_{it} and unit offer p_{it} to maximize its own benefits. This is translated into the following optimization problems as Formula (6).

$$\text{Max } R_{it} = p_{it} q_{it} - C_{it}(q_{it}) \quad \text{s.t. } 0 \le q_{it} \le W_i \tag{6}$$

In the stage of $[t, t+1]$, the actual electricity price sold by the generator is $P_{it} = P_{it} + k(Q_t)$, and the biggest revenue problem of the single generator is expressed as Formula (7).

$$\max R_{it} = (p_{it} + k(Q_t))q_{it} - C_{it}(q_{it}) \quad \text{s.t. } 0 \le q_{it} \le W_i \tag{7}$$

In the stage of $[t, t+1]$, the electricity generator maximizes the overall income through the centralized trading platform, and its problem can be expressed as Formula (8).

$$\max_{q_{1t}, q_{2t} \cdots q_{mt}} \sum_{i=1}^{m} ((p_{it} + k(Q_t))q_{it} - C_{it}(q_{it})) \quad \text{s.t. } 0 \le q_{it} \le W_i \tag{8}$$

3.3 Dynamic Pricing Strategy of Dual Electricity Producers in the Competitive Environment

3.3.1 Pricing Analysis and Strategies of Dual Electricity Producers in Single Stage Game

1) **Pricing analysis and modeling of dual electricity producers in single stage game**

Due to the different scale and capacity of competing enterprises, the order of players often exists in the actual competition process. Stackelberg game model is introduced to divide the electricity producers participating in centralized trading into two types. One is the electricity market leader, who firstly provides q_{1t} of planned electricity supply to the centralized trading platform in every $[t, t + 1]$ period. In each $[t, t + 1]$ period, the leader determines his own electricity supply according to his planned electricity supply q_{1t}. Since q_{2t} is determined by q_{1t}, q_{2t} is the response function of q_{1t}. Let's say that $q_{2t} = S(q_{1t})$. Assuming that the demand behavior of electricity users is the same in each period, and the proportion of choosing leaders is α_t, the proportion of choosing followers is $(1 - \alpha_t)$, and $\alpha_t > 0.5$.

It can be inferred from the balance analysis in Sect. 3.1 that different electricity producers may have different price adjustment mechanisms in order to achieve the supply-demand balance within the market share of each electricity producer. There is a certain correlation between leader and follower of the electricity supply because the follower knows leader' quotation and supply, namely $q_{2t} = S(q_{1t})$. The association is follower and leader who applies the same price adjustment mechanism to implement its own supply and demand balance function, namely $k(q_{1t}) = k(q_{2t}) = k(Q_t)$. Formula (9) can be derived according to Formula (4).

$$\frac{d_1 - q_1}{\alpha ne_j} = \frac{d_2 - q_2}{(1 - \alpha)ne_j} = \frac{D_t - Q_t}{ne_j} \tag{9}$$

In function (9), $d_1 = \alpha n(N_j(t) - e_j p_{1t})$, $d_2 = (1 - \alpha)n(N_j(t) - e_j p_{2t})$, $D_t = d_1 + d_2$, $Q_t = q_1 + q_2$. Simplify the function (9) to obtain Formula (10) as follows.

$$Q_t = \frac{q_{1t}}{\alpha_t} = \frac{q_{2t}}{1 - \alpha_t} \quad D_t = \frac{d_{1t}}{\alpha_t} = \frac{d_{2t}}{1 - \alpha_t} \tag{10}$$

The return function of leaders and followers is as follows.

$$R_{it} = (p_{it} + k(Q_t))q_{it} - C_{it}(q_{it}) \tag{11}$$

In function (11), $0 \leq q_{it} \leq W_i$, $C_{it}(q_{it}) = \frac{a_i}{2}q_{it}^2 + b_i q_{it}$, $i = 1, 2$,

$$Q_t = q_{1t} + q_{2t}, p_{it} = a_i q_{i1} + b_i$$

$$k(Q_t) = \frac{D_t - Q_t}{ne_j} = \frac{d_{1t} - q_{1t}}{\alpha ne_j} = \frac{d_{2t} - q_{2t}}{(1 - \alpha)ne_j}$$

2) **Pricing strategy of dual electricity producers in single period game**

Firstly, the follower adjusts the electric quantity q_{2t} according to the leader's electric quantity q_{1t} and determine the quotation to maximize his own profit in the second stage. After price adjustment, the trading platform obtains P_{2t} and gains R_{2t} according to Formula (11). The optimization order of follower is calculated as Formula (12).

$$q_{2t} = S(q_{1t}) = \frac{n(N_{(t)} - eb_2) - q_1}{2 + ne_j a_2} \tag{12}$$

The leader predicts that the stalker chooses according to $S(q_{1t})$ and maximizes R_{1t} in the first stage. To sum up, the pricing and quantitative strategies of leaders are as follows.

$$P_{1t}^* = a_1 q_1^* + b_1 + \frac{L - Mq_1^* - Nq_2^*}{ne_j} \quad q_{1t}^* = \frac{n(FG - H)}{a_1 ne_j F + 2(F - 1)}$$

Follower pricing and quantitative strategies are follows.

$$P_{2t}^* = a_2 q_2^* + b_2 + \frac{L - M q_1^* - N q_2^*}{ne_j} \quad q_{2t}^* = S(q_{1t}^*) = \frac{nH - q1*}{F}$$

F, G, H, L, M and N are defined as follows.

$$F = 2 + ne_j a_2 \quad G = N(t) - e_j b_1 \quad H = N_{(t)} - e_j b_2$$

3.3.2 Dynamic Pricing Analysis and Strategy of Multi-stage Game Considering User's Strategic Choice

The strategic choice behavior of users should be considered since there are rational users in the actual market. This paper divides the electricity users into active users and loyal users. Active users will choose electricity producers according to their own maximization interests at the expense of paying the switching cost S. Loyal customers will not switch. Assuming that electricity users are in the same area, the proportion of active users among users of different generators is the same. With the gradual liberalization of the electricity market, the share of active users is higher than that of loyal users. In other words, the proportion of active users is assumed to be V and V > 50% [8].The transfer costs of the active users who choose the leader and the follower are respectively as s_1 and s_2. The market share of the leader is α_0 at the beginning. Due to the existence of active users, the market share of the leader and the follower (α_t, $1 - \alpha_t$) during the competition has four scenarios.

In the [t, t + 1] phase, the cost paid by active users is as Formula (13).

$$U_{it} = P_{it}^*(N_j(t) - e_j P_{it}^*) \tag{13}$$

If another electricity supplier is selected, its cost is as Formula (14).

$$U_{it}' = P_{(3-i)t}^*(N_j(t) - e_j P_{it}^*) + s_i \tag{14}$$

If $U_{it} > U_{it}'$, active users switch generators in the next period. If $U_{it} \leq U_{it}'$, the original generator is maintained. Active users who choose leaders and electricity suppliers in this period choose electricity suppliers in the next period in the following ways, as shown in Table 1.

Table 1. Active user selection behavior matrix

	$U_{2t} \leq U_{2t}'$	$U_{2t} > U_{2t}'$
$U_{1t} \leq U_{1t}'$	Leader, follower	Leader, leader
$U_{1t} > U_{1t}'$	Follower, follower	Follower, leader

The market shares of leaders and followers in the next period under different conditions are shown in Table 2.

Table 2. Next generation market share matrix

$\alpha_{t+1}, 1 - \alpha_{t+1}$	$U_{2t} \leq U'_{2t}$	$U_{2t} > U'_{2t}$
$U_{1t} \leq U'_{1t}$	$\alpha_0, 1 - \alpha_0$	$v + \alpha_0(1 - v), (1 - \alpha_0)(1 - v)$
$U_{1t} > U'_{1t}$	$\alpha_0(1 - v), 1 - \alpha_0(1 - v)$	$(1 - \alpha_0)v + \alpha_0(1 - v), \alpha_0 v + (1 - \alpha_0)(1 - v)$

Substitute α_{t+1} into Eq. (13) and (14) to obtain the optimal pricing $P_{it+1}{}^*$ and electricity supply $q_{it+1}{}^*$ of each generator in the next stage $[t + 1, t + 2]$.

In order to analyze the influence of equilibrium strategy on the overall profit of the electricity generation side in the competitive environment, the validity of the subgame refined Nash equilibrium should be considered. Let $V_{dec}(q_{1t}, q_{2t})$ represent the total profit of electricity generation companies under equilibrium conditions in competition and $V_{int}(q_{1t}, q_{2t})$ represent the maximum profit of electricity generation companies after integration in the absence of competition. The validity of equilibrium is defined as follows.

$$\eta = \frac{R_{dec}(q_{1t}, q_{2t})}{R_{int}(q_{1t}, q_{2t})} \tag{15}$$

$$R_{dec}(q_{1t}, q_{2t}) = \sum_{t=1}^{4} \sum_{i=1}^{2} (p_{it}^* + k(Q_t))q_{it}^* - C_{it}(q_{it}^*) \tag{16}$$

$$R_{int}(q_{1t}, q_{2t}) = \max_{q_{1t}, q_{2t}} \sum_{t=1}^{4} \sum_{i=1}^{2} ((\frac{\partial C_{it}(q_{it})}{\partial q_{it}} + k(Q_t))q_{it} - C_{it}(q_{it}))$$

$$\text{s.t.}\ \ 0 \leq q_{it} \leq W_i$$

$$q_{1t}/\alpha_0 = q_{2t}/(1 - \alpha_0) \tag{17}$$

To sum up, due to the multi-stage game involved, the solution of the model is complicated. The electricity generator strategy can be obtained by using MATLAB. If the equilibrium validity is close to or over 1, it indicates that the strategy is effective.

4 Numerical Simulation and Analysis

Suppose there are two different electricity producers on the centralized trading platform which are the leader and the follower respectively according to their size and market share. Firstly, the leader puts forward q_1 and offer p_1, and the follower puts forward q_2 and offer p_2 accordingly. The centralized trading platform calculates the demand D_t at this stage according to the producer's quotation and electricity consumption. The balance between supply and demand is realized by adjusting the electricity price. The leader and the follower determine the electricity supply and quotation for the next period according to their market share and the current price adjustment function. Related parameters are set as follows: $a_1 = 0.3$, $a_2 = 0.4$, $b_1 = 1.2$, $b_2 = 1.2$, $n = 5$, $s_1 = 100$, $s_2 = 80$,

$W_1 = 600\text{MWh}$, $W_2 = 500\text{MWh}$, $e_j = 0.8$, 1, 1.2, $v = 0.55$, 0.75, 0.95, $\alpha_0 = 0.6$, 0.7, 0.8. In order to simplify the calculation, this paper studies the pricing problem of dynamic game from 6 a.m. to 18 p.m. on workday. The day is divided into four periods including morning (6–8), morning (9–11), noon (12–14) and afternoon (15–18). For each transaction, the quantity of electricity and price of electricity are determined on an hourly basis. The supply quantity and price of electricity are the same for each hour in each time period. The electricity consumption of multiple large electricity users in the same area in each period is from reference [16]. The simplified N_j (t) discrete data from is as shown in Table 3.

Table 3. N_j (t) Discrete data

Time frame t/h t/h	6–8	9–11	12–14	15–18
$N_j(t)/W$	110	200	250	300

4.1 Comparative Analysis of Dynamic Pricing Strategies

Under the condition of $\alpha_0 = 0.6$, $v = 0.6$, the optimal solution of each generation company in each period without considering competition and considering competition and consumers' strategic choice behavior is obtained. The results are shown in Table 4. It can be seen that the pricing of each generation is lower than that of the single-period optimization problem and the electricity supply also increases in the multi-period dynamic game. Although the income of the two generation companies in each period declines, the electricity price is greatly reduced and the supply is increased. The result is that the benefit of electricity users is improved and the higher demand is met. The effectiveness of the equilibrium solution is 0.85, that is the difference between the equilibrium return in the competitive environment and the maximum return in the integration environment is small. It proves that the competition mechanism is effective.

4.2 Sensitivity Analysis of Key Parameters

4.2.1 Influence of Price Demand Elasticity on Equilibrium Solution

The influences of user's price sensitivity on equilibrium are shown in Fig. 2 and Fig. 3 when e is 0.8, 1.0, and 1.2 respectively. It can be seen from the figure that the equilibrium electricity supply and equilibrium electricity price of the two electricity producers in each period decrease with the increasing of price demand elasticity of large electricity users. When the user has unit elasticity, the price of the two generators is the same. When there is no flexibility, leaders' price is higher than followers.

This is because the leader has a larger market share and is more affected by changes in the quantity of demand than the follower. The change of the price elasticity of demand is small, the price is more significant, leaders improve the electricity price gains from higher than the revenues increased demand with the increase of the price elasticity of

Table 4. Comparison between single-period optimal pricing strategy and dynamic game pricing strategy

Optimal Solution in Single-period						
		Time frame	t=1	t=2	t=3	t=4
Leader		q_{it}/MWh	120.26	219.53	274.68	329.82
		p_{it}/¥/MWh	87.39	158.53	198.05	237.57
		R_{it}/¥	8196.04	27309.46	42753.39	61644.02
Follower		q_{it}/MWh	82.61	150.79	188.67	226.55
		p_{it}/¥/MWh	85.87	155.76	194.58	233.41
		R_{it}/¥	5629.60	18758.01	29365.97	42341.35
Revenue/time frame		R_t/Yuan	13825.64	46067.47	72119.36	103985.36
Total revenue		R_{int}/Yuan	**235997.84**			
Equilibrium Solution in Multi-stage Dynamic Games						
	Time frame		t=1	t=2	t=3	t=4
Leader	q_{it}/MWh		148.90	271.80	340.08	408.35
	p_{it}/¥/MWh		73.01	132.70	165.73	198.76
	R_{it}/¥		6258.09	20965.79	32822.27	47324.82
Follower	q_{it}/MWh		110.08	200.94	251.42	301.90
	p_{it}/¥/MWh		72.37	131.54	164.28	197.02
	R_{it}/¥		5411.25	18114.47	28358.49	40888.71
Revenue/time	R_t/Yuan		11669.33	39080.26	61180.76	88213.54
Total revenue	R_{dec}/Yuan		**200143.89**			
Equilibrium validity			**0.84807509**			

demand, demand is affected by the price one by one, leader, reduce the loss of the demand is higher than the lower electricity loss, thus greatly reduce the electricity price to reduce revenue losses.

4.2.2 The Influence of the Leader's Initial Market Share on the Equilibrium Solution

It shows the influence of the leader's initial market influence on the equilibrium in Fig. 4 and Fig. 5 when α_0 is 0.55, 0.75 and 0.95 respectively. It can be seen that the leader's market share does not affect the electricity supply in each period, but has a small influence on the pricing. The higher the market share is, the higher the electricity price will be. This makes both leaders and followers willing to compete through trading platforms.

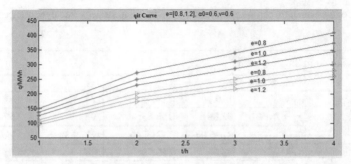

Fig. 2. q_{it} curve: e = [0.8, 1.2], $\alpha0 = 0.6$, v = 0.6

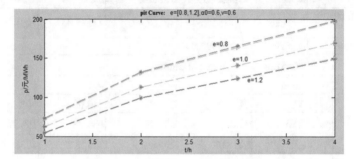

Fig. 3. P_{it} curve: e = [0.8, 1.2], $\alpha0 = 0.6$, v = 0.6

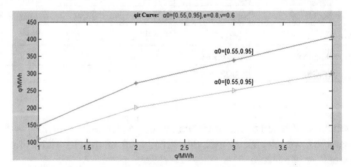

Fig. 4. q_{it} curve: $\alpha0 = [0.55, 0.95]$, e = 0.8, v = 0.6

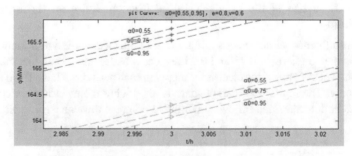

Fig. 5. P_{it} curve: $\alpha0 = [0.55, 0.95]$, e = 0.8, v = 0.6

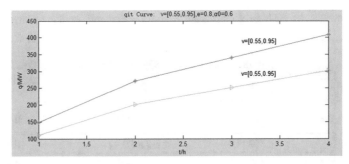

Fig. 6. q_{it} curve: v = [0.55, 0.95], e = 0.8, α0 = 0.6

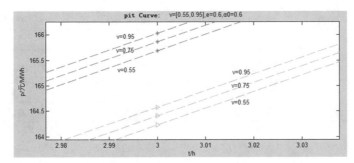

Fig. 7. P_{it} curve: v = [0.55, 0.95], e = 0.6, α0 = 0.6

4.2.3 Influence of Market Share Ratio of Active Users on Equilibrium Solution

Figure 6 and Fig. 7 show the influences of market share ratio of active users on equilibrium solution when V is set at 0.55, 0.75 and 0.95 respectively. It can be seen from the figure that the equilibrium electricity supply of the two electricity producers remains unchanged when the proportion of active users decreases gradually. At the same time, the equilibrium price decreases slightly.

4.3 Comparative Analysis of Income Under Different Parameter

The equilibrium income and equilibrium effectiveness are calculated under different parameter settings. The key parameter and results are shown in Table 5. It is used to study the impact of key parameters on equilibrium.

The table shows that the equilibrium effectiveness at 0.8 above, the optimal value of competition environment and integration environment were similar, namely in the electric electricity market to introduce the competition mechanism is effective. With the increasing of price elasticity of demand, the equilibrium return under dynamic competition and the optimal return under integration environment are all decreasing, but the equilibrium effectiveness increases. With the increase of the leader's initial market share, the equilibrium return and optimal return decrease, and the equilibrium effectiveness increases. When the leader's initial market share is large, the equilibrium effectiveness approaches 1.With the increase of the proportion of active users, the equilibrium returns

Table 5. Comparative analysis of returns under single-period optimization and dynamic game

Parameter	Value	Equilibrium Revenue $R_{dec}/¥$	Optimal Revenue $R_{int}/Yuan$	Equilibrium Validity
e	0.8	200143.89	235997.84	0.8481
	1	150323.96	177099.32	0.8488
	1.2	123831.38	138940.27	0.8913
α_0	0.55	200205.09	236064.24	0.8481
	0.75	199960.27	230089.36	0.8691
	0.95	199715.46	207129.65	0.9642
v	0.55	200064.44	235997.84	0.8477
	0.75	200382.22	235997.84	0.8491
	0.95	200700.01	235997.84	0.8504

under dynamic competition gradually increase. In the integrated environment, the optimal returns remain unchanged and the equilibrium effectiveness increases because active users' choice behavior is not taken into account.

5   Conclusions

This paper studies the dynamic pricing strategy of electricity generation side of supply chain. The results show that:

1) The competition mechanism introduced into the electric power direct marketing can effectively reduce the electricity price and improve the electricity supply.
2) The price regulation mechanism set by the trading platform can achieve the balance of supply and demand in the electricity market.
3) Balanced electricity supply has nothing to do with the market share of electricity producers.
4) With the increase of sensitivity, the electricity price will gradually decrease and always be lower than the optimal electricity price in the integrated environment.
5) Large electricity users are willing to enter the competitive market to reduce the unit electricity purchase cost.
6) Equilibrium returns decrease with the increase of the elasticity of user price demand and the leader's initial market share, increase with the increase of active users' market share.

Acknowledgment. This project is supported by Industry-University Collaborative Education Project of the Ministry of Education (202102489014).

References

1. Philpott, A.B., Pettersen, E.: Optimizing demand-side bids in dayahead electricity markets. IEEE Trans. Electr. Syst. **21**(2), 488–498 (2006)
2. Bu, S., Yu, F.R., Liu, P.X.: Dynamic pricing for demand-side management in the smart grid. In: 2011 IEEE Online Conference on Green Communications (GreenCom), pp. 47–51. IEEE (2011)
3. Kanellou, E., Mastakas, O., Askounis, D.: A game theory approach towards real-time pricing in future electricity markets. Int. J. Decis. Support Syst. **4**(3), 217–234 (2021)
4. Dai, Y., Gao, Y.: Dynamic pricing decision based on distributed generation system. Syst. Eng. **34**(2), 70–75 (2016). (in Chinese)
5. Zhu, G., Shan, Y., Lao, Y., Song, X., Zhang, L., Wei, Y.: Research on the pricing strategy of thermal power plants in the electricity spot market-analysis based on short-term bidding game model. Price Theory Pract. (06), 92–96+180 (2020). (in Chinese)
6. Chen, Z., Xiao, J., Jing, Z., Zhang, X., Zhang, H., Leng, Y.: A zonal transmission pricing approach for electricity market in China. Power Syst. Technol. **41**(07), 2124–2130 (2017). (in Chinese)
7. Cong, Y., Zhang, L., Tao, W.: Quantitative evaluation method of transmission pricing mechanism under circumstance of electricity spot market. Proc. Chin. Soc. Electr. Eng. **40**(21), 6925–6936 (2020)
8. Lv, K., Hu, H., Wu, C.: Paying customers to switch in electrical retail market-based on multi-type clients and differential switching cost. Syst. Eng.-Theory Pract. **12**, 2644–2655 (2012). (in Chinese)
9. Zhang, W., Xiao, B., Li, S.: Pricing and market competitiveness of electricity retailers under new strategy of electric power system reform. Smart Power **46**(09), 45–52 (2018). (in Chinese)
10. Kimes, S.E.: Yield management: a tool for capacity-constrained service firms. J. Oper. Manag. **8**(4), 348–363 (1989)
11. Chopra, S.: Supply Chain Management (Version 7). China Renmin University Press (2021)
12. Kincaid, W.M., Darling, D.A.: An inventory pricing problem. J. Math. Anal. Appl. **7**, 183–208 (1963)
13. Liu, H., Bi, W.: Optimal dynamic pricing for multi-products with consumers' reference effects and strategic behavior. Chin. J. Manag. Sci. 1–9 (2021). https://doi.org/10.16381/j.cnki.iss n1003-207x.2019.1248. (in Chinese)
14. Hou, F., Zhai, Y., Hu, Y.: Dynamic pricing strategy for perishable products in competitive markets with customer inertia. Oper. Res. Manag. Sci. **29**(09), 179–185 (2020). (in Chinese)
15. Liu, L., Xia, M.: Research and suggestion on reform of domestic electric pricing mechanism. Res. Dev. **1**, 133–136 (2015). (in Chinese)
16. Chen, X., Yu, Y., Xu, L.: Linear supply function equilibrium with demand side bidding and transmission constrain. Proc. Chin. Soc. Electr. Eng. **24**(8), 17–23 (2004)

Competitiveness of China Railway Express Node Cities Based on AHP

Chen Chen[✉] and Qiuhui Li

School of Business Administration, Wuhan Business University, Wuhan 430056, China
cc3221@qq.com

Abstract. The China Railway Express (CR Express) is not only an important practice of the Belt and Road, but also an important component of the international trade logistics channel. However, with the rapid increase in opening cities, problems such as insufficient supply of goods, fewer return trains, and fierce price competition have become more and more obvious. It is necessary to analyze the level of competitiveness of node cities in network. Through the analysis of each city, development strategies can be formulated separately to ensure the healthy and sustainable development of the trains. This paper establishes the competitiveness index of the node cities of CR Express, and evaluate the city competitiveness base on AHP. The results show that the logistics competitiveness of the 33 hub cities of CR Express is quite different, showing obvious hierarchical and hierarchical phenomena. According to the ranking of competitiveness, cities can be roughly divided into three categories. The overall economic development level of the first category of cities is good, and the second category has considerable advantages in a certain aspect of urban competitiveness, and the third category of cities has average competitiveness.

Keywords: China railway express · Node city competitiveness · Analytic hierarchy process

1 Introduction

The CR Express train is a fast cargo compartment that uses container freight trains to travel from China to Europe. Currently, the CR Express train has three channels: West, China and China. The total number of trains in China and Europe has exceeded 40,000, reaching 41,008, 31 domestic cities with more than 100 trains, and 168 cities in 23 European countries. However, with the increasing number of domestic cities, the problems of low efficiency of train return, insufficient supply of goods, and price war of large proportion of freight subsidies are becoming more and more obvious. It is necessary to analyze the competitiveness of the cities in the macroscopic view. According to the development needs of the cities and the specific conditions of the cities, the precise analysis and planning of the city division strategies should be carried out to ensure the healthy and sustainable development of the cities in China and Europe.

Many scholars apply it to the port competitiveness evaluation, besides, there are many studies on the urban development and logistics development, and different scholars have

© The Author(s), under exclusive license to Springer Nature Switzerland AG 2022
Z. Hu et al. (Eds.): ICCSEEA 2022, LNDECT 134, pp. 186–193, 2022.
https://doi.org/10.1007/978-3-031-04812-8_16

established various models for the evaluation by using various methods. In 1980, overseas scholars had put forward the concept of "competitiveness". James and Chris on the basis of constructing the evaluation index system of regional logistics competitiveness, used the method of subjective distribution value to evaluate the development level of regional logistics competitiveness [1]. Ekici O and others measured the level of logistics development competitiveness in Turkey by means of neural network analysis from six indicators of infrastructure, customs, service level, timeliness, transport and national logistics performance [2]. The research on the level of competitiveness of domestic urban logistics development is becoming more and more mature. Dai Ying, Li Qin and Song Han established a comprehensive evaluation capability system of goods distribution centers composed of 28 three-level indicators by gradually shrinking indicators and deleting duplicate indicators, and adopted a multi-index comprehensive evaluation model to evaluate the competitiveness of CR Express logistics [3]. Wei Songbo used the factor analysis method to measure the integrated level of logistics in central Europe and China based on the growth level theory, the hub-and-spoke network theory, and the logistics integration theory. The social network analysis method is used to estimate the logistics development level of the cities in China and Europe, and the relationship analysis and structure analysis of the logistics network in cities in China and Europe are studied.

2   Competitiveness of CR Express Node Cities Model

2.1   AHP

The analytic hierarchy process (AHP) (Analytic Hierarchy Process, AHP) combines quantitative and qualitative methods by layering decision factors and then relative treatment, which can be applied to multiple fields. Moreover, AHP is easy to combine with other methods to bring the advantages of multiple methods to solve various decision problems. Luo [4] Apply AHP to the selection of chemical engineering courses. As an important method of environmental analysis, AHP can objective part of the subjective process, so it is suitable to make strategic and policy choices. References [5–7] select AHP methods to educational and vocational guidance, mobile health adoption and village government evaluation. Therefore, AHP methods can be applied in many fields, and it also very suitable for urban competitiveness evaluation, which has multiple layers and crosses and is difficult to describe quantitatively. The AHP-based evaluation of urban competitiveness in CR Express includes the following steps.

1) Build a hierarchical model. Constructing Competitiveness Evaluation Indexes of Node Cities Based on Characteristics of CR Express rope Classes.
2) Construct a decision matrix. The judgment matrix is formed by comparing the elements of each level. This is the key to determine the relative importance of the lower element to the upper element. The judgment matrix satisfies the following requirements. The evaluation index of competitiveness and the evaluation grade between node cities are determined by quantitative and expert survey methods.

$$
b_{ij} = \begin{bmatrix} b_{11} & b_{12} & \ldots & b_{1n} \\ b_{21} & b_{22} & \ldots & b_{2n} \\ \ldots & \ldots & \ldots & \ldots \\ b_{n1} & b_{n2} & \ldots & b_{nn} \end{bmatrix} \tag{1}
$$

3) Hierarchical single sorting and consistency check. The ranking of competitiveness among cities and individual cities is determined, and the inconsistency of results is ensured within the allowable range. The eigenvalues of the calculation matrix are the weights of lower-layer factors, and λ_{max} is calculated, as shown in formula 2.

$$CR = \frac{CI}{RI} \qquad (2)$$

Where, RI is a random consistency index, and $CI = \frac{\lambda_{max}-n}{n-1}$, when $CR < 0.1$, the consistency test is passed.

4) Hierarchical order and consistency tests. The competitiveness ranking of urban nodes in the CR Express is determined and the inconsistency of the results is ensured within the allowable range. The overall consistency is calculated by formula 3.

$$CR = \frac{\sum CI(j)a_j}{\sum RI(j)a_j} \qquad (3)$$

When $CR < 0.1$, index passed the consistency check.

2.2 Evaluation Indicators of City Competitiveness in CR Express

The evaluation index of urban node competitiveness involves many contents. Combined with the research theme of this paper and considering the objective conditions of data processing, the evaluation index system of urban logistics development level of CR Express class node is constructed, which includes 10 indexes of 3 categories.

2.2.1 Economic Development Index

Urban economic development is an important material basis depending on urban functions. The level of economic development is one of the important factors that influence the development of urban logistics, and the level of logistics development is closely related to the level of urban economic development [8]. The consumption level of urban residents with a good economic development level is bound to be high, so the high demand makes the development of logistics have a foundation. This paper selects GDP per capita, foreign trade volume, the added value of tertiary industry, and the total social consumer goods as indicators to evaluate the level of urban economic development. GDP per capita is the gross domestic product of a country or region divided by the number of resident population in a specific accounting period. The result can be used to determine the level of economic development of a region. Foreign trade volume refers to the sum of the value of a country's foreign trade in goods and the value of service trade, which reflects one of the important indicators of the foreign trade scale of a country or region. The added value of the tertiary industry is the value added by circulation and service cycle (generally annual) over the previous liquidation cycle. The total retail sales of social commodities are the total sales of consumer goods sold by various departments of national economy to urban and rural residents and social groups, agricultural production materials sold to farmers, and consumer goods sold directly to non-agricultural residents. This indicator reflects the demand scale of logistics services in each industry. The larger the total retail sales amount, the higher the demand for logistics services.

2.2.2 Logistics Service Capability Index

Logistics service capability is an important basis for logistics service providers to provide logistics service level and gain competitive advantage. The logistics service capability of a city refers to the ability of the production logistics enterprises and commercial logistics enterprises in a city to provide important value-added for the supply chain by saving their costs. This paper introduces the logistics industry added value, freight turnover, fixed assets investment, freight volume and railway carrying capacity to measure the logistics service capacity. Freight volume refers to the actual quantity of goods delivered by a transportation enterprise within a certain period, which basically reflects the scale and capacity of local logistics operations. Cargo turnover refers to the total transportation volume calculated by the actual completion of various transportation modes and transportation distance in a certain period of time. The turnover of goods is divided into the turnover of various types of goods completed by various transportation modes, such as railway, highway, water, air and pipeline, which can comprehensively reflect the production results of transportation. The heat of logistics development relates to the investment of logistics capital, which reflects the financial performance of supporting the logistics service capacity of cities, the government's support for the development of logistics industry, and is the core index of logistics industry development. The railway carrying capacity is the maximum freight transport volume that can be completed in a certain line section (or section) according to the locomotives, rolling stock and personnel provided in a certain period under the condition of certain technical equipment and running organization method. It reflects the adaptation of momentum and kinetic energy in this region.

2.2.3 Informatization Index

Logistics informatization refers to the management activity that enterprises can control the flow of goods by means of modern information technology, thus reducing the cost and improving the benefit. Informationization is the soul of modern logistics and the inevitable requirement and cornerstone of modern logistics development. The degree of informationization of a city can reflect the level of logistics development of a city. This paper uses the number of Internet users and the number of mobile phone users as the evaluation indicators to evaluate the city's informationization level. The number of Internet subscribers refers to the number of subscribers registered with the Chinese Internet through telecom enterprises in a region at the end of the reporting period, and the number of mobile phone subscribers refers to the total number of subscribers accessing the mobile phone websites in a region through mobile phone exchanges and occupying mobile phone numbers. These two indicators generally reflect the improvement of mobile communication in an area [9].

2.3 Hierarchy Mode

According to the index system, city node competitiveness hierarchy model is shown in Fig. 1.

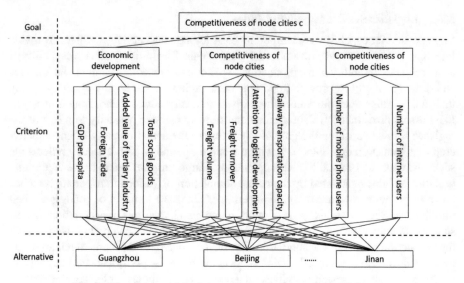

Fig. 1. Hierarchy model of city node competitiveness in CR Express

2.4 Competitiveness Indicator Judgment Matrix

Considering previous studies data, experts opinion, theoretic analysis and case study, the importance of competitiveness index is evaluated, and the competitiveness index judgment matrix is constructed according to the scale of 1–9.

2.5 Competitiveness Level Decision Matrix Between Cities

The indicators designed include per capita GDP, foreign trade volume, added value of tertiary industry, total consumer goods, freight traffic volume, freight turnover, railway carrying capacity, mobile phone users and internet users. The data comes from the statistical yearbook of each city and the statistical bulletin of national economic and social development published by the city information statistics website. The qualitative index (logistics development heat) is obtained by the expert score method. The index data of 33 hub nodes are collected respectively, and the data comparison between two cities is made by using Excel to construct the judgment matrix of city competitiveness level.

2.6 Hierarchical Sorting and Consistency Testing

According to the calculation results of the judgment matrix, the hierarchical ordering and consistency check are carried out. Consistency ratio is calculated by Formula 4.

$$CR = \frac{CI}{RI} \tag{4}$$

Where, RI is a random consistency index, and $CI = \frac{\lambda_{\max} - n}{n - 1}$, when $CR < 0.1$, the consistency test is passed. After calculation, all the judgment matrices of each layer of the index system constructed in this paper have passed the satisfactory consistency test. Hierarchical single-sorting is to obtain the relative importance weight of a single layer to the previous layer through normalization and weighting. The total ranking of levels is to calculate the weight of the relative importance of all factors of a level to the highest level.

3 Case Study of the Competitiveness of Cities in CR Express

According to the calculation steps of AHP, the collected urban economic data and expert scores are substituted for calculation, and the competitiveness rankings of 33 cities are obtained as shown in Table 1.

According to the evaluation results, the hub nodes are divided into three categories: Guangzhou, Chongqing, Beijing, Dalian, Shenzhen, Wuhan and Tianjin, Ningbo, Chengdu and Suzhou.

The first-class node city, Guangzhou, has an absolute advantage over other cities in terms of freight volume and freight turnover, so it has the highest score and the city's comprehensive competitiveness. The second type of hub cities are divided into two categories: one is the cities with higher scores on the index with larger weight, Dalian has a relative advantage in freight volume, and the other has its own competitive advantage, with higher scores on most indexes, good development in all aspects, and high comprehensive competitiveness of cities. Tianjin has an absolute advantage in railway transport capacity, logistics development heat is good, Chongqing has a better information development degree, Beijing and Shenzhen have an absolute advantage in economic development, and Wuhan has a relative advantage in logistics service capacity. The third type of hub nodes are generally developed in all aspects, not only in the weight of the index is low, but also in the weak position of other indexes, without its own competitive advantage, the city comprehensive competitive level is average.

According to the objective results of city competitiveness, the first and second types of hub nodes with strong competitiveness, namely Guangzhou, Chongqing, Beijing, Dalian, Shenzhen, Wuhan and Tianjin, can be considered as alternative hub nodes of the CR Express class train. The central node is used as the goods distribution center to drive the development of other cities. At the same time, Guangzhou, as the hub node with the maximum freight volume and the maximum freight turnover, has sufficient supply of goods, and can reasonably be distributed to the surrounding node cities, which can not only improve its own freight turnover efficiency, but also bring goods sources to the surrounding node cities, thus realizing mutual benefit and win-win. Wuhan, a hub with relative advantages in logistics service capability, can serve as a center for goods collection and transportation, to integrate resources, reduce repeat lines, and reduce logistics cost. For Chongqing, a hub node with high informatization, we should make full use of its advantages for innovative development. We can build a hub node logistics information sharing platform to facilitate real-time monitoring and tracking of goods status and improve customer satisfaction. We can also create a language conversion system to reduce communication barriers between different language countries in China and Europe. Improve transportation efficiency. As for hubs such as Beijing and Shenzhen, which have relatively good economic development, we should actively explore

Table 1. City competitiveness ranking

Ranking	City	Competitiveness	Ranking	City name	Competitiveness
1	Guangzhou	0.1119	18	Xiamen	0.0236
2	Chongqing	0.0624	19	Qingdao	0.0229
3	Beijing	0.0534	20	Alashan Pass	0.0224
4	Dalian	0.0480	21	Yiwu	0.0223
5	Shenzhen	0.0475	22	Jinan	0.0209
6	Wuhan	0.0459	23	Dongguan	0.0208
7	Tianjin	0.0393	24	Lianyungang	0.0196
8	Ningbo	0.0350	25	Shenyang	0.0182
9	Chengdu	0.0348	26	Urumqi	0.0175
10	Suzhou	0.0345	27	Harbin	0.0169
11	Nanjing	0.0316	28	Erenhot	0.0164
12	Hangzhou	0.0296	29	Lanzhou	0.0162
13	Zhengzhou	0.0296	30	Manzhouli	0.0146
14	Xi'an	0.0285	31	Qinzhou	0.0145
15	Changsha	0.0258	32	Yingkou	0.0144
16	Hefei	0.0239	33	Ulanqab	0.0136
17	Holgos	0.0238			

the source of goods, which is the driving force for the sustained development of China and Europe, encourage enterprises to explore overseas markets, establish cooperation with logistics enterprises in European countries, and attract the source of goods suitable for railway transportation with quality and accurate services. For the third type of hub nodes with poor development in each aspect, we should strengthen the construction of logistics infrastructure, make use of their respective location advantages, and form strategic alliance with the hub nodes with strong competitiveness, so as to promote their own development.

4   Conclusion

This paper constructs a competitiveness evaluation system of CR Express hub nodes, and selects 10 indicators of GDP per capita, foreign trade volume and the added value of the tertiary industry from three aspects: economic development level, logistics service capability and information technology. By using the analytic hierarchy process (AHP) qualitative and quantitative analysis, the competitiveness ranking of a hub node is obtained and classified, and different suggestions are given for different hub nodes.

The ranking of the city competitiveness of the CR Express will help to manage the CR Express more reasonably. In the follow-up research, the network planning of the CR

Express can be optimized and laid out according to the competitiveness of the city, so as to promote the CR Express reasonable development.

Acknowledgment. This project is supported by Teaching Research Projects of Wuhan Municipal Education Bureau (2021091).

References

1. Book Binder, J.H., Tam, C.S.: Comparison of Asian and European logistics systems. Int. J. Phys. Distrib.-Logist. Manage. **33**(1), 35–58 (2003)
2. Ekici, O., Kabak, O., Ulengin, F.: Linking to compete: logistics and global competitiveness interaction. Transp. Policy **48**(22), 117–128 (2016)
3. Ying, D., Qin, L., Han, S.: Research on the optimal layout of CR Express rope Banlie domestic goods distribution centers under the Belt and Road initiative[J]. J. Chongqing Univ. Technol. (Soc. Sci.) **32**(10), 66–75 (2018). (in Chinese)
4. Luo, W.: On the school-enterprise cooperation development plan of logistics management in private universities. PR Mag. (07), 223–227 (2019). (in Chinese)
5. Essaid, E.H., Abdellah, A., Mohamed, E.H.: Using FAHP in the educational and vocational guidance. Int. J. Mod. Educ. Comput. Sci. **10**(12), 36–43 (2018)
6. Farhad, L., Kimia, F., Nasrin, B.: An analysis of key factors to mobile health adoption using Fuzzy AHP. Int. J. Inf. Technol. Comput. Sci. **12**(2), 1–17 (2020)
7. Vickky, L., Ema, U.: Decision support system performance-based evaluation of village government using AHP and TOPSIS methods: Secang sub-district of Magelang regency as a case study. Int. J. Intell. Syst. Appl. **10**(4), 18–28 (2018)
8. Wei, S.: Research on the structure and correlation of urban logistics network in CR Express. Chang'an University, Xi'an (2018)
9. Sun, Y.: Research on urban logistics competitiveness of central plains urban agglomeration based on entropy weight TOPSIS model. Zhejiang University of Technology, Hangzhou (2019)
10. Cai, R.: Research on the model of central and central-europe class and the optimization of central hub layout. Wuhan University of Technology, Wuhan (2018)

Design and Implementation of Cold Chain Logistics Temperature Measurement and Control System Based on LabVIEW

Zhang GengE[1,2(✉)], Huang Fang[1], and Shen Jialun[1]

[1] Nanning University, Nanning 530200, Guangxi, China
78704108@qq.com
[2] University of Sabah, Sabah, Malaysia

Abstract. With the progress of science and technology and the development of refrigeration technology, the global agricultural cold chain logistics industry has made great progress. The average annual growth rate of China's cold storage capacity and the number of refrigerated vehicles is increasingly raised. Closely monitoring and strictly controlling the temperature of refrigerated trucks plays a vital role in reducing the rate of cargo loss. Based on the analysis of the automatic virtual instrument function of LabVIEW software, this paper puts forward the system design scheme of virtual instrument, completes the hardware design of switching power supply and circuit board, front and rear panel design and software design, and verifies the effectiveness of the system through experimental tests. The system can realize the temperature measurement and control of cold chain logistics, and provides research value for improving the operation efficiency of cold chain logistics.

Keywords: Virtual instrument · Refrigerated truck · Temperature measurement and control system

1 Introduction

Cold chain logistics is an vital foundation to support the development of large-scale industrialization of agriculture, promote agricultural transformation and increase farmers' income, and help rural revitalization [1]; It is an important guarantee to improve the quality and safety system of fresh agricultural products "from farmland to dining table and from branches to tip of tongue", and improve the quality control ability of vaccines and other pharmaceutical products in the whole process. It is also an significant means to reduce postpartum loss of agricultural products and waste of food circulation [2]. With the progress of science and technology, the development of refrigeration technology and the rapid development of China's agricultural cold chain logistics industry, during the 13th Five Year Plan period, the average annual growth rate of China's cold storage capacity and refrigerated vehicle ownership exceeded 10% and 20% respectively. By 2020, the scale of cold chain logistics market has exceeded 380 billion, and the number of refrigerated vehicles has exceeded 280000, which is about 2.4 times and 2.6 times

Z. Hu et al. (Eds.): ICCSEEA 2022, LNDECT 134, pp. 194–205, 2022.
https://doi.org/10.1007/978-3-031-04812-8_17

respectively at the end of the 12th Five Year Plan [3]. The statistics of refrigerated vehicle market in recent years are shown in Table 1.

Table 1. Refrigerated vehicle market ownership

	2015	2016	2017	2018	2019	2020
Number of refrigerated vehicles (10,000)	9.34	11.50	14.00	18.00	21.47	28.7
Year on year growth%	23.1	23.13	21.7	28.6	19.3	33.5

Table 1 shows that the average growth rate of the market ownership of refrigerated vehicles in China in recent years is more than 20%, with a growth rate of more than 30% (33.5%) in 2020, a new high in recent years. It shows that the development momentum of China's refrigerated vehicle market is stronger and has great development potential. Therefore, more and more attention has been paid to the quality safety and risk assessment of cold chain [4]. New methods and technologies of safety management have also been introduced [5–7]. Real time monitoring and strict control of the temperature of refrigerated vehicles and ensuring the quality of goods play a vital role in reducing the loss rate of cold chain goods [8].

With the development of science and technology, various advanced detection methods and data processing methods have been widely used in different fields [9–11]. Many advanced technologies and means are also used in the management and control of cold chain and logistics system [12]. Simulation technology, embedded operating system and virtual instrument have shown outstanding advantages in the field of logistics and cold chain [13, 14]. Under this background, this paper discusses the important problem of cold chain logistics temperature control based on LabVIEW.

2 LabVIEW Function Analysis

LabVIEW software can be regarded as an automatic virtual instrument, a graphical based program editor, which runs in the form of a borderless block to automatically generate the whole program [15]. At present, most of them have been widely used in various types of embedded application monitoring systems such as engineering simulation, data acquisition, instrument quality control, computer measurement and analysis and engineering data video display [16, 17]. It directly uses the control software function of a virtual display on the computer mobile display to automatically simulate various traditional virtual instrument signals, and then accurately uses a variety of operation methods and signal forms to express the technical results of analog detection of various signal data output by the instrument: a powerful equipment software control function that directly uses the computer, We can directly realize the direct operation, analysis and data processing of various signals: a device software directly using the interface of single machine I/O system. It can directly complete the operation, acquisition, analysis and conditioning of various signals, so that we can directly complete the process automation test of various signal detection and control functions [18, 19]. LabVIEW development environment integrates all tools needed by engineers and scientists to quickly build various applications, which can help engineers and scientists solve problems, improve productivity and innovate [20, 21]. The virtual instrument solution is shown in Fig. 1.

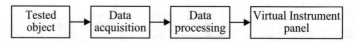

Fig. 1. Application scheme of virtual instrument

2.1 Temperature Sensor

Inside the temperature sensor there is a semiconductor thermistor. The lower its temperature, the greater the resistance will be; the higher the temperature, the smaller the resistance. The temperature sensor is installed in the refrigerator car and is in direct contact with the air in the refrigerator car. Thus, the air temperature in the refrigerator car can be accurately measured.

2.2 Data Acquisition

Data acquisition card is a very key component in the measurement system. Its main functions include: analog input, analog output, digital I/O, counter/computer, etc. [22]. In this paper, the USB-6216 data acquisition card of Ni company is selected, which is NI-USB multi-functional data acquisition. The shape design is shown in Fig. 2.

Fig. 2. Data acquisition card

The main performance indexes of the selected data acquisition card are introduced as follows [23].

(1) Analog input resolution: 16 bits (65536);
(2) Number of analog input channels: 8 differential/16 single ended;

(3) Analog output: 2 channels (update rate 833 ks/s);
(4) Analog output resolution: 16 bits (65536);
(5) Counter/timer: 2 channels (32 bits);
(6) Maximum sampling rate: 400 ks/S;
(7) Digital I/O port: 24 (digital trigger).

3 Design of Water Temperature Measurement and Control System Based on LabVIEW

3.1 Hardware Design

3.1.1 Enclosure Type Switching Power Supply

The main purpose of switching power supply is a small and portable switching power supply rectifier conversion circuit equipment, which generally mainly includes switching shell power capacitor, switching power supply capacitor, power supply rectifier transformer and power supply rectifier conversion circuit. Generally, it can be roughly divided into two types: control AC voltage output type and DC voltage input type. Switching power supply control refers to the automatic control of relevant opening or closing through the full application of various modern electronic signal processing sensors, or advanced electronic signal processing sensor technologies such as personal computers, A switching power supply control element which can make the switching current output normally and stably [24]. One of the two central components of switching power supply is switching pulse power supply automatic width modulation (PWM) and microcontroller PIC and PVMOSFET.

Social development is changing with each passing day. With the continuous progress of modern electronic technology, the technology of switching power supply is constantly updated and developed, and the technology is constantly improved. Switching power supply embodies high efficiency, energy saving, convenience and low cost, which is what we human beings have been pursuing. Therefore, switching power supply is widely used in various electronic products and equipment. Obviously, switching power supply has become an indispensable part of electronic devices. The enclosure switching power supply used in the experimental design is shown in Fig. 3. The parameters are as follows:

(1) DC output range: 5 V, 0.5–8 A/12 V, 0.2–3.5a/12 v, 0.0–1.0 A (rated output: 5v5a/12v2.8a/- 12v0.5a);
(2) Output voltage accuracy: +2%/+6%/+5%;
(3) Efficiency: 77%;
(4) Input voltage range: 88–264 Vac/120–373 vdc (withstand 300 VAC high voltage input for 5 s without damage);
(5) Voltage adjustment range: 5 V: 4.75–5.5 V;
(6) Operating temperature: −25–70 °C.

Fig. 3. Enclosure type switching power supply

3.1.2 Circuit Board Design

The resistance signal of the temperature sensor is converted into a voltage signal through a self-made circuit. As shown in Fig. 4, the power supply is connected at mark G, the voltage chip Ref02 stably outputs 5 V voltage, the engine water temperature sensor is connected at mark Rx, and the real-time resistance value of the sensor is obtained, indicating that LabVIEW is connected at U_1 and U_2. At this time, U_1 can be calculated from the voltage formula (1), Then U_2 is calculated from the voltage formula (2), and finally the potential difference is calculated from $U_1 - U_2$. After simplification, formula (3) is obtained (Fig. 5).

$$U_1 = 5 * (Rx/Rx + 2K) \tag{1}$$

$$U_2 = 5 * (150/150 + 2K) \tag{2}$$

$$Rx = U_1 - U_2 = ((86000 * u + 30000)/(200 - 43 * u))/1000 \tag{3}$$

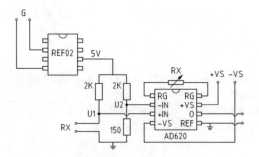

Fig. 4. Circuit diagram

Fig. 5. Circuit board

3.2   Software Design

3.2.1   Front Panel Design

In the software start page and startup window, select "New VI" to enter a software design working environment based on virtual instrument application. The overall design appearance and daily operation of the software of virtual instrument and its application program completely imitate the real real instruments, such as oscilloscope and multimeter. Input function information through input function to obtain information from a user network interface or other technical means, and then transmit these input information through display in an e-mail or some form or directly to other electronic files or computers.

LabVIEW application consists of three main parts: foreground board, program block diagram, "track" program icon/software patch panel. The front panel refers to an operation control interface based on user programming based on VI control program. It refers to an interactive output control port based on input and output sensors based on VI control program. Generally, we can directly create the program by using a display and control interface based on input port controls and output sensors. It is necessary to obtain or automatically convert the generated display information to control the structural block diagram of the display instrument program in real time.

The front panel of the experimental system consists of physical channel configuration selection, real-time calculation result display, table lookup method value, data storage path, waveform diagram showing resistance and temperature and Boolean button. It enables people to operate conveniently and presents the change of temperature value in a more intuitive way. In the front panel, the replacement channel can be configured in the physical channel, and the sampling mode, number of points and rate can be set. The real-time calculation results are displayed and adjusted to adapt to different temperature sensors. The existing temperature values are digitally displayed, and the temperature change curve is displayed in the form of waveform diagram. The front panel is shown in Fig. 6.

3.2.2   Design of Rear Panel of Temperature Measurement and Control System

The code design in the program block diagram adopts a graphical programming code design method, which can be used to design, implement, manage and control each object of the whole program block diagram.

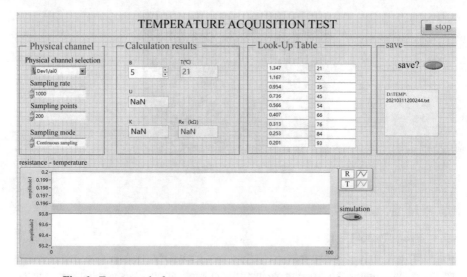

Fig. 6. Front panel of temperature sensor measurement and control system

The object of the program block diagram mainly includes a terminal of a connector and two nodes constituting the program block diagram. The wiring interface terminal is a wiring object on the current control panel. The main functions of the nodes in the program block diagram include a function, VI, express VI and its structure. Figure 7 shows the program block diagram of the rear panel of this design. Its main purpose is to enter the voltage collected in the experiment into a cycle, calculate the real-time temperature signal through the formula, and finally save the collected data information in the form of text document to the computer folder.

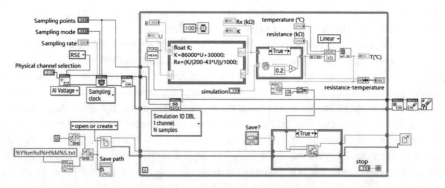

Fig. 7. Program block diagram of temperature measurement and control system

(1) Temperature data acquisition module

Temperature data acquisition is the first step in the detection of the whole system. It is to collect the resistance signal sent by the temperature sensor and the

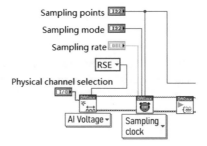

Fig. 8. Data acquisition module

voltage signal sent after circuit conversion, fixedly select its physical channel, set the sampling rate, sampling points and sampling mode, and start the voltage signal acquisition, as shown in Fig. 8.

(2) Temperature data acquisition calculation and waveform display module

Take an average value of a large number of voltage signals obtained through conversion in the previously designed circuit, bring the value into formula (4), calculate and obtain the K value, and then substitute the obtained K value into formula (5). After calculation, finally obtain the resistance value of the water temperature sensor of the experimental vehicle under the current condition. After obtaining the resistance value, carry it in with the obtained resistance value, In the linear equation between the resistance value of the temperature sensor and the corresponding temperature, as shown in Fig. 9, the corresponding temperature value is finally obtained. As shown in Fig. 9.

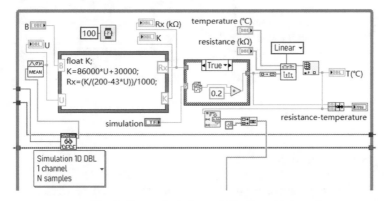

Fig. 9. Data calculation and display module

$$K = 86000 * u + 30000 \tag{4}$$

$$Rx = (K/(200 - 43 * u))/1000 \tag{5}$$

It can be seen from Table 2 that a linear relationship between resistance value and temperature of the sensor.

(3) Data storage

The module saves the obtained experimental data in a file. A new document will be created to store the data every time the experiment is opened, which improves the accuracy and accuracy of the data and enables people to more intuitively observe the change of temperature value, as shown in Fig. 10.

Table 2. Linear relationship between resistance value and temperature of the sensor

R (KΩ)	T (°C)
1.347	21
1.167	27
0.954	35
0.736	45
0.566	54
0.407	66
0.313	76
0.253	84
0.201	93

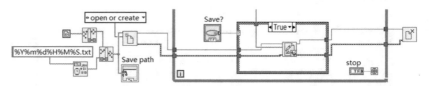

Fig. 10. Data storage module

4   Experimental Test

Before the whole experiment, we need to study the relationship between the resistance value of the temperature sensor and the temperature value.

At the beginning of the experiment, the switching power supply is connected to the data acquisition card, which is connected to the computer LabVIEW program for data acquisition.

The front panel of LabVIEW will display the real-time temperature change value and the waveform diagram of resistance and temperature change. As shown in Fig. 11. The computer desktop will also generate a document recording the experimental data, recording all the data of this experiment.

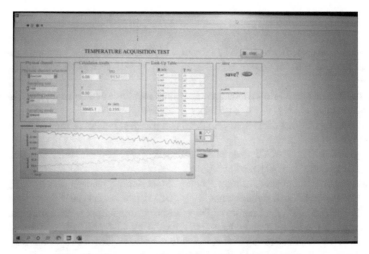

Fig. 11. System front panel experimental data display

Sort out and analyze the data saved in the whole experiment, and change the system to more intuitively show the corresponding relationship between the output change of the resistance signal of the temperature sensor and the temperature signal. When the temperature is lower or higher than the specific temperature, it can give an alarm in time.

5 Conclusion

Facing the urgent demand of cold chain logistics temperature control, this paper analyzes the storage capacity of logistics refrigerated truck, and puts forward that strictly controlling the temperature of refrigerated truck plays a vital role in reducing the rate of cargo loss. Through the research on the function of virtual instrument, a measurement and control system for real-time monitoring the temperature of refrigerator truck and ensuring the quality of goods is designed by using LabVIEW software. The designed measurement and control system can monitor the temperature in real time and provide guarantee for cold chain logistics.

In this paper, the temperature measurement and control system is designed from both qualitative and quantitative aspects. The measurement and control system designed by virtual instrument plays a time-saving and efficient role, and provides a practical and effective scheme to solve the problem of cargo damage in cold chain logistics during refrigerated truck transportation.

Acknowledgment. The project is financially supported by: (1) the special research project on private higher education of Guangxi Education Science Planning in 2021 (2021zjy667), (2) China ASEAN International Joint Key Laboratory of comprehensive transportation, the district level undergraduate teaching reform project in 2021 (2021jgb428) and (3) the second batch of teaching reform project of specialized creative integration program of Nanning University (2020xjzc02).

References

1. Bottani, E., Casella, G., Nobili, M., Tebaldi, L.: Assessment of the economic and environmental sustainability of a food cold supply chain. IFAC PapersOnLine **52**(13), 367–372 (2019)
2. Xie, R.: Current situation and development countermeasures of cold chain logistics in China. Logist. Technol. **33**(21), 1–3 + 7 (2014). (in Chinese)
3. Cui, Z.: Cold chain logistics: review in 2019 and prospect in 2020. China Logist. Procure. **1**, 23–24 (2020). (in Chinese)
4. Zhang, Q., Chen, Z.: HACCP and the risk assessment of cold-chain. Int. J. Inf. Eng. Electron. Bus. (IJIEEB) **2**(15), 67–71 (2011)
5. Saqaeeyan, S., Rismantab, A.: A novel method in food safety management by using case base reasoning method. Int. J. Educ. Manag. Eng. (IJEME) **10**, 48–54 (2015)
6. Kakelli, A.K., Aju, D.: An Internet of Thing based agribot for precision agriculture and farm monitoring. Int. J. Educ. Manag. Eng. (IJEME) **08**(08), 33–39 (2020)
7. Ding, J., Wang, X.: Food safety testing technology based on the spectrophotometer and ARM. Int. J. Intell. Syst. Appl. (IJISA) **10**, 48–54 (2011)
8. Pi, S.: Research on temperature control system of cold chain logistics based on RFID technology. Chang'an University, Xi'an (2013). (in Chinese)
9. Sinitsyn, R.B., Yanovsky, F.J.: Copula ambiguity function for wideband random radar signals. In: 2011 IEEE International Conference on Microwaves, Communications, Antennas and Electronic Systems, COMCAS 2011, Tel-Aviv, Israel, vol. 11, no. 5, pp. 7–9 (2011)
10. Bokal, Z.M., Sinitsyn, R.B., Yanovsky, F.J.: Generalized copula ambiguity function application for radar signal processing. In: Microwaves, Radar and Remote Sensing Symposium, MRRS-2011 – Proceedings, pp. 313–316 (2011)
11. Sinitsyn, R.B., Yanovsky, F.J.: Acoustic noise atmospheric radar with nonparametric copula based signal processing. Telecommun. Radio Eng. **71**(4), 327–335 (2012)
12. Xia, Z., Han, Y.: Application of computer logistics simulation technology in enterprise decision-making. China Water Transp. (Acad. Edn.) **06**, 155–156 (2006). (in Chinese)
13. Sun, H., Wu, X., Chen, Q.: Application and research of embedded operating system in remote laboratory construction. Aviat. Manuf. Technol. **10**, 91–94 (2005). (in Chinese)
14. Qiao, W.: Virtual instrument technology and its application and prospect. Electr. Drive Autom. **04**, 6–9+13 (2007). (in Chinese)
15. Qi, X., Zhou, J., Jiao, J.: Introduction and typical examples of LabVIEW 8.2 Chinese version, vol. 07. People's Posts and Telecommunications Press, Beijing (2008). (in Chinese)
16. Li, Z., Qi, Y.: Design of temperature acquisition system based on LabVIEW. Mach. Manuf. **55**(11), 86–87 + 98 (2017). (in Chinese)
17. Zuo, Z., Zheng, B., Wu, J., Chen, G.: Design of a water temperature monitoring system based on LabVIEW. Comput. Knowl. Technol. **6**(34), 9922–9923 (2010). (in Chinese)
18. Han, J.: Design of water temperature and water level monitoring system based on LabVIEW. China Sci. Technol. Inf. **04**, 74 (2012). (in Chinese)
19. Patel, D.M., Shah, A.K.: LabView based control system design for water tank heater system. Trends Electr. Eng. **7**(3), 31–40 (2017)
20. He, X., Zhu, S., Qin, B., et al.: Design of remote temperature acquisition system based on LabVIEW. J. Hunan Univ. Technol. **27**(06), 89–93 (2013). (in Chinese)
21. Mao, H., Chen, Z., Fang, J., Liu, S.: Photoacoustic spectrum detection system based on LabVIEW. J. Zhejiang Norm. Univ. (Nat. Sci. Edn.) **34**(03), 277–280 (2011). (in Chinese)
22. Zeng, Q., Jiang, W., Liu, H.: Design of real-time temperature acquisition system based on LabVIEW and DS18B20. Autom. Appl. **07**, 92–94 (2015). (in Chinese)

23. Yang, C.: Design of measurement and control system for sequential turbocharged diesel engine test bench based on virtual instrument. Harbin Engineering University, Harbin (2007). (in Chinese)
24. Wang, S.: Design of high efficiency switching power supply for multimedia equipment. Heilongjiang Sci. **5**(09), 295 (2014). (in Chinese)

Perfection of Computer Algorithms and Methods

Recognition of Potholes with Neural Network Using Unmanned Ground Vehicles

Maksym Alpert and Viktoriia Onyshchenko$^{(\boxtimes)}$

National Technical University of Ukraine "Igor Sikorsky Kyiv Polytechnic Institute", 37 Peremohy Avenue, Kyiv 03056, Ukraine
v.onyshchenko@kpi.ua

Abstract. Nowadays, with the rapid development of science and technology, many methods of object recognition have appeared. The most well-known of them are used to obtain more accurate information about the recognized object. Drones are usually used to obtain information in the locality. Modern drones can be used to collect information about the objects that surround us. Neural networks can process this information much more effectively than humans can. In this article is discussed a method to recognize of potholes with neural network using unmanned ground vehicles. This method will help to avoid traffic emergencies. Unmanned ground vehicles allow to conduct research in unfavorable conditions and hard-to-reach places, and a neural network with object recognition will allow to quickly indicate the safest path. Different known libraries of objects recognition such as Keras API and Tensorflow are compared in this article. Keras is a neural network library while TensorFlow is the open-source library for a number of various tasks in machine learning. TensorFlow provides both high-level and low-level APIs while Keras provides only high-level APIs. In addition, a single deep neural network called SSD (Single Shot Multibox Detector) is chosen to use with Tensorflow. This library is integrated into a website for controlling self-made unmanned ground vehicle and using object recognition of potholes. All these object recognition methods are described and analyzed in this paper. Main advantages and disadvantages of these methods are considered too. New approach of applying neural network to recognize the potholes is proposed in this work. An experiment has been conducted that shows the accuracy of potholes recognition. The results of testing object recognition are shown in data tables. Appropriate graphs demonstrate that the recognition accuracy is dependent to the number of potholes in the image.

Keywords: Machine learning · Neural network · UGV · Tensorflow

1 Introduction

Many different companies developed automated drones for delivering products or exploring the area. Such ground drones like Amazon Scout [1], REV-1 [2] are good for delivering products. These ground vehicles also have neural network integrated to their firmware. However, these drones have a number of important disadvantages:

Z. Hu et al. (Eds.): ICCSEEA 2022, LNDECT 134, pp. 209–220, 2022.
https://doi.org/10.1007/978-3-031-04812-8_18

Case 1: Overweight with different sensors.
Case 2: Well-known ground-based mobile platforms have a software part, which is a separate solution in most cases, and it is hard to integrate some new features like recognition of specific object.
Case 3: Drones from the example cannot recognize potholes.

To solve these problems, it was decided to develop our own unmanned ground vehicle, which is based on standard frameworks and can be developed quickly and conveniently maintained in the future. The control system will be created in the form of a web application that supports neural network. In addition, it allows using any microprocessor system that supports browsers as a remote control (Fig. 1).

Unmanned ground vehicles (UGVs) are robotic systems that operate on land under the control of a human operator or autonomously. They are used for a wide range of civilian tasks, especially in dangerous or unfavorable conditions for humans, as well as for tasks that are complex or have a high risk to life.

UGV can be controlled remotely using a portable or stationary control station, or operate independently. Autonomous UGVs can move between pre-defined points or in space to perform their task. UGVs operating can collect information in non-mapped changing environments and build a map of their environment using methods such as simultaneous localization and mapping. Artificial intelligence and machine learning can also help them adapt to their environment.

UGVs allow conducting research in unfavorable conditions and hard-to-reach places, and a neural network with object recognition will allow to quickly indicating the safest path. Own UGV was developed based on Raspberry Pi for solution to these problems (Fig. 1).

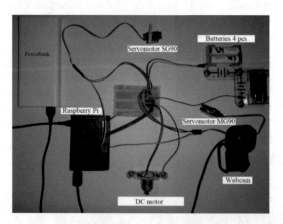

Fig. 1. Model of an UGV

Different known methods of objects recognition will be described and analyzed in this paper. Main advantages and disadvantages of these methods will be considered too. New approach of applying neural network to recognize the potholes is proposed in this work.

2   Choosing Method of Recognition

Let's conduct analysis and comparison with others well-known recognition methods that use a neural network.

In the article [3], a modified component diagram was developed that contains infrastructure elements to ensure the operation of a neural network. The Keras API was selected for implementation. Keras is a leading open-source library designed for creating neural networks and machine learning projects.

The approach of working and interacting with a neural network, which is described in the article [3], is not very successful. Interaction with the camera takes place through a script in the file server.py using the OpenCV library, which is executed via the SSH (Secure Shell) protocol. This approach is not secure, and the performance of such a solution is quite low. So, object recognition with Raspberry Pi takes a lot of resources and as a result, there is no more than 1–2 frames/second at the output.

The article [4] considers the construction of a software system based on a low-power microcontroller, but this solution can be used as an auxiliary solution for connecting an actuator and a basic controller containing a neural network system.

In the article [5] the authors propose to use Raspberry Pi and MPU-9250 sensor module to detect anomalies such as potholes, bumps etc. A Participatory Sensing system based on raspberry pi is designed and developed to detect and record road surface anomalies that are measured by the inertial measurement unit (IMU) and GPS sensor. The also developed an android application that is designed to alert the user from an upcoming high intensity rough area on the road.

In the paper [6] authors use transfer learning method that can use the pre-training model to solve the problem of insufficient target task data.

In the paper [7] authors focus on the lane navigation which has an important part of the AV movement on the road. Here lane decision making is optimized by using deep learning techniques in creating a Neural Network model that focuses on generating steering commands by taking an image the road mapped out with lane markings.

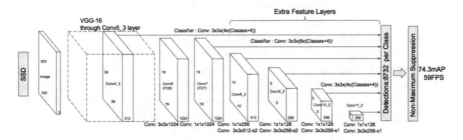

Fig. 2. Architecture of SSD [8]

In the article [8], Wei Liu, Dragomir Anguelov and others proposed a new method for detecting objects in images using a single deep neural network called SSD (Single Shot Multibox Detector). This method allows detecting objects in images using a single deep neural network. SSD discretizes the output space of bounding blocks into a set of blocks for different image formats and scales. During forecasting, the network generates estimates of the presence of each category of objects in each block by default and makes adjustments for the block to better match the shape of the object. In addition, the network combines forecasts from several feature maps with different resolutions for natural processing of objects of different sizes. Object detection, localization, and classification tasks are performed in one direct network pass. The SSD architecture is shown in Fig. 2.

3 Technology of Training and Adaptation of a Neural Network for UGV

The UGV software system includes a neural network-based object recognition module to assist the operator in the process of automated UGV control. The general structure of an experimental UGV with a neural network is shown in Fig. 3. Let us describe the purpose of each element of the UGV circuit:

Case 1: The UGV remote controller is used to transmit information, Audio, and video to a human operator.
Case 2: The UGV microcomputer is used to process information received from sensors, cameras, and sensors.
Case 3: The UGV communication devices are used for communicating with actuators.
Case 4: UGV actuators are used to change the position of the UGV in space.
Case 5: The system module of image capture is used to analyze the image from the video camera.
Case 6: The neural network is used to recognize objects from the image capture system module.
Case 7: Software elements for processing operator signals and information are used to obtain recognition results for the human operator.

Consider a practical implementation of a neural network for control of UGV. As an example, let us take potholes that occur on the UGV path.

To solve this problem, about 200 images of potholes were collected from the Internet [9]. About 50 images were used for testing, and the rest were used for part of the training. All the images for testing can be found in "test" folder. All images were taken with daylight and about 1 m above the ground. The next step is to annotate the resulting images. To do this, let us use LabelImg program. LabelImg is a tool for annotating graphic images. The resulting annotations are saved as XML files in PASCAL VOC format. The following are samples of potholes (Fig. 4). Figure 5 shows an annotated image using LabelImg.

When working with large data sets, using a binary file format for storing data can have a significant impact on the performance of the import pipeline (images) and, as a result, on the training time of the model. Binary data takes up less disk space, copies less time, and can be read from disk much more efficiently. TFRecord is suitable for these tasks. However, pure performance is not the only advantage of this file format. TFRecord is optimized for use with TensorFlow in several variants. The TFRecord format makes

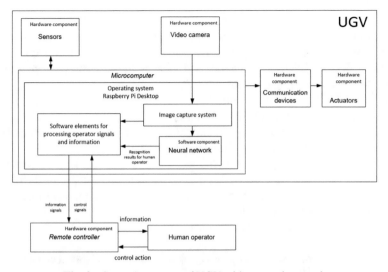

Fig. 3. General structure of UGV with a neural network

Fig. 4. Photos of potholes before processing

Fig. 5. Photos of potholes after processing

it easy to combine various sets of data and easily integrates with the data import and preprocessing functions provided by the library. In particular, for sets of data that are too big can be fully stored in memory, this is an advantage. Only the data that is currently needed is loaded from disk and then processed. Convert the XML files that were received after the stroke in LabelImg to TFRecord format.

To convert XML files to TFRecord, first convert them to CSV using a Python script xml_to_csv.py, which can be found at the link [10].

XML files in the "images/training" and "images/test" folders are converted to two CSV files, one for training and one for testing the model.

From the repository [10], let us use the script generate_tfrecord.py. One can use this script to generate TFRecord files. To do this, one needs to make adjustments to it. In the class_text_to_int(row_label) method, the name of the object being investigated and id (if row_label = = 'potholes': return 1) must be added.

Run the following scripts in the console:

Case 1: python generate_tfrecord.py --csv_input = data/train_labels.csv --output_path = data/train.record [10].
Case 2: python generate_tfrecord.py --csv_input = data/test_labels.csv --output_path = data/test.record [10].

To get a potholes detector, one can use a pre-trained model and then use transfer learning to study a new object, or study new objects completely from scratch. The advantage of transfer learning is that learning can be much faster, and the data one may need is much less. For this reason, it was chosen to conduct transfer learning. TensorFlow has quite a few pre-trained models with available checkpoint files, as well as configuration files.

The ssd_mobilenet_v1_pets configuration file was used for this task.config [11], which will be adjusted according to the model. In the configuration file, one needs to set the number of objects under study (num_classes). in fine_tune_checkpoint, specify the path to the folder where the previously trained model is located. A placemark map also has to be created, which is basically a dictionary containing the ID and name of the classes one wants to detect. Such a file will have the pbtxt extension.

Let us go directly to training the model. First, we need to download the repository [11] and copy all the files from the legacy folder to the object_detection folder and go to this folder. The following script will start training the model:

python   train.py   --train_dir   =   training/   --pipeline_ config_path = training/ssd_mobilenet_v1_pets.config --logtostderr

Let us check how well the model works. To do this, export the output graph. In "models" object_detection", there is a script "export_inference_graph.py".

python   export_inference_graph.py   --input_type   image_tensor   --pipeline_ config_path   training/ssd_mobilenet_v1_pets.config   --trained_checkpoint_ prefix training/model.ckpt-**** --output_directory new_graph, where **** checkpoint number.

A custom model has been created that can detect the required object. This model has the output format*.pb.

In order to compare the process of training a neural network, it was decided to use the Google Colab service. This service allows writing and executing Python code in the browser. Also in Google Colab, you can easily change the runtime environment and perform training on a remote GPU.

For local training of its own model, the proprietary Intel Core i7 4770 CPU was used.

The training process was performed on a remote GPU and a local CPU. The following is a comparative description of the neural network learning process.

4   Comparison of the Neural Network Learning Process

TensorBoard provides the visualization and tools you need to experiment with machine learning:

Case 1: Track and visualize metrics such as loss and accuracy.
Case 2: Visualization of the model graph (operations and layers).
Case 3: View histograms of weights, offsets, or other tensors as they change over time.
Case 4: Designing attachments in a smaller space.
Case 5: Display images, text, and audio data.
Case 6: Profiling TensorFlow applications.

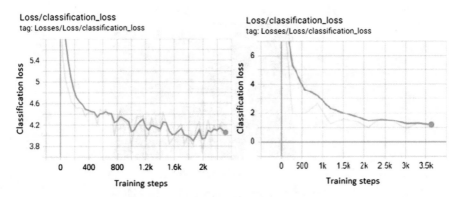

Fig. 6. Loss of object classification

Classification loss describes how much the result for object classification obtained by the neural network differs from the expected result, i.e. it indicates the amount of error that the model made during the forecast. When analyzing an image, the loss of classification indicates whether the bounding box class matches the predicted class. The loss of classification is shown in Fig. 6.

Figure 6 shows that classification loss of potholes decreases due to the quantity of training steps while training neural network.

5   Test Results of Object Recognition

Let's prepare for the experiment. We will need a web-camera and a tablet. The web-camera will be broadcasting the image with potholes from the tablet to the browser and our pre-trained neural network will detect potholes with a certain percentage of recognition (Fig. 7).

Fig. 7. Web-camera and tablet

Fig. 8. Web-interface of potholes recognition app with example

Let's take 10 random photos of potholes on the roads. The UGV camera has a resolution of 640 * 480. Photos are colored and the number of potholes in the photos is from 1 to 13 (Fig. 8). The website is located on localhost: 4200 address. Web-interface consists of camera window and 3 buttons. "Toggle Webcam" button turns on webcam in the browser, "Detect pit" turns on potholes recognition that is based on Sect. 3 in this article.

The next step is to activate the potholes recognition camera. After the camera recognizes the image, let's record the results in Table 1.

For example, there are 13 potholes on the image_10.jpg and trained neural network recognizes only 7 potholes (Fig. 8).

The number of experiments depends on the difficulty of recognizing potholes. If there are more than 2 potholes in the image, the experiment for this image will be repeated (Table 2).

Table 1. Neural network recognition results

Experiment number	Photo number, short description of the photo	Recognition results, characteristics
1	image_1.jpg, availability 1 potholes	Recognized 1 potholes
2	image_2.jpg, availability 2 potholes	Recognized 2 potholes
3	image_3.jpg, availability 1 potholes	Recognized 1 potholes
4	image_4.jpg, availability 6 potholes	Recognized 4 potholes
5	image_5.jpg, availability 5 potholes	Recognized 4 potholes
6	image_6.jpg, availability 1 potholes	Recognized 1 potholes
7	image_7.jpg, availability 2 potholes	Recognized 2 potholes
8	image_8.jpg, availability 1 potholes	Recognized 1 potholes
9	image_9.jpg, availability 2 potholes	Recognized 2 potholes
10	image_10.jpg, availability <13 potholes	Recognized 7 potholes

The results for Table 2 are based on the quantity of experiments for different images with different quantity of potholes.

Now let's count the averaged recognition characteristics. For example in experiment 4 from Table 1 we detected only 4 potholes of 6. It means that the percentage of undetected potholes is 0%.

Table 2. Neural network recognition results depended on number of experiments

Experiment number	Number of experiments	Recognition percentage	Averaged recognition characteristics
1	1	image_1.jpg, 88–92%	90%
2	2	image_2.jpg, 80–90%	85%
3	3	image_3.jpg, 97–99%	98%
4	3	image_4.jpg, 0–90%	60%
5	4	image_5.jpg, 0–94%	75,2%
6	1	image_6.jpg, 93–99%	96%
7	2	image_7.jpg, 83–91%	87%
8	1	image_8.jpg, 97–99%	98%
9	2	image_9.jpg, 84–88%	86%
10	6	image_10.jpg, 0–88%	33,85%

Averaged recognition characteristics for the 4th experiment $= \frac{0+0+90+90+90+90}{6} = 60\%$, where 6 – is the total quantity of potholes.

The generalization of the processing results is summarized in the graphs shown in Fig. 9 and Fig. 10.

Figure 9 shows the total quantity of potholes in red. The identified potholes are marked in green. If the total quantity of potholes in the image does not exceed 4, the neural network recognizes all potholes. If the total quantity of potholes in the image is greater than 4, the accuracy of recognition by the neural network decreases, as shown in Table 2.

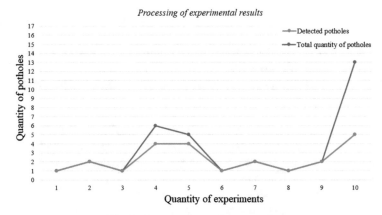

Fig. 9. Processed experiment results

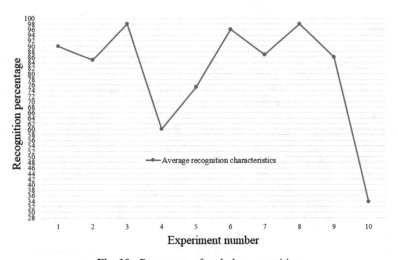

Fig. 10. Percentage of potholes recognition

Figure 10 shows the average characteristics of potholes recognition. If the total number of potholes increases during the experiment, the recognition percentage decreases. The need to increase the number of experiments for image recognition is due to the number of potholes in the image. Increasing the number of experiments improves recognition results.

6 Summary and Conclusion

Different known methods of objects recognition were described and analyzed in this paper. Main advantages and disadvantages of these methods were considered too. New approach of applying neural network to recognize the potholes was proposed in this work.

Based on the results of experiments, the following conclusions were obtained:

1) If the total number of potholes in the image is more than 4, the accuracy of recognition by the neural network decreases;
2) If the total number of potholes increases during the experiment, the recognition percentage decreases.

The number of image recognition experiments depends on the number of potholes in the image. Increasing the number of experiments improves recognition results.

The ambient light also affects the accuracy of image recognition results.

In further research, will be considered a variant with fully automatic control without a human operator. A swarm of unmanned ground and aerial vehicles will be created based on game theory to make quick decisions how to overcome obstacles either potholes or trees.

References

1. Scott, S.: Meet scout. US About Amazon, 23 January 2019. Retrieved from https://www.abo utamazon.com/news/transportation/meet-scout
2. Ai, R.: Three-wheeled delivery robot sees increased use in Michigan. New Atlas, 12 May 2020. https://newatlas.com/robotics/rev-1-delivery-robot/
3. Alpert, S., Alpert, M., Katin, P., Litvinova, N.: Software and hardware infrastructure of a terrestrial autonomous platform with the elements of artificial intelligence. Math. Mach. Syst. 1, 24–31 (2021). https://doi.org/10.34121/1028-9763-2021-1-24-31
4. Katin, P.: Development of variant of software architecture implementation for low-power general purpose microcontrollers by finite state machines. EUREKA Phys. Eng. 3, 49–54 (2017). https://doi.org/10.21303/2461-4262.2017.00361
5. Ahmad, M., WaqarRaza, Omer, Z., Asif, M.: A participatory system to sense the road conditions. Int. J. Eng. Manuf. (IJEM) 7(3), 31–40 (2017). https://doi.org/10.5815/ijem.2017.03.04
6. Zheng, D., Lia, H., Yin, S.: Action recognition based on the modified two-stream CNN. Int. J. Math. Sci. Comput. (IJMSC) 6(6), 15–23 (2020). https://doi.org/10.5815/IJMSC.2020.06.03
7. Tahir, N.M., Batureb, U.I., Abubakar, K.A., Baba, M.A., Yarima, S.M.: Image recognition based autonomous driving: a deep learning approach. Int. J. Eng. Manuf. (IJEM) 10(6), 11–19 (2020). https://doi.org/10.5815/ijem.2020.06.02
8. Liu, W., et al.: SSD: single shot multibox detector. In: Leibe, B., Matas, J., Sebe, N., Welling, M. (eds.) ECCV 2016. LNCS, vol. 9905, pp. 21–37. Springer, Cham (2016). https://doi.org/10.1007/978-3-319-46448-0_2
9. Alpert, M.: Images for training neural network. Maksym Alpert - Google Drive, 18 April 2021. https://drive.google.com/drive/folders/1QuDjrNaJH374VnbZ8IxBKJX-8Tb AWdaE. Accessed 2022
10. Tran, D.: Raccoon Detector Dataset. GitHub (2018). https://github.com/datitran/raccoon_d ataset
11. Tensorflow. GitHub. https://github.com/tensorflow/models/tree/master/research/object_det ection/samples/configs

Mobile-YOLO: A Lightweight and Efficient Implementation of Object Detector Based on YOLOv4

ChunZhi Wang[✉], Xin Tong, Rong Gao, and LingYu Yan

School of Computing, Hubei University of Technology, Wuhan 430068, China
chunzhiwang@mail.hbut.edu.cn

Abstract. To address the problems of large number of parameters and high computational complexity in current target detection models, we propose a lightweight target detection model based on YOLOv4 to simple the network structure and reduce parameters, which makes it be suitable for developing on the mobile and embedded devices. In our model, we first use the lightweight network Mobilenetv3 as the backbone feature extraction network, and fuse the depth-separable convolution in the SPP and PANet structure to compress the number of parametres of the model to one-sixth of the original; then the coordinate attention mechanism and adaptive spatial feature fusion mechanism are combined to improve the extraction and fusion capability of the backbone network and Neck module for image features; The SoftPool pooling method and FReLU activation function are also introduced in MobileNetv3 to reduce the information loss of feature maps and improve the ability of the network to capture the spatial information of images. The experimental results on the Pascal VOC dataset show that the improved lightweight target detection model based on YOLOv4 achieves 87.33% accuracy with a lower number of parameters, and reaches the best results in the comparison experiments.

Keywords: Target detection · Lightweight neural network · Cooperative attention mechanism · Adaptive spatial feature fusion

1 Instruction

As one of the important research tasks in the field of computer vision, target detection has been widely used in various fields such as intelligent monitoring and intelligent driving. Traditional target detection methods use sliding window technology to screen the target area, extract image features through artificially designed algorithms, and finally use a classifier to determine the target category. With the continuous development of deep learning technology, target detection technology based on convolutional neural networks has gradually replaced traditional target detection algorithms, and has achieved significant improvements in accuracy and speed. At present, there are mainly two types of algorithms in the field of target detection. One is two-stage target detection algorithm, such as Mask-RCNN, Faster-RCNN, etc. This type of algorithm first generates a series

© The Author(s), under exclusive license to Springer Nature Switzerland AG 2022
Z. Hu et al. (Eds.): ICCSEEA 2022, LNDECT 134, pp. 221–234, 2022.
https://doi.org/10.1007/978-3-031-04812-8_19

of target candidate frames, and uses convolutional neural networks to classify and predict the candidate frames, the computational complexity is relatively high [1]; The other is single-stage target detection algorithm, such as YOLO and SDD series of algorithms, this type of algorithm directly predict the category and location of the target object through the convolutional neural network, which has a faster detection speed [2].

In order to extract richer image features and achieve better target detection results, the number of convolutional neural network layers in the target detection model is gradually deepened, and the parameter amount and calculation cost of the model are also significantly increased; Meanwhile, in order to improve the portability of the equipment and reduce the computing cost, the storage resources and computing resources of mobile devices such as mobile phones, unmanned aerial vehicles, unmanned vehicles, etc. are often very limited, and large-scale target detection networks cannot be deployed [3]. Therefore, it is of great significance to compress the parameter and calculation amount of the target detection model based on the convolutional neural network, reduce the use of memory and computing resources, and enable it to be deployed to low energy consumption and low computing power devices.

Based on the above requirements, in this paper, we propose an improved lightweight target detection model based on YOLOv4: To reduce the amount of parameter calculation of the model, we combine the lightweight network MobileNetv3 and the deep separable convolution; Then, the Coordinate attention and adaptive spatial feature fusion mechanism are introduced into the backbone network and PANet module to improve the extraction and fusion capabilities of the backbone network and neck structure for image features; We also use the SoftPool pooling layer and the FReLU activation function to reduce the loss of the sampling information of the feature map of the MobileNetv3 network, and enhance the network's ability to extract image spatial information.

2 Related Work

In order to make the target detection model more suitable for embedded devices with low energy consumption, the research and application of lightweight models are of great significance. One of the effective methods for lightweight design of target detection model is to use lightweight convolutional neural network as the backbone network [4]. Separate convolution, channel shuffling and other methods can reduce the amount of parameters in the convolution process and improve the calculation speed of the network [5].

Commonly used lightweight networks include: SqueezeNet, Xception, ShuffleNet series, MobileNet series, etc. SqueezeNet uses 1×1 ordinary convolution and grouped convolution as the basic structure of the network to reduce the number of input channels of the image and the pooling operation in the network [6]; Xception improves the convolution structure of Inceptionv3, which improves the accuracy of the network while reducing the amount of parameters [7]; ShuffleNetV1 uses grouped convolution to reduce the amount of network parameters and reduces the loss of channel information caused by grouped convolution [8]; MobileNetv1 proposes the deep separable convolution, which greatly improves the operating speed of the network, but causes a large loss of image information. Therefore, MobileNetv2 adds an inverted residual (Inverted

Residuals) module and a linear bottlenecks (Linear Bottlenecks) module to MobileNetv1 [9], which reduces image information loss and memory usage [10]; MobileNetv3 mainly uses the network search algorithm [11], and at the same time introduces the SE module on the basis of MobileNetv2, which has improved the running speed and accuracy of the network [12].

In recent years, many scholars have proposed a design method for target detection models based on lightweight networks. Literature [13] uses grouped convolution and deep separable convolution to improve the backbone network of SSD, and uses bidirectional feature fusion, which has achieved better results than the original model on the Pascal VOC data set; Literature [14] uses MobileNetv2 as the backbone network of YOLOv3, combined with the CoordAttention attention mechanism, to achieve high accuracy in the detection of remote sensing data sets; Literature [15] proposed a GS-YOLO model based on YOLOv4, replacing the backbone network with the lightweight network GhostNet, and using channel pruning to further reduce the parameters and calculations of the model, but the accuracy of the model has decreased; In [16], the author of YOLOv4 proposed a lightweight model YOLOv4-tiny, which reduces the number of feature layers and prediction heads of the backbone network, and uses FPN as a feature extraction network, which greatly reduces the amount of model parameters, but the accuracy has decreased.

It can be seen from the research of the above-mentioned scholars that, introducing a lightweight network or modifying the convolution structure of the network into the existing target detection model can reduce the amount of model parameters and calculations, However, although the methods in literature [15] and literature [16] reduce the amount of model parameters, the accuracy of the model is also reduced. In order to avoid this problem, in this paper, we first lighten the target detection model, then combine Coordinate attention, adaptive spatial feature fusion, FReLU and SoftPool to improve the model's ability to process the features of the input image. On the whole, our model can improve the detection accuracy of the model while reducing the amount of parameters.

3 Mobile-YOLO

3.1 Network Structure

In this paper, we propose a lightweight target detection model based on YOLOv4. We use lightweight convolutional neural networks and deep separable convolutions to reduce the amount of model parameters, and combine Coordinate attention (collaborative attention mechanism) and ASFF (adaptive spatial feature fusion mechanism), FReLU activation function and SoftPool pooling to improve the accuracy of the model. The model consists of the backbone network, feature extraction module and YOLO Head. The backbone network is the lightweight network MobileNetv3; The feature extraction module is composed of SPP, PANet and adaptive feature fusion modules. The overall network structure design is shown in Fig. 1. In the process of model training and testing, the input picture (416 × 416) is first extracted through the lightweight network MobileNetv3, and the number of output channels of the backbone network's output feature layer are (13, 13, 1024), (26, 26, 512) and (52, 52, 256); The feature layer with the number of output channels (13, 13, 1024) is the input feature layer of SPP, after three convolution and pooling operations, it enters the PANet structure. The feature layer with the number of output channels (26, 26, 512) and (52, 52, 256) directly enters the PANet structure; In the PANet module, upsampling, downsampling and five-time convolution blocks are used for feature fusion. The three output feature layers are then subjected to adaptive feature fusion through the ASFF module, and finally the three output feature layers of the ASFF structure are passed to YOLO Head for prediction.

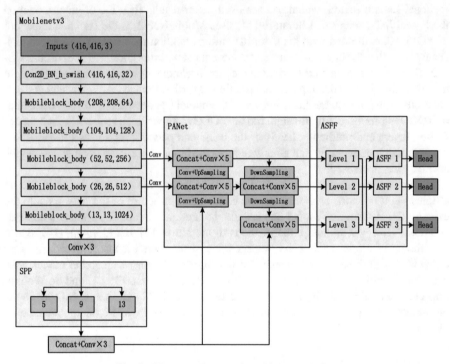

Fig. 1. The overall network architecture design

3.2 Lightweight Design of the Network

3.2.1 Lightweight Backbone Network

The backbone network of YOLOv4 is CSPDarknet53, which is composed of five large residual blocks and contains a total of 53 convolutional layers, with a large amount of parameters and a high amount of calculation. In this paper, we use the lightweight network MobileNetv3 as the backbone network, reduce the number of parameters of the backbone network by the deep separable convolution. We also use the residual inversion structure and the SE attention mechanism to improve the backbone network's ability to extract image features. On the premise of a small reduction in the accuracy of the model, the parameter amount of the model and the amount of calculation in the training process are significantly reduced.

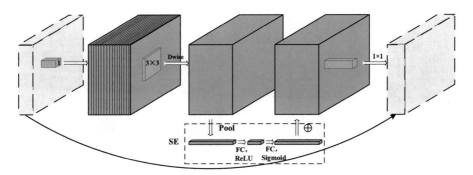

Fig. 2. Backbone network's neck structure

The improved backbone network neck structure is shown in Fig. 2, which is composed of deep separable convolution, residual inversion structure and SE attention module. We first use 1×1 convolution to increase the dimension of the input feature map, and then perform 3×3 depth separable convolution to achieve information integration between different channels. After the depth separable convolution, we integrate the SE attention module and use global pooling to compress the features of the image. Then the weights of different channels in the feature map are obtained through two fully connected layers connected to the ReLU and Sigmoid activation functions, and the different weights are multiplied by the feature layer to obtain the output feature layer of the SE module. Finally, 1×1 convolution is used to reduce the dimensionality of the feature map, and the output feature layer of the backbone network is obtained.

3.2.2 Lightweight Neck Structure

The Neck structure of YOLOv4 is divided into two parts: SPP (Spatial Pyramid Pooling) and PANet (Path Aggregation Network). The calculation parameters of the Neck structure mainly come from the three-time convolution blocks, the five-time convolution blocks and the 3×3 convolution in the down-sampling process. In order to reduce the amount of parameters of the Neck structure, we use the depth separable convolution proposed in MobileNetv1 to replace the ordinary 3×3 convolution in the three-time convolution blocks and the five-time convolution blocks. The left and right images of Fig. 3 respectively show the improvement of the three-time convolution block and the five-time convolution block:

In summary, we have made light-weight improvements to the backbone network and Neck structure of YOLOv4 respectively. The model obtained by improving the backbone network is denoted as YOLOv4-Mbv3, and the model obtained by improving the Neck structure is denoted as YOLOv4-Mbv3-PdNet. As shown in Table 1, the data shows that after the lightweight improvement of the YOLOv4 backbone network, the parameter amount of the model is significantly reduced, and the calculation amount is reduced by about one-half. On this basis, we make lightweight improvements to the Neck structure of YOLOv4. The amount of model parameters is reduced to one-sixth of the original, and the amount of calculations is reduced to one-ninth of the original.

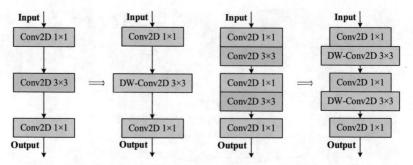

Fig. 3. Three-time and five-time convolution block structure improvement

Table 1. Comparison of model params and the amount of operations

Method	Backbone	Params size	GFLOPs
YOLOv4	CSPDarknet53	244.29M	29.95
YOLOv4-Mbv3	MobileNetv3	152.55M	13.36
YOLOv4-Mbv3-PdNet	MobileNetv3	44.74M	3.82

3.3 Feature Extraction and Fusion Module Design

3.3.1 Coordinate Attention

The backbone network MobileNetv3 uses the SE attention mechanism. Although it can improve the network's ability to extract image information, SE only converts the input feature map into a single feature vector through 2D global pooling, ignoring the important location information in the feature map [17]. In this paper, we add the Coordinate attention mechanism to the first layer of MobileNetv3 and the three-time convolution block in the SPP module to embed the location information of the image into the channel attention, which reduces the information loss of the feature map and improve the network's ability to capture image spatial information.

In the Coordinate attention module, we first factorize the global pooling as formulated in Eq. (1) into spatial information encoding operations in different directions:

$$z_c = \frac{1}{H \times W} \sum_{i=1}^{H} \sum_{j=1}^{W} x_c(i, j) \tag{1}$$

Specifically, given the input X, we use two spatial extents of pooling kernels (H, 1) or (1, W) to encode each channel along the horizontal coordinate and the vertical coordinate, respectively. Thus, the output of the c-th channel at height h can be formulated as:

$$z_c^h(h) = \frac{1}{W} \sum_{0 \le i \le W} x_c(h, i) \tag{2}$$

Similarly, the output of the c-th channel at width w can be formulated as:

$$z_c^w(w) = \frac{1}{H} \sum_{0 \le i \le H} x_c(j, w) \tag{3}$$

Then the spatial information is weighted and fused. The structure is shown in Fig. 4. First, we use the average pooling to encode each channel of the input feature map, perform feature aggregation along two directions to obtain feature maps in two directions. Then we use the cascade operation and 1×1 convolution to perform feature mapping on information in different directions, decompose the feature map into two different tensors along the spatial dimension, then two 1×1 convolutions are used to make the tensor have the same number of channels. Finally, the horizontal and vertical attention weights are obtained through the Sigmoid activation function to weight and fuse the image features.

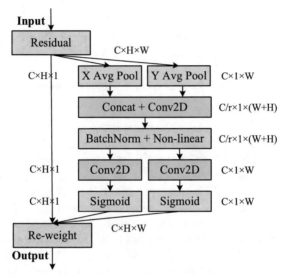

Fig. 4. The structure of coordinate attention

As shown in Fig. 5, in this paper, we add Coordinate attention to the backbone network and the three-time convolution block respectively: Adding Coordinate attention after the h_Swish activation function of the first layer of the MobileNetv3 network is more conducive to the backbone network to obtain the shallow features of the input image; Adding Coordinate attention after the depth separable convolution in the three-time convolution block of the SPP module can enable the SPP module to extract information from features of different scales more accurately, which is conducive to subsequent feature fusion operations.

3.3.2 Adaptive Feature Fusion Module

The feature fusion module of YOLOv4 is PANet, which integrates the features of the input image through the two paths of "top-down" and "bottom-up", making the low-level information of the network easier to spread. However, PANet cannot make full use of the information of feature maps of different scales. In this article, we add ASFF (Adaptive Spatial Feature Fusion) after the PANet module to form a new adaptive feature fusion module. By calculating the weight parameters of feature fusion, the information

First layer   **Three-time convolution**

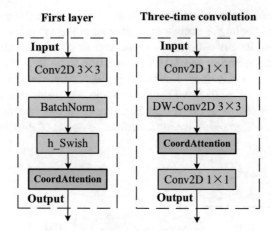

Fig. 5. The integration of the Coordinate attention

of feature maps of different scales is adaptively fused, so that the network can filter the image features and retain useful information for combination.

As shown in Fig. 6, X1, X2, and X3 are the three output feature layers of the backbone network. After up-sampling, down-sampling and five-time convolution blocks in the PANet structure, the three fused feature layers Level1, Level2, and Level3 are used as the input feature layers of ASFF for further feature fusion. In the ASFF structure, if Level1 remains unchanged, we first use 1×1 convolution on Level2 and Level3 respectively, adjust their channel numbers to be the same as Level1, and then use upsampling to make the dimensions of Level2 and Level3 the same as Level1. The three feature layers with the same number of channels and dimensions are denoted as Level1, resized-Level2, and resized-Level3. Then we use 1×1 convolution for Level1, resized-Level2, and resized-Level3 to obtain the weight parameters α, β, and γ, Finally, Level1, resize_Level2, and resize_Level3 are multiplied by α, β, and γ respectively and then added to obtain the final fusion feature layer ASFF1; the calculation method of ASFF2 and ASFF3 is the same as that of ASFF1. The calculation method is as follows:

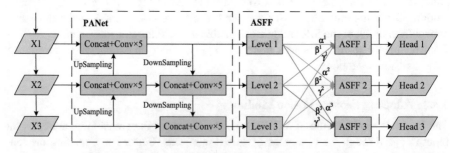

Fig. 6. Adaptive spatial feature fusion module architecture

$$y_{ij}^l = \alpha_{ij}^l \cdot x_{ij}^{1 \to l} + \beta_{ij}^l \cdot x_{ij}^{2 \to l} + \gamma_{ij}^l \cdot x_{ij}^{3 \to l} \tag{4}$$

In the formula, y_{ij}^l represents the output feature layer obtained through the ASFF module. When the value of l is 1, 2, and 3, it represents ASFF1, ASFF2, and ASFF3 respectively; $x_{ij}^{n \to l}$ represents the feature vector obtained by converting the number of channels of feature layer n into the same as feature layer l; α_{ij}^l, β_{ij}^l, γ_{ij}^l represent the weight values of three different feature layers. Finally, we use the Softmax function to make them satisfy the formula as follows [18]:

$$\alpha_{ij}^l + \beta_{ij}^l + \gamma_{ij}^l = 1 (\alpha_{ij}^l, \beta_{ij}^l, \gamma_{ij}^l \in [0, 1]) \tag{5}$$

The adaptive feature fusion module proposed in this paper combines ASFF adaptive feature fusion on the basis of PANet two-way feature fusion. The improved module learns the weight parameters of different feature layers, fully integrates the three output feature layers of the backbone network, and improves the training and testing effects of the model.

3.3.3   Activation Function and Pooling Layer Improvement

In order to further improve the performance of the model, we also use FReLU and SoftPool to improve the ReLU activation function and AvgPool (average pooling) in the backbone network MobileNetv3, which strengthen the network's ability to capture image spatial information and reduce the information loss of feature maps in the pooling process.

First, we replace part of the ReLU in MobileNetv3 with FReLU (Funnel ReLU), the calculation formula of the activation function is changed from y = max(x, 0) to y = max(x, T(x)), and the scope of the activation function changes from 1 dimension to 2 dimension, compared with ReLU, the value of FReLU depends on the spatial context of each pixel, rather than the pixel itself, and is better for visual tasks such as object detection and instance segmentation.

On the other hand, the AvgPool used in the backbone network may cause loss of information in the feature map. In this paper, we use SoftPool to replace AvgPool in MobileNetv3, introduce softmax exponential weighting to retain the basic attributes of the input feature map, and then use the activation method to amplify the stronger features, which can retain more image information during the pooling process. As shown in Table 2, we change the ReLU in the first six layers of MobileNetv3 to the FReLU activation function, and use SoftPool in the last layer of the network. In the improved MobileNetv3 structure, the first six layers including MNeck use the FReLU activation function, and the h_Sigmoid activation function is used in the last nine-layer network; After the 2 × 2 convolutional layer, SoftPool pooling is used to reduce the size of the feature map. In this way, we can strengthen the network's ability to capture image spatial information and reduce the information loss of feature maps in the pooling process.

Table 2. Improved MobileNetv3's structure

	Input	Operator	out	SE	NL	s
	$208^2 \times 16$	Mneck 3×3	16	-	**FReLU**	1
	$208^2 \times 16$	Mneck 3×3	24	-	**FReLU**	2
	$104^2 \times 24$	Mneck 3×3	24	-	**FReLU**	1
	$104^2 \times 24$	Mneck 5×5	40	√	**FReLU**	2
2	$52^2 \times 40$	Mneck 5×5	40	√	**FReLU**	1
	$52^2 \times 40$	Mneck 3×3	80	-	HS	2
3	$26^2 \times 80$	Mneck 3×3	80	-	HS	1
	$26^2 \times 80$	Mneck 3×3	112	√	HS	1
	$26^2 \times 112$	Mneck 3×3	112	√	HS	1
	$26^2 \times 112$	Mneck 5×5	160	√	HS	2
2	$13^2 \times 160$	Mneck 5×5	160	√	HS	1
	$13^2 \times 960$	Conv2D 2×2	960	-	-	1
	$13^2 \times 960$	**SoftPool 7×7**	-	-	-	1

4   Experiment and Result Analysis

4.1   Experimental Environment and Data Set

The experiments in this paper use the Pytorch deep learning framework to build a target detection network. The computer operating system is Win10, the GPU is NVIDIA GeForce GTX1070 8G, the PyTorch version is 1.6, and the Python version is 3.8.2. The data set used in the experiment is the Pascal VOC2007+2012 data set, which contains a total of 16,551 pictures, including 20 categories of objects: people, animals (cats, dogs, cows, horses, sheep, birds), vehicles (cars, buses) Cars, bicycles, motorcycles, boats, airplanes, trains) and household items (chairs, tables, sofas, televisions, water cups, potted plants) [19]. We use 10% of the pictures as the test set, and 90% of the pictures as the training set and the validation set. The ratio of the training set to the validation set is 9:1.

4.2   Ablation Experiment

In this paper, we propose an improved lightweight target detection model based on YOLOv4. The improvement method is mainly divided into two major steps: (1) Lightweight improvement of the network: reduce the amount of network parameters and calculations from both the backbone network and the convolution structure; (2) Improve the accuracy of the model: Introduce the Coordinate attention mechanism in the three-time convolution blocks in the backbone network and SPP module; Add the ASFF adaptive feature fusion mechanism after the PANet feature fusion module; Replace the ReLU activation function in the backbone network MobileNetv3 with the FReLU activation function; Introduce SoftPool pooling in the last layer of the backbone network. We compare the parameters and calculations of the lightweight YOLOv4-mbv3 and the original YOLOv4 model in the previous section, in order to verify the effectiveness of the improved method in step (2), in this section, we use ablation experiments to compare the improvement effects of different methods on the model.

Table 3. Comparison of ablation experiments

YOLOv4-Mbv3	Ca	ASFF	FReLU	SoftPool	Params	map%
√					44.74M	82.80
√	√				48.50M	84.63 (+1.83)
√		√			76.10M	85.56 (+2.76)
√			√		47.73M	86.02 (+3.22)
√	√			√	44.74M	84.45 (+1.65)
√	√	√			76.77M	86.74 (+3.94)
√	√	√	√		79.76M	87.20 (+4.40)
√	√	√	√	√	79.76M	**87.33 (+4.53)**

As shown in Table 3, Ca means the introduction of the Coordinate attention mechanism, ASFF means the introduction of the adaptive feature fusion mechanism, FReLU means the improvement of the activation function, and SoftPool means the improvement of the pooling layer; the evaluation indicators used in the experiment are Params (memory occupied by parameters) and map (The accuracy of the model). The introduction of the Coordinate attention mechanism slightly increases the amount of network parameters and improves the network accuracy by 1.83%; The introduction of the ASFF adaptive feature fusion mechanism increases the amount of parameters by about 30M, but the accuracy of the model is significantly improved To 85.56%. Using FReLU to improve the activation function increases the amount of parameters by about 3M and improves the network accuracy by 3.22%; The introduction of the SoftPool pooling layer improves the accuracy of the model by 1.65% without increasing the amount of network parameters. After comparing the effects of the four improved methods, we performed cumulative experiments on the above improved methods, that is, adding ASFF, FReLU, and SoftPool improvements on the basis of the introduction of the Coordinate attention mechanism. The accuracy of the model gradually increased, and the final model parameter amount is 79.76M, the accuracy reaches 87.33%. Compared with the YOLOv4-mbv3 model obtained by lightening YOLOv4, the model accuracy is increased by 4.53%, although the parameter amount is increased by about 35M, but compared with the original YOLOv4, the parameter amount is reduced by two-thirds.

4.3 Performance Comparison of Different Models

In order to verify the performance of the lightweight target detection model proposed in this paper, we record the lightweight target detection model obtained by improving YOLOv4 as Mobile-YOLO, and compare it with five good target detection models proposed in recent years. We ensure that the experimental environment and the number of input image channels are the same, then compare the model's Params size (parameter amount), GFLOPs (calculation amount) and map (detection accuracy on the VOC2007+2012 data set).

Table 4. Performance comparison of different models

Method	Image size	Params size	GFLOPs	map%
YOLOv4	416	244.29M	29.95	84.42
YOLOv4-tiny	416	22.74M	3.43	80.93
YOLOv3-ASFF	416	236.32M	33.05	80.06
CenterNet	416	124.61M	22.15	82.95
Efficientdet-d4	512	78.97M	10.57	83.63
YOLOX-l	640	206.79M	32.88	86.63
Mobile-YOLO (ours)	416	79.76M	11.97	**87.33**

As shown in Table 4, compared with the original YOLOv4, our mobile-YOLO reduces the amount of parameters and calculations by about two-thirds, and the accuracy of the model has increased from 84.38% to 87.33%; Compared with the lightweight model YOLOv4-tiny proposed by the author of YOLOv4, although the amount of parameters and calculations of the algorithm in this paper are higher, the accuracy is improved by 6.4%. Compared with YOLOv3-ASFF and CenterNet, our model achieves better results in terms of model weight and accuracy; Compared with Efficientdet-d4, the parameter amount is almost the same, but the accuracy of our model is significantly improved; Compared with YOLOX-l, the accuracy of Mobile-YOLO is 0.7% higher, and the amount of parameters and calculations are less. On the whole, the improved target detection model in this paper has achieved good detection results while reducing the amount of network parameters and calculations.

5   Conclusion

In order to design a lighter and more efficient network to make the target detection model more suitable for low-energy embedded devices, this paper proposes a lightweight target detection model based on YOLOv4, which improves the model's performance on the basis of reducing the amount of parameters and the amount of calculations. The improvement idea of the model is divided into two major steps: (1) Design a lightweight network structure: Combine the lightweight network MobileNetv3 and deep separable convolution to improve the backbone network and Neck structure, The parameter amount of the model is reduced to one-sixth of the original YOLOv4; (2) Improve the accuracy of the model: Introduce Coordinate attention mechanism in the backbone network and Neck structure, introduce an adaptive feature fusion mechanism in the PANet feature fusion module, strengthen the network's ability to extract and integrate image features. We also use the SoftPool and the FReLU activation function to improve the backbone network, reduce the information loss of the feature map, and improve the network's ability to acquire image spatial information. The experimental results on the Pascal VOC data set show that the parameter amount of the target detection model in this paper is one third of that of YOLOv4. Our model achieves an accuracy of 87.33% while achieving a lightweight network, which is more suitable for target detection tasks in a low energy consumption and low computing power environment. In the next step, we will continue to study the method of model lightweight, improve the accuracy of the model at a lower cost, and design a more lightweight and efficient target detection model.

Acknowledgment. This research is supported in part by National Natural Science Foundation of China (61772180) and Key R & D plan of Hubei Province (2020BHB004, 2020BAB012).

References

1. Rahman, A.K.M.F., Raihan, M.R., Islam, S.M.M.: Pedestrain detection in thermal images using deep saliency map and instance segmantaion. Int. J. Image Graph. Signal Process. (IJIGSP) **13**(1), 40–49 (2021). https://doi.org/10.5815/ijigsp.2021.01.04
2. Ren, S., He, K., Girshick, R., et al.: Faster R-CNN: Towards real-time object detection with region proposal networks. Adv. Neural. Inf. Process. Syst. **28**, 91–99 (2015)
3. Long, D.T.: A lightweight face recognition model using convolutional neural network for monitoring students in e-learning. Int. J. Mod. Educ. Comput. Sci. (IJMECS) **12**(6), 16–28 (2020). https://doi.org/10.5815/ijmecs.2020.06.02
4. Lan, R., Sun, L., Liu, Z., et al.: MadNet: a fast and lightweight network for single-image super resolution. IEEE Trans. Cybern. **51**(3), 1443–1453 (2020)
5. Zhang, L., Piao, Y., Liu, Y.: Detection method of robot grasp based on lightweight network. In: Journal of Physics: Conference Series, vol. 1920, no. 1, p. 012113. IOP Publishing (2021)
6. Iandola, F.N., Han, S., Moskewicz, M.W., et al.: SqueezeNet: AlexNet-level accuracy with 50x fewer parameters and <0.5 MB model size. arXiv preprint arXiv:1602.07360 (2016)
7. Chollet, F.: Xception: deep learning with depthwise separable convolutions. In: Proceedings of the IEEE Conference on Computer Vision and Pattern Recognition, pp. 1251–1258 (2017)
8. Zhang, X., Zhou, X., Lin, M., et al.: ShuffleNet: an extremely efficient convolutional neural network for mobile devices. In: Proceedings of the IEEE Conference on Computer Vision and Pattern Recognition, pp. 6848–6856 (2018)
9. Yao, T., Zhang, Q., Wu, X., et al.: Image recognition method of defective button battery base on improved MobileNetV1. In: Wang, Y., Li, X., Peng, Y. (eds.) Image and Graphics Technologies and Applications. IGTA 2020. Communications in Computer and Information Science, vol. 1314, pp. 313–324. Springer, Singapore (2020). https://doi.org/10.1007/978-981-33-6033-4_24
10. Sakharkar, Y.A., Singh, M., Kumar, K.A., Aju, D.: A reinforcement learning-based offload decision model (RL-OLD) for vehicle number plate detection. Int. J. Eng. Manuf. (IJEM) **11**(6), 11–18 (2021). https://doi.org/10.5815/ijem.2021.06.02
11. Wang, H., Bhaskara, V., Levinshtein, A., et al.: Efficient super-resolution using MobileNetV3. In: Bartoli, A., Fusiello, A. (eds.) Computer Vision – ECCV 2020 Workshops. ECCV 2020. Lecture Notes in Computer Science, vol. 12537, pp. 87–102. Springer, Cham (2020). https://doi.org/10.1007/978-3-030-67070-2_5
12. Howard, A., Sandler, M., Chu, G., et al.: Searching for mobilenetv3. In: Proceedings of the IEEE/CVF International Conference on Computer Vision, pp. 1314–1324 (2019)

13. Feng, Y., Zhang, S., Wu, X.: Research on lightweight design of SSD network for target detection. Signal Process **36**(5), 756–762 (2020)
14. Li, Y., Wang, J., Lu, L., et al.: A lightweight real-time target detection model for remote sensing images. Adv. Lasers Optoelectron. **58**(16), 1615007 (2021)
15. Fu, H., Wang, P., Li, X., Lu, Z., Di, R.: A lightweight network model for moving target recognition. J. Xi'an Jiaotong Univ. **55**(07), 124–131 (2021)
16. Wang, C.Y., Bochkovskiy, A., Liao, H.Y.M.: Scaled-YOLOv4: scaling cross stage partial network. In: Proceedings of the IEEE/CVF Conference on Computer Vision and Pattern Recognition, pp. 13029–13038 (2021)
17. Jiang, Z., Zhao, L., Li, S., et al.: Real-time object detection method based on improved YOLOv4-tiny. arXiv preprint arXiv:2011.04244 (2020)
18. Liu, S., Huang, D., Wang, Y.: Learning spatial fusion for single-shot object detection. arXiv preprint arXiv:1911.09516 (2019)
19. Adarsh, P., Rathi, P., Kumar, M.: YOLOv3-tiny: object detection and recognition using one stage improved model. In: 2020 6th International Conference on Advanced Computing and Communication Systems (ICACCS), pp. 687–694. IEEE (2020)

Program Module of Cryptographic Protection Critically Important Information of Civil Aviation Channels

Anna Ilyenko[(✉)] and Sergii Ilyenko

National Aviation University, Kyiv 03058, Ukraine
ilyenko.a.v@nau.edu.ua

Abstract. Given the ever-growing statistics of cyber-attacks on civil aviation worldwide, after in-depth analysis and study of this issue, the authors highlighted the current state of cyber security and organization of protection of ground-to-air and air-to-air fleet of aircraft in operation. Airlines of Ukraine, as well as to consider in detail the world experience of aircraft leaders such as Airbus and Boeing. The authors propose test software for the protection of aircraft communication channels designed for the Antonov Design Bureau to ensure the integrity and confidentiality of the transmission of information in order to prevent the implementation of cyber-attacks. Proposed test software use electronic digital signature technology based on elliptic curves algorithm in combination with symmetric encryption and allows today reduces the quantitative value of the average key generation time by an average of 36%, which requires 12 times less load on the computer system and increases the amount of memory to store the keys in the civil aviation communication channels of Antonov aircraft and provides authentication of aircraft and fully protect every information message they transmit. This solution for Antonov Design Bureau aircraft is new, has not been certified on serial aircraft and has no analogues.

Keywords: Cybersecurity · Aviation · Integrity · Confidentiality

1 Introduction

Today the civil aviation industry, as a critical object of international transport, has faced the challenge of increasing cyber threats, which are increasingly difficult to detect, control and neutralize. Every year, the number of cyberattacks on the aircraft itself during its life cycle, as well as on aviation infrastructure (airports, airfields, air navigation centers, etc.) increases in arithmetic progressions. Considering the modern aircraft as an object of cyber-attacks, it should be noted that such an aircraft is in fact one large computer-integrated complex of all automated systems interaction and complexes involved in the aircraft maintenance. An extensive network of aircraft controllers provides interoperability and control of various avionics systems. And it is the primary object of cybernetic influence from the outside (control systems, navigation, sounding, monitoring, information communication and entertainment systems) (Fig. 1). Critical information (which

© The Author(s), under exclusive license to Springer Nature Switzerland AG 2022
Z. Hu et al. (Eds.): ICCSEEA 2022, LNDECT 134, pp. 235–247, 2022.
https://doi.org/10.1007/978-3-031-04812-8_20

reliability can lead to unpredictable consequence) includes engine telemetry settings, flight parameters, data on the aircraft location in space and time, speed, and so on. Substitution or interception of this critical information can not only provoke catastrophic situations with a high decommissioning probability, but also endanger the health and lives of a significant number of people. It should also be stated that the number of threat vectors aimed at compromising aviation organizations increased. Therefore, the relevance of this research is dictated by the challenge of time and ensuring the safety of air transport worldwide [3–6].

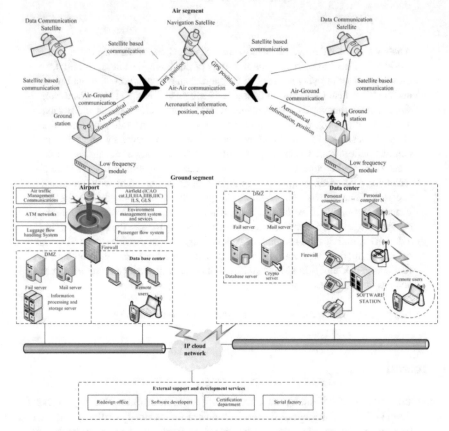

Fig. 1. Architecture of modern civil aviation communication channels

The purpose of this article is to develop a test software module for the procedure of ensuring the critical information integrity and authenticity of civil aviation communication channels in the Antonov family. This decision is a challenge and a requirement for safe modern aircraft of the Antonov family operation. Nowadays there is no analogue to the proposed approach to the organization of civil aviation communication channels of Ukrainian-made aircraft protection. Based on the purpose, the task of this scientific article: 1) to study the current state of cybersecurity of civil aviation in Ukraine and the world; 2) to investigate cryptographic methods of information protection and practical approaches to ensuring the integrity and confidentiality of information; 3) to develop test software module for control of message protection of Antonov family aircraft; 4) test and implement the cryptographic module and to substantiate the expediency of its use to protect messages in the information networks of aircraft of the Antonov family.

The scientific novelty is due to the solution of the problem of information protection, in particular information stored, processed and transmitted in modern information networks of airplanes of the Antonov family on the basis of ensuring confidentiality and integrity. The protection procedure for instant messaging has been improved, based on the use of a combined cryptosystem with elliptic curves algorithm and AES, which allowed to ensure the integrity and confidentiality of data and authentication of aircraft.

2 Related Works

Today on-board wired and wireless avionics aircraft devices have access to the route construction system and program algorithms for flight on the route using aircraft controls in different flight modes (modern on-board computer FMC-Flight Management Computer controlled by the CDU display – Central Display Unit). Unauthorized reprogramming of the route (including hacking methods) can lead to intentional or unintentional damage of the data and/or systems that are important (critical) for the safe aircraft maintenance [3, 4, 9]. The most critical cyber threats to the civil aviation industry are those that affect the safety of aviation activities - namely, those that can cause, in the worst case, even catastrophic consequences. This can be the theft of employee's personal data of various structural enterprises and aviation industry and passengers divisions. And intrusion into aeronautical (satellite, aeronautical) computer networks to carry out powerful large-scale cyberattacks. To such cyber-attacks the authors include: phishing attacks; non-compliance with regulations; data loss or theft; insider threats; distributed denial of service (DDoS) attacks. These attacks are aimed at serious reputational, financial and moral losses in order to significantly compromise the information airports networks, air traffic control centers and directly the operators of the aircraft themselves. In the works [8, 9] authors identify the most common objects of cyber-attacks during the modern aircraft maintenance are: air traffic control system (ATC); onboard IP networks; interface devices for both ground equipment and the aircraft itself; office technology platforms, etc. As more and more passenger aircrafts provide access to the Internet on board during the flight, it is possible to gain unauthorized access to any control aircraft systems through the network injection via Wi-Fi. Also, access to on-board control systems can be done through on-board entertainment and information systems by using different interfaces (touch screens, USB-interfaces, Wi-Fi, GSM-communication).Therefore, the process of information security in the civil aviation industry is a complex concept that should include: computer networks of airports and airfields; information channels for interaction between aircraft; information networks of the aircraft itself; own databases [7–10]. In the works [8–10] authors identify that new types of aircraft use TCP/IP technology for systems that connect both domain aircraft and cabin interfaces in a way that makes the aircraft virtually a network domain server. The architecture of this air network allows to connect to external systems and networks, such as wireless transmission and service systems, satellite communications (SATCOM), e-mail, the Internet and more. The main advantage of using the TCP/IP protocol is the ability to transfer information to the aircraft without the use of media. The use of this approach leads to vulnerabilities and external threats, which can lead to unauthorized access and affect the operation of the functional avionics systems of modern aircraft. Unauthorized access to the aircraft avionics at any stage of the modern aircraft network will violate the confidentiality, integrity and availability of data, which is likely to create extreme conditions for the aircraft maintenance, justify the use of airline cyber security [3, 4, 8, 9, 11–15].

The authors of this article identify four main areas of an integrated approach to ensuring the aircraft cybersecurity: 1. regular security assessment of the onboard aircraft equipment, which is to verify compliance with regulatory requirements; 2. Application of architectural constructing principles of the secure operating systems for designing aircraft information infrastructure; 3. use of classic means of cybersecurity, adapted to the civil aviation industry requirements (means of identification and authentication, access control, encryption, electronic-digital signature); 4. Regular scheduled and unscheduled risks and vulnerabilities assessment of the aircraft information infrastructure. In this article authors focus on one of the promising areas of civil aviation communication channels protection based on the encryption of modern aircraft information, both foreign design bureaus (Boeing and Airbus) and the state enterprise Antonov. Since 2020, due to the air traffic crisis caused by the COVID-19 pandemic, an increase in air traffic has been recorded, in which the Antonov division and its airline are successfully competing. The following types of transport Antonov aircrafts include: AN-22 Antey, AN-124 Ruslan, AN-225 Mriya, AN-74, as well as passenger aircraft AN-148, AN-158, etc. This approach allows to provide confidentiality, integrity and authenticity of the critical information source. The world's aircraft design bureaus encrypt information to protect personal workers and service personnel. The aircraft and ground systems communication by air is another area where the use of ciphers is acquired and is a vital control. All critical information and users of information exchange must be authenticated, and the aircraft must be encrypted for protection against eavesdropping, deletion and modification [14–18].

Today, the world's leading aircraft design bureaus, such as Boeing and Airbus, use modern cryptographic methods to ensure the confidentiality and authenticity of critical information for the organization of secure ground-to-air, air-to-air, and ground-to-ground communication channels. Boeing and Airbus are world leaders in developing and implementing public key certificate distribution policies that define the use of digital certificates to authenticate a source of critical information. Namely the process of interaction between ground stations and aircraft. Also, providing the integrity of critical information through electronic digital signatures and symmetric encryption to ensure the confidentiality and identification of the final participants in the information exchange between aerospace domains. In works [1, 2] the directions of use of cryptographic transformations for maintenance of protection of aviation communication channels are offered that allows to assert that the Ukrainian regulations do not regulate accurate methods of counteraction to cyberattacks of aircraft designed by Antonov Design Bureau. Table 1 summarizes the use of cryptographic means of critical information protection for the organization of civil aviation communication channels "ground-to-air", "air-to-air", "ground-to-ground" [1, 2, 16–20, 25–28].

Table 1. Comparison of cryptographic means using by leading aircraft design bureaus

Cryptographic function	Boeing	Airbus
Symmetric encryption	3DES	3DES/AES
Hashing	SHA-1	SHA-1/SHA-256
Electronic digital signature	RSA-2048/ECDSA-224	RSA-2048/ECDH-224
Distribution of cryptographic keys	RSA-2048/ECDH-224	RSA-2048/ECDH-224

Modern aircraft have dozens of connected subsystems that transmit critical telemetry and control data to each other. Most confidential and aeronautical information, that is critical, is transmitted over unsecured open communication channels. The issue of secure communication channels for aircraft designed by Antonov Design Bureau, as well as promising aircraft developments and aircraft in maintenance, is a very important and will be covered in this article [21–24].

3   Justification of the Choice of Cryptographic Methods for the Implementation of Protection

3.1   Proposed Mechanism for Civil Aviation Security of Modern Aircraft Antonov Family

Therefore, the mechanism of civil aviation communication channels protection of modern aircraft of the Antonov family, according to the authors, should follow the classic combined scheme, which speeds up encryption and decryption operations and ensures the three cryptographic methods operation of three services - symmetric encryption, asymmetric encryption and electronic digital signature. The implementation of this approach will further solve the problem of authentication "ground-to-air", "air-to-air", to ensure the integrity and confidentiality of information, namely navigation data transition between the aircraft and the ground station.

The task of the source authentication and information integrity is realized through the use of electronic digital signature procedure based on RSA or ECDSA algorithms that support the PKCS7 standard and use SHA256 hashing. These electronic-digital schemes (EDS) use X509 containers that encapsulate the token and the public key of the asymmetric pair (RSA and EC). It is also proposed to use symmetric encryption to encrypt the confidential information "ground-to-air" and "air-to-air" during transmitting, as well as to encrypt private parts of asymmetric pairs stored in the database and used for mutual authentication. Due to modern methods of symmetric encryption, the most optimal are SP-network schemes. So to solve the problem of symmetric encryption, it is proposed to use the AES-128 algorithm in the encryption mode of the ciphertext blocks (CBC). The goal of this approach is to use asymmetric encryption to form an encrypted symmetric session key, which encrypts the content of the message. In the RSA scheme, the session random key is encrypted by exponentiation of encrypting exponent (public key) in the final field. As a result the application creates a standard PKCS7-container EnvelopedData, which describes service information on encryption algorithms for the recipient – parameters of the combined encryption scheme. The EC scheme uses the ECIES (ecliptic curve integrated encryption scheme), which is widely used in modern encryption schemes, since only EDS (ECDSA) and key exchange protocol (ECDH) are standard service EC keys are used with. To ensure an acceptable level of cryptocurrency, RSA keys have a size of 2048 bits, EC keys use the standard elliptic curve secp256r1, which is recommended by NIST and is widely used in modern EDS circuits. Next, let's consider the possibility of two schemes functioning.

RSA scheme. This scheme combines such cryptographic algorithms as 3DES, SHA-1 and RSA-2048.

Stage 1. Key generation. Firstly, two random numbers of 2048 bits are generated (p, q); checking them for mutual simplicity; calculation of the Euler function $\varphi(n) = (p-1)(q-1), n = pq$; obtaining a static encryption exponent e = 65537; calculating the private exponent $de \equiv 1 (\mathrm{mod}\varphi(n))$, and then further pair forming (d, n) and (e, n) as private and public key respectively.

Stage 2. Encryption. Firstly, a random session key is generated k_E (256 біт), AES-256 is selected as the session key; then the content encryption of the message using $C = 3DES(k_E, m)$. Next, exponent e = 65537 is used and n from the public key of the remote user and the session key encryption $k_{ENC} = k_E^e(\mathrm{mod}n)$ and the formation of PKCS # 7 container EnvelopedData with the fields corresponding X509 certificate and (k_{ENC}, C).

Stage 3. Decryption. Receiving container PKCS#7 EnvelopedData from k_{ENC} and C; decription of the session symmetric key k_E, using the private key d and n: $k_E = k_{ENC}^d(\mathrm{mod}n)$ and the final decryption of the message $m = AES256(k_E, C)$.

Stage 4. EDS formation. Receiving m (binary PKCS # content 7 of the EnvelopedData container) and calculation of a hash for the message $h = SHA(m)$. EDS is calculated using the private key d and n and TSP flag to the TSP server and inserting an additional attribute from TimeStampToken to the PKCS#7 container and additional X509 TSP server certificates: $s = h^d(\mathrm{mod}n)$ and $tsp = SHA(s)$ and further packing (s, m) to the PKCS#7 container SignedData and X509 certificate of the current user.

Stage 5. EDS validation. Obtaining PKCS#7 container SignedData (s, m). Using exponent e = 65537 and n of the remote user public key, calculation $h' = s^e(\mathrm{mod}n)$ and checking $h' = SHA(m)$ are processed if the TimeStamp attribute is present in the PKCS#7 container: validation of the certificate chain for the TSP server and checking the SHA256 (s) in the TimeToken attribute.

Authors propose to use ECIES scheme in two variations such as combination of elliptic curves cryptography (ECC) and symmetric algorithm. First way, scheme combines such cryptographic algorithms as AES-256, SHA-1 and ECDSA-secp256r1. The second way to obtain system-wide parameters is to use the state standard DSTU 4145-2002, which presents elliptic curves with key length 163 bits, which are suitable for use in cryptographic transformations. Let's show the use of proposed scheme of cryptographic transmission in aviation channels of Antonov aircrafts on the example of an algorithm ECDSA.

In the general case, the process of constructing system-wide parameters for cryptographic transformations in groups of elliptic curves points involves the following actions: field selection; random selection of elliptic curve coefficients from the elements of the selected field; calculating the order of the elliptic curve; checking the suitability of this curve for use in cryptographic applications; selection of the base point and calculation of its order; checking the suitability of using the base point for cryptographic transformations.

Stage 1. Key generation. Random number generation $r \in [1; n - 1]$ and public key calculation: $Q = rG$, where Q - public key; r - private key.

Stage 2. Encryption. Random number generation $r \in [1; n - 1]$; random number packing $R = rG$; calculation of the general secret using the public key of the remote user K_B, $S = P_x$; $P = (P_x, P_y) = rK_B$; calculation of session symmetric key and MAC key $k_e \| k_m = KDF(S \| S_1)$; message encryption $m = AES256(k_E, C)$; MAC calculation: $d = MAC(k_m; c \| S_2)$. Domains elliptic curve parameters—(n, G), public keys $K_a = k_a G$, $K_b = k_b G$, additional information S_1, S_2.

Stage 3. Decryption. The following steps are required to decrypt in ECIES: receiving $R \| c \| d$; calculation of the general secret using a private key k_b; calculation of the general secret $S = (P_x)$, $P = (P_x, P_y) = rk_b$ $P = k_b R = rk_b G$; calculation of encryption key and MAC key $k_e \| k_m = KDF(S)$; fulfillment of the condition $d = MAC(k_m; c \| S_2)$; message decryption $m = AES256(k_E, C)$.

Stage 4. EDS formation. Receiving m (binary container with $R \| c \| d$) and private key d; hash calculating $h = SHA256(m)$, sampling the left bits $z = h$ according to the bit length of the curve n; ephemeral key generation $f \in [1; n - 1]$; EC point calculation $(x_1, y_1) = fG$; calculation of $r = x_1 (\mathrm{mod}\, n)$; $r = f^{-1}(z + rd_A)(\mathrm{mod}\, n)$ by TSP flag: $tsp = SHA256(s)$ to TSP server and insert additional attribute from TimeStampToken to PKCS7 container and additional X509 TSP server certificates Packing (r; s; m) to PKCS#7 container SignedData.

Stage 5. EDS validation. Obtaining PKCS # 7 SignedData container (r, s, m) and user public key Q_A; verification of r and s $r \in [1; n - 1], s \in [1; n - 1]$; calculation of $u_1 = zs^{-1}(\mathrm{mod}\, n)$ and $u_2 = rs^{-1}(\mathrm{mod}\, n)$; calculation of $(x_1, y_1) = u_1 G + u_2 Q_A$; verification of $r \equiv x_1 (\mathrm{mod}\, n)$ if there the TimeStamp attribute in the PKCS#7 container: validation of the certificate chain for the TSP server and verification of the SHA256(s) presence in the TimeToken attribute.

The KDF (Key Derivation Function) generates a pair of secret keys based on a secret value S, optional S_1, and the MAC function generates lossy information, which serves to verify the integrity of the transmitted data using k_m and optional S_2. Thus, the ECIES scheme provides encryption and verification of data integrity.

The ECIES scheme is not provided by the Enveloped Data container standard. So the internal content for ECIES is a type of Octet String in ASN1, consisting of a triplet $R \| c \| d$: R integrates a secret random number from which a shared secret is formed and only the owner (remote interlocutor) of the private key knows it; c - encrypted message with a common secret (a point on the elliptic curve that can be obtained from

R and represents the coordinate x; d is used to verify the integrity of the received crypto message (this is a control bits sequence obtained by using MAC.

The CMS Signed Data container is reciprocal for the two schemes, since the standard provides a signature generation with both RSA and ECDSA algorithms.

Thus, the following procedure is used in the proposed scheme of forming a container with a message: the message is encrypted with a random symmetric key (in this case - AES, 128 bits in CBC mode), which is encrypted with the public key of the remote party or is the coordinate x of the common secret; enveloped-data is formed with the information from the previous step or Octet String triplet $R\|c\|d$; enveloped-data (or triplet $R\|c\|d$) is embedded in signed-data and signed by the sender's private key; the sender's certificate is also attached. As a result, the remote party knows the random symmetric key, and anyone can verify the signature because the CMS has an attached certificate. Optionally, after user activates the TSP flag from the main menu of the application, a request is additionally generated for the TSP server with the SHA256 digest for the message signature. As a result of a network operation with an HTTP POST request with a special header (Content type: application/timestamp-query), a timeStampToken object appears, which is inserted into the CMS container of SignedData format with all the necessary certificates to check the additional attribute.

To summarize, it can be said that the proposed solution for the civil aviation communication channels protection uses cryptographic tools, namely a combined (symmetric and asymmetric) cryptosystem for encrypting and signing information messages; symmetric cryptography as an auxiliary for private keys encryption in the local SQLite database of the ground station; password hashing and "salting" of the local user (a ground air navigation station or an aircraft can be considered as the user) from which a key is formed for symmetric fields encryption with private keys in the device database; protected pseudo-random number generator, etc. ASN.1 is used to describe the abstract data syntax, cryptograms are packed according to PCKS#7, and its subset CMS. It describes a standardized structure that combines the protected data and information needed to work properly with them. Among the additional information added to the CMS-containers the following can be distinguished: information about the hashing and signature algorithm; information about the encryption algorithm and its recipients, who will be able to decrypt symmetric keys to decrypt the content of the whole message, user certificates (subscribers), additional chains of certificates (such as root certificates inclusion), revoked certificate lists. There are two types of CMS in crypto containers: Enveloped Data and Signed Data to enable encryption of the message and remote party verification by the party receiving the message. There is no centralized server that identifies users and stores information about messages. The authors suggest using the Google Firebase service, which personalizes a user through FirebaseToken and allows the test application to create RealTime Database, which is a NoSQL database, and stores data in JSON format and can be used as an application for data storage. Thus it provides pseudo-centralized functions in the current architecture. The network interaction between applications is a dual model implemented by the Firebase cloud data base. When a device with an application is online, an event occurs and the device can communicate with the cloud database. The constructed cloud cloud data base scheme provides creation of a branch with the personalized device identifiers (FirebaseToken) in a cloud data base.

While writing a message, application passes a cryptocontainer with data to a branch with the FirebaseToken ID from the X509 container (from the Subject Alternative Name field). Thus, the application does not directly create sockets for network operations, open TCP or UDP ports for listening, and so on.

3.2 Testing of the Offered Software Solution

The test software for the protection of civil aviation communication channels was made with Java programming language, which implements the following functionality: code navigation; integration with Concurrent Versions System (CVS); editing assembly files; SDK manager; View software level designer; software function debugger; profiler; application debugging console; attaching to log files on the device; memory debugger; virtual device manager; device file navigator; APK file shaper and loader. The shaper uses the standard gradle build system version 3.4.1, which allows you to add dependencies through editing. The software uses the Bouncy Castle Java library for cryptographic functions, which encapsulates both standard high-level cryptographic APIs and low-level proprietary APIs for more efficient access to functionality, and is released under the MIT license. The application uses the company's cloud database instead of its own centralized intermediary server infrastructure between users of the application.

Testing of the application local functions is the easiest vector for testing the application because the visual parts reflect the data states that can be clearly seen and require a single device. Therefore, the functionality of adding a user to the certificate, generating asymmetric key pairs and embedding them in the certificate is illustrative and interdependent with the number of elements, displayed in the certificate and the contact snippets. In addition, you can make sure that the certificate is generated correctly by clicking on it to trigger its export, and then look at the ASN1 structure and its contents. After this testing, some conclusions can be obtained: the local database has the correct structure; user entities and certificates are correctly added to the local database; the cryptographic component for generating asymmetric key pairs, creation of a certificate and exporting it, and symmetric private key encryption fields work correctly; the selection window of certificate exchange method also works as expected.

Testing of sending messages can be done as follows: after the procedure of adding a user and his visual appearance on the contacts tab, it is possible to click on it and proceed to the screen with messages related to this contact and the user of the application. While clicking on a contact at first, the user's certificate (and with it the private key) must be selected from the list of his certificates, which will be used to communicate with this contact. After this selection, all messages must be signed with a private key that corresponds to the public key of the certificate. There is a Firebase Console for remote cloud database viewing and the content that has changed in it. If the application is working properly after encrypting, signing the message and storing it in the local database, the message model should appear in the cloud database in a separate path for the remote device. After the message appears on the screen and in the cloud database console, which was just written by the local user, the following conclusions can be obtained: the table of message databases works correctly because messages were added; the send button and content field also work as expected; sending messages to the cloud database works correctly; the cloud database has a planned structure that corresponds to the structure from the design stage; cryptographic component for the CMS type enveloped-data and signed-data generation, i.e. encryption and signing of messages works as expected. In addition, from the cloud database console you can see the ASN1 message structure and its components. Testing of receiving messages from contacts follows the same scenario as sending testing from the current user. However, another device (sender) is added to the test. The accurate application operation involves receiving messages if the application is running and the device is online. Independently of the screen which the user uses (the screen of contacts and certificates or the screen of messages) the application must process the reception of messages correctly. In addition, the application may receive messages at any time from an unknown contact. The user must be notified through a mandatory dialogue with the confirmation of an unknown user. As the application performs the functions of the messenger and its main set of functions work correctly (receiving and sending messages, with other auxiliary stages, including: encryption, storage in the local database, interaction with the cloud database, etc.), which was proved during testing, it can be stated that the application corresponds to the theme of the project and the requirements described in the task.

A comparative information of expediency of cryptographic methods implementation of civil aviation communication channels is given in Fig. 2. As a result of the proposed method, a software implementation of the procedure for the encryption, decryption, formation and verification of a digital signature ih the aviation communication channels of Antonov aircrafts. The algorithms were tested in the Crypto++ 5.6.0 software environment on a dual-core Intel Core 2.83 GHz processor running Windows 10 64 bit x64. The main aim of performing the experiment is to compare algorithms on the same grounds and draw a conclusion about procedure of encryption, decryption, formation and verification of electronic digital signature. During the experiment the algorithms where constructed in such a way that the same parameters are almost everywhere equivalent in value, which provides clarity of comparison. All algorithms where compared by using some experiment specification such as: K - length of the key and time of encryption, decryption, formation and validation of electronic digital signature. After experiment we received results which show important difference between proposed method of the electronic digital signatures procedure using a group of points of the elliptic curves with providing the possibility of ensuring the integrity and confidentiality of information.

The following conclusions drawn from the obtained results: First of all, if we take into account the speed of the algorithm, namely the execution of operations, the use of the ECIES scheme reduces the quantitative value of the average key generation time by an average of 36%, although the encryption and EDS procedure is inferior in speed, but the decryption procedure is reduced to 1.09 times, and the validation procedure – up to 4.9 times. Secondly, with the same 128-bit security level, the cryptographic key is reduced from 2048 bits to 163 bits for the ECIES scheme, which requires 12 times less load on the computer system and increases the amount of memory to store the keys. Therefore, it is better to use the ECIES scheme, which is based on the use of a combined cryptosystem with elliptical cryptography algorithms and AES, to organize the civil aviation communication channels protection of Antonov aircraft.

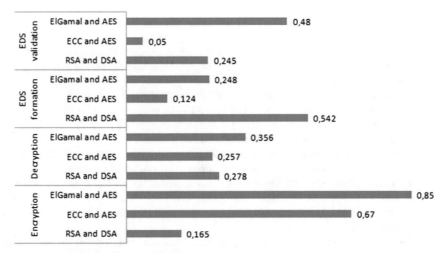

Fig. 2. Average time (ms) of cryptographic operations of implemented schemes in civil aviation communication channels

4 Conclusion

Thus, in this article authors provide a complete description of the proposed schemes and the proposed solution for the civil aviation communication channels protection in the critical information transmission. The proposed software module for civil aviation communication channels protection ensures information confidentiality and integrity by the use of cryptographic technologies including the use of combined cryptosystem with elliptical cryptography algorithms and AES-256. It can be said that this proposed approach is unique for modern civil aviation communication channels of Antonov aircraft and much more economical, because aviation communication channels have limited computing resources. Proposed test software module for Antonov aircrafts allows reduces the quantitative value of the average key generation time by an average of 36%, which requires 12 times less load on the computer system and increases the amount of memory to store the keys. This software module can be further integrated in the civil aviation communication channels of Ukraine and can provide authentication of Antonov aircrafts. Our future research will focus on creating and implementing other encryption algorithms based on elliptic curves using strong elliptic curves over finite fields of simple order.

References

1. Boeing Commercial Airline PKI Basic Assurance certificate policy (2013)
2. Airbus Public Key Infrastructure. Certificate Policy (2018)
3. Sampigethaya, K., Poovendran, R., Shetty, S., Davis, T., Royalty, C.: Future E-enabled aircraft communications and security: the next 20 years and beyond. Proc. IEEE **99**, 2040–2055 (2011)
4. Sampigethaya, K., Poovendran, R., Bushnell, L.: Secure operation, control, and maintenance of future e-enabled airplanes. Proc. IEEE **96**(12), 1992–2007 (2008)

5. De Cerchio, R., Riley, C.: Aircraft systems cyber security. In: IEEE/AIAA 30th Digital Avionics Systems Conference, pp. 1C3-1–1C3-7. IEEE (2011)
6. Mahmoud, M., Larrieu, N., Pirovano, A.: A performance-aware public key infrastructure for next generation connected aircrafts. In: 29th Digital Avionics Systems Conference, pp. C. 3-1-3–C. 3-16. IEEE (2010)
7. Robinson, R., et al.: Impact of public key enabled applications on the operation and maintenance of commercial airplanes. In: 7th AIAA ATIO Conference, 2nd CEIAT International Conference on Innovation and Integration in Aero Sciences, p. 7769 (2007)
8. Sampigethaya, K., Poovendran, R., Bushnell, L.: A framework for securing future eEnabled aircraft navigation and surveillance. In: AIAA Infotech Aerospace Conference, p. 1820 (2009)
9. Robinson, R., et al.: Electronic distribution of airplane software and the impact of information security on airplane safety. In: Saglietti, F., Oster, N. (eds.) SAFECOMP 2007. LNCS, vol. 4680, pp. 28–39. Springer, Heidelberg (2007). https://doi.org/10.1007/978-3-540-75101-4_3
10. Hulínská, Š., Kraus, J.: Fatigue risk management system in aviation. Risks Bus. Process. 174–180 (2016)
11. Mahmoud, M., Larrieu, N., Pirovano, A.: Aeronautical communication transition from analog to digital data. A network security survey. Comput. Sci. Rev. 11, 1–29 (2014)
12. Gaurav, D., Gaurav, C., Vikas, S., Ilsun, Y., Choo, K.K.R.: Cyber security challenges in aviation communication, navigation, and surveillance. Comput. Secur. 112, 102516 (2022)
13. Nobles, C.: Cyber threats in civil aviation. In: Emergency and Disaster Management: Concepts, Methodologies, Tools, and Applications, pp. 119–141. IGI Global (2019)
14. Kazmirchuk, S., Anna, I., Sergii, I.: Digital signature authentication scheme with message recovery based on the use of elliptic curves. In: Hu, Z., Petoukhov, S., Dychka, I., He, M. (eds.) ICCSEEA 2019. AISC, vol. 938, pp. 279–288. Springer, Cham (2020). https://doi.org/10.1007/978-3-030-16621-2_26
15. Kazmirchuk, S., Ilyenko, A., Ilyenko, S.: The improvement of digital signature algorithm based on elliptic curve cryptography. In: International Conference on Computer Science, Engineering and Education Applications, pp. 327–337 (2020)
16. Kazmirchuk, S., Ilyenko, A., Ilyenko, S.: Improved gentry's fully homomorphic encryption scheme: design, implementation and performance evaluation (2020)
17. Korchenko, O., Vasiliu, Y., Gnatyuk, S.: Modern quantum technologies of information security against cyber-terrorist attacks. Aviation 14(2), 58–69 (2010)
18. Vysotska, O., Davydenko, A.: Keystroke pattern authentication of computer systems users as one of the steps of multifactor authentication. In: Hu, Z., Petoukhov, S., Dychka, I., He, M. (eds.) ICCSEEA 2019. AISC, vol. 938, pp. 356–368. Springer, Cham (2020). https://doi.org/10.1007/978-3-030-16621-2_33
19. Hu, Z., Dychka, I., Onai, M., Zhykin, Y.: Blind payment protocol for payment channel networks. Int. J. Comput. Netw. Inf. Secur. 6(11), 22–28 (2019)
20. István, V.: Construction for searchable encryption with strong security guarantees. Int. J. Comput. Netw. Inf. Secur. 5(11), 1–10 (2019)
21. Goyal, R., Khurana, M.: Cryptographic security using various encryption and decryption method. Int. J. Math. Sci. Comput. 3(3), 1–11 (2018)
22. Ilyenko, A., Ilyenko, S., Kvasha, D.: The current state of the cybersecurity of civil aviation of Ukraine and the world. Electron. Prof. Sci. Edn. Cybersecur. Educ. Sci. Tech. 1(9), 24–36 (2020). (in Ukrainian)
23. Stolzer, A.J., Goglia, J.J.: Safety Management Systems in Aviation. Routledge, London (2016)
24. Moir, I., Seabridge, A., Jukes, M.: Civil Avionics Systems. Wiley, Hoboken (2013)
25. Moir, I., Seabridge, A.: Aircraft Systems: Mechanical, Electrical, and Avionics Subsystems Integration. Wiley, Hoboken (2011)
26. Zakharchenko, V., et al.: System efficiency of programmed operation of avionics (2018)

27. Ilyenko, S., et al.: Functional automated systems and complexes of aircraft: textbook. Manual (2019)
28. Zakharchenko, V., et al.: Methods and means of ensuring the reservation of avionics (2020)

Automated License Plate Recognition Process Enhancement with Convolutional Neural Network Based Detection System to Improve the Accuracy and Reliability of Vehicle Recognition

Yakovlev Anton[(✉)] and Lisovychenko Oleh

Igor Sikorsky Kyiv Polytechnic Institute, Peremohy prosp. 37, Kyiv 03056, Ukraine
liferunner@gmail.com

Abstract. This paper addresses the issue of automatic plate recognition in road safety systems. Practical observations show that a large number of road safety systems in production still use OCR-only approach for plate recognition tasks. Full scale experiment has been conducted by the authors with the Axis P1344 high-speed camera unit, installed on highway and an OCR-only based detection system, which confirmed instability of accuracy and detection time under volatile environmental conditions, such as lighting, camera-caused image issues, weather, mechanical issues, etc. A method and algorithm were proposed for custom data training and automated plate recognition with pre-trained convolutional neural network (CNN) based detection system Yolo v5 as an intermediate logical layer of detection. Methods for image classification, dataset formation and training automation were used with a custom dataset of 2000 environmentally variable images. The efficiency of automated plate recognition in various conditions was verified with developed software using a set of 20 000 input images with a plate detection precision value more than 98% with average detection time 0,6s. The results obtained have practical application as an opportunity to improve the plate recognition process accuracy with plate-only region detection by CNN detection system, which is especially important when detecting in volatile environmental conditions.

Keywords: Plate recognition · Machine learning · Pattern recognition · Road safety · YOLOv5

1 Introduction

Vehicle recognition subsystem is an automated information technology system that automatically supplies road traffic safety (RTS) management systems with vehicle-specific identity information from road events-specific captured data [1]. Processed information transformed from raw graphical representation obtained with capturing devices (high-speed camera units) to per-system understandable representation. The process reduces human resource cost, human error and directly influences road safety growth [2, 3].

© The Author(s), under exclusive license to Springer Nature Switzerland AG 2022
Z. Hu et al. (Eds.): ICCSEEA 2022, LNDECT 134, pp. 248–259, 2022.
https://doi.org/10.1007/978-3-031-04812-8_21

As per constant vehicle count growth on roads annually [4] road safety systems role becomes vital for further sustainable development. This statement is specifically relevant in the era of substantial development of computer systems and software engineering technology when any common solution can be revised with modern approaches.

From a functional point of view, a vehicle recognition subsystem is a detection system. Structurally it is a component of the system of RTS management. It consists from a camera unit, communication channels, and software with intelligence functions [5]. Camera unit is installed within controlled sections of the roads and aimed at the way it will capture an area with one or multiple vehicles. Depending on conditions the camera unit is supplied with additional equipment like speed sensor, additional illumination spotlight, etc. Once a system predefined road event occurs, the camera captures the controlled area. On further steps the image is being processed with the software in accordance with system architecture (centralized or distributed). As a result all vehicle license plates should be detected, recognized and processed [6]. The OCR approach used for license plate recognition, as any other detection approach, has its own risk of accuracy degradation. One of the reasons for accuracy degradation are camera-related issues like focusing, exposure, light, etc. Problems from the described area are easily solved by camera unit adjustment from one side and image post processing from the other side. However, due to limited OCR detecting capabilities under various environmental conditions a systems performance, in terms of accuracy and time cost, using OCR-only based approach could degrade. This leads to overall system reliability degradation as vehicle recognition is one of the key stages of road events processing in RTS management systems [7].

The influence of various factors on license plate detection accuracy considered in [8]. A probable way to achieve mentioned above factors influence reduction is the creation of automated recognition systems with neural networks and deep learning approaches used for license plate recognition [9–13].

Major research objectives lie in the field of convolutional neural network based detection system YOLOv5 usage as an intermediate stage of license plate recognition to achieve better precision value, time cost and computational cost compared to existing approaches. Numerous researches are focused on OCR and/or YOLOv3 usage for detection purposes, when YOLOv5 delivers better performance characteristics and requires additional research within in a license plate recognition with pattern recognition capabilities process. Custom dataset collection, classification, formation and training processes automation with modern software engineering approaches provides a way to higher detection precision as larger and variable train datasets could be formed. Task-specific input training dataset forming factor importance highlighted as a way to achieve better precision and timing characteristics.

2   Accuracy and Reliability Factors of Automated License Plate Recognition (LPR) Process Review Focused Literature Sources Review

Automated detection systems, including RTS, reliability and accuracy increasing issues are under active consideration in numerous research works, like [14–23].

Therefore, the paper [14] focuses on Iterative Threshold Segmentation (ITS) Algorithm for vehicle number plate recognition techniques. It is stated that iteration based threshold used to separate image foreground and background. Outdoor images with a vehicle pictured were used for testing. Obtained automatic threshold processing of the experimental images results showing the validity of the method. This is important study of LPR process enhancement but a set of images used for experiments and accent on its variety is sufficient to make this study valuable enough for dramatic precision growth.

The article [15] discusses the problem of Image Processing approach for Automatic Number Plate Recognition System. An efficient approach proposed by the authors for automatic number plate recognition, which states that input graphical data with a vehicle pictured is processed with a sequence of stages. Primarily input data is filtered with iterative bilateral approach. Secondly equalization of adaptive histogram performed. During third and further steps number plate extraction achieved by image subtraction, boundary box analysis and detection of Sobel vertical edge, morphological operation, image thresholding\binarization. As authors states a method being proposed shows notable results in conditions of digital noise, in cases of filmed area low illumination, blurry images, sufficient contrast as well as for under exposed or overexposed images. A methods proposed in the research are effective within LPR task solving within mentioned improvement factors context, but requires sufficient computational abilities as well as highly productive software engineering technologies involved.

The article [16] touches on a solution for the Automatic LPR problem. It is discussed that OCR-based approach is erroneous within specific problems like fonts variability, image segmentation by thresholding, etc. Each case from above is reviewed with a "problem-solution" approach. For instance, lack of letter character templates problem proposed to be solved through the additional templates addition usage. Constant development approach for OCR database is working solution but it is limited due to variability and unpredictability of letters distortion in real world, also in some cases, OCR approach is slow and cannot be suitable for real time applications, which LPR process is in some cases.

The article [17] is a critical study for LPR systems as well as new technological approaches evaluation for LPR systems. Most useful methods of detection were discussed and evaluation performed for MATLAB-specific methods with results shown. Case study conducted with general LPR process steps described. Most of described methods lies in field of image processing which is disadvantaged with slow performance, also no full-scale experiments with representative enough and variable enough input data quantity.

The authors of article [18] proposing RL-OLD model which implements reinforcement learning based unloading decisions. The model is used for high precision plate detection and recognition with computer resources optimal consumption declared. Effective edge computing utilization performed by a model for various kinds of license plates detection. Plate individual characteristics defines either detection-specific computation is performed within local system or cloud system, depending on complexity of detection process. High level of accuracy, marginal data loss and minimum delays has been achieved by proposed approach. Proposed approach defines flexible architecture for local and global deployments depending on tasks solved by the system. However, proposed architecture mass deployment could be cost ineffective in some cases comparing to similar solutions. Yolo v3 for license plate patterns region detection is outdated and later versions of CNN-based detection systems should be considered.

The article [19] explored a present-day approach for an LPR task solution. Multiple sieve-like steps proposed from image obtaining to plate letters output as ASCII characters. These steps are detection of a vehicle region, detection of a plate region, segmentation of plate characters and a recognition of characters within segments. Authors stating that a level of 100% accuracy was reached for a detection of license plate and 97.5% for stage of recognizing characters. In addition, it is stated that the method being proposed overcomes methods proposed last years in time and accuracy performance factors. Proposed methods implements latest technology in every field of LPR detection process stage. However, authors did not provide information on evaluation datasets used during experiments within the method proposed and lead to mentioned above precision levels.

Automated tracking system studied in article [20] for vehicle tracking. An experimental real-time computer vision method used for LPR task solving within the security level enforcement during vehicle transitions within closed institution. The researcher's result showed 96% rate for monitoring within optimal criteria of operation. Proposed system experimental approach showing high success rates with computer vision involved. This can be overcome with additional learning approach for computer vision training.

The article [21] considered a VMR approach, which stands for Vehicle Manufacturer Recognition. A CNN system usage considered for precise definition. The results obtained show that strong generalizations are peculiar to the algorithm is insufficient in classification accuracy. Result proves that the average accuracy of the class will be higher than with other methods usage, which proves convolutional neural networks are better at vehicle manufacturer recognition with a results of a comparative tests. This study proves good potential in VMR approach, which could be additional logical layer in LPR problem solving. This study requires more experiments for various input data to obtain the scope of its usage within problem being solved.

The article [22] studied methodology of a license plate embedding. This methodology is used for large volumes of accurate Mercosur license plate images generations that leads to supervised LPD training. Methodology being considered proves an accuracy level at 95% and average time cost at 40 ms. Attempt to higher precision by dataset size with variable image is good for precision growth, but retrain on failures could supply with comparatively higher precision values.

The article [23] describes the LPR problem solution in the video stream by optimal training parameters selection and training dataset transformation. As a result CNN detector efficiency growth is 30.5% comparing to initial value. This is achieved with a Yolo v3 usage and should be revised with recent technology promising higher precision and time cost effectiveness.

The standard [24] describes requirements for two systems: "VIDEOKONTROL-Rubezh" [25] and PDR «CASCADE» [26]. Every mentioned RTS management system involves license plate recognition. Concurrently the standard explicitly states that license plate recognition should use OCR approach with guaranteed accuracy level not lower than 90%.

The above review demonstrates that proposed solutions do not consider accuracy and reliability increasing approaches of the RTS management system with a custom real world data training from one side and Yolo v5 CNN detection system on another. The closest solution [25, 26], which is operated on, involves OCR only approach which is insufficient under various environmental factors. Accordingly, the development of the method, an algorithm and a software for an automated license plate recognition system, considering the environmental variability, is an existent technical and scientific problem.

The processes of automated license plate recognition under capturing environment variability are the object of this research.

Method, algorithm and software for the automated license plate recognition with capturing environment variability is the subject of this research.

The purpose of the research can be stated as improve the reliability and accuracy of automated LPR systems in capturing environment variability.

This is achieved by a custom dataset D_c obtained from the real world using Axis P1344 high-speed camera unit, aimed at a specific area of busy highway, custom dataset automated classification method $C_c(n)$ and adopted usage of Yolo v5 CNN-based detection system, pre trained with custom data, as an intermediate detection layer.

3 Automated License Plate Recognition Method and Algorithms

During the research, authors employed the following methods: mathematical modeling methods, a full-scale experiment methods, an analysis, synthesis and automatic control systems methods.

Method of task specific custom dataset D_c collection uses camera unit, software and storage hardware. With the captured frame i equation of the process is determined as following

$$D_c = \sum_{i=1}^{n} f(i)$$

Once the needed dataset is obtained it requires Yolo v5 detection system specific annotated dataset D_a for every single image i_c, and every single classification region per image r_c. This process involves images itself and processes classified region coordinated projection to annotation files. Yolov5 Region coordinates supplied with coordinates x_c_n, y_c_n, w_n, h_n, normalized by the dimensions of the image. Having annotated (or classified) dataset D_a, unclassified image from custom dataset i and classification region r equation of the process is determined as following

$$D_a = \sum_{i=1}^{n} f(i) \sum_{r=1}^{m} f(r)$$

As classified dataset D_a is collected and all required for training information is present a detection system learning process performed. As a result of learning process train data D_t is generated, equation of the process is determined as following

$$D_t = \sum_{i_c=1}^{n} f(i_c)$$

Train data is used to recognize license plate region(s) R_d on captured image i_d as following

$$R_d = f(i_d)$$

Once R_d is obtained OCR component is used to recognize license plate letters character data L_d as following

$$L_d = f(R_d)$$

4 Experiment

4.1 Full-scale Experiment with Yoloanno Application, yPlateReco Software and Axis P1344 IP Camera

The authors conducted an experiment with developed Yoloanno application, yPlateReco software and an Axis P1344 IP Camera installed on a 3 lane highway.

Figure 1 shows image capturing system during operation.

Figure 1(a) shows Axis P1344 IP Camera installation parameters like distances and angles to capturing object (vehicle).

Figure 1(b) shows a sample image (1280 × 800 pixels), captured with an installed camera unit. Array of images obtained with camera unit used for custom dataset collection, Yolo v5 detection system train on custom dataset and further license plate region recognition.

Full-scale modeling with developed software and a proposed method for automated license plate recognition were evaluated for efficiency and operability of algorithms.

Figure 2 demonstrates Yoloanno application [30] user interface used for custom dataset annotation (classification). Training dataset was created as a result of classification with the Yoloanno application. Detection system training dataset consists of image files and corresponding Yolo v5 annotation text files. A set of 2000 images was classified. Obtained a custom classified dataset that was used as an input for Yolo v5 training interface with next hyperparameters: image size - 640, batch size - 16, epoch - 300. Train process produces a custom data weight file that is further used for recognition.

Figure 3 presents a recognition result with obtained on previous stage weight file.

A set of environmentally variable 20000 images obtained by the capturing camera unit was used in experiment to determine detection accuracy value of plate region detection. The results of detection for all images and images with plate present only are shown in Table 1.

The above results indicate that, for plate recognition tasks the accuracy is higher as stated for OCR-only approaches. Accuracy growth shown is 5–8% higher depending on input data (Fig. 4).

A processing time cost factor Δt was considered in comparison to the "OCR-only" approach Δt_{OCR} for an all image area and the proposed "Plate region detection" approach Δt_{RD} combined with "Detected region OCR" approach Δt_{RO}. The results of 1500 images processing within all Δt factor approaches are shown in Table 2.

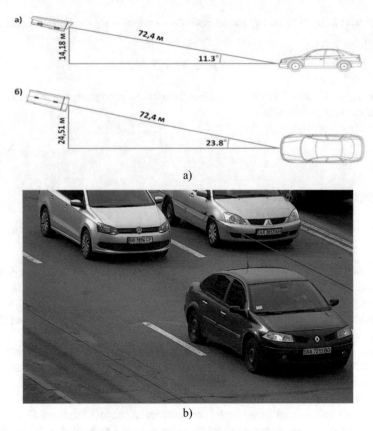

a)

b)

Fig. 1. Experiment with Axis P1344 camera unit

Fig. 2. Main application window of Yoloanno classification tool with Yolo v5 format support

a)

b)

```
0  0.223438  0.27  0.06875  0.025
0  0.842578  0.236875  0.0648438  0.02625
0  0.907422  0.97625  0.0804688  0.0325
```

c)

Fig. 3. Custom weight file usage recognition results: a,b - graphical representation, c - text representation (normalized xywh)

Table 1. Result distribution for testing dataset input on license plate region detection

Type of dataset	Correct detection (A) [%]	Semi-correct detection (B) [%]	False detection (C) [%]
All images	98,25	1,11	0,64
Plate present only	95,57	3,19	1,24

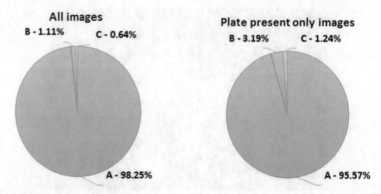

Fig. 4. Result distribution for testing dataset input on license plate region detection: **A** - correct detection, **B** - Semi-correct detection, **C** - false detection. Correct detection percentage growth is a positive trend.

Table 2. Time cost for testing dataset input recognition with various approaches

Approach	t_{min}[sec.]	t_{max}[sec.]	t_{avg}[sec.]
Δt_{OCR}	0,592	19,969	3,989
Δt_{RO}	0,333	2,125	0,517
Δt_{RD}	0,555	0,867	0,651
$\Delta t_{RD} + \Delta t_{RO}$	0,888	2.992	1,168

The above results indicate that plate region recognition with custom pre-trained CNN-based detection system Yolo v5 combined with reduced to plate only region OCR shows better performance results on time factor Δt than the OCR only approach on an all area of the processed image on the same set of images and a hardware.

5   Conclusion

A method, algorithm and software proposed for an automated LPR system allowed to remarkably improve the reliability and accuracy of the road safety system overall in variable environment conditions. The existing solutions incorporated to the RTM systems such as "VIDEOKONTROL-Rubezh", PDR «CASCADE» and others, matching a standard, provide an OCR-only approach when it comes to license plate recognition, which is not enough under various environmental conditions. The results of a full-scale experiment with the Axis P1344 camera unit, various capturing conditions like lighting, camera-caused image issues, weather, mechanical issues shows that recognition accuracy and reliability deteriorates significantly. Such behavior is unacceptable in RTM systems as critical information could be missing or the system can be deceived by unscrupulous drivers. Recognition deterioration occurs because the OCR mechanism is sensitive to various character features like geometry, saturation, font, letter-background contrast, etc., which are hard to keep matching requirements in the real world. The scientific novelty of results obtained resides in design features theoretical justification of the original automated license plate recognition subsystem in road safety systems, consisting in constant, an additional logical detection layer added with use of pre-trained with custom dataset CNN-based detection system to crop plate recognition area from all image to plate-only region with further OCR recognition of detected region. Custom train data usage for pattern recognition covers a large amount of probable environmental influences on how a license plate is captured with a camera unit. The obtained results notably improve time expenses and accuracy of the license plate recognition process. The practical effect of the obtained results lies in the development and production of the automated license plate recognition original system within RTM with addition of CNN-based detection system for recognition process improvement. Therefore, reliability and accuracy of the LPR that uses CNN in general under environmental variability can be reached. As an outcome, different kinds of uncertainty in input capturing data can be reduced, which proves the practical value of the results obtained in overall. The perspective of further research will lie in improving the variability of input training dataset for larger deviation cases coverage and detection time expenses reduction.

References

1. ISO. 2021. ISO 39001:2012 - Road traffic safety (RTS) management systems—Requirements with guidance for use (2021). https://www.iso.org/standard/44958.html
2. Ranking EU progress on road safety, 15th Road Safety Performance Index Report, June 2021 (2021). https://etsc.eu/wp-content/uploads/15-PIN-annual-report-FINAL.pdf
3. European regional status report on road safety 2019, ISBN 978 92 890 5498 0 (2019). https://apps.who.int/iris/bitstream/handle/10665/336584/9789289054980-eng.pdf
4. The European Automobile Manufacturers' Association: NEW PASSENGER CAR REGISTRATIONS, EUROPEAN UNION. 16 September 2021 (2021). https://www.acea.auto/files/20210916_PRPC_2107-08-FINAL.pdf
5. Shan, D., Ibrahim, M., Shehata, M., Badawy, W.: Automatic license plate recognition (ALPR): a state-of-the-art review. IEEE Trans. Circuits Syst. Video Technol. 23(2), 311–325 (2013). https://doi.org/10.1109/TCSVT.2012.2203741. S2CID 206661467
6. Massoud, M.A., Sabee, M., Gergais, M., Bakhit, R.: Automated new license plate recognition in Egypt. Alexandria Eng. J. 52, 319–326 (2013). https://doi.org/10.1016/j.aej.2013.02.005
7. Lin, C.-H., Lin, Y.-S., Liu, W.-C.: An efficient license plate recognition system using convolution neural networks. In: 2018 IEEE International Conference on Applied System Invention (ICASI). IEEE (2018). https://doi.org/10.1109/icasi.2018.8394573
8. Rhead, M., Gurney, R., Ramalingam, S., Cohen, N.: Accuracy of automatic number plate recognition (ANPR) and real world UK number plate problems. In: 2012 IEEE International Carnahan Conference on Security Technology (ICCST). IEEE (2012). https://doi.org/10.1109/ccst.2012.6393574
9. Koval, V., Turchenko, V., Kochan, V., Sachenko, A., Markowsky, G.: Smart license plate recognition system based on image processing using neural network. In: Second IEEE International Workshop on Intelligent Data Acquisition and Advanced Computing Systems: Technology and Applications, 2003 Proceedings (2003). https://doi.org/10.1109/idaacs.2003.1249531
10. Fahmy, M.M.M.: Automatic number-plate recognition: neural network approach. In: Proceedings of VNIS'94 - 1994 Vehicle Navigation and Information Systems Conference (1994). https://doi.org/10.1109/vnis.1994.396858

11. Shashirangana, J., et al.: License plate recognition using neural architecture search for edge devices. Int. J. Intell. Syst. (2021). https://doi.org/10.1002/int.22471
12. Shashirangana, J., Padmasiri, H., Meedeniya, D., Perera, C.: Automated license plate recognition: a survey on methods and techniques. IEEE Access **9**, 11203–11225 (2021). https://doi.org/10.1109/access.2020.3047929
13. Shrivastava, S., Singh, S.K., Shrivastava, K., Sharma, V.: CNN based automated vehicle registration number plate recognition system. In: 2020 2nd International Conference on Advances in Computing, Communication Control and Networking (ICACCCN) (2021). https://doi.org/10.1109/icacccn51052.2020.9362737
14. Akther, M., Ahmed, M., Hasan, M.: Detection of vehicle's number plate at nighttime using iterative threshold segmentation (ITS) algorithm. Int. J. Image Graph. Signal Process. **5**(12), 62–70 (2013). https://doi.org/10.5815/ijigsp.2013.12.09
15. Kaur, S.: An automatic number plate recognition system under image processing. Int. J. Intell. Syst. Appl. **8**(3), 14–25 (2016). https://doi.org/10.5815/ijisa.2016.03.02
16. Ramshankar, Y., Deivanathan, R.: Development of machine vision system for automatic inspection of vehicle identification number. Int. J. Eng. Manuf. **8**(2), 21–32 (2018). https://doi.org/10.5815/ijem.2018.02.03
17. Mie Aung, M.: Study for license plate detection. Int. J. Image Graph. Signal Process. **11**(12), 39–46 (2019). https://doi.org/10.5815/ijigsp.2019.12.05
18. Sakharkar, Y.A., Singh, M., Kumar, K.A., Aju, D.: A reinforcement learning based offload decision model (RL-OLD) for vehicle number plate detection. Int. J. Eng. Manuf. (IJEM) **11**(6), 11–18 (2021). https://doi.org/10.5815/ijem.2021.06.02
19. Pirgazi, J., Sorkhi, A.G., Kallehbasti, M.M.P.: An efficient robust method for accurate and real-time vehicle plate recognition. J. Real-Time Image Proc. **18**(5), 1759–1772 (2021). https://doi.org/10.1007/s11554-021-01118-7
20. Calitz, A., Hill, M.: Automated license plate recognition using existing university infrastructure and different camera angles. African J. Inf. **12**(2), Article 4 (2020). https://digitalcommons.kennesaw.edu/ajis/vol12/iss2/4
21. Ma, L., Zhang, Y.: Research on vehicle license plate recognition technology based on deep convolutional neural networks. Microprocessors Microsyst. **82**, 103932 (2021). https://doi.org/10.1016/j.micpro.2021.103932
22. Silvano, G., et al.: Synthetic image generation for training deep learning-based automated license plate recognition systems on the Brazilian Mercosur standard. Des. Autom. Embed. Syst. **25**(2), 113–133 (2020). https://doi.org/10.1007/s10617-020-09241-7
23. Schegolihin, Y., Mitrohin, M., Sazykina, V., Semenkin, M.: Gradual labeling of the training set to improve the efficiency of image detection by a neural network on the example of license plate recognition. In: 2021 29th Conference of Open Innovations Association (FRUCT) (2021). https://doi.org/10.23919/fruct52173.2021.9435602
24. DSTU (State standard of Ukraine) 8809:2018 Metrology. Traffic control devices with photo and video recording functions. Remote vehicle speed meters, remote space-time parameters of vehicle location meters. Metrological and technical requirements (2018). (Ukr.)
25. VIDEOKONTROL-Rubezh, complex. Ollie.com.ua (2020). http://www.ollie.com.ua/videocontrol/index.html. (Ukr.)
26. USIT, Produkciya - USI. Ukrsi.com.ua (2020). http://ukrsi.com.ua/products/. (Ukr.)
27. Yampolskyi, L.S., Lisovichenko, O.I., Oliynyk, V.V.: Neurotechnologies and Neurocomputer Systems. Dorado-Druk, Kyiv (2016). (Ukr.)
28. Joshi, P., Escrivá, D., Godoy, V.: OpenCV by Example. Packt Publishing, Birmingham (2016)
29. Ultralytics/yolov5. GitHub (2020). https://github.com/ultralytics/yolov5
30. Yakovlev, A.: AntonYakovlev/Yoloanno. GitHub (2020). https://github.com/AntonYakovlev/Yoloanno

31. Yakovlev, A., Lisovychenko, O.: An approach for image annotation automatization for artificial intelligence models learning. Adapt. Syst. Autom. Control **1**, 32–40 (2020). https://doi.org/10.20535/1560-8956.36.2020.209755

32. Omar, N., Sengur, A., Al-Ali, S.G.S.: Cascaded deep learning-based efficient approach for license plate detection and recognition. Expert Syst. Appl. **149**, 113280 (2020). https://doi.org/10.1016/j.eswa.2020.113280

33. Kumar Sahoo, A.: Automatic recognition of Indian vehicles license plates using machine learning approaches. In: Materials Today: Proceedings (2020). https://doi.org/10.1016/j.matpr.2020.09.046

34. Jamtsho, Y., Riyamongkol, P., Waranusast, R.: Real-time license plate detection for non-helmeted motorcyclist using YOLO. ICT Exp. **7**(1), 104–109 (2020). https://doi.org/10.1016/j.icte.2020.07.008

35. Qian, Y., et al.: Spot evasion attacks: adversarial examples for license plate recognition systems with convolutional neural networks. Comput. Secur. **95**, 101826 (2020). https://doi.org/10.1016/j.cose.2020.101826

36. Silva, S.M., Jung, C.R.: Real-time license plate detection and recognition using deep convolutional neural networks. J. Visual Commun. Image Represent. **71**, 102773 (2020). https://doi.org/10.1016/j.jvcir.2020.102773

37. Slimani, I., Zaarane, A., Al Okaishi, W., Atouf, I., Hamdoun, A.: An automated license plate detection and recognition system based on wavelet decomposition and CNN. Array, **8**, 100040 (2020). https://doi.org/10.1016/j.array.2020.100040

38. Xiang, H., Zhao, Y., Yuan, Y., Zhang, G., Hu, X.: Lightweight fully convolutional network for license plate detection. Optik **178**, 1185–1194 (2019). https://doi.org/10.1016/j.ijleo.2018.10.098

39. Björklund, T., Fiandrotti, A., Annarumma, M., Francini, G., Magli, E.: Robust license plate recognition using neural networks trained on synthetic images. Pattern Recognit. **93**, 134–146 (2019). https://doi.org/10.1016/j.patcog.2019.04.007

40. Kessentini, Y., Besbes, M., Ammar, S., Chabbouh, A.: A two-stage deep neural network for multi-norm license plate detection and recognition. Expert Syst. Appl. **136**, 159–170 (2019). https://doi.org/10.1016/j.eswa.2019.06.036

41. Asif, M.R., Qi, C., Wang, T., Fareed, M.S., Raza, S.A.: License plate detection for multi-national vehicles: an illumination invariant approach in multi-lane environment. Comput. Electr. Eng. **78**, 132–147 (2019). https://doi.org/10.1016/j.compeleceng.2019.07.012

42. Rao, W., Wu, Y. J., Xia, J., Ou, J., Kluger, R.: Origin-destination pattern estimation based on trajectory reconstruction using automatic license plate recognition data. Transp. Res. Part C: Emerg. Technol. **95**, 29–46 (2018). https://doi.org/10.1016/j.trc.2018.07.002

43. Li, H., Wang, P., You, M., Shen, C.: Reading car license plates using deep neural networks. Image Vision Comput. **72**, 14–23 (2018). https://doi.org/10.1016/j.imavis.2018.02.002

44. Wang, R., Sang, N., Wang, R., Jiang, L.: Detection and tracking strategy for license plate detection in video. Optik **125**(10), 2283–2288 (2014). https://doi.org/10.1016/j.ijleo.2013.10.126

45. Öztürk, F., Özen, F.: A new license plate recognition system based on probabilistic neural networks. Procedia Technol. **1**, 124–128 (2012). https://doi.org/10.1016/j.protcy.2012.02.024

46. Yuren, D., Shi, W., Liu, C.: Research on an efficient method of license plate location. Phys. Procedia **24**, 1990–1995 (2012). https://doi.org/10.1016/j.phpro.2012.02.292

47. Kocer, H.E., Cevik, K.K.: Artificial neural networks based vehicle license plate recognition. In: Procedia Computer Science, vol. 3, pp. 1033–1037. Elsevier (2011). https://doi.org/10.1016/j.procs.2010.12.169

Modern Requirements Documentation Techniques and the Influence of the Project Context: Ukrainian IT Experience

Denys Gobov[1(✉)] and Inna Huchenko[2]

[1] National Technical University of Ukraine "Igor Sikorsky Kyiv Polytechnic Institute", Kyiv, Ukraine
d.gobov@kpi.ua
[2] National Aviation University, Kyiv, Ukraine
inna.huchenko@npp.nau.edu.ua

Abstract. The success of development and testing activities depends on the quality of the requirement engineering and business analysis deliverables. The most important part of them is requirement and design specification. Many specification and modeling techniques may be used alternatively or in conjunction with others to accomplish a particular business analysis document. Business analysts should select the most appropriate techniques based on the project context, their previous experience, considering budget and time restrictions. The paper aims to analyze the current practices of requirement specification in modeling and define dependencies between project attributes and techniques. We conducted a survey study involving 328 specialists from Ukrainian IT companies and a series of interviews with experts to interpret the survey results. The results can be used to enhance educational programs for IT specialists, taking into account the needs of Ukrainian companies. A set of statistically significant dependencies between project context and requirements documentation techniques allows developing a business analysis approach recommendation system.

Keywords: Software requirements engineering · Business analysis · Specification techniques · Project context

1 Introduction

Requirements Analysis and Design Definition knowledge area describe the tasks that business analysts perform to structure and organize requirements discovered during elicitation activities, specify and model requirements and designs, validate and verify information, identify solution options that meet business needs and estimate the potential value that could be realized for each solution option [1]. Tasks from this knowledge area are part of the core business analysis cycle [2]. A business analysis practitioner has to define the deliverables that will be produced and the business analysis techniques that may be utilized. The purpose of deliverables is to provide stakeholders with the appropriate level of detail about the project to understand the information it contains and use it as an input for their activities (e.g., software architecture design, implementation, testing, etc.). The final step of specification and modeling is a requirement architecture creation. A requirements architecture fits the individual models and specifications together to ensure that all of the requirements form a single whole that supports the overall business objectives and produces a useful outcome for stakeholders [3].

Industrial guidelines and empirical studies define multiple specification and modeling techniques that have proven themselves in practice. A thorough understanding of the variety of techniques available, the project stakeholders, and their preferences regarding the presentation and format of business analysis documents, regulatory and contractual constraints assists the business analyst in adapting to a particular project context [4]. A set of predefined requirement specifications and modeling techniques can be created based on a selected collection of standard viewpoints across an industry or organization. But usually, this template does not strictly prescribe particular techniques. It means that selecting appropriate techniques remains the responsibility of the business analyst. Its result profoundly impacts the project plan, required costs, and resources. This study was conducted to analyze the current practices of business analysts from Ukrainian IT companies and IT departments in non-IT companies using the specification and modeling techniques. The main research objectives are: to define current trends in requirements specification and modeling techniques in Ukrainian companies and to check statistically significant dependencies between project context and requirements specification/modeling techniques. The paper is structured as follows. Section 2 contains a review of the related requirements analysis activities and surveys regarding requirement engineering and business analysis, guidance on their use. Section 3 is devoted to the structure of the questionnaire and survey results. Section 4 concludes the paper with a discussion of the study findings and future work.

2 Related Literature Review

Many studies were conducted to analyze the practical and theoretical aspects of business analysis technique selection and provide some guidance on their use. Dieste and Juristo proposed a framework for selecting elicitation techniques based on contextual attributes of the elicitation process. They established the adequacy values of each technique for each attribute value [5]. Two groups of students were involved in the experiment, and practitioners did not participate. Soares and Cioquetta [6] studied eight user requirement documenting techniques. They assessed there based on the proposed characteristics (Human readable, Independent towards methodology, Identify and represent types of requirements, etc.). Studies have been conducted to analyze the local specifics of requirement analysis and modeling techniques in software development projects. Jarzębowicz and Połocka [7] conducted a survey study involving 42 Polish IT industry professionals, asking them to select techniques applicable to different projects. Jarzębowicz and Sitko also investigated the practice of using requirements documentation techniques in an Agile environment in the context of the Polish IT industry [8]. The number of respondents who passed verification was 69. Ali, Rafiq, and Majeed investigated requirement engineering practices in small and medium companies of Pakistan [9] and confirmed that Use Case, Data Flow Diagram, and Class Diagram were widely used for requirement elicitation and analysis. Fatima and Mahmood conducted a survey study among professionals working in the IT industry to compare the requirement engineering in traditional and agile companies [10]. Large-scale surveys in requirement engineering areas were conducted by the NaPiRE initiative, including but not limited to specification techniques usage [11]. A set of research is devoted to using modern agile-oriented requirements specification techniques such as User Stories and Story Mapping [12, 13]. The main limitations of all surveys mentioned above are the low number of participants that do not allow statistically correct results assessment and students involvement instead of experienced practitioners. Some researchers built requirement elicitation techniques selection guidance using machine learning algorithms [14, 15]. These models have strict dependencies on gathered data used for model training.

In order to investigate current practices in requirement specification and modeling area, we created a long list of techniques based on the business analysis and requirement engineering bodies of knowledge published by the most recognizable international institute and associations, namely: [1, 3] by the International Institute of Business Analysis (IIBA), [16] by the Project Management Institute (PMI), [17] by the International Requirement Engineering Board (IREB), and [18] by British Computer Society (BCS). The analysis of these sources gives us a set of seventeen techniques: Acceptance and Evaluation Criteria, Activity Diagrams, Business Model Canvas, Business Process Models, Class/Entity-Relationship Diagrams, Data Dictionary, Data Flow Diagrams (DFD), Goal Models, Prototypes (High-fidelity and Low-fidelity), Natural Language / Informal (plain) Text, Roles and Permissions Matrix, Sequence Diagrams, State Machines, Functional Decomposition, Use Cases, Use Case Diagrams, User Stories, User Journey Map/User Flow Diagrams.

3   Survey Study

The survey's target group consists of Ukrainian IT professionals, mainly business analysts and other roles involved in business analysis or requirements engineering activities. The overall number of survey participants is 328. Details are described in [19], focusing on elicitation techniques. This article briefly describes the questionnaire design and concentrates on its Analysis and Design section.

The questionnaire basis was taken from the NaPIRE initiative [11] and reworked considering such sources as [1, 3, 16–18].

3.1   Project Context Survey Section

The section "General information" was intended to gather the following information about survey respondent and their project:

- Project size.
- The primary industrial sector of the current project. The set of industrial sectors was taken from [11] and reworked to domain areas within which services are offered by most of the Ukrainian IT Companies.
- Company type: IT or non-IT. The separation was made among Outstaff, Outsource, and Product for IT companies.

- Company size.
- Class of systems or services such as business, embedded, scientific software, etc.
- Team distribution (co-located or dispersed).
- Role in the Project.
- Experiencein a business analyst (BA) or a requirements engineer (RE) role.
- Certifications.
- Way of working in the project (adaptive versus predictive)
- Project category for most of the participant's projects (e.g., greenfield engineering).
- BA/RE activities in which the respondent is usually involved.

3.2 Requirements Analysis and Design Section

Within the given questions category, we were interested in requirements documentation usage, in particular, section/subsections in business analysis deliverables, requirements specification and modeling techniques, specification templates, and classes of non-functional requirements. All questions except question about templates were multiple-choice with a predefined set of answers. Question regarding templates was formulated in the following way: "Do you use templates for requirements specification?" and survey participants could select the following options:

- Company templates based on the best practices
- Standards
- We do not use templates
- Own, customized for the project
- Other

3.3 Data Transformation

The results of the Q1 section analysis (project factors and participants' background) were described in [19]. The data transformations were conducted for further log-linear analysis on multiple associations identification and Chi-square analysis [20] for pairs of categorical variables.

Firstly, records with zero experience in the BA role and a few with the "I don't know" answer about team distribution were removed. Those were the outliers. The number of records for further analysis was decreased to 324.

Secondly, some answer options were merged to have the expected count for variables combinations more than 5, namely:

- Experience value "Up to 1 year" was merged with "1–3 years" and resulted in "Up to 3 years".
- Ways of working were split into three groups instead of 5: agile, hybrid, and plan-driven.Company sizes were merged to get three groups instead of 5: up to 200 members, 201- 1500, and over 1500 members.
- Outsource and Outstaff were merged into one answer option for Company type questions.

3.4 Survey Results

The following results were received based on respondents' answers:

- 18,9% of respondents use requirements documentation as part of the contract. In 53,25% of cases, documentation is considered during customer acceptance.
- Business analysis documents more often include functional (69%), business (63%) and non-functional (60%) requirements, glossary (62%), and other sections. The less used sections are cost-benefit analysis – 7,74%, deployment specifics – 13%, success metrics – 14,86%. The problem statement is considered by 31,27% of respondents, stakeholder analysis – only by 22%, and technical interfaces – by 24,77% (Fig. 1).

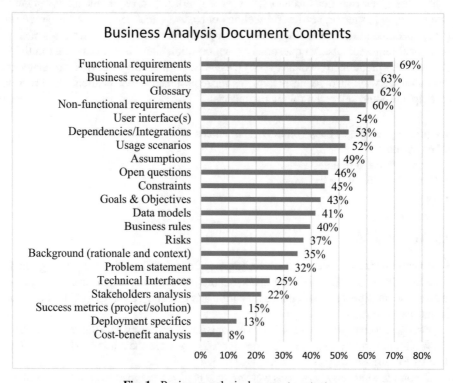

Fig. 1. Business analysis document contents

- The following specification and modeling techniques are rarely used: Goal Models – 1,24%, Business Model Canvas – 12,38%, Data Dictionary – 23,22%. The most popular techniques are User Stories – 79,26%, Use Cases – 65,63%, and Activity Diagrams – 63,16% (Fig. 2). The list of most popular techniques coincides with results in [7, 8].

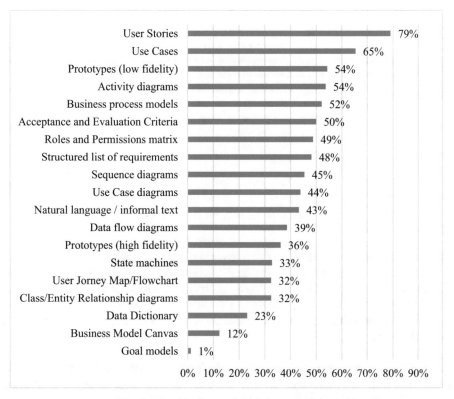

Fig. 2. Specification and modeling techniques

- Regarding the types of NFRs specified in the document, the following results were obtained: 9,29% of respondents don't specify NFRs at all; portability is considered by 14,55%; safety – by 17,03%. The most popular NFRs are usability – mentioned by 64,07%, security – 60,06%, performance efficiency – 57,28% (Fig. 3).
- As for template usage, 50,15% of respondents use company templates, 4,64% - standards, 8,36% - own templates, and 36,84% answered that they do not use templates at all.

We used a hierarchical log-linear model to determine the primary associations between the "factor-specification/modeling technique" pairs of variables in this study. Other interaction effects (for example, relationships between three variables, etc.) are outside the scope of this research. The SPSS [21] program was used to conduct the log-linear analysis. Hierarchical model selection was executed for each specification/modeling technique, along with additional background considerations from Q1. Two-way connections with a significance level greater than 0.05 (p < 0.05) were examined in partial associations tables. After that, the simple Chi-Square test for association was done for each two-way association found while hierarchical model selection.

The Chi-Square test's basic assumptions were examined and confirmed, including:

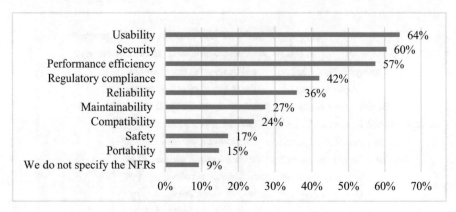

Fig. 3. Non-functional requirements

1. Both variables are categorical, and in our case, they are nominal.
2. Observations in groups/levels inside the variable are independent. As mentioned above, data were transformed to classify observations into mutually exclusive classes.
3. All cells should have expected counts greater than 5. Crosstabs procedure in the SPSS produces the expected count value for each variable, so it is easy to check this assumption.

The result of the P-value calculation, which should be less than 0.05 considering 0,95 confidence level, is presented in Table 1. These results allow us to make the conclusion about statistical significance dependencies for found pairs.

Table 1. Chi-square test results for pairs "Factor – Specification/modeling technique"

Factor – Specification/modeling technique	p-value	Details
Company Type –User Stories	0	The technique is used widely within IT Outsource/Outstaff companies, less actively – in IT Product companies, and rarely in non-IT companies. Moreover, the gap between "Yes"/"No" answers is minimal for in-house development
Project Category – User Stories	0.025	User Stories are more prevalent in greenfield engineering and reengineering projects than user interface engineering or product/platform customization

(continued)

Table 1. (*continued*)

Factor – Specification/modeling technique	p-value	Details
Ways of Working –User Stories	0.002	Almost 9% of respondents working in agile projects don't use User Stories. The technique is unlikely to be used in plan-driven methodologies, with a high probability of being hybrid and actively used in agile projects
Company Type –User Journey Map/Flowcharts	0.023	The technique is used mainly within IT Outsource/Outstaff companies, less actively – in IT Product companies, and rarely in non-IT companies. However, even in IT Outsource/Outstaff companies, only 65% of survey participants use the mentioned techniques
Project Category –User Journey Map/Flowcharts	0.008	The technique is used mostly in greenfield engineering and user interface engineering projects. Only 20–26% of survey participants, who work in reengineering or product/platform customization, use this technique
Project Category –Use Case Diagrams	0.023	This diagram is actively used in greenfield engineering projects. For other project categories, it is rather not used
Experience – State Machines	0	State machine diagrams are used almost 50/50 after five years of experience in business analysis. Respondents with experience up to 3 years use this technique in 24,3% of cases, and those with 3–5 years of experience – in 19, 2% of cases
Project Category –Sequence Diagrams	0.008	Sequence diagrams are popular mostly in greenfield engineering, less in reengineering and product/platform customization, and rarely in user interface engineering

(*continued*)

Table 1. (*continued*)

Factor – Specification/modeling technique	p-value	Details
Project Category –Roles and Permissions Matrix	0	This technique is more likely to be used in greenfield engineering projects
Experience –Roles and Permissions Matrix	0.014	The probability of using the mentioned technique is higher if the business analyst is more experienced
Company Size –High-fidelity Prototypes	0.004	The technique is used 50/50 in companies up to 200 specialists and around 30% of cases within larger companies
Company Type –DFD	0.047	These diagrams are used less in non-IT companies. Within IT companies, only around 40% of respondents use the given technique
Experience – DFD	0	Participants with more experience in business analysis tend to use these diagrams more often
Team Distribution – Activity Diagram	0	The technique is much more popular within distributed teams

The influence of industry, team size, class of systems, and project category factors on the specification techniques selection was not confirmed.

4   Conclusion

Data about software development project context, requirement specification and modeling activities were gathered via survey. The survey structure was built based on the worldwide known industry standards. Three hundred twenty-eight specialists from Ukrainian companies (mainly business analysts and product owners) took part in the survey. Most used requirements specification and modeling techniques were defined: User Stories, Use Cases, Activity Diagrams, Low-fidelity Prototypes, Business Process Models, Acceptance and Evaluation Criteria, Roles and Permissions Matrix, and Sequence Diagrams. In most cases, requirement specifications contained the following non-functional requirements: usability (64%), security (60%), and performance efficiency (57%). 50% of the participants answered that they used templates developed and approved by the company. It may indicate that companies are trying to standardize and increase the maturity of their business analysis processes. About 37% of respondents do not use templates. Business analysis documents more often include functional (69%), business (63%) and non-functional (60%) requirements, glossary (62%), the least common sections are success metrics (15%), deployment specific (13%), and cost-benefit analysis (8%).

The hypothesis "project factors influence requirements specification/modeling technique" was formulated and checked via statistical methods. After survey data cleansing, the log-liner hierarchical model selection algorithm was used for defining the two-way associations between project factors/participants' background and usage of the specification/modeling techniques. A set of statistically significant dependencies was found for the following techniques: Acceptance and Evaluation Criteria, Activity Diagrams, Data Flow Diagrams, High-fidelity Prototypes, Roles and Permissions Matrix, Sequence Diagrams, State Machines, User Stories, User Journey Map/Flowcharts, Use Case Diagrams. The following project factors have been identified as influencing the technique selection: Company Type, Company Size, Project Category, Ways of Working, Practitioner Experience, Team Distribution.

The study results analysis described in this paper is limited by two-way associations only, which were additionally checked with the Chi-Square test. The list of techniques used in the survey is not exhaustive. Nevertheless, it includes the most popular specification and modeling techniques used in practice. Several directions for future research can be considered, such as studying current business analysis and requirement engineering practices in other countries, finding dependencies and formulating recommendations for techniques selection in regards to the whole set of business analysis activities.

References

1. International Institute of Business Analysis. A guide to the business analysis body of knowledge (BABOK Guide), ver. 3. IIBA, Toronto (2015)
2. Gobov, D., et al.: Approaches for the concept "business analysis" definition in IT projects and frameworks. In: CEUR Workshop Proceedings, vol. 2711, pp. 321–332 (2020)
3. International Institute of Business Analysis. A Core Standard A Companion to A Guide to the Business Analysis Body of Knowledge (BABOK® Guide) ver. 3. IIBA, Toronto (2017)
4. Rehman, T., Khan, M., Riaz, N.: Analysis of requirement engineering processes, tools/techniques and methodologies. Int. J. Inform. Technol. Comput. Sci. (IJITCS) 5(3), 40–48 (2013). https://doi.org/10.5815/ijitcs.2013.03.05
5. Dieste, O., Juristo, N.: Systematic review and aggregation of empirical studies on elicitation techniques. IEEE Trans. Softw. Eng. 37(2), 283–304 (2011). https://doi.org/10.1109/tse.201 0.33
6. dos Santos Soares, M., Cioquetta, D.S.: Analysis of techniques for documenting user requirements. In: Murgante, B., et al. (eds.) Computational Science and Its Applications – ICCSA 2012. LNCS, vol. 7336, pp. 16–28. Springer, Heidelberg (2012). https://doi.org/10.1007/978-3-642-31128-4_2
7. Jarzębowicz, A., Połocka, K.: Selecting requirements documentation techniques for software projects: a survey study. In: Proceedings of Federated Conference on Computer Science and Information Systems (FedCSIS), pp. 1189–1198. IEEE (2017). https://doi.org/10.15439/201 7F387
8. Jarzębowicz, A., Sitko, N.: Communication and documentation practices in agile requirements engineering: a survey in polish software industry. In: Wrycza, S., Maślankowski, J. (eds.) Information Systems: Research, Development, Applications, Education. Lecture Notes in Business Information Processing, vol. 359, pp. 147–158. Springer, Cham (2019). https://doi.org/10.1007/978-3-030-29608-7_12
9. Ali, W., Rafiq, A., Majeed, M.: Requirements engineering in software houses of Pakistan. Int. J. Mod. Educ. Comput. Sci. 6(9), 47–53 (2014). https://doi.org/10.5815/ijmecs.2014.09.07
10. Fatima, T., Mahmood, W.: Requirement engineering in agile. Int. J. Educ. Manag. Eng. 9(4), 20 (2019)

11. Fernandez, D., Wagner, S.: Naming the pain in requirements engineering: a design for a global family of surveys and first results from Germany. Inf. Softw. Technol. **57**, 616–643 (2015). https://doi.org/10.1016/j.infsof.2014.05.008

12. Perkusich, M., et al.: Intelligent software engineering in the context of agile software development: a systematic literature review. Inf. Softw. Technol. **119**, 106241 (2020). https://doi.org/10.1016/j.infsof.2019.106241

13. Ochodek, M., Kopczyńska, S.: Perceived importance of agile requirements engineering practices–a survey. J. Syst. Softw. **143**, 29–43 (2018). https://doi.org/10.1016/j.jss.2018.05.012

14. Darwish, N., Mohamed, A., Abdelghany, A.: A hybrid machine learning model for selecting suitable requirements elicitation techniques. Int.J. Comput. Sci. Inf. Secur. **14**(6), 1–12 (2016)

15. Gobov, D., Huchenko, I.: Influence of the software development project context on the requirements elicitation techniques selection. In: Hu, Z., Petoukhov, S., Dychka, I., He, M. (eds.) Advances in Computer Science for Engineering and Education IV. LNDECT, vol. 83, pp. 208–218. Springer, Cham (2021). https://doi.org/10.1007/978-3-030-80472-5_18

16. Project Management Institute. The PMI Guide to BUSINESS ANALYSIS. PMI, Newtown Square, Pennsylvania (2017)

17. Pohl, K.: Requirements Engineering: Fundamentals, Principles, and Techniques. Springer, Cham (2010)

18. Paul, D., et al.: Business analysis, 3rd edn. BCS, The Chartered Institute for IT (2014)

19. Gobov, D., Huchenko, I.: Requirement elicitation techniques for software projects in Ukrainian IT: an exploratory study. In: Proceedings of the Federated Conference on Computer Science and Information Systems, pp. 673–681. IEEE (2020).https://doi.org/10.15439/2020f16

20. Yates, D., et al.: The Practice of Statistics, 1st edn. W.H. Freeman, New York (1999)

21. Morgan, G., et al.: IBM SPSS for Introductory Statistics: Use and Interpretation. Routledge, London (2019)

The Problems and Advantages of Using Non-separable Block Codes

Yaroslav Klyatchenko$^{(\boxtimes)}$, Oxana Tarasenko-Klyatchenko, Georgiy Tarasenko, and Oleksandr Teslenko

National Technical University of Ukraine "Igor Sikorsky Kyiv Polytechnic Institute", 37, Peremohy, Kyiv 03056, Ukraine
k_yaroslav@ukr.net

Abstract. The growth of digital telecommunication technologies that utilize more modern methods the efficiency and reliability of data transfer is improved. These include the methods of data encoding, based on the artificial introduction of redundancy, that allows the recipient's part not only to detect distortion of transmitted data but also form the correct values. The proposed technique is based on using non-separable codes. Using FPGA allows creating universal coders and decoders for 8-bit non-separable codes. These devices could be configured for any of these codes with any decimal digits (non-separable codes do not have a section for information and verification). This allows increasing the transmission speed of binary-coded decimal words by 25%.

Keywords: Error-correcting code · Boolean functions · Sphere packing · Hamming code · Non-separable block codes

1 Introduction

Nowadays, there are many examples of the use of code families that code data in blocks and have a wide range of practical applications - error correction, for example. Also, there are many parallels between error correcting codes and certain types of lattice sphere packings [1]. It is known [2], that in the case of block n-digit codes base 2 to correct the t distortions it is required and sufficient that $d \geq 2t + 1$, where the where d - is the smallest Hamming distance between any codewords. The theoretical speed of the error correcting code is determined as $R = log_2 N_k/n$, where N_k - is a number of codewords in a code.

A theory and practice of building block codes, where the clear separation between informational and additional bits does exist, and additional bits are created based on the informational ones, was developed the most.

Examples of such codes are Hamming code [3] and Bose-Chaudhuri-Hocquenghem codes [4]. An advantage to such codes is the analytical determination of allowed combinations and comparatively simple realisation of encoders and decoders. At the same time, the speed of separable codes could be maximal not entirely with all values of n. For example, for the Hamming code with $d = 3$ the maximally possible speed is attained for the values of n from series 7, 15, 31… It can be assumed, that for other values of n, the speed could be raised using the non-separable codes. In the non-separable block codes, as compared to the separable ones, in the code' codewords the division of digits of code to informational and additional is absent.

It is proven theoretically in the paper [5] that for 8-digit non-separable codes the maximum possible value of codewords is 20. At the same time, the 8-digit Hamming code could only have 16 codewords. Respectively, the value of the Hamming code speed is $R = 4/8 = 0.5$. In the case with the non-separable codes $R = log_2 20/8 = 0.54024$.

2 The Statement of the Research Problem

Better indicators of the theoretical speed of non-separable codes prompt to conduct research in the following directions:

- determine the contents of codewords in non-separable codes;
- create encoders and decoders for non-separable codes.

It has to be pointed out, that the general theory of bulldog with retaining the advantages on the speed of both non-separable codes and their encoders and decoders is absent at present. Separate classes of non-seperable codes do exist, for example, The Hadamard codes, but such codes do not enable the advantages over separable codes in terms of speed.

Taking into account the peculiarities of the implementation of specialized devices using integrated technology, the most advantageous is the use of the technology of field-programmable gate array (FPGA) [3, 6–9].

3 Codewords' Contents Determination in Non-separable Codes

To solve the problem of finding the maximal code of non-distributing block correction codes we should formulate the problem within the context of graph theory.

The elements from $N = 2n$ multitude of all binary numbers could be represented as a Hamming graph node, where two nodes are adjacent if the Hamming distance between respective codes is no less than d.

It is apparent that the maximal code is a maximum clique of Hamming graph. To search for the maximum cliques the Bron-Kerbosch algorithm is used [10]. But, according to [11], the task of finding the maximum clique of graph in the general case is NP-complete. Thus several algorithms are proposed for separate cases [6, 12, 13] that showcase decent results.

Further improvement of the Bron-Kerbosch algorithm is as follows. The notion of equivalency of non-separable error correcting code is used. It is apparent, that for every non-separable code an equivalent code does exist with a null value in all the codeword digits. The notion of the distance between two codewords is used as the minimal value of the distance between any codeword of two codes. If two codes with n-bit are found, the distance between which is not less than $d-1$, then it is easy to create the $n + 1$-bit code on their basis, with the distance between the codewords no less than d.

Since the Hamming codes are optimal provided $n = 7$ (densely packed [4]), hence this enables for significant simplification of the 8-bit codes search for 20 codewords using the respective modification of the Bron-Kerbosch algorithm. The modification goes as follows. Two non-equivalent 7-bit Hamming code are selected. Their turn-by-turn reduction is conducted (removal of some codewords or the other) until the distance between the shortened codes becomes no less than 2. Since the maximal number of codewords in an 8-bit non-separable code is known, then the result is determined with an aggregate number of 20 codewords in the shortened 7-bit codes.

It has to be pointed out, that according to [5] the maximal number of codewords in 9-bit codes cannot exceed 40 and such a code could be easily obtained in accordance with the described modified algorithm.

4   Implementation of 8-bit Non-separable Block Codes

The maximum number of 20 codewords of 8-bit non-separable code determines the following opportunities for application - using numbers in a positional numeral system base 20 or binary-coded decimal numbers.

Let's examine the creation of encoders and decoders for binary-coded decimal numbers in the packed format.

Let the binary-coded decimal words be presented in a packed format, where each half code has two decimal digits and the binary representation of a decimal number is separated into 5-bit corteges (Fig. 1).

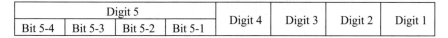

Digit 5				Digit 4	Digit 3	Digit 2	Digit 1
Bit 5-4	Bit 5-3	Bit 5-2	Bit 5-1				

Fig. 1. Cortege of decimal numbers

To create non-separable code sequences, we will use a pair of corteges that look like "bit 5 – i", where the digit i = 1, 2, 3, 4. Each of these pairs has 20 different values and can be converted into one code word of non-separable code. At the same time there can be a problem of assignment to codewords of a code of pairs of corteges. The encoder implementation consists of an implementation of eight Boolean functions with five variables. The LUT of modern FPGA can be configured to implement any Boolean function of 6 or any two Boolean functions of 5 variables. Thus, for the implementation of the encoder, 4 LUTs are enough for any 8-bit non-separable code, with any assignment of the codewords of the pairs of corteges. Boolean functions of the encoder are not defined in case of distortion of input data. From the viewpoint of hardware costs (by the number of LUTs), the addition of a definition of Boolean functions can be arbitrary. To control some possible distortions of input data in the.encoder, it is possible to realize additional two Boolean functions that would signal the presence of distortions and in the case of using determined Boolean functions to take into account the most probable distortion.

The implementation of the decoder is more complicated. One should implement at least 5 Boolean functions consisting of 8 variables. It is also advisable to implement a sixth Boolean function that would have the value of "0" for correct data (or corrected for single distortions) and a value of "1" for detected double distortions. For the implementation of any Boolean function of 8 variables, the structure in Fig. 2 is presented, where block 1 is marked for 6-LUT, and block 2 is a multiplexer, implemented by FPGA. Thus, for the implementation of the decoder for any 8-bit non-separable code, it needs 20 (or 24) LUT and 15 (18) multiplexers are required for any 8-bit non-dividing code, with any assignment of the words of corteges pairs.

Let us consider the possibility of reducing hardware costs in separate implementations of the decoder. Let the corresponding data transformations be present in Table 1, where y8-y1 are digits of the codecs and z1-z5 are digits of data corresponding to pairs of corteges for five decimal digits (Fig. 1).

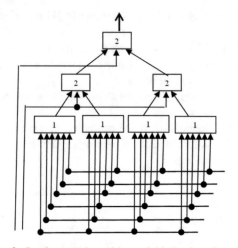

Fig. 2. Implementation arbitrary 8-bit Boolean function

When implementing a device, keep in mind the following: if the data appears on decoder input, for example, in accordance with Table 2, the data "10111" and "0" must appear on output.

Thus Boolean functions are defined on 180 combinations of inputs, and others are not defined. All other combinations correspond to double distortion. Therefore, the sixth function is completely defined, on other combinations of values of variables it equals "1". This allows for arbitrary determination of the first five functions in order to reduce hardware costs.

Minimization of Boolean functions on the basis of all functions of six or less variables consists of the use of the decomposition method [14–16]. In the case of fully defined Boolean functions between subsidiary functions, there is a relation of equality, which is the relation of equivalence. This allows for a simple definition of the decomposition coefficients. In the case of incompletely defined Boolean functions, there is a compatibility relation between subsidiary functions that is not an equivalence relation. Therefore, to determine the coefficients of decomposition, let us use the following algorithm, which is

Table 1. Example of data transformations

y_8	0	0	0	0	0	0	0	0	0	0	0	0	1	1	1	1	1	1	1	1
y_7	0	0	0	0	0	0	1	1	1	1	1	1	0	0	0	0	0	1	1	1
y_6	0	0	0	1	1	1	0	0	0	1	1	1	0	0	1	1	1	0	1	1
y_5	0	0	1	0	0	1	0	1	1	0	1	1	0	1	0	1	1	0	0	0
y_4	0	0	1	1	1	0	1	0	0	0	1	1	1	1	0	0	1	1	0	1
y_3	0	1	0	0	1	1	1	0	1	0	0	1	1	0	0	0	1	0	1	0
y_2	0	1	0	1	0	1	1	1	0	0	1	0	0	1	1	0	1	1	1	0
y_1	0	1	1	0	1	0	0	0	1	1	1	0	0	0	1	0	1	1	0	0
z_5	1	0	0	0	1	0	0	0	1	1	0	1	0	1	1	0	1	1	1	0
z_4	0	0	0	0	0	0	0	1	1	0	1	1	0	0	0	0	0	0	0	0
z_3	1	1	0	1	1	1	0	0	0	0	0	0	0	1	0	1	1	0	0	0
z_2	1	1	1	1	1	0	0	0	0	0	0	0	0	0	0	0	0	1	1	1
z_1	1	1	0	0	0	0	0	0	0	0	1	1	1	0	1	1	1	1	0	1

Table 2. Example of determining decoder boolean functions

y_8	y_7	y_6	y_5	y_4	y_3	y_2	y_1	z_6	z_5	z_4	z_3	z_2	z_1
0	0	0	0	0	0	0	0	0	1	0	1	1	1
1	0	0	0	0	0	0	0	0	1	0	1	1	1
0	1	0	0	0	0	0	0	0	1	0	1	1	1
0	0	1	0	0	0	0	0	0	1	0	1	1	1
0	0	0	1	0	0	0	0	0	1	0	1	1	1
0	0	0	0	1	0	0	0	0	1	0	1	1	1
0	0	0	0	0	1	0	0	0	1	0	1	1	1
0	0	0	0	0	0	1	0	0	1	0	1	1	1
0	0	0	0	0	0	0	1	0	1	0	1	1	1

a modification of the well-known Quine–McCluskey algorithm to minimize disjunctive normal form (DNF).

Algorithm for determining the coefficients of decomposition of incompletely determined Boolean functions:

- Stage 1. Perform the gluing of all primitive and created subsidiary functions. Two child functions are glued together if they are compatible. When gluing primitive subsidiary functions, it is possible to create a new subsidiary function, which was absent among the initial ones.

- Stage 2. Implementation of the absorption and creation of a shortlist of subsidiary functions. Some subsidiary function absorbs another if it is defined and is equal to another one for all sets of variables where this other function is defined. The absorbed subsidiary function is deleted from the list of subsidiary functions.
- Stage 3 Generation of the dead-end lists of subsidiary functions and selection of a dead-end list with a minimum number of them. The dead-end list is created by removing redundant extra-child functions.

If in the first stage (the step of gluing) no new subsidiary functions are created, then the decomposition factor will be equal to the maximum number of pairwise incompatible initial child functions, which simplifies the algorithm.

According to the considered algorithm, a program was developed for minimizing the method of decomposition of incompletely defined Boolean functions.

A device was created (Fig. 3) for decoding the non-separable 8-bit code for binary-coded decimal numbers, according to Table 1. In Fig. 3, the digits 5_j denote the variables y_j ($j = 1, 2,..., 8$), and the numbers 6_i are variables z_i. ($i = 1, 2,..., 5$). Output 6_6 implemented the signal of double distortion.

The device for decoding the correction of non-conductive 8-bit codes consists of unit 1 for determining an error signal and unit 2 for detecting signals on the first and fifth outputs of the device, unit 3 for detecting signals on the second and third outputs of the device, and unit 4 for detecting signals on the fourth output of the device. The Boolean function for the determination of the double error is not minimized and is realized with the use of the Shannon expansion in accordance with the structure on Fig. 3.

The device works as follows. When the device enters signals values that do correspond to one codeword in Table 1 or signals the values that do correspond to codeword with single-bit error, then the outputs of the device implemented the correct value of BCD number. Thus, at the output of unit 2_4 the value of bit z_1 from Table 1 is implemented. At the output of unit 3_4 - z_2, 3_5 - z_3, 4_2 - z_4, 2_5 - z_5, and at the output of block 1_7 there is value "0". On any other combination of input signals at the output of unit 1_7, the value "1" is realized as a mark of the presence of double bit error in input data. Thus, the device corrects single-bit errors and detects some double-bit errors.

Units of the device, according to the given tables, can be implemented by any of the known engineering techniques. In the case of using an FPGA to implement the device, you need 16 6-LUT and 3 multiplexers, which is one and a half times less, unlike the universal implementation.

Let the sequence of n packaged BCD words be transmitted. In the case of using an 8-bit Hamming code, you need to transmit $2n$ bytes (one byte per decimal number). When using an 8-bit non-separable code, it is necessary (according to Fig. 1) to transmit 8n/5 bytes. The transfer rate of the entire sequence increases by 25%.

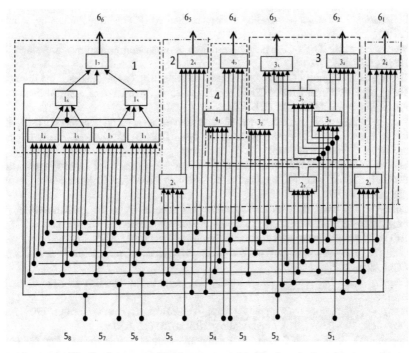

Fig. 3. Structure of 8-bit non-separable block code decoder

5 Conclusions

Therefore, the increase in data transmission speed for non-separable codes requires using complex algorithms of finding maximal cliques and complex algorithms of shared decomposition of Boolean functions. This increases the time to find a solution compared to separable codes, although increases the speed in operation.

The use of the proposed modification of the Bron-Kerbosch algorithm enables simplification of the non-separable codes to search based on the codes with less bitness.

The use of the proposed algorithm of shared decomposition of non-fully determined Boolean functions enables to reduce the expenses of LUT FPGA by 1.5 times to implement a decored as put against the Shannon expansion. At the same time, the speed of transmission of the whole decimals stream is increased by 25%.

Further research is concerned with analysing non-separable codes provided the codes' basis larger than 2, using the Reed-Solomon codes for comparison.

References

1. MacWilliams, F.J., Sloane, N.J.A.: The Theory of Error Correcting Codes
2. Blahut, R.E.: Theory and Practice of Error Control Codes. Addison-Wesley, Reading
3. Repka, M., Varchola, M.: Correlation power analysis using measured and simulated power traces based on hamming distance power Model - Attacking 16-bit integer multiplier in FPGA. IJCNIS **7**(6), 10–16 (2015)

4. Berlekamp, E.R.: Algebraic Coding Theory, revised World Scientific Publishing, Singapore (2014)
5. Viazovska, M.S.: The sphere packing problem in dimension 8. Ann. Math. Second Series **185**(3), 991–1015 (2017)
6. Teslenko, O., Shapoval, I.: Non-separable block codes. In: System Analysis and Information Technology 18th International Conference SAIT 2016, Kyiv, Ukraine, May 30–June 2 2016, Proceedings - ESC\IASA "NTUU\KPI", 440 p. (2016)
7. Rani, A., Grover, N.: Area & power optimization of asynchronous processor using Xilinx ISE & Vivado. Int. J. Inform. Eng. Electron. Bus. (IJIEEB) **10**(4), 8–15 (2018). https://doi.org/10.5815/ijieeb.2018.04.02
8. Mahabub, A.: Design and implementation of a novel complete filter for EEG application on FPGA. Int. J. Image Graph. Sig. Proc. (IJIGSP) **10**(6), 22–30 (2018). https://doi.org/10.5815/ijigsp.2018.06.03
9. Cunțan, C.D., Baciu, I., Osaci, M.: Studies on the necessity to integrate the FPGA (Field Programmable Gate Array) circuits in the digital electronics lab didactic activity. IJMECS **7**(6), 9–15 (2015). https://doi.org/10.5815/ijmecs.2015.06.02
10. Bron, C., Kerbosch, J.: Algorithm 457: finding all cliques of an undirected graph. Commun. ACM **16**(9), 575–577 (1973)
11. Cormen, T.H., Leiserson, C.E., Rivest, R.L., Stein,C.:Introduction to Algorithms,Third Edition
12. Johnston, H.C.: Cliques of a graph-variations on the Bron-Kerbosch algorithm. Int. J. Parallel Prog. **5**(3), 209–238 (1976). https://doi.org/10.1007/BF00991836
13. Eppstein, D., Löffler, M., Strash, D.: Listing all maximal cliques in sparse graphs in near-optimal time. In: Cheong, O., Chwa, K.-Y., Park, K. (eds.) ISAAC 2010. LNCS, vol. 6506, pp. 403–414. Springer, Heidelberg (2010). https://doi.org/10.1007/978-3-642-17517-6_36
14. Grover, N., Soni, M.K.: Simulation and optimization of VHDL code for FPGA-based design using simulink. IJIEEB **6**(3), 22–27 (2014). https://doi.org/10.5815/ijieeb.2014.03.04
15. Tarasenko, V., Teslenko, O.: Implementation of operations in finite fields on a one-dimensional cascade of constructive modules. Sys. Res. Inf. Technol. **2**, 7–27 2006
16. Klyatchenko, Y., Tarasenko, G., Tarasenko-Klyatchenko, O., Tarasenko, V., Teslenko, O.: Optimization of processor devices based on the maximum indicators of self-correction. In: Hu, Z., Petoukhov, S., Dychka, I., He, M. (eds.) Advances in Computer Science for Engineering and Education ICCSEEA 2018. Advances in Intelligent Systems and Computing, vol. 754, pp. 380–390. Springer, Cham (2019). https://doi.org/10.1007/978-3-319-91008-6_38

Parallelization of Algorithms by the OpenCL Library for Solving a System of Linear Algebraic Equations

O. S. Sharatskiy, I. M. Kuzmenko[✉], and A. I. Burenok

National Technical University of Ukraine "Igor Sikorsky Kyiv Polytechnic Institute", Kyiv, Ukraine
i.m.kuzmenko@kpi.ua

Abstract. The problem of accelerating the solution of a system of linear algebraic equations (SLAE) with the use of parallel calculations is considered. The object of research is the numerical methods of SLAE calculation by different processors.

The calculation time of the task of parallelization of the algorithms for calculating the system of algebraic equations using the OpenCL library is estimated. The paper describes the algorithms for the numerical solution of SLAE with the estimation of the time complexity of algorithms that perform parallel calculations.

The parallelization is performed on the CPU and GPU, depending on the settings specified for the OpenCL library. The running time of an algorithm running on a graphics processor is less than on a central one, regardless of the type of algorithm and the size of the coefficient matrix. The Gaussian iteration time deviated from the parabola, which was obviously due to the inhomogeneity of the equation system and the GPU load. It is established that the time of execution of iterations by the Gaussian method can sometimes deviate significantly from the parabolic one. Prospects for further research may be to use this library to parallelize valid mathematical calculations with the predominant use of GPUs.

Keywords: Numerical methods · System of linear algebraic equations · Parallel calculation · OpenCL library

1 Introduction

The numerical solution of mathematical physics problems is reduced to a system of linear algebraic equations and its solution requires modern methods and algorithms. At present, multithreaded and parallel calculation methods are widely used, in particular when using numerical methods.

The object of study is the numerical methods of SLAE calculation by CPU and GPU devices.

Algorithms for numerical methods for calculating systems of equations are time-consuming. The use of parallel calculations reduces the time complexity of the algorithms on which numerical analyses are based. However, this requires efficient partitioning of the problem, depending on the system of equations. Therefore, to improve the speed, the task is divided into threads automatically, using the OpenCL library.

The subject of study is the minimization of the calculation time consuming for numerical analysis of SLAE using a parallel calculations that were implemented automatically.

The purpose of the work is to minimize calculation time of a system of linear algebraic equations by using a parallel calculation.

Developed and implemented software reduces the calculation time of SLAE calculation algorithms by the Gaussian method and by the method of simple iterations. Based on the OpenCL computing library the software is realized and the influence of parallel calculations on the solution of systems of linear algebraic equations for a hydrodynamic problem is investigated. Studies have shown that the time of the algorithm, which runs on a graphics processor is less than on the central, almost regardless of the type of algorithm and the size of the matrix of coefficients.

2   Problem Statement

On the bases of literature review known that the parallel algorithm for Gauss–Seidel method [1] reduces the calculation in 5–6 times and authors describe the parallel algorithm. Algorithm listed in [2] distributes computational tasks between processors based on the floating-point rounding mode required for the task. PARFES [3] is a parallel sparse direct solver for systems of linear equations obtained using the finite element method in mechanics problems. In [4] the value of parallel processing in solving a system of linear equations is reviewed.

The OpenFOAM open source library solves the problems of numerical calculation of fluid dynamics with automatic parallelization [5]. However, the system does not have a graphical interface, and using the OpenMPI library allows only processors to calculate the task. There are similar systems, but they also cannot perform calculations on a graphics processor.

In [6] describes two naive parallel algorithms based on row block and row cyclic data distribution and put special emphasis on presenting a third parallel algorithm based on the pipeline technique. Further, there given an implementation of the pipelining technique in OpenMP interface.

Authors in [7] analyze the bottlenecks and propose minor extensions to current graphics architectures which would improve their effectiveness for solving problems.

The paper [8] examines more efficient algorithms that make the implementation of large matrix multiplication on upcoming GPU architectures more competitive. In work [9] authors modified Backpropagation Algorithm and Boltzmann Machine Algorithm with CUDA parallel matrix multiplication and discovered that the planned strategies achieve a quick training of Deep Neural Networks. A multithreading approach for the text classification process has been built with the algorithm using GPU parallelism using CUDA [10].

The paper [11] describes the matrix multiplication algorithm on dual core 2.0 GHz processor with two parallel threads.

Despite the above-mentioned research results, no attempt has been made to analyze the performance of threads for the use of parallel calculations reduces the calculation time of the algorithms with using the OpenCL library for numerical analyses. Therefore, to minimize calculation time, the task is divided into threads automatically, using the OpenCL library.

The purpose of this work is to parallelize the algorithms for the numerical solution of the system of linear equations with the estimation of the calculation time of parallel calculation on CPU and compute units of the GPU.

3 Methods of Solving

3.1 Solving of a SLAE Using Gaussian Method

We have a system of n inhomogeneous equations with n unknowns.

$$\begin{cases} a_{11}x_1 + a_{12}x_2 + \dots + a_{1n}x_n = b_1 \\ a_{21}x_1 + a_{22}x_2 + \dots + a_{2n}x_n = b_2 \\ \dots \\ a_{n1}x_1 + a_{n2}x_2 + \dots + a_{nn}x_n = b_n \end{cases} \tag{1}$$

To reduce the matrix of the system to a triangular form, subtract first equation from the second, multiplied by the corresponding $c_{21} = \frac{a_{21}}{a_{11}}$. Then in the same way subtract the first equation from the third, fourth, etc., with $c_{i1} = \frac{a_{i1}}{a_{11}}$. These operations - subtraction of equations are parallelized using OpenCL. As a result, we get

$$\begin{cases} a_{11}x_1 + a_{12}x_2 + \dots + a_{1n}x_n = b_1 \\ 0x_1 + a_{22}^{(2)}x_2 + \dots + a_{2n}^{(2)}x_n = b_2^{(2)} \\ \dots \\ 0x_1 + a_{n2}^{(2)}x_2 + \dots + a_{nn}^{(2)}x_n = b_n^{(2)} \end{cases}$$

where $a_{ij}^{(2)} = a_{ij} - c_{i1}a_{1j}$, $b_i^{(2)} = b_i - c_{i1}b_1$. This eliminates all elements of the first column below the main diagonal. Subtract from the new third line the second, multiplied by $c_{32} = \frac{a_{32}^{(2)}}{a_{22}^{(2)}}$. Do a similar operation with other lines 4, 5, ... n, i.e. subtract from each i-th line the second, multiplied by $c_{i2} = \frac{a_{i2}^{(2)}}{a_{22}^{(2)}}$. We get the system

$$\begin{cases} a_{11}x_1 + a_{12}x_2 + \dots + a_{1n}x_n = b_1 \\ 0x_1 + 0x_2 + \dots + a_{2n}^{(3)}x_n = b_2^{(3)} \\ \dots \\ 0x_1 + 0x_2 + \dots + a_{nn}^{(3)}x_n = b_n^{(3)} \end{cases}$$

Carrying out similar actions with the third, fourth, etc. equations, in the n-1 step with the execution of parallel calculations, we obtain a triangular matrix. The accuracy of the result was determined with $\varepsilon < 10^{-7}$ using single-precision floating-point format.

3.2 Solving of a SLAE Using the Fixed-Point Iteration Method

With a large number of unknowns in a linear system, approximate numerical methods are used in contrast to the Gaussian method, which gives an exact solution. In particular, the fixed-point iteration method. For system (1) we denote.

$$\frac{b_i}{a_{ii}} = \beta_i, \quad \frac{-a_{ij}}{a_{ii}} = \alpha_{ij}$$

when $i \neq j$ and $\alpha_{ij} = 0$ when $i = j$, $(i, j = \overline{1, n})$. Denote the matrices.

$$\alpha = \begin{bmatrix} \alpha_{11} & \cdots & \alpha_{1n} \\ \vdots & & \vdots \\ \alpha_{n1} & \cdots & \alpha_{nn} \end{bmatrix}; \beta = \begin{bmatrix} \beta_1 \\ \vdots \\ \beta_n \end{bmatrix}$$

the system can be written in matrix form $X = \beta + \alpha x$. For the initial (zero) approximation we take the column of constant terms $x^{(0)} = \beta$. Next, we consistently build matrices-columns $x^{(1)} = \beta + \alpha x^{(0)}$ (first approximation), $x^{(2)} = \beta + \alpha x^{(1)}$ (second approximation) and so on. If the sequence of approximations $x^{(0)}, x^{(1)} \ldots, x^{(k)}, \ldots$ has limit $x = x^{(k)}$, it is the solution of the system. This method of successive approximations is called the fixed-point iteration method. Multi-threaded calculations included calculating the sum above.

Write the formulas of approximations in explicit form

$$\begin{cases} x_i^{(0)} = \beta_i \\ x_i^{(k+1)} = \beta_i + \sum_{j=1}^{n} \alpha_{ij} x_i^{(k)} \\ (i = \overline{1, n}, k = 0, 1, 2 \ldots) \end{cases}$$

The iterative process must be continued until the condition $\left| x_i^{(k+1)} - x_i^{(k)} \right| \leq \varepsilon$ is met, where $\varepsilon = 0,001$ is the accuracy of the computational process.

4 A Parallelization of Algorithms for Solving Linear Systems on CPU and GPU

4.1 A Parallelization and OpenCL library

Initially, C# multithreading was used to numerically solve Eq. (1), but the approach showed a low speed at two-cores CPU. Therefore, to solve linear systems OpenCL library was used - allowing acceleration of similar calculations, performing operations in parallel on modern processors and video cards, with a choice of one or more devices [12]. The OpenCL library dynamically defines a free processor, including multi-core CPUs and GPUs, and provides parallelism at the instruction and data levels. This scales the performance of the software depending on the user's available hardware.

The implementation of the Gauss method using OpenCL is presented below. The main function sets the OpenCL environment and controls the execution flow.

```
// creating program and OpenCL context
ComputeContext computeContext = new ComputeContext(-1, new ComputeCon-
textPropertyList(ComputePlatform.Platforms[platform]), null, IntPtr.Zero);
// here we specify OpenCL program and device(s) to use
ComputeProgram computeProgram = new ComputeProgram(context,
File.ReadAllText("kernel.c")).Build(new
List<ComputeDevice>{ComputePlatform.Platforms[platform].Devices[device]}, "",
null, IntPtr.Zero);

ComputeKernel computeKernel = computeProgram.CreateKernel("gaussian1");

ComputeBuffer<float> computeBuffer = new ComputeBuff-
er<float>(computeContext, 9L, equation);

// passing memory parameters to kernels
computeKernel.SetMemoryArgument(0, computeBuffer);
computeKernel.SetValueArgument(1, size);
ComputeCommandQueue computeCommandQueue = new ComputeCommand-
Queue(computeContext, ComputePlatform.Platforms[platform].Devices[device], 0L);

// main loop for gaussian method
for (int i = 1; i < size; i++) {
  //setting iteration number or line to subtract
  computeKernel.SetValueArgument(2, i);
  int num = size - i;
  int num2 = size - i + 1;
  // waiting for OpenCL to compute
  computeCommandQueue.Execute(computeKernel, new long[2], new long[] {
num2,num }, null, null);
  computeCommandQueue.Finish();
}
// passing matrix back to program
computeCommandQueue.ReadFromBuffer<float>(computeBuffer, ref equation, true,
null);

// calculating unknows
float[] x = new float[size];
for (int j = size - 1; j >= 0; j--) {
    x[j] = equation[j * (size + 1) + size] / equation[j * (size + 1) + j];
    for (int k = size - 1; k > j; k--)
        x[j] -= equation[j * (size + 1) + k] * x[k] / equation[j * (size + 1) + j];
}
```

Matrix of coefficients is stored in equation in plain form. It is passed to device memory, transformed in triangular form and passed back to calculate unknowns.

The OpenCL program looks as follows. It takes a matrix array, unknowns count and current iteration, which determines which row is subtracted now. OpenCL program is executed on the compute block in parallel and subtracts each element multiplied by c_i simultaneously. Number of elements to compute is specified by the main program.

```
__kernel void gaussian1(__global float * v, int size, int k) {

    int cols = size - k + 1;
    int i = k + get_global_id(0);
    int j = k + get_global_id(1);

    float t = v[j * (size + 1) + k - 1] / v[(k - 1) * (size + 1) + k - 1];
    v[j * (size + 1) + i] -= t * v[(k - 1) * (size + 1) + i];

}
```

Main diagonal elements are not changed as they are used to compute other elements and are known to be 1 after that.

4.2 Testing of the Algorithm

The testing of the algorithm was performed as follows: the input module, receiving the boundary conditions and the compute grid size, builds the linear system, which it passes to the solution module. The solution module solves a system with unknowns using Gaussian method and a fixed-point method. Algorithms for these methods include thread parallelization using the OpenCL library. First, the memory is allocated for the matrix, which will be solved on the computing blocks in parallel and is loaded the data, and then performs the corresponding number of iterations. After the calculations, the output module loads the data back, returns an array of unknowns, and saves the results to a file.

Using OpenCL, the linear system was solved for 100, 400, 900, ..., 10 000 unknowns for $10 \times 10, 20 \times 20, 30 \times 30,..., 100 \times 100$ compute grid by the Gaussian method and the fixed point iteration method.

Table 1 shows the dependence of the system resolution time (in ms) for different numbers of unknowns (left column).

Calculations were performed on a i7-9700K CPU with theoretical single floating-point performance 114 GFLOPS and on RTX 2070 GPU with a performance 7.5 TFLOPS.

When compared, the time complexity of the Gaussian algorithm is less than fixed-point, provided that in both cases ε did not exceed 10^{-7} for Gaussian and 0.001 for fixed-point in Table 1.

The right two columns show a relative performance boost with parallel calculation.

Table 1. The dependence of the system resolution time (in ms) for various numbers of unknowns

Number of unknowns	Gaussian CPU	Gaussian GPU	Gaussian single thread	Fixed-point CPU	Fixed-point GPU	Fixed-point single thread	Gaussian single thread\Gaussian CPU	Fixed-point single thread \fixed-point GPU
100	6	26	1	43	102	8	0,17	0,19
400	17	50	80	177	170	493	4,71	2,79
900	100	60	995	571	500	5 603	9,95	9,81
1600	508	278	5 636	5 431	1 395	31 810	11,09	5,86
2500	2 144	971	21 218	22 685	3 048	120 374	9,90	5,31
3600	6 759	2 792	63 073		6 707	336 542	9,33	
4900	15 876	7 133	159 903		16 510	669 941	10,07	
6400	35 469	16 254	358 900		32 789	1 261 119	10,12	
8100	70 602	32 912	717 214		59 038	1 736 103	10,16	
10000	130 385	65 216	1 367 215		134 025	2 815 393	10,49	

5 Results

Figures 1a, b show the dependence of the time of one iteration for Gaussian method (in ms) on the number of iterations for the system with $80 \times 80 = 64,000$ unknowns. The number of iterations is placed on the X axis, the time of one iteration is placed on the Y axis.

Calculations were performed on a i7-9700K CPU with theoretical single floating-point performance 114 GFLOPS and on RTX 2070 GPU with a performance 7.5 TFLOPS.

Theoretically, the time of one iteration for Gaussian method should decrease quadratically, because a system of N unknowns requires $N-1$ iterations, each consisting of $(N-i) (N-i-2)$ operations, where i is the iteration number from 1 to $N-1$ inclusive. However, in practice, the time of execution of iterations can deviate significantly from the theoretical to a lesser or greater extent (Fig. 1a–b for 6400 unknowns), regardless of processor power.

Obviously, this distribution is due to the heterogeneity of the system at working with floating-point arithmetic.

From Figs. 1a, 1b it can be seen that the time of one iteration for Gaussian method of CPU exceeds this of GPU. Integral value of the iterations time for the system by the Gaussian method with parallel calculation on the CPU is 37.2 s, on the GPU 17.9 s for the above examples (Fig. 1a, 1b) respectively.

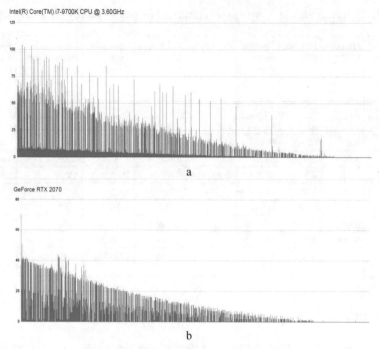

Fig. 1. a. Iterations time for i7-9700K, b. Iterations time for RTX 2070.

6   Conclusion

A system of numerical calculation of the velocity field was developed using C++ and parallel calculation with OpenCL. In the Gaussian method by performing the parallel calculation, we obtain a triangular matrix. The method of fixed-point iteration is used to calculate the sum. The parallel calculation is performed on CPUs and GPUs, depending on the settings specified for the OpenCL library.

The novelty of obtained results in the next - the time of an algorithm running on a graphics processor is less than on a central one, almost regardless of the type of algorithm and the size of the matrix.

The iteration time in the Gaussian method should decrease quadratically, however, it deviates significantly from the quadratic, due to the heterogeneity of the system and float point arithmetic.

The practical significance is that the OpenCL library dynamically defines a free processor scales the performance of the software and provides threads at the instruction and data levels.

Prospects for further research are to study all methods for SLAE calculation with parallel calculation computing and OpenCL library use.

References

1. Koester, D.P., Ranka, S., Fox, G.C.: A parallel Gauss-Seidel algorithm for sparse power system matrices. In: SuperComputing 1994, pp. 184–193 (1994)
2. Kolberg, M., Cordeiro, D., Bohlender, G., Fernandes, L.G., Goldman, A.: A Multithreaded verified method for solving linear systems in dual-core processors. In: Workshop on State-of-the-Art in Scientific and Parallel Computing (PARA 2008), p. 9. Trondheim, Norway (2008). ffhal-00542920f
3. Fialko, S.: Parallel direct solver for solving systems of linear equations resulting from finite element method on multi-core desktops and workstations. Comput. Math. Appl. **70**(12), 2968–2987 (2015)
4. Dixon, L.C.W.: A review of parallel methods for solving sets of linear equations and their application within optimization algorithms. Calcolo **25**, 21–36 (1988). https://doi.org/10.1007/BF02575745
5. https://www.openfoam.com/
6. Michailidis, P.D., Margaritis, K.G.: Parallel direct methods for solving the system of linear equations with pipelining on a multicore using OpenMP. J. Comput. Appl. Math. **236**(3), 326–341 (2011)
7. Thompson, C.J., Hahn, S., Oskin, M.: Using modern graphics architectures for general computing: a framework and analysis. In: Proceedings of the 35th IEEE/ACM International Symposium Microarchitecture, pp. 306–317 (2002)
8. Larsen, E.S., McAllister, D.: Fast matrix multipliers using graphics hardware. In: Proceedings of the High Performance Networking and Computing Conference, p. 55 (2001)
9. Dharmajee Rao, D.T.V., Ramana, K.V.: Accelerating training of deep neural networks on GPU using CUDA. Int. J. Intell. Syst. Appl. (IJISA) **11**(5), 18–26 (2019). https://doi.org/10.5815/ijisa.2019.05.03
10. Chatterjee, S., Jose, P.G., Datta, D.: Text classification using SVM enhanced by multithreading and CUDA. Int. J. Mod. Educ. Comput. Sci. (IJMECS) **11**(1), 11–23 (2019). https://doi.org/10.5815/ijmecs.2019.01.02
11. Dash, Y., Kumar, S., Patle, V.K.: Evaluation of performance on open MP parallel platform based on problem size. Int. J. Mod. Educ. Comput. Sci. (IJMECS) **8**(6), 35–40 (2016) https://doi.org/10.5815/ijmecs.2016.06.05
12. Kruger, J., Westermann, R.: Linear algebra operators for GPU implementation of numerical algorithms. ACM Trans. Graph. **22**, 908–916 (2003)

Mathematics and Software for Building Nonlinear Polynomial Regressions Using Estimates for Univariate Polynomial Regressions Coefficients with a Given (Small) Variance

Alexander Pavlov$^{(\boxtimes)}$ ⓘ, Maxim Holovchenko ⓘ, Iryna Mukha ⓘ, and Kateryna Lishchuk ⓘ

National Technical University of Ukraine "Igor Sikorsky Kyiv
Polytechnic Institute", Kyiv 03056, Ukraine
`pavlov.fiot@gmail.com, {ma4ete25,lishchuk_kpi}@ukr.net,`
`mip.kpi@gmail.com`

Abstract. The presented results address the building of multivariate nonlinear regressions based on experimental data and using them in various special-purpose information systems. In contrast to the well-known methods for building multivariate polynomial regressions given by a redundant representation using an active experiment, we claim that the proposed methodology and algorithms for its implementation allow finding estimates for the coefficients at nonlinear terms of multivariate regression with a variance sufficiently small for practice and a small number of tests. We show that the solution of this problem reduces to the sequential building of univariate polynomial regressions. Each regression yields corresponding strictly nondegenerate systems of linear equations with deterministic coefficients. The variables in the systems of linear equations are the coefficients at the terms of the nonlinear polynomial regression. The right parts of the systems are estimates for the coefficients at nonlinear terms of univariate regression found based on one pre-built set of normalized orthogonal polynomials of Forsythe which guarantee a given (small) value of their variance. This way, we assure efficient finding of coefficients at nonlinear terms of the multivariate polynomial regression. In conclusion, we give the software architecture that implements the proposed methodology and algorithms for building a multivariate polynomial regression given by a redundant representation.

Keywords: Multivariate polynomial regression · Normalized orthogonal polynomials of Forsythe · Redundant representation · Active experiment · Software architecture · Information systems

1 Introduction

1.1 Related Works Review Substantiating the Theoretical and Practical Relevance of the Results We Present

Regression analysis is used today in many areas of science, technology, economy, finance (e.g., [1–4]). In particular, the areas include data mining [5], near-infrared spectroscopy

[6], forensic dating [7], multi-element quantitative analysis of soils [8], medical diagnostics [9], image filtering [10], predictive models design in various fields of agriculture and meteorology [11], etc. In most cases, multivariate linear regressions are used (e.g., [6–8, 11]) because multivariate nonlinear and, in particular, polynomial regressions face theoretical and practical difficulties at their adequate building. Thus, for example, well-known methods—a stepwise technique for adding and removing polynomial terms [12], neural networks [10], group method of data handling [13], genetic algorithms [14], hierarchical approximation [10]—do not allow in the general case to estimate the coefficients of the nonlinear terms of multivariate polynomial regression (MPR) with guaranteed small variances of the estimates. This paper addresses the solution to this problem. We develop a methodology, and algorithms for its implementation, to estimate the coefficients for nonlinear terms of MPR given by a redundant representation. Our approach uses estimates of univariate polynomial regressions (UPRs) coefficients with a given (small) variance and solving nondegenerate systems of linear equations which variables are the coefficients at nonlinear terms of the MPR. Under a redundant representation, we mean that the expression for the MPR may contain nonlinear terms absent in true regression expression. Hence, the solution to this problem is extremely relevant and significant in theoretical terms and for practical use in designing modern intelligent information systems for various purposes.

1.2 General Principles for Solving the Problem

The methodology presented in this work and the mathematical and algorithmic support for an MPR building from its redundant representation are based on previous results [15–21]. The following is the main difference between the proposed methodology and the algorithms that implement it from the known methods (e.g., [6, 22, 23]). Based on an active experiment, the solution of the problem is reduced to building sequential UPRs. The coefficients at nonlinear terms of the UPRs have a given (small) variance. Each of the UPRs yields a nondegenerate system of linear equations that ensure the finding of all coefficients for nonlinear terms of multivariate regression with acceptable for practice accuracy. Finding the remaining coefficients is reduced to the standard procedure of estimating the coefficients of multivariate linear regression (MLR) based on an active experiment.

1.3 The Structure of the Paper

Section 2 presents well-known theoretical results used to solve the problem. In Sect. 3, we give a formal statement of the problem of estimating unknown coefficients of MPR given by a redundant representation. Section 4 presents the methodology and algorithms for its implementation which reduce the problem of estimating the coefficients of nonlinear terms of MPR to a sequential solution of the problem of estimating the coefficients of nonlinear terms of UPR and the solution of corresponding nondegenerate systems of linear equations. In Sect. 5, we present a generalized methodology for finding estimates with a given variance for coefficients at nonlinear terms of an MPR given by a redundant representation with the use of an active experiment. Section 6 provides an illustrative example. Section 7 shows the software architecture that implements the presented mathematical support for estimating the coefficients for nonlinear terms of MPR. Section 8 contains conclusions about the results we obtained.

2   General Theoretical Foundations

Hereinafter, we will use the following well-known theoretical results set out, e.g., in [24], which allow estimating of coefficients of UPR using normalized orthogonal polynomials of Forsythe (NOPFs).

Suppose that a UPR has the form

$$Y(x) = \theta_0 + \theta_1 x + \ldots + \theta_r x^r + E, \tag{1}$$

where x is a scalar deterministic input, E is a random variable with an arbitrary distribution, zero mean and limited variance σ^2. We have the results of experiments $(x_i \to y_i, i = \overline{1,n})$, where $y_i = \theta_0 + \theta_1 x_i + \ldots + \theta_r x_i^r + \delta_i, i = \overline{1,n}$; δ_i is an unknown realization of the random variable E. We need to find estimates for the coefficients $\hat{\theta}_j, j = \overline{0,r}$.

Let us convert the model (1) to the following model (2):

$$Y(x) = \sum_{j=0}^{r} w_j Q_j(x) + E \tag{2}$$

where

$$Q_j(x) = q_{j0} + q_{j1}x + \ldots + q_{jj}x^j, j = \overline{0,r}, \tag{3}$$

is the j-th NOPF. Coefficients of the NOPF are found by the known recurrent procedure. Estimates for the coefficients $w_j, j = \overline{0,r}$, are found by the formula

$$\hat{w}_j = \sum_{i=1}^{n} y_i Q_j(x_i), j = \overline{0,r}, \tag{4}$$

and the estimates $\hat{\theta}_j, j = \overline{0,r}$, are then found by $\hat{w}_j, j = \overline{0,r}$, as follows:

$$\hat{\theta}_j = \hat{w}_r q_{rj} + \ldots + \hat{w}_j q_{jj}, j = \overline{0,r}, \tag{5}$$

the accuracy of the estimate $\hat{\theta}_j, j = \overline{0,r}$, is evaluated by the formula

$$M\hat{\theta}_j = \theta_j, \operatorname{Var}\left(\hat{\theta}_j\right) = \sigma^2 \sum_{i=r}^{j} q_{ij}^2. \tag{6}$$

3   The Problem Statement: Building an MPR Given by a Redundant Representation

Let MPR be given by a redundant representation [15, 19, 20, 24]

$$y(\overline{x}) = \sum_{\forall (i_1,\ldots i_t) \in K} \sum_{\forall (j_1,\ldots j_t) \in K(i_1,\ldots i_t)} b_{i_1 \ldots i_t}^{j_1 \ldots j_t} (x_{i_1})^{j_1} \cdot (x_{i_2})^{j_2} \ldots (x_{i_t})^{j_t} + E \tag{7}$$

where $\bar{x} = (x_1 \ldots x_m)^\top$ is the deterministic vector of input variables, E is a random variable with $ME = Q_j$, $\mathrm{Var}(E) = \sigma^2 < \infty$.

The following active experiment can be carried out in this case. We can input arbitrary values of the input variables x_j in the range $[x_{j_{min}}, x_{j_{max}}]$, $j = \overline{1, m}$. We assume that

$$\bigcap_{j=1}^{m} [x_{j_{min}}, x_{j_{max}}] = [a, b] \neq \{\emptyset\}. \tag{8}$$

Values of all the coefficients $b_{i_1 \ldots i_t}^{j_1 \ldots j_t}$ are unknown, and some of the nonlinear terms may equal to zero. We need to exclude from (7) terms with zero coefficients and estimate the values of all non-zero coefficients.

4 Methodology for an MPR Building Based on a Redundant Representation and Algorithms that Implement It

4.1 UPRs Building

Calculation of NOPF coefficients is a rather cumbersome computational procedure in the general case. In addition, accuracy of the coefficients $\hat{\theta}_j$, $j = \overline{0, r}$ significantly depends on the choice of the interval $[a, b]$ containing the values of the scalar variable x_i, $i = \overline{1, n}$ (formula (6)). Computational experiments [15, 19, 20] have shown that the best interval has the form $[x_1 = -|x_n|, x_n]$, $x_n > 1$. We have shown the following accuracy of finding the coefficients $\hat{\theta}_j$, $j = \overline{0, 5}$, within the interval $[-50, 50]$ for $n = 10$ and the maximum degree of the polynomial $r = 5$, according to formula (6): $\mathrm{Var}\left(\hat{\theta}_0\right) = \sigma^2 \cdot 0.37$; $\mathrm{Var}\left(\hat{\theta}_1\right) = \sigma^2 \cdot 2.32 \cdot 10^{-3}$; $\mathrm{Var}\left(\hat{\theta}_2\right) = \sigma^2 \cdot 1.8 \cdot 10^{-6}$; $\mathrm{Var}\left(\hat{\theta}_3\right) = \sigma^2 \cdot 4.72 \cdot 10^{-9}$; $\mathrm{Var}\left(\hat{\theta}_4\right) = \sigma^2 \cdot 2.61 \cdot 10^{-13}$; $\mathrm{Var}\left(\hat{\theta}_5\right) = \sigma^2 \cdot 4.47 \cdot 10^{-16}$ where $\sigma^2 = \mathrm{Var}(E)$.

If values of x_i, $i = \overline{1, n}$, belong to an interval $[a, b]$ where a and b have the same sign, then the estimates $\hat{\theta}_j$, $j = \overline{0, r}$, are worse than those for an interval symmetric to zero. Let us consider the three types of intervals: (1) $[a, b]$, $a < b \leq -1$; (2) $[-|b|, b]$, $b > 1$; (3) $[a, b]$, $1 \leq a < b$, for given a, b, n. Thus, the following is desirable for each interval of these three types:

1. To find three sets of NOPFs with a given precision in advance. This problem can be solved with the following practical procedure. Using the recurrent procedure, find $Q_j(x)$, $j = \overline{0, r}$. Set arbitrary functions $y(x) = \theta_0 + \theta_1 x + \ldots + \theta_r x^r$ with given coefficients $\theta_0, \ldots, \theta_r$. Suppose $E = 0$. Estimate $\hat{\theta}_j$, $j = \overline{0, r}$, by the set of NOPF data $\{x_i, y_i\}$ by formulas (4), (5) and compare $\hat{\theta}_j$ with the known θ_j, $j = \overline{0, r}$. If necessary, refine the NOPFs coefficients until a given accuracy is reached.
2. To be able to recalculate the found NOPFs for an interval $[ka, kb]$ with a simple computational procedure. k here is an arbitrary positive number. This problem was solved in [19]: suppose that q_{jl}, $\forall j = \overline{0, r}$, $l = \overline{0, j}$, are the coefficients of NOPFs calculated by x_i, $i = \overline{1, n}$. If z_i, $i = \overline{1, n}$, where $\forall i\ z_i = kx_i$, $k > 0$, then the NOPFs coefficients q_{jl}^z, $\forall j = \overline{0, r}$, $l = \overline{0, j}$, have the form

$$q_{jl}^z = \frac{1}{k^j} q_{jl}; \tag{9}$$

$$\text{Var}\left(\hat{\theta}_j^z\right) = \sigma^2 \sum_{l=r}^{j} \left(\frac{1}{k^j} q_{jl}\right)^2 = \left(\frac{1}{k^j}\right)^2 \text{Var}\left(\hat{\theta}_j\right), \quad j = \overline{0, r}. \tag{10}$$

Thus, the NOPFs found for the interval $[a, b]$ and the numbers $a = x_1 < x_2 < \ldots < x_n = b$ are efficiently recalculated by formula (9) for the interval $[ka, kb]$ $\forall k > 0$ for numbers $ka = z_1 = kx_1 < z_2 = kx_2 < \ldots < z_n = kx_n = kb$ where $[ka, kb]$ is an interval in which an active experiment can be carried out.

3. To design a new active experiment if some values of the variance of estimates $\text{Var}\left(\hat{\theta}_j^z\right), j = \overline{2, r}$, do not satisfy the given accuracy (we suppose that the value of σ^2 or its upper bound is known). The experiment is carried out on the interval $[ka, kb]$ and the numbers $z_1, \ldots, z_n$, for which we found the NOPFs by (10). This will allow us to reach the given accuracy for the estimates $\hat{\theta}_j^z, j = \overline{2, r}$, of the unknown coefficients $\theta_j, j = \overline{2, r}$, using only the previously found set of NOPFs. This problem in general formulation was solved in [20]: let the input of the object be a numerical sequence $z_1, \ldots, z_n, z_1, \ldots, z_n \ldots, z_1, \ldots, z_n, n > r$, where the sequence of the numbers $z_1, \ldots, z_n$ repeats l times. Let us denote $Z = (z_1, \ldots, z_n)$; $Y = \left(\frac{\sum_{k=1}^{l} y_{k1}}{l}, \frac{\sum_{k=1}^{l} y_{k2}}{l}, \ldots, \frac{\sum_{k=1}^{l} y_{kn}}{l}\right)$; $Z' = (z_{11}, z_{21}, \ldots, z_{l1}, z_{12}, z_{22}, \ldots, z_{l2}, \ldots, z_{1n}, z_{2n}, \ldots, z_{ln})$ where $z_{ki} = z_i \forall k = \overline{1, l}, \forall i = \overline{1, n}$; $Y' = (y_{11}, y_{21}, \ldots, y_{l1}, \ldots, y_{1n}, y_{2n}, \ldots, y_{ln})$ where y_{ij} is the output value of the object at the input value z_{ij}. Let $Q_j(z), j = \overline{0, r}$, be the NOPFs built on the numerical sequence Z, and $Q'_j(z)$ be the NOPFs built on the numerical sequence Z'. Then [20] $Q'_j(z) = Q_j(z)/\sqrt{l}, j = \overline{0, r}$ (for the values of $z = z_i = z_{ki}, k = \overline{1, l}$). From this it follows [20] that the estimates obtained from the experiments (Z, Y) and (Z', Y') are equal. In other words, the estimates $\theta_j, j = \overline{0, r}$, obtained from the experiment (Z, Y) correspond in fact to the following UPR problem:

$$Y(z) = \theta_0 + \theta_1 z + \ldots + \theta_r z^r + E_1, ME_1 = 0, Var(E_1) = \sigma^2/l.$$

By choosing the appropriate l, we can achieve the given accuracy of the estimates $\hat{\theta}_j, j = \overline{2, r}$.

4. With the appearance of the theoretical result [21], the problem of estimating the coefficients with a small variance for nonlinear terms of a UPR has a solution with a small number of tests. It was shown in [21] that for any interval $[c, d]$, in which an active experiment can be carried out, the variance of coefficient estimates is found from a set of NOPFs built in an arbitrary segment $z_1 = -\hat{b} < z_2 < \ldots < z_n = \hat{b}$, $\hat{b} > 1$. In the example shown in Sect. 6, at $\hat{b} = 50$, number of experiments $n = 10$, variances of coefficients at nonlinear terms 2...5 vary in the range from $D\widehat{\theta_2} = DE \cdot 1.8 \cdot 10^{-6}$ to $D\widehat{\theta_5} = DE \cdot 4.47 \cdot 10^{-16}$.

Remark 1. We can exclude division operations when estimating the coefficients at nonlinear terms of the original UPR using estimates for the coefficients at nonlinear terms of virtual UPR found from the above set of NOPFs. For that, we need [21] to set the parameter a in the form $1/k, k > 1$, preferably an integer or a fractional number with a finite number of decimal places.

Remark 2. The work [21] contains the results of statistical tests, the reliability of which follows from the structural analysis of the matrix A using the 3-sigma rule. The tests showed that the use of the NOPFs set (Sect. 6) gives the results acceptable for practice in the case when [21] $DE \in [1, 2], r = 3, n = 10, x_1 \in [0, 1], d - c = 50$. Then, the values of θ_2 and θ_3 are found with an average accuracy of up to 10^{-2} and 10^{-3} measurement units, respectively.

4.2   Estimating the Coefficients at Nonlinear Terms of MPR

The set of terms of MPR has the following form, according to (7):

$$\left\{ b_{i_1 \ldots i_t}^{j_1 \ldots j_t} \left(x_{i_1} \right)^{j_1} \cdot \left(x_{i_2} \right)^{j_2} \ldots \left(x_{i_t} \right)^{j_t} \right\} \tag{11}$$

$$\forall (i_1, \ldots i_t) \in K, \forall (j_1, \ldots j_t) \in K(i_1, \ldots i_t)$$

where $\forall b_{i_1 \ldots i_t}^{j_1 \ldots j_t}$ are unknown coefficients.

Using one of the three available sets of NOPFs, we construct NOPFs for $j = \overline{0, r}$,

$$r = \max_{\forall (i_1, \ldots i_t) \in K, \forall (j_1, \ldots j_t) \in K(i_1, \ldots i_t)} (j_1, \ldots j_t) \tag{12}$$

by formulas (9) with the values of $a = z_1 < z_2 < \ldots < z_n = b$ on the interval $[a, b]$ defined in (8).

Remark 3. If the result from [21] is used to build UPRs, then the NOPFs set for the interval $\left[-\hat{b}, \hat{b} \right]$ should be used.

4.3   Methodology and Algorithms for Its Implementation Which Guarantee Finding All Coefficients at Nonlinear Terms of MPR

The proposed methodology and the algorithms that implement it are as follows.

1.1.   First, select from the redundant representation of the MPR in an arbitrary order, one by one, the nonlinear terms, each of which contains at least one variable to a power greater than or equal to two. Let x_j be such a variable. Then carry out an active experiment at the following values of the input variables: $x_{ji} = x_i, x_i \in [a, b]$ or $x_i \in \left[-\hat{b}, \hat{b} \right], i = \overline{1, n}; x_{li} = x_{lfix}^1, i = \overline{1, n} \, \forall l \neq j$, where x_{lfix}^1 is a fixed value of the variable x_l when building the first UPR in order.

Remark 4. If the interval $[a, b]$ has been chosen, then, if necessary, carry out an experiment with repeated data.

The MPR turns into a univariate regression, the coefficients of which are generally a linear weighted sum of some unknown coefficients at the nonlinear terms of the MPR, possibly including the constant. Using the selected set of NOPFs, estimate with a given accuracy (by the value of variance) the coefficients of the UPR at the variable to a power greater than or equal to two. Sequentially consider all equations containing one unknown coefficient at the nonlinear terms of the MPR and find their values. Further, consider, one by one, the linear equations containing at least two unknown coefficients of the MPR. Let the first of them contain $P_1 \geq 2$ unknown coefficients of the MPR. Carry out $P_1 - 1$ active experiments on the following input data: $\{x_{ji} = x_i, x_i \in [a, b]$ or $x_i \in \left[-\hat{b}, \hat{b}\right], i = \overline{1, n}; x_{li} = x^s_{l\text{fix}}, i = \overline{1, n} \; \forall l \neq j\}, s = \overline{2, P_1}.$

The sets of numbers $\{x^s_{l\text{fix}} \forall l \neq j\}, s = \overline{1, P_1}$, must be such that the corresponding system of P_1 linear equations with P_1 variables was not degenerate. Obviously, by virtue of the rule for building a univariate regression (only one input variable changes, the rest of the variables take fixed values), such sets of numbers always exist. Repeat the described procedure for all remaining linear equations containing unknown coefficients of the MPR considering the previously found coefficients for nonlinear terms of the MPR.

Remark 5. The previously built UPRs can be used when building systems of linear equations to find the unknown coefficients of the MPR.

1.2. Select the next nonlinear term of the MPR with at least one variable to a power greater than or equal to two and not yet found the value of the coefficient. Repeat the procedure described in step 1.1 considering the coefficients of the MPR already found in step 1.1. Calculate, in a finite number of steps, the values of all coefficients for nonlinear terms of the MPR containing at least one variable to a power greater than or equal to two.

2. Sequentially, one by one, consider all nonlinear terms of the MPR containing only variables to the power of one. In this case, an active experiment that reduces MPR to UPRs has the following design. At least two, but not more than r (see (12)) variables take the same values $x_i, i = \overline{1, n}, x_i \in [a, b]$ or $x_i \in \left[-\hat{b}, \hat{b}\right]$, the rest of the variables in all experiments take fixed values. Further, execute the procedure described in steps 1.1, 1.2.

Remark 6 (To Steps 1.1, 1.2, 2). The choice of nonlinear terms of the MPR in an arbitrary order in steps 1.1, 1.2, and 2 guarantees finding all coefficients for nonlinear terms of the MPR. But we recommend choosing them in a way (under the given redundant representation of the MPR) that minimizes both the number of UPRs built and the dimensions of the linear equations to be solved.

3. After execution of steps 1.1, 1.2, 2, the problem of finding the remaining coefficients of the MPR has been reduced to the problem of estimating unknown coefficients of MLR by ordinary least squares in an active experiment [22, 23].

5 A Generalized Methodology for Finding Estimates with a Given Variance for Coefficients at Nonlinear Terms of an MPR Given by a Redundant Representation Using an Active Experiment

1. For a given interval $[a, b]$, find a set of NOPFs that guarantees estimating the coefficients for nonlinear terms of UPRs with a given variance.
2. Sequentially carry out, based on the theoretical results presented in Sect. 4, a series of active experiments in a given order. Each of the experiments turns an MPR into UPRs with estimates for the coefficients at its nonlinear terms, the estimates have a given variance. The UPRs yield the systems of strictly nondegenerate linear equations, their variables are the coefficients at the nonlinear members of the MPR, and the right-hand sides contain the found estimates at the nonlinear terms of UPRs. Section 4 proves that the sequential fulfillment of the series of active experiments in the proposed order guarantees estimating the coefficients at all the nonlinear terms of the MPR given by a redundant representation.

The results obtained are strict (see Sect. 4) and are confirmed by the results of statistical tests (see Remark 2 in Sect. 4). The example below effectively illustrates the possibility of practical use of the proposed methodology and algorithms for its implementation.

6 Illustrative Example: Estimating the Coefficients for Nonlinear Terms of an MPR

Let us specify the redundant representation of an MPR as follows:

$$Y(x_1, x_2, x_3, x_4)$$
$$= \theta_0 + \theta_1 x_1 + \theta_2 x_2 + \theta_3 x_3 + \theta_4 x_4 + \theta_5 x_1 x_2 + \theta_6 x_1 x_2 x_4 + \theta_7 x_1 x_3 x_4 + \theta_8 x_1^2 x_3$$
$$+ \theta_9 x_1 x_3^2 + \theta_{10} x_2^2 x_3^2 + \theta_{11} x_1^4 x_2 x_3^2 + \theta_{12} x_3^2 + E$$

where $ME = 0$, $\mathrm{Var}(E) = 2500$, E has a normal distribution.

We used a part of the $C++$ *boost* extension library to generate values of the random variable E.

We use the following realizations of the random variable E: 54.436785298584844; 36.65484440268196; 39.37525282158212; 17.06828734421936; 32.91653994279295; −6.102376584160259;   −36.63056533846738;   57.61740273439848; −2.4409080323190033; 1.5231642970171395.

The true values of the coefficients are: $\theta_0 = 1$, $\theta_1 = 0$, $\theta_2 = 2$, $\theta_3 = 0$, $\theta_4 = 3$, $\theta_5 = 4$, $\theta_6 = 5$, $\theta_7 = 6$, $\theta_8 = 7$, $\theta_9 = 0$, $\theta_{10} = 8$, $\theta_{11} = 9$, $\theta_{12} = 10$.

Thus, the true description of the MPR does not have the terms with the coefficients $\theta_1, \theta_3, \theta_9$.

Remark 7. We will use the true values of the coefficients $\theta_j, j = \overline{0, 12}$, to:

- implement a virtual active experiment;
- illustrate the efficiency of the proposed method of an MPR coefficients estimation.

Preliminary Stage. For the values of the scalar variable given in Table 1, we have found exact coefficients of the NOPFs up to the fifth degree (Table 2, see (3)).

Table 1. The values of the scalar variable x

x_1	x_2	x_3	x_4	x_5	x_6	x_7	x_8	x_9	x_{10}
-50	-38.888	-27.778	-16.666	-5.555	5.555	16.666	27.778	38.888	50

Remark 8. All values of the scalar variable must be non-zero.

Table 2. Found coefficients of the NOPFs

	Q_0	Q_1	Q_2	Q_3	Q_4	Q_5
q_{j0}	0.3162277660 1683794	0	−0.359030465 25330308508	0	0.3760102939 1503510256	0
q_{j1}		0.0099087567 581727680483	0	−0.023723738 974404719567	0	0.0407333774 2261610065
q_{j2}			0.0003525085 350084986334 4	0	−0.001293696 619986197477 5	0
q_{j3}				0.0000131169 489449199365 89	0	−0.000067433 562532976568 096
q_{j4}					5.1116893449 436679076× × 10^{-7}	0
q_{j5}						2.1143486041 566852428× × 10^{-8}

According to (9), the variances of estimates of the univariate regression coefficients $\hat{\theta}_j, j = \overline{2, 5}$, range from $\mathrm{Var}\left(\hat{\theta}_2\right) = \mathrm{Var}(E) \cdot 1.8 \cdot 10^{-6}$ to $\mathrm{Var}\left(\hat{\theta}_5\right) = \mathrm{Var}(E) \cdot 4.47 \cdot 10^{-16}$ which is acceptable. The preliminary stage has been completed.

Step 1 is implemented for the term $\theta_8 x_1^2 x_3$ of the MPR. We carry out a virtual active experiment for the following values of the input variables: $x_{1i} = x_i, i = \overline{1, 10}$ (see Table 1); $x_{2i} = x_{2fix}^1 = 3, i = \overline{1, 10}$; $x_{3i} = x_{3fix}^1 = 7, i = \overline{1, 10}$; $x_{4i} = x_{4fix}^1 = 11, i = \overline{1, 10}$. As a result, our MPR model will take the form of a UPR:

$$Y(x) = \overbrace{\left(\theta_0 + \theta_2 x_{2fix}^1 + \theta_3 x_{3fix}^1 + \theta_4 x_{4fix}^1 + \theta_{10}\left(x_{2fix}^1\right)^2\left(x_{3fix}^1\right)^2 + \theta_{12}\left(x_{3fix}^1\right)^2\right)}^{b_0^1}$$

$$+ \overbrace{\left(\theta_1 + \theta_5 x_{2fix}^1 + \theta_6 x_{2fix}^1 x_{4fix}^1 + \theta_7 x_{3fix}^1 x_{4fix}^1 + \theta_9\left(x_{3fix}^1\right)^2\right)}^{b_1^1} x + \overbrace{\left(\theta_8 x_{3fix}^1\right)}^{b_2^1} x^2$$

$$+ \overbrace{0}^{b_3^1} x^3 + \overbrace{\left(\theta_{11} x_{2fix}^1\left(x_{3fix}^1\right)^2\right)}^{b_4^1} x^4 + \overbrace{0}^{b_5^1} x^5 + E$$

Remark 9. The superscript of the coefficient $b_j^i, j = \overline{0, 5}$, is the order number of the built UPR ($i = 1$ at step 1).

According to the found coefficients of the NOPFs (Table 2) and (4), (5), we obtain results of the virtual active experiment: $\hat{b}_2^1 = 49.03141$; $\hat{b}_4^1 = 1{,}322.999990725$.

Remark 10. $\hat{b}_3^1 = \hat{b}_5^1 = 0$, accurate to thousandths.

We find the estimates for the coefficients $\hat{\theta}_8$ and $\hat{\theta}_{11}$ by $\hat{b}_2^1$ and $\hat{b}_4^1$:

$$\hat{\theta}_8 = \frac{\hat{b}_2^1}{x_{3fix}^1} = 7.004 \ (\theta_8 = 7); \ \hat{\theta}_{11} = \frac{\hat{b}_4^1}{x_{2fix}^1 \left(x_{3fix}^1\right)^2} = 9 \ (\theta_{11} = 9).$$

Step 2 is implemented for the term $\theta_{10} x_2^2 x_3^2$ of the MPR. We carry out a virtual active experiment for the following values of the input variables: $x_{2i} = x_i, i = \overline{1, 10}$ (see Table 1); $x_{1i} = x_{1fix}^2 = 5, i = \overline{1, 10}$; $x_{3i} = x_{3fix}^2 = 11, i = \overline{1, 10}$; $x_{4i} = x_{4fix}^2 = 13, i = \overline{1, 10}$. The MPR becomes a UPR in which $b_2^2 = \theta_{10}\left(x_{3fix}^2\right)^2$.

According to the found coefficients of the NOPFs (Table 2) and (4), (5), we obtain the results of the virtual active experiment: $\hat{b}_2^2 = 968.03141$ and

$$\hat{\theta}_{10} = \frac{\hat{b}_2^2}{\left(x_{3fix}^2\right)^2} = 8 \ (\theta_{10} = 8)$$

Step 3 is implemented for the term $\theta_9 x_1 x_3^2$ of the MPR. We carry out a virtual active experiment for the following values of the input variables: $x_{3i} = x_i, i = \overline{1, 10}$ (see

Table 1); $x_{1i} = x_{1\text{fix}}^3 = 11, i = \overline{1, 10}$; $x_{2i} = x_{3\text{fix}}^3 = 13, i = \overline{1, 10}$; $x_{4i} = x_{4\text{fix}}^3 = 17, i = \overline{1, 10}$. As the result, the MPR becomes a UPR in which

$$b_2^3 = \theta_9 \cdot x_{1\text{fix}}^3 + \theta_{10} \cdot \left(x_{2\text{fix}}^3\right)^2 + \theta_{11} \cdot \left(x_{1\text{fix}}^3\right)^4 \cdot x_{2\text{fix}}^3 + \theta_{12}.$$

The unknown coefficients are θ_9 and θ_{12}. According to the found coefficients of the NOPFs (Table 2) and (4), (5), we obtain the result of the virtual active experiment: $\hat{b}_2^3 = 1.71436 \cdot 10^6$. We have got a single equation with two unknowns:

$$\hat{b}_2^3 = \theta_9 \cdot x_{1\text{fix}}^3 + \hat{\theta}_{10} \cdot \left(x_{2\text{fix}}^3\right)^2 + \hat{\theta}_{11} \cdot \left(x_{1\text{fix}}^3\right)^4 \cdot x_{2\text{fix}}^3 + \theta_{12}$$

To obtain the second linear equation, we carry out a virtual active experiment for the following values of the input variables: $x_{3i} = x_i, i = \overline{1, 10}$ (see Table 1); $x_{1i} = x_{1\text{fix}}^4 = 13, i = \overline{1, 10}$; $x_{2i} = x_{3\text{fix}}^4 = 19, i = \overline{1, 10}$; $x_{4i} = x_{4\text{fix}}^4 = 23, i = \overline{1, 10}$. The second equation is

$$\hat{b}_2^4 = \theta_9 \cdot x_{1\text{fix}}^4 + \hat{\theta}_{10} \cdot \left(x_{2\text{fix}}^4\right)^2 + \hat{\theta}_{11} \cdot \left(x_{1\text{fix}}^4\right)^4 \cdot x_{2\text{fix}}^4 + \theta_{12}$$

where $\hat{b}_2^4 = 4.88683 \cdot 10^6$.

Let us solve the system of linear equations

$$\begin{cases} \theta_9 \cdot x_{1\text{fix}}^3 + \hat{\theta}_{10} \cdot \left(x_{2\text{fix}}^3\right)^2 + \hat{\theta}_{11} \cdot \left(x_{1\text{fix}}^3\right)^4 \cdot x_{2\text{fix}}^3 + \theta_{12} = \hat{b}_2^3; \\ \theta_9 \cdot x_{1\text{fix}}^4 + \hat{\theta}_{10} \cdot \left(x_{2\text{fix}}^4\right)^2 + \hat{\theta}_{11} \cdot \left(x_{1\text{fix}}^4\right)^4 \cdot x_{2\text{fix}}^4 + \theta_{12} = \hat{b}_2^4. \end{cases}$$

We know θ_{10} and θ_{11}, so it is a system of two linear algebraic equations with two unknowns θ_9 and θ_{12}. Its solution is: $\hat{\theta}_9 = 0$ ($\theta_9 = 0$) and $\hat{\theta}_{12} = 10.003$ ($\theta_{12} = 10$).

Step 4 is implemented for the term $\theta_5 x_1 x_2$ of the MPR. We carry out a virtual active experiment for the following values of the input variables: $x_{1i} = x_{2i} = x_i, i = \overline{1, 10}$ (see Table 1); $x_{3i} = x_{3\text{fix}}^5 = 37, i = \overline{1, 10}$; $x_{4i} = x_{4\text{fix}}^5 = 10, i = \overline{1, 10}$. As the result, the MPR becomes a UPR in which

$$b_2^5 = \theta_5 + \theta_6 \cdot x_{4\text{fix}}^5 + \theta_8 \cdot x_{3\text{fix}}^5 + \theta_{10} \cdot \left(x_{3\text{fix}}^5\right)^2.$$

The unknown coefficients are θ_5 and θ_6. According to the found coefficients of the NOPFs (Table 2) and (4), (5), we obtain the result of the virtual active experiment: $\hat{b}_2^5 = 11,265.552807758$. We have got a single equation with two unknowns:

$$\hat{b}_2^5 = \theta_5 + \theta_6 \cdot x_{4\text{fix}}^5 + \hat{\theta}_8 \cdot x_{3\text{fix}}^5 + \hat{\theta}_{10} \cdot \left(x_{3\text{fix}}^5\right)^2$$

To obtain the second linear equation, we carry out a virtual active experiment for the following values of the input variables: $x_{1i} = x_{2i} = x_i, i = \overline{1, 10}$ (see Table 1); $x_{3i} = x_{3\text{fix}}^6 = 21, i = \overline{1, 10}$; $x_{4i} = x_{4\text{fix}}^6 = 50, i = \overline{1, 10}$. The second equation is

$$\hat{b}_2^6 = \theta_5 + \theta_6 \cdot x_{4\text{fix}}^6 + \hat{\theta}_8 \cdot x_{3\text{fix}}^6 + \hat{\theta}_{10} \cdot \left(x_{3\text{fix}}^6\right)^2$$

where $\hat{b}_2^6 = 3{,}929.27152675$.

Having solved the system of two equations with two unknowns, we obtain $\hat{\theta}_5 = 4.02$ ($\theta_5 = 4$) and $\hat{\theta}_6 = 5.001$ ($\theta_6 = 5$).

Step 5 is implemented for the term $\theta_7 x_1 x_3 x_4$ of the MPR. We carry out a virtual active experiment for the following values of the input variables: $x_{1i} = x_{3i} = x_{4i} = x_i$, $i = \overline{1, 10}$ (see Table 1); $x_{2i} = x_{2\text{fix}}^7 = 34$, $i = \overline{1, 10}$. The MPR becomes a UPR in which $b_3^7 = \theta_7 + \theta_8 + \theta_9$. According to the found coefficients of the NOPFs (Table 2) and (4), (5), we obtain the result of the virtual active experiment: $\hat{b}_3^7 = 13.00585$.

Thus, $\hat{b}_3^7 = \theta_7 + \theta_8 + \theta_9$ and $\hat{\theta}_7 = 6.001$ ($\theta_7 = 6$).

We have estimated all coefficients for nonlinear terms of the MPR given by a redundant representation with sufficient accuracy. Finding the estimates for the coefficients $\theta_j, j = \overline{0, 4}$, has been reduced to the MLR problem which, in the case of an active experiment, is solved efficiently by ordinary least squares if Var(E) is small.

7 The Software Architecture for MPR Coefficients Estimation Using an Active Experiment

Let us show the software architecture in the form of a universal library that implements the above presented mathematical support for estimating the values of unknown coefficients of an MPR based on the results of an active experiment. The object model of the library packages is shown in Fig. 1.

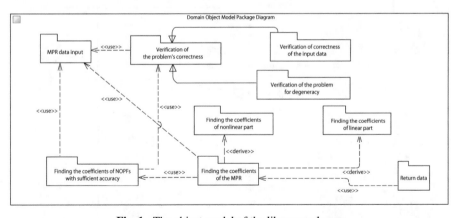

Fig. 1. The object model of the library packages

The functionality of the library from the user's point of view is shown in Fig. 2.

Figure 3 shows an example of using the library described above in abstract software with a layered architecture.

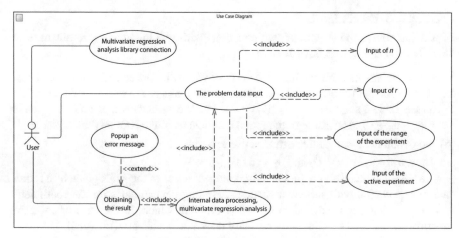

Fig. 2. The library's functional

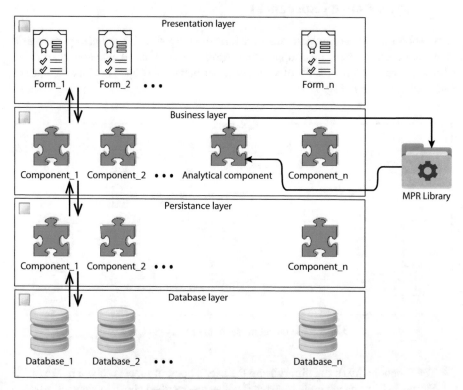

Fig. 3. The library usage example

8 Conclusions

1. Based on the analysis of the state-of-the-art related works, we have substantiated the relevance and significance, both in theoretical and practical terms, of building and using multivariate nonlinear (in particular, polynomial) regressions in various fields of science, technology, economics, and finance. We have shown that the existing well-known techniques for multivariate polynomial regression building do not ensure estimation of the coefficients at its nonlinear terms with small variance. This significantly narrows their scope. Thus, the solution of the problem of building a multivariate polynomial regression given by a redundant representation, using the results of a small number of active experiments with sufficiently small variances of coefficient estimates at its nonlinear terms, is extremely relevant both in theoretical terms and in terms of expanding their practical use to design modern intelligent various purpose information systems.

2. We have shown that the solution of this problem is reduced to the sequential building of univariate polynomial regressions and the solution of nondegenerate systems of linear equations. The variables of the equations are the unknown coefficients at nonlinear terms of the redundant representation of the multivariate polynomial regression. Thus, the proposed methodology and the algorithms of its implementation, for building a multivariate polynomial regression given by a redundant representation, using the results of an active experiment, are fundamentally different from all known methods for solving such a problem.

3. We have shown that only one previously found set of normalized orthogonal Forsythe polynomials is needed to estimate the coefficients at nonlinear terms of univariate polynomial regressions with a given small variance (which are the coefficients at the right-hand sides of linear equations). Thus, the accuracy of estimating the coefficients at nonlinear terms of a multivariate polynomial regression depends only on the accuracy of estimating the nonlinear terms of univariate polynomial regressions, since we have shown that the corresponding systems of linear equations are strictly nondegenerate, and the value of their determinant is in the general case a function of parameters, which values can be taken from a given interval.

4. We have shown that the calculation of the remaining coefficients of multivariate polynomial regression is reduced to the standard multivariate linear regression problem, which is solved efficiently using an active experiment.

5. We have given an example of estimating coefficients at nonlinear terms of a multivariate polynomial regression given by a redundant representation that contains four independent deterministic scalar variables and eight nonlinear terms. The example illustrates the effectiveness of the proposed method.

6. We have shown the architecture of the software that implements the algorithms for multivariate nonlinear polynomial regression given by a redundant representation. The software can be effectively used in special-purpose information systems.

References

1. Yu, L.: Using negative binomial regression analysis to predict software faults: A study of Apache Ant. Int. J. Inf. Technol. Comput. Sci. (IJITCS) **4**(8), 63–70 (2012). https://doi.org/10.5815/ijitcs.2012.08.08
2. Shahrel, M.Z., Mutalib, S., Abdul-Rahman, S.: PriceCop–price monitor and prediction using linear regression and LSVM-ABC methods for e-commerce platform. Int. J. Inf. Eng. Electron. Bus. (IJIEEB) **13**(1), 1–14 (2021). https://doi.org/10.5815/ijieeb.2021.01.01
3. Satter, A., Ibtehaz, N.: A regression based sensor data prediction technique to analyze data trustworthiness in cyber-physical system. Int. J. Inf. Eng. Electron. Bus. (IJIEEB) **10**(3), 15–22 (2018). https://doi.org/10.5815/ijieeb.2018.03.03
4. Isabona, J., Ojuh, D.O.: Machine learning based on kernel function controlled Gaussian process regression method for in-depth extrapolative analysis of Covid-19 daily cases drift rates. Int. J. Math. Sci. Comput. (IJMSC) **7**(2), 14–23 (2021). https://doi.org/10.5815/ijmsc.2021.02.02
5. Sinha, P.: Multivariate polynomial regression in data mining: methodology, problems and solutions. Int. J. Sci. Eng. Res. **4**(12), 962–965 (2013)
6. Kalivas, J.H.: Interrelationships of multivariate regression methods using eigenvector basis sets. J. Chemom. **13**(2), 111–132 (1999). https://doi.org/10.1002/(SICI)1099-128X(199903/04)13:2%3C111::AID-CEM532%3E3.0.CO;2-N
7. Ortiz-Herrero, L., Maguregui, M.I., Bartolomé, L.: Multivariate (O)PLS regression methods in forensic dating. TrAC Trends Anal. Chem. **141**, 116278 (2021). https://doi.org/10.1016/j.trac.2021.116278
8. Guo, G., Niu, G., Shi, Q., Lin, Q., Tian, D., Duan, Y.: Multi-element quantitative analysis of soils by laser induced breakdown spectroscopy (LIBS) coupled with univariate and multivariate regression methods. Anal. Methods **11**(23), 3006–3013 (2019). https://doi.org/10.1039/C9AY00890J
9. Nastenko, E.A, Pavlov, V.A, Boyko, A.L., Nosovets, O.K: Mnogokriterialnyi algoritm shagovoi regressii [Multi-criterion step-regression algorithm]. Biomedychna inzheneriya i tekhnolohiya **3**, 48–53 (2020). https://doi.org/10.20535/2617-8974.2020.3.195661 (in Russian)
10. Sergeev, V.V., Kopenkov, V.N., Chernov, A.V.: Comparative analysis of function approximation methods in image processing tasks. Comput. Opt. **26**, 119–122 (2004). (in Russian)
11. Babatunde, G., Emmanuel, A.A., Oluwaseun, O.R., Bunmi, O.B., Precious, A.E.: Impact of climatic change on agricultural product yield using k-means and multiple linear regressions. Int. J. Educ. Manag. Eng. (IJEME) **9**(3), 16–26 (2019). https://doi.org/10.5815/ijeme.2019.03.02
12. Vaccari, D.A., Wang, H.K.: Multivariate polynomial regression for identification of chaotic time series. Math. Comput. Model. Dyn. Syst. **13**(4), 395–412 (2007). https://doi.org/10.1080/13873950600883691
13. Ivahnenko, A.G.: Modelirovanie Slojnyh Sistem. Informacionnyi Podhod [Complex Systems Modeling. Informational Approach]. Vyshcha shkola, Kyiv (1987). (in Russian)
14. Jackson, E.C., Hughes, J.A., Daley, M.: On the generalizability of linear and non-linear region of interest-based multivariate regression models for fMRI data. In: 2018 IEEE Conference on Computational Intelligence in Bioinformatics and Computational Biology (CIBCB), pp. 1–8 (2018). https://doi.org/10.1109/CIBCB.2018.8404973
15. Pavlov, A.A., Chekhovskii, A.V.: Postroenie mnogomernoi polinomialnoi regressii. Aktivnyi eksperiment [Multidimensional polynomial regression construction. Active experiment]. Syst. Res. Inf. Technol. **2009**(1), 87–99 (2009). (in Russian)

16. Pavlov, A.A., Chekhovskii, A.V.: Postroenie mnogomernoi polinomialnoi regressii. Aktivnyi eksperiment s ogranicheniiami [Multidimensional polynomial regression construction. Active experiment with limitations]. Bull. Natl. Tech. Univ. "KhPI". Ser.: Syst. Anal. Control Inf. Technol. **4**, 174–186 (2009). (in Russian)
17. Pavlov, A.A., Chekhovskii, A.V.: Svedenie zadachi postroeniia mnogomernoi regressii k posledovatelnosti odnomernykh zadach [Reducing the problem of constructing multivariate regression to a sequence of one-dimensional problems]. Visnyk NTUU KPI Inform. Oper. Comput. Sci. **48**, 111–112 (2008). (in Russian)
18. Pavlov, A.A., Holovchenko, M.N.: Postroenie odnomernoi i mnogomernoi polinomialnoi regressii po izbytochnomu opisaniiu s ispolzovaniem aktivnogo eksperimenta [Univariate and multivariate polynomial regression construction from a redundant representation using an active experiment]. Bull. Natl. Tech. Univ. "KhPI". Ser.: Syst. Anal. Control Inf. Technol. **1**(3), 9–13 (2020). https://doi.org/10.20998/2079-0023.2020.01.02. (in Russian)
19. Pavlov, A.A., Kalashnik, V.V.: Rekomendacii po vyboru zony provedeniya aktivnogo eksperimenta dlya odnomernogo polinomialnogo regressionnogo analiza [Recommendations on the selection of the area of the active experiment for the one-dimensional polynomial regression analysis]. Visnyk NTUU KPI Inform. Oper. Comput. Sci. **60**, 41–45 (2014). (in Russian)
20. Pavlov, A.A., Kalashnik, V.V., Kovalenko, D.A.: Postroenie mnogomernoi polinomialnoi regressii. Regressiya s povtoryayuschimisya argumentami vo vhodnyh dannyh [Construction of a multi-dimensional polynomial regression. Regression with repetitive arguments in the input data]. Visnyk NTUU KPI Inform. Oper. Comput. Sci. **62**, 57–61 (2015). (in Russian)
21. Pavlov, A.A.: Estimating with a given accuracy of the coefficients at nonlinear terms of univariate polynomial regression using a small number of tests in an arbitrary limited active experiment. Bull. Natl. Tech. Univ. "KhPI". Ser.: Syst. Anal. Control Inf. Technol. **2**(6), 3–7 (2021). https://doi.org/10.20998/2079-0023.2021.02.01
22. Draper, N.R., Smith, H.: Applied Regression Analysis, 3rd edn. Wiley & Sons, New York (1998). https://doi.org/10.1002/9781118625590
23. Bolshakov, A.A., Karimov, R.N.: Metody obrabotki mnogomernykh dannykh i vremennykh riadov [Methods of multidimensional data and time series processing]. Goriachaia liniia–Telekom, Moscow (2007). (in Russian)
24. Zgurovsky, M.Z., Pavlov, A.A.: The four-level model of planning and decision making. In: Zgurovsky, M.Z., Pavlov, A.A. (eds.) Combinatorial Optimization Problems in Planning and Decision Making: Theory and Applications, Studies in Systems, Decision and Control, 1st edn., vol. 173, pp. 347–406. Springer, Cham (2019). https://doi.org/10.1007/978-3-319-989 77-8_8

Adaptive Routing Method in Scalable Software-defined Mobile Networks

Yurii Kulakov, Sergii Kopychko, and Iryna Hrabovenko[⊠]

National Technical University of Ukraine "Igor Sikorsky Kyiv Polytechnic Institute", 37 Peremohy Ave., Kyiv 03056, Ukraine

iryna.hrabovenko@gmail.com

Abstract. Mobile computer networks are characterized by frequent changes in topology that may lead to link loss between network nodes. Hence, such characteristic makes the routing process more complex, especially in networks of large size. Clustering helps to simplify the routing process when the mobile network size increases. In this paper, we propose a centralized method of route formation that excludes the recalculation of routing information on previously formed sections of the route. In the proposed method, first, clusters are constructed by the k-means clustering algorithm. To transfer data, inter-cluster and intra-cluster types of communication are used to reduce routing time complexity and routing overhead. Then modified wave distance routing algorithm is utilized to build a set of disjoint paths in every cluster that belongs to the inter-cluster route. As a result, the proposed routing algorithm implementation in the SDN controller eliminates the main drawback of the distance vector routing algorithm, namely, the problem of counting to infinity. The simulation results confirm the effectiveness of the proposed method in terms of route reliability and stability due to the cluster-based approach and the formation of additional paths from all intermediate nodes in the direction to the final node.

Keywords: Software-defined networking · Multipath routing · Streaming algorithm · Traffic balancing

1 Introduction

Mobile ad hoc networks (MANETs) usually consist of multiple wireless devices that create a self-organizing network. The node's transmission range in such a type of network is limited, and the environment is highly dynamic [1]. Regarding the continuous topology changes, routing algorithms should quickly respond to these changes, so the delay stays unnoticeable for the users. Therefore, improving the existing and creating new routing algorithms for mobile computer networks is still in demand [2].

One of the most effective routing techniques to guarantee stability and load balancing in mobile networks is to use clustering methods. Different clustering methods apply according to predefined conditions such as network topology or characteristics of nodes [3]. In this paper, the division of the network into clusters presents the k-means

Z. Hu et al. (Eds.): ICCSEEA 2022, LNDECT 134, pp. 304–313, 2022.
https://doi.org/10.1007/978-3-031-04812-8_26

algorithm, which refers to distributed clustering methods and helps create a two-level routing structure.

This article proposes to use SDN technology to increase the speed of transferring the routing information [4, 5]. In the SDN architecture, the control plane places on a separate device called a controller. As the controller helps efficiently exchange service information between the nodes of the data level, it is the most considerable part of SDN architecture [6–8].

In considered solutions authors propose to combine SDN architecture with cluster-based routing to enhance centralized mobile networks scalability. However these solutions don't ensure alternative paths generation for data transmission in cases when nodes leave cluster or links between the nodes is lost, which doesn't eliminate main path recalculation from the source node.

In response to frequent topology changes, the objective of this paper is to provide stable routing and dynamic reconfiguration procedures. Hence, this work integrates multipath routing taking into account the delay of transmission links, clustering technique to maintain scalability, and SDN architecture to decrease network overload.

2 Literature Review

Modern mobile computer networks are characterized by the large size and a varied composition of equipment. This complicates the process of managing this type of network, in particular, routing and traffic engineering. Among characteristic features of mobile networks there are frequent topology changes and rapid scalability. Therefore, high-quality support for data transmission in networks with such a fast and high data transmission rate is an extremely important task.

In order to solve these problems, the paper [9] investigates various methods and approaches to optimize data traffic in the network using SDN. The focus is on issues such as quality of service (QoS), load balancing and congestion control. Compared to traditional network, the main advantage of load balancing in SDN is that it is done centrally in SDN controller. This contributes to more efficient load balancing strategy. The SDN controller updates routing information for SDN switches by updating their routing tables in order to select the optimal route in terms of minimizing energy consumption and channel congestion. Compared to distributed traffic engineering and balancing methods, the centralized method eliminates the need to exchange service information between network switches. The paper [10] proposes a multipath routing algorithm that allows to increase the network performance by 10–15% due to reducing the volume of service packets. It allows reducing energy consumption by about 41% and increases the maximum utilization of communication channels by 60% in comparison with distributed methods of traffic engineering and balancing [8]. In addition, ping latency in network is reduced by 5–10%, and the number of control packets is reduced by 60–70% [11].

In turn, the use of multipath routing provides quick route rebuilding process after failures of links and network switches [12].

In paper [13], a method of centralized formation of routing information is proposed, which excludes the reconstruction of paths on the intermediate sections of the route.

In work [14] a method of traffic balancing is proposed, which, due to the centralized method of generating routing information in SDN controller and the use of multipath routing, simplifies traffic reconfiguration procedure and ensures the most uniform network load. Formation of routing information is carried out using wave routing algorithm [15]. At the same time, paths are formed from all intermediate nodes to the final node. This makes it possible to simplify the procedure for building routes between intermediate nodes of previously formed routes.

3   Proposed Routing Method in Large Mobile Computer Networks

This paper proposes the method of routing in a cluster-based network using SDN architecture. The given approach helps to implement an effective network functioning using a centralized managing type. Moreover, it contributes to an acceleration of routes construction and reconstruction processes.

The architecture of the proposed method consists of three levels of a software-defined network. At the application level, centralized routing algorithm functions. And the SDN controller uses this algorithm for processing the data received from local network controllers. It should be noted that with the help of the SDN controller it is quite easy to reconfigure the cluster structure in case of nodes exiting. The central controller is located at the control level where it performs interaction between application and data transmission layers. The data transmission level consists of grouped mobile nodes in so-called clusters. Each mobile node has one of three possible states. It can either be a cluster head boundary node or an ordinary member of the cluster. Each cluster head has a built-in local controller and is connected to the central controller. This organization allows local controllers to provide information about any changes in data transmission level supporting the global network overview by the central controller.

The division of the network into clusters is carried out by the k-means method, which refers to partitioning clustering methods. The method is based on a random or heuristic method of selecting k initial representatives. Every network object is then assigned to the nearest cluster. Then recalculation of a new representative for each cluster, using an average value of nodes within it, takes place.

Using the idea of the k-means method, the graph of the network is divided into k cluster domains. Each node is distributed to a cluster with the closest average location. The cluster head is elected based on the closest location to the center of the cluster. Cluster nodes that have connections with nodes of another cluster are established as boundary nodes that transfer inter-cluster information. All other nodes stay ordinary cluster members.

The proposed routing method has a two-level structure because it consists of intra-cluster and inter-cluster levels of routing. If the source and destination nodes are located in different clusters, then the combined routing method is used to find the path between them. Otherwise, only intra-cluster routing takes place.

The formation of paths between clusters using a modified multipath OSPF algorithm carries out in a central network controller.

Within clusters, the modified wave distance-vector algorithm operates to construct multiple paths. That makes it possible to maintain a set of paths among several cluster nodes. As paths between two remote nodes of the cluster build, the paths between its internal nodes are built as well. And at every next step of the algorithm, the next hop towards the destination node is found, till the complete route will be generated.

4   Routing Algorithm in Cluster-Based Mobile Networks

4.1   Route Building Algorithm

While designing routes the delay of transmitting routing information through communication links is considered as a metric. The given approach makes it possible to build several routes with minimum allowable delay and choose the most optimal for data transmission. The processes of route building and packets transmission are carried out in accordance with the algorithm presented below.

Pseudocode of the algorithm

Notations:
V_i, X_i, U_i: cluster members
V_j, X_j, U_j: boundary nodes
V_{ch}, X_{ch}, U_{ch}: cluster heads
V_{rt}: routing table of V_{ch}
L_i: optimal routing path
S_{rt}: routing table built by SDN controller
$W_{k+1} = \{V_i \mid i = 1, \dots m\}$: set of nodes adjacent to $W_k = \{V_i \mid i = 1, \dots m\}$
$T_i\{V_n, V_i, d_i\}$: distance vector of two adjacent nodes
d_k, D_l: link and route delays

```
1: begin
2: V_i would sent data to Ui;
3: V_i sends request to V_ch;
4: if (V_ch received request from V_i) then
5:     if (V_rt maintains L_i) then
6:         go to 20;
7:     else
8:         if (U_i ∈ N(V_ch)) then
9:             go to 12;
10:        else (V_ch sends request to SDN controller) then   /* inter-cluster routing */
11:            SDN controller builds and sends Srt to V_ch, X_ch, U_ch
12:                begin   /* inner-cluster routing */
13:                k = 0;   /* initializing k step */
14:                for k = k + 1 step 1 form W_k+1 = {V_i | i = 1,...m}
15:                    if W_k+1 = ∅ then go to 19
16:                    for i = 1 step 1 to m calculate T_i{V_n, V_i, d_i}
17:                        if d_k > D_l then D_l = d_k
18:                    go to 14;
19:                end
20: V_ch sends L_i to V_i;
21: end
```

Due to the presented algorithm, we consider the two cases. Either both initial and final nodes are in the same cluster or different clusters. In the first case, the cluster head *Vch* of starting cluster builds all possible routes between the nodes of its cluster, then stores them in its routing table. Next, it transmits the best-found route to starting node *Vi*, respectively, then the process of transmitting information on the specified route *Li* to destination node *Ui* starts. If nodes are in different clusters, first, the inter-cluster routing takes place with the help of a central network controller. Next, an intra-cluster routing process takes place within the clusters of the generated inter-cluster route. Thus, since routing information generation and updating is centralized - it significantly reduces the time of route formation.

The process of building paths within each cluster takes place in the local cluster controller sequentially starting from the initial node $V_i \in W_{k+1}$ and $V_j \in W_k$ of adjacent sets. Adjacent sets of nodes are the sets $W_{k+1} = \{V_i\}$ and $W_k = \{V_j\}$ with common links $L_{i,j}$, where $V_i \in W_{k+1}$ and $V_j \in W_k$. The formation of paths begins with the source node V_s at $i = 1$. In this case, the set $W_1 = \{V_s\}$ and the set $W_2 = \{V_i\}$ – is the set of nodes adjacent to the node V_s. Then, for the nodes $V_i \in W_2$ adjacent to the node V_s, routing tables are formed in the direction of the node V_d. During the next wave of routing, the building of routes from nodes $V_i \in W_{k+1}$ to nodes $V_j \in W_k$ continues. The path formation process lasts until all paths between nodes V_s and V_d are found. After algorithm completion, we will have generated tables of routes from nodes $V_i \in W_{k+1}$ to V_d.

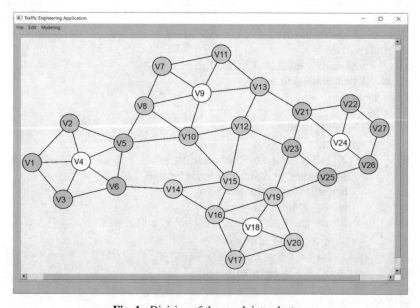

Fig. 1. Division of the graph into clusters

4.2 Routing Algorithm Simulation

Figure 1 shows the division of the graph into clusters using the k-means method into the designed simulation application (where weights of delay on the links were set in the process of graph creation).

The network graph consists of four clusters each cluster has a corresponding number - *CL1*, *CL2*, *CL3,* and *CL4*. Nodes *V4*, *V9*, *V18*, and *V24* are cluster heads of the formed clusters. On the example of a given graph, we consider the route computation between nodes *V1* ∈ *CL*1 and *V27* ∈ *CL*4. Since these nodes are located in different clusters, routes between the clusters *CL1* and *CL4* will, first, be built based on the clusters relationship table (Table 1). In this case, two paths are computed:

P1: = CL1 > CL2 > CL4 and P2: = CL1 > CL3 > CL4.

Table 1. Clusters relationships table

	CL1	CL 2	CL3	CL 4
CL 1	-	1	1	0
CL 2	1	-	1	1
CL 3	1	1	-	1
CL 4	0	1	1	-

Intra-cluster building process of disjoint paths between boundary nodes of the formed inter-cluster route begins after computing the inter-cluster routes.

For each cluster, a table is formed, including delay weights between cluster nodes as well as delays on the links with an adjacent cluster. The adjacency nodes table of CL1 is presented in Table 2. It is constructed similarly for all other clusters.

Table 2. The adjacency table of the *CL1*

	V1	V2	V3	V4	V5	V6	V8	V10	V14
V1	-	0.1	0.2	0.2					
V2	0.1	-		0.4	0.3				
V3	0.2		-	0.1		0.4			
V4	0.2	0.4	0.1	-	0.3	0.2			
V5		0.3		0.3	-	0.2	0.4	0.7	
V6			0.4	0.2	0.2	-			0.8
V8					0.4		-		
V10					0.7			-	
V14						0.8			

Then, for all adjacent nodes of selected inter-cluster paths, intra-cluster paths nodes tables of distance vectors are computed.

The computation of distance vectors starts from the gateway node with a minimum delay to the node of the adjacent cluster. In this case, the delay between adjacent gateway nodes *V5* and *V8* is 0.4, and between *V6* and *V14* is 0.8. Therefore, the *V5* node is selected as the starting node. Firstly, a set of nodes $W1 = \{V5\}$ is formed. Next, based on Table 3, the adjacent set of nodes $W2 = \{V2, V4, V6\}$ is formed. The next step is to build routing tables for the set of nodes $W2 = \{V2, V4, V6\}$ to the set of nodes $W1 = \{V5\}$ taking into account the delay of the path to node *V5*. Table 3 shows an example of the routing table of node *V4*.

Table 3. The table of distance vectors of *V4* at the first step

Cluster number	Destination node	Adjacent node	Delay
CL1	V5	V5	0.3

At the next step, routing tables are built from the set of nodes $W3 = \{V1, V3\}$ to node *V5* through adjacent nodes of the set $W2 = \{V2, V4, V6\}$.

Then the path is calculated from *V1* to *V5* through *V6*.

Table 4 shows an example of the routing table of *V1*.

Table 4. Tables of distance vectors of *V1*

Cluster number	Destination node	Adjacent node	Delay
CL1	V5	V2	0.4
CL1	V5	V4	0.5
CL1	V5	V3	0.8

In this case, route $V1 > V2 > V5$ has a lower delay.

5   Results and Discussions

After the intra-cluster routing tables building process completes, nodes *V4*, *V9*, *V18*, and *V24* compute disjoint routes between gateway nodes of their clusters, taking into account the path metrics.

The final calculation step is the complete transmission route between the nodes *V1* and *V27*. The results of the algorithm simulation are shown in Fig. 2.

From the table of results, first, we may state that the algorithm calculated three disjoint paths in the first and second clusters of the route and two disjoint paths in the fourth cluster. Secondly, to form the optimal path from each routing table are selected paths with lesser delay and hops count.

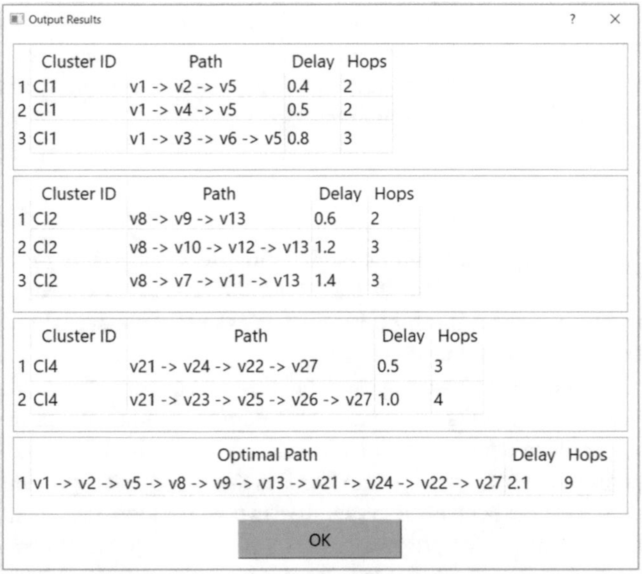

	Cluster ID	Path	Delay	Hops
1	Cl1	v1 -> v2 -> v5	0.4	2
2	Cl1	v1 -> v4 -> v5	0.5	2
3	Cl1	v1 -> v3 -> v6 -> v5	0.8	3

	Cluster ID	Path	Delay	Hops
1	Cl2	v8 -> v9 -> v13	0.6	2
2	Cl2	v8 -> v10 -> v12 -> v13	1.2	3
3	Cl2	v8 -> v7 -> v11 -> v13	1.4	3

	Cluster ID	Path	Delay	Hops
1	Cl4	v21 -> v24 -> v22 -> v27	0.5	3
2	Cl4	v21 -> v23 -> v25 -> v26 -> v27	1.0	4

	Optimal Path	Delay	Hops
1	v1 -> v2 -> v5 -> v8 -> v9 -> v13 -> v21 -> v24 -> v22 -> v27	2.1	9

Fig. 2. The results of algorithm simulation

Since in the mobile networks there are situations of link loss between the cluster nodes, we will consider the route reconfiguration process in case node *V21* leaves cluster *CL4*. The content of routing tables and the optimal route after topology changes as shown in Fig. 3.

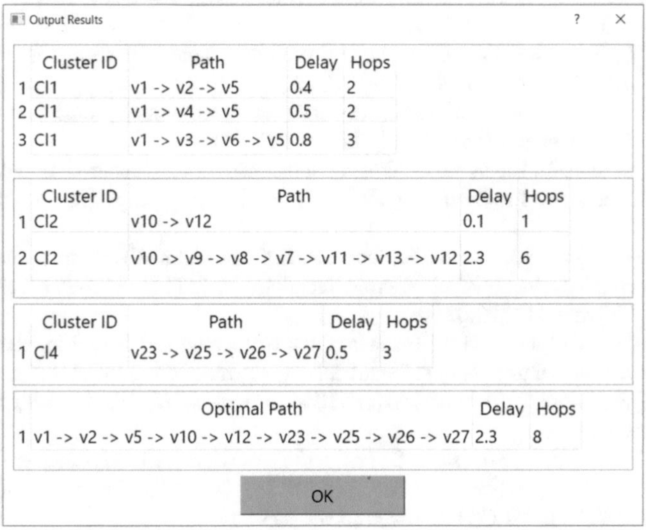

	Cluster ID	Path	Delay	Hops
1	Cl1	v1 -> v2 -> v5	0.4	2
2	Cl1	v1 -> v4 -> v5	0.5	2
3	Cl1	v1 -> v3 -> v6 -> v5	0.8	3

	Cluster ID	Path	Delay	Hops
1	Cl2	v10 -> v12	0.1	1
2	Cl2	v10 -> v9 -> v8 -> v7 -> v11 -> v13 -> v12	2.3	6

	Cluster ID	Path	Delay	Hops
1	Cl4	v23 -> v25 -> v26 -> v27	0.5	3

	Optimal Path	Delay	Hops
1	v1 -> v2 -> v5 -> v10 -> v12 -> v23 -> v25 -> v26 -> v27	2.3	8

Fig. 3. Route reconfiguration results

Comparing the results of Table 2 and Table 3, we may see how the routes within *CL2* and *CL3* become reconfigured, so the total route and its metrics have changed. The given results demonstrate the effectiveness of the proposed routing method due to its flexibility in terms of dynamic network structure.

6 Conclusion

The paper proposes a method of traffic engineering, which, by taking into account the peculiarities of SDN organization, in particular, the presence of a central controller, makes it possible to reduce the time it takes to calculate the set of the routes, thus simplifying the procedure of traffic balancing.

Given the importance of ensuring the reliability and quality of data transmission in modern large mobile networks, this paper considers the method of traffic engineering considering the delay in communication channels. A distinctive feature of the proposed algorithm against the known routing algorithms is that it supports dynamic route reconfiguration, therefore there is no need to spend extra time for route recalculation. Another significant characteristic of the proposed method is that the choice of data transmission route considers the quality of service metrics, which reduces the likelihood of choosing the shortest but suboptimal route.

In further works, when building a route, we consider predicting the load based on the principle of similarity and choosing the path taking into account the change in route delay.

References

1. Hinds, A., et al.: A review of routing protocols for mobile ad-hoc networks (MANET). Int. J. Inf. Educ. Technol. **3**(1), 1 (2013)
2. Rezaee, M., Yaghmaee, M.: Cluster based routing protocol for mobile ad hoc networks. INFOCOMP J. Comput. Sci. **8**(1), 30–36 (2009)
3. Anupama, M., Sathyanarayana, B.: Survey of cluster based routing protocols in mobile adhoc networks. Int. J. Comput. Theory Eng. **3**(6), 806 (2011)
4. Isong, B., Kgogo, T., Lugayizi, F.: Trust establishment in SDN: controller and applications. Int. J. Comput. Netw. Inf. Secur. (IJCNIS) **9**(7), 20–28 (2017). https://doi.org/10.5815/ijcnis.2017.07.03
5. Sahoo, K.S., Mishra, S.K., Sahoo, S., Sahoo, B.: Software defined network: the next generation internet technology. Int. J. Wirel. Microw. Technol. (IJWMT) **7**(2), 13–24 (2017). https://doi.org/10.5815/ijwmt.2017.02.02
6. Abdullah, M.Z., Al-awad, N.A., Hussein, F.W.: Evaluating and comparing the performance of using multiple controllers in software defined networks. Int. J. Mod. Educ. Comput. Sci. (IJMECS) **8**, 27–34 (2019). https://doi.org/10.5815/ijmecs.2019.08.03. In MECS http://www.mecs-press.org
7. Singh, K.S., Kumar, N., Srivastava, S.: PSO and TLBO based reliable placement of controllers in SDN. Int. J. Comput. Netw. Inf. Secur. (IJCNIS) **2**, 36–42 (2019). https://doi.org/10.5815/ijcnis.2019.02.05. In MECS http://www.mecs-press.org
8. Gamess, E., Tovar, D., Cavadia, A.: Design and implementation of a benchmarking tool for openflow controllers , Int. J. Inf. Technol. Comput. Sci. (IJITCS) **10**(11), 1–13 (2018). https://doi.org/10.5815/ijitcs.2018.11.01. http://www.mecs-press.org/

9. Matnee, Y.A., Abooddy, C.H., Mohammed, Z.Q.: Analyzing methods and opportunities in software-defined (SDN) networks for data traffic optimizations. Int. J. Recent Innov. Trends Comput. Commun. (IJRITCC) **6**(1), 75–82 (2018). http://www.ijritcc.org

10. Rajasekaran, K., Balasubramanian, K.: Energy conscious based multipath routing algorithm in WSN. Int. J. Comput. Netw. Inf. Secur. (IJCNIS) **1**, 27–34 (2016). https://doi.org/10.5815/ijcnis.2016.01.04. In MECS http://www.mecs-press.org/

11. Kumar, P., Dutta, R., Dagdi, R., Sooda, K., Naik, A.: A programmable and managed software defined network. Int. J. Comput. Netw. Inf. Secur. (IJCNIS) **12**, 11–17 (2017). https://doi.org/10.5815/ijcnis.2017.12.02. In MECS http://www.mecs-press.org/. Accessed Dec 2017

12. Moza, M., Kumar, S.: Analyzing multiple routing configuration. Int. J. Comput. Netw. Inf. Secur. (IJCNIS) **5**, 48–54 (2016). https://doi.org/10.5815/ijcnis.2016.05.07. In MECS http://www.mecs-press.org/. Accessed May 2016

13. Kulakov, Y., Kohan, A., Kopychko, S.: Traffic orchestration in data center network based on software-defined networking technology. In: Hu, Z., Petoukhov, S., Dychka, I., He, M. (eds.) Advances in Computer Science for Engineering and Education II, ICCSEEA 2019. Advances in Intelligent Systems and Computing, vol. 938, pp. 228–237. Springer, Cham (2020). https://doi.org/10.1007/978-3-030-16621-2_21

14. Kulakov, Y., Kopychko, S., Gromova, V.: Organization of network data centers based on software-defined networking. In: Hu, Z., Petoukhov, S., Dychka, I., He, M. (eds.) Advances in Computer Science for Engineering and Education, ICCSEEA 2018. Advances in Intelligent Systems and Computing. vol. 754. pp. 447–455. Springer, Cham (2019). https://doi.org/10.1007/978-3-319-91008-6_45

15. Kulakov, Y., Kogan, A.: The method of plurality generation of disjoint paths using horizontal exclusive scheduling. Adv. Sci. J. **10**, 16–18 (2014). https://doi.org/10.15550/ASJ.2014.10. ISSN 2219–746X

Big Data Methods in Learning Analytics System by Using Dask Cluster Computer Framework

Fail Gafarov[✉] and Lilija Khairullina

Institute of Computational Mathematics and Information Technologies, Kazan Federal University, Kremlyovskaya Street 18, Kazan 420008, Russia
fgafarov@yandex.ru

Abstract. Large datasets are generated and stored in different commercial and state informational systems. These datasets contain detailed information about the dynamics of processes for long periods of time, and can be used for statistical analysis and for predictive modelling in decision making systems. The analysis of such large volumes of data needs to develop a data processing pipelines based on BigData methods by using computational clusters. In this work we present a data processing framework based on using Python language based Dask library for parallel computing in cluster computing systems and high-level Pandas library for subsequent data processing (correlation and regression analysis). By using the developed framework, we performed a large-scale empirical study of the educational environment parameter's influence on the school student's academic performance, based on the analysis of the large datasets stored in the "Electronic education in the Republic of Tatarstan" system for time period from 2015 till 2020 year.

Keywords: BigData · Educational data mining · Dask · Correlation analysis · Regression analysis

1 Introduction

The technological development leads to exponential growth in the amounts of generated data, giving to organizations more opportunities to grow and improve the quality of their services. Every year, there has been an increase in the number of organizations that are starting to use BigData technologies and machine learning tools to improve the efficiency in decision making, customer experience personalizing, business processes optimization based on big data analysis [1]. BigData technologies allow to process and to analyze a large amount of unstructured data, and can lead to new discoveries in various fields of science [2].

This work is devoted to the use of Big Data methods and technologies for the analysis of data describing educational process, collected in the State Information System "Electronic Education in the Republic of Tatarstan" for the period from 2015 to 2020. This database contains more than two billion information units, including information on the progress of more than a million students and the professional activities of more

Z. Hu et al. (Eds.): ICCSEA 2022, LNDECT 134, pp. 314–323, 2022.
https://doi.org/10.1007/978-3-031-04812-8_27

than 120,000 teachers. To study the educational environment parameter's influence on the school student's academic performance we have to perform a statistical analysis of this raw data.

The main contributions of this work:

- For the first time we demonstrated the possibility of using the Dask distributed computing framework [3] in educational analytics.
- Analytical framework that allows performing a high-performance analysis of large educational data sets has been developed, by using Dask framework.
- A large scale analysis of big data from the State Information System "Electronic Education in the Republic of Tatarstan" was carried out by using the methods of mathematical statistics.
- Conclusions, that characterize the influence of the educational environment on the academic success of school students and the professional success of school teachers, have been obtained.

The framework developed in this work can be quickly expanded and expanded by additional modules implementing new specific data analysis methods (for example, machine learning methods, etc.). One of the immediate applications is the analysis of the effects, caused by school's transition to distance learning because of COVID, on the academic performance of school students. It should also be noted that the proposed framework can be reconfigured for use for big data analysis in other areas (for example, in medicine, or in different state informational systems).

2   Related Works

Currently, there is a growing interest of researchers in the fields related to educational analytics and Educational data mining (EDM), and especially in the use of big data methods. A systematic literature review of educational data mining in mathematics and science education is presented in paper [4]. The paper [5] describes common sources of data, key objectives of EDM and recent findings in this field. In their study authors used datasets provided by the Australian Bureau of Statistics for analyzing the patterns of success (and failure) of students and providing insight into the possible steps that could be taken to improve outcomes. Currently, the intellectual analysis of educational big data and the use of machine learning technologies [6] is considered in the education industry as a critical technology that will make the most important transition to evidence-based educational policy and evidence-based pedagogy, as well as individualize the educational process through taking into account personal characteristics [7], educational and socio-economic factors influencing the effectiveness of the teacher's professional activity and the student's performance [8]. For example, the multimedia technology method of teaching has impacted positively on the improvement and interest of secondary school students in Mathematics [9]. By using paired sample t-test analysis authors showed that there was a significant difference in the mean score for the multimedia method than the tradition method. It should be noted that in modern literature, much attention is paid to the use of mathematical methods for identification factors affecting academic performance [10]. For studying the influence of educational system parameters on academic performance correlation analysis, factorial, taxometric analysis and other methods of statistical analysis are used [11].

Big data methods are very important in educational systems, and can lead to very important improvements [12]. High performance analysis of educational data by using Big data methods opens up new prospects for creating a new educational experience as an early vocational guidance, new adaptive educational trajectories, for control of professional trajectories, openness and transparency of education [13].

Apache Spark is the most commonly used tool for dealing with big data [14]. Hadoop and Spark-based scalable algorithms addressing the frequent itemset mining problem in the Big Data analytics, by using detailed theoretical and experimental comparative analyses, presented in the review [15]. In paper [16] author showed that Apache Spark highly outperforms MapReduce, and as the data grows Spark becomes more reliable and fault tolerant, and increasing of the number of blocks on the HDFS, also increases the run-time of both the MapReduce and Spark programs. The work [17] presents a recommender system built with help of Apache Spark, based on HPC implementation of the bisecting KMeans clustering algorithms. Researcher performed detailed analysis of KMeans and Bisecting KMeans clustering algorithm's performance on Apache Spark environment.

Dask is a newer system for dealing with big data, and has been actively gaining popularity recently. It is a new flexible parallel computing library for high-performance data analytics, designed primarily to provide scalability and extensibility to existing Python packages (like NumPy, SciPy, Pandas and others), and it can run locally or scaled to run on a cluster. Dask allows running a Phyton programs in parallel mode with minimal code changes, taking advantage of the full computing power [3]. Dask also provides a real-time dashboard that highlights key metrics of user processing tasks such as project progress, memory consumption, and more [18]. In paper [19] it had been demonstrated the applicability of Dask to producing simple and interpretable rules for highlighting anomalies in application log data to scale and in a distributed environment. Authors implemented the k-means algorithm to separate anomalies from normal events, and a gradient tree boosting classification model to produce the interpretable meaningful rationale rule set for generalizing its application to a massive number of unseen events.

Parallel implementation, of data mining techniques research results on school record analytics, by using Spark platform was presented in paper [20]. This work is devoted to finding frequent patterns on academic records for all students of Universidad Michoacana (UMSNH) from 2005 to 2016, and for searching relevant frequent pattern subsets. Authors reported the superior performance achieved by parallel implementation compared to non-parallel versions of the application. In study [21] was presented a Rapid Miner application to assist in processing student's performance data. By using three models (Decision Tree Algorithm model, Naive Bayes and K-Means) authors showed how it is possible to predict the level of student's performance for increasing decision making system's results.

3   The Architecture of Proposed System

One of the best tools for intellectual data analysis in Python is the Pandas library. But Pandas does not support parallel processing mechanisms, and therefore this library does not use full advantage of the modern cluster-based data processing capabilities. One of the effective solutions of this problem can be done by using the Dask library for processing big data sets [18].

By using Dask and Pandas packages, we have built an information and analytical system for processing and studying big data sets containing information about activities of teacher, school administration and student's performance. The anonymized data, describing different entities (grades, lesson topics, timetable, information about teachers and students) was obtained as separate csv of xml files. The raw data was loaded into data structures called Dask DataFrames. It is a large parallel DataFrame composed of many smaller Pandas DataFrames, split along the index, and one Dask DataFrame operation triggers many operations on the constituent Pandas DataFrames.

Because the raw data has been scattered in different files it is often necessary to use the operations of merging (merging) entities by some key. We used the Dask DataFrame.merge() method to merge data frames obtained by loading different data files. Grouped, reduced and aggregated DataFrames obtained from raw datasets by using Dask methods subsequently were processed by using mathematical statistics methods of the Pandas, Numpy, SciPy libraries. In our study the school student's training level was used (calculated as the mean annual mark in each grade for each subject) as an indicator of academic performance.

In order to obtain probabilistic and statistical models to describe the educational process, we used the methods of mathematical statistics (Spearman's correlation coefficient, Parson's $\chi 2$ test, Student's t-test, and one-way ANOVA test). High-level view of the analytical system is presented in Fig. 1.

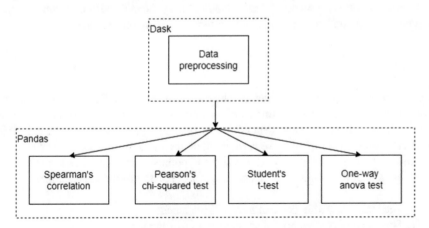

Fig. 1. The system architecture

$$t = \frac{m_1 - m_2}{\sqrt{s^2(\frac{1}{n1} + \frac{1}{n2})}}$$

Correlation analysis methods were implemented into system by calculation of Spearman's correlation coefficient to determine the relationship between different variables.

To test the hypothesis about the equality of the general means of two independent samples the classic Student's t-test had been implemented. In this formula m_1 and m_2 are the means of the two groups being compared, s- is the pooled standard error of the two groups, and n1 and n2 are the number of observations in each of the groups. To test the hypothesis about the correspondence of the empirical distribution to the assumed theoretical distribution Pearson's goodness-of-fit test is included by using the formula

$$\chi^2 = \sum_{i=1}^{n} \frac{(O_i - E_i)^2}{E_i}$$

where O_i – is the number of observations of type i, n- the total number of observations, E_i -is the expected (theoretical) count of type i, asserted by the null hypothesis. We used the chi-squared statistic's p-value by comparing the value of the statistic to a chi-squared distribution.

The one-way analysis of variance (ANOVA) was implemented to investigate the significance of differences between the means in different groups [22]. The analytical system has been deployed on a high-performance computing was carried out on a cluster consisting of 4 virtual machines with 32 GB of RAM and 16 computing cores. As already mentioned above, in computational cluster-based processing of big data, Apache Spark [13] based tools are also quite often used. However, Dask has advantages over Spark, which are important for our project. Dask is smaller, lighter and simpler than Spark, and flexible as Pandas with more power to compute on a cluster based computational systems, and fully based on Python. It works well on a single machine to make use all of cores on laptop and processes larger- than-memory data. Therefore, program scripts are easy to debug on personal computers before deploying on a cluster.

4   Results

In this section we present main results obtained by using analytical system presented above. Firstly, we analyzed the impact of teachers' education on the average grades of their students. The study about dependence of teacher's mean marks on the university where he received his diploma, we carried out on the basis of the Pearson's criterion of goodness, which is most often used to test the hypothesis that a certain sample belongs to the theoretical distribution. In this case, we based on the following hypotheses:

- H0: The University affects mean marks of the teacher.
- H1: The University does not affect mean marks of the teacher

We used the significance probability (p-value) is a measure in statistical hypothesis testing, and assumed that if the p-value < 0.05, then the null hypothesis is not refuted. Top 5 universities with the largest number of teachers were taken for the analysis, student's mark had been divided into 5 clusters by using k-means method. A separate analysis was made for primary school teachers (grades 1–4) and subject teachers (grades 5–11). The values of p-value for several subject teachers (Mathematics, Russian language, Biology, Russian literature) are presented in the Table 1.

Table 1. Pearson's chi-square test p-values for subject teachers

Years	Math	Russ. lang	Biology	Russ. lit
2015–2016	0.056	0.003	0.060	0.022
2016–2017	0.000	0.069	0.091	0.033
2017–2018	0.001	0.000	0.143	0.371
2018–2019	0.000	0.000	0.376	0.266

By using Pearson's chi-square test's p-values, we divided the subjects into two groups (the first group- teacher's education plays a role, the second group- the teacher's education does not plays a significant role). The first group includes: mathematics, Russian language, geography, history, social studies, English; the second group includes: physical culture, biology, Russian literature, physics, chemistry, IT, the Tatar language, technology, life safety, Tatar literature, music, fine arts.

Next, by using the results correlation analysis we concluded that primary school teachers (grades 1–4) have a positive correlation in mean marks with teacher experience time (the longer times the teaching experience of a primary school teacher, the higher the quality of student knowledge). Also, we investigated the correlation between age of the teachers and the average grade obtained by his students. In elementary school teacher we obtained a weak positive correlation in mathematics, Russian language, English and weak negative correlations in technology and music. For secondary school teachers, we obtained a weak negative correlation in Mathematics, Physics, Geography, History, Computer science, Life safety, Technology, Arts. (See Table 2, for some subjects). This means that the young teachers give high marks to their students than older colleagues.

Based on a comparative analysis of the average grades given by male and female teachers (by using the Student's t-test), it is shown that in primary school, male teachers in physical education are less successful than female teachers, and among subject teachers, men are more successful than women (see Table 3).

Table 2. Spearman's correlation coefficient between average marks and teacher's age (5–11 grades)

Years	Math	Physics	Eng. lang	Comp. sci
2015–2016	−0.056**	−0.083**	0.02	−0.097**
2016–2017	−0.048**	−0.07*	−0.042*	−0.09**
2017–2018	−0.079**	−0.088**	0.055**	−0.074**
2018–2019	−0.076**	−0.105**	0.037*	−0.097**
2019–2020	−0.1**	−0.08**	0.056**	−0.125**

Analysis of variance (ANOVA-test) was applied to determine the effect of the teacher's qualification category on academic performance. Here the null hypothesis

Table 3. Comparison of average marks (grades 5–11): male teachers- female teachers (Student's t-test for independent samples, * - p-value < 0.05, ** - p-value < 0.005)

Years	Math	Russ. lang	Biology
2015–2016	3.77–3.71**	3.81–3.76*	4.06–3.99**
2016–2017	3.79–3.72**	3.82–3.76*	4.06–3.97**
2017–2018	3.82–3.74**	3.78–3.76	4.03–3.98**
2018–2019	3.86–3.76**	3.82–3.76	4.06–3.98**
2019–2020	3.9–3.81**	3.78–3.75	4.06–3.99**

states that the mean for all groups is equal. The p-value was used to decide whether an alternative hypothesis can be accepted or not. If the p-value is less than 0.05, we reject the null hypothesis in favor of the alternative: this means that at least one group mean is significantly different. The teachers grade qualification category grouped into four categories:

- No qualification category
- First qualification category
- Second qualification category
- Highest qualification category

The charts (see Fig. 2) show the distribution of grades by subject for secondary school. The box plot shows the median, lower and upper quartiles, and the minimum and maximum values of the sample.

The p-value obtained from the ANOVA analysis is significant ($p < 0.05$) for all subject, and therefore, we conclude that there are significant differences between marks for different qualification categories of the teacher.

The data that support the findings of this study are available from The Ministry of Education and Science of the Republic of Tatarstan. Restrictions apply to the availability of these data, which were used under license for this study. Data are available from the corresponding author with the permission of The Ministry of Education and Science of the Republic of Tatarstan.

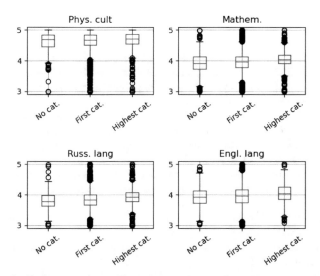

Fig. 2. Boxplots of student's mean marks grouped by categories of teachers for primary school

5 Conclusions

In this work we developed a data processing framework for a large-scale empirical study of school student's academic success and teacher's professional activity based on HPC computing library for distributed computing in Python. Here we for the first time, the possibilities of using Dask for the analysis of large volumes of educational data has been demonstrated. By using the framework, we performed statistical analysis of big data stored in the system "Electronic education in the Republic of Tatarstan" for time period from 2015 till 2020 year, and we found interesting regularities regarding the educational trajectory of school students and the professional activity of teachers. The proposed system provides new opportunities in educational analytics and will allow researchers and school management staff to conduct a quick and effective intellectual data analysis stored in educational system.

The system, reported in this can be easily scaled up depending on the volumes of the studied data, and it is also possible to add new modules for effective data preprocessing for more complex data processing algorithms. In this work, we have built a static computing cluster, what is one of the limitations of our system. For a more efficient use of computing resources, it is also desirable to use Kubernetes [23] for elastic deployment of a computing system in a cloud. In the future we will extend the system by machine learning tools, as well as by using the Kubernetes technology for elastic deploying in Amazon or Google clouds.

Acknowledgment. The research was supported by the Russian Science Foundation under grant № 22-28-00923, https://rscf.ru/project/22-28-00923/.

References

1. Elgendy, N., Elragal, A.: Big data analytics: a literature review paper. Lect. Notes Comput. Sci. **8557**, 214–227 (2014). https://doi.org/10.1007/978-3-319-08976-8_16
2. Gandomi, A., Haider, M.: Beyond the hype: Big data concepts, methods, and analytics. Int. J. Inf. Manage. **35**(2), 137–144 (2015)
3. Rocklin, M.: Dask: parallel computation with blocked algorithms and task scheduling. In: Proceedings of the 14th Python in Science Conference, Austin, pp. 130–136 (2015)
4. Shin, D., Shim, J.: A, systematic review on data mining for mathematics and science education. Int. J. Sci. Math. Educ. **19**, 639–659 (2021)
5. Alom, B.M.M., Courtney, M.: Educational data mining: a case study perspectives from primary to university education in Australia. Int. J. Inf. Technol. Comput. Sci. (IJITCS) **10**(2), 1–9 (2018)
6. Luan, H., et al.: Challenges and future directions of big data and artificial intelligence in Education. Front. Psychol. **11**, 2748 (2020)
7. Lodge, J.M., Corrin, L.: What data and analytics can and do say about effective learning. npj Sci. Learn **2**, 1–2 (2017)
8. Buniyamin, N., bin Mat, U., Arshad, P.M.: Educational data mining for prediction and classification of engineering students achievement. In: 2015 IEEE 7th International Conference on Engineering Education (ICEED), pp. 49–53 (2015)
9. Dominic-Ugwu, B.: Ogwueleka Francisca Nonyelum, the assessment of multimedia technology in the teaching of mathematics in secondary schools in Abuja-Nigeria. Int. J. Mod. Educ. Comput. Sci. (IJMECS) **11**(6), 8–18 (2019)
10. Ganorkar, S.S., Tiwari, N., Namdeo, V.: Analysis and prediction of student data using data science: a review. In: Zhang, Y.-D., Senjyu, T., SO–IN, C., Joshi, A. (eds.) Smart Trends in Computing and Communications: Proceedings of SmartCom 2020. SIST, vol. 182, pp. 443–448. Springer, Singapore (2021). https://doi.org/10.1007/978-981-15-5224-3_44
11. Simonacci, V., Gallo, M.: Statistical tools for student evaluation of academic educational quality. Qual. Quant. Int. J. Methodol. **51**(2), 565–579 (2017). https://doi.org/10.1007/s11135-016-0425-z
12. Park, Y.-E.: Uncovering trend-based research insights on teaching and learning in big data. J. Big Data **7**, 1–17 (2020). https://doi.org/10.1186/s40537-020-00368-9
13. Baig, M.I., Shuib, L., Yadegaridehkordi, E.: Big data in education: a state of the art, limitations, and future research directions. Int. J. Educ. Technol. High. Educ. **17**, 1–23 (2020)
14. Salloum, S., Dautov, R., Chen, X., Peng, P.X., Huang, J.Z.: Big data analytics on apache spark. Int. J. Data Sci. Analytics **1**, 145–164 (2016). https://doi.org/10.1007/s41060-016-0027-9
15. Apiletti, D., Baralis, E., Cerquitelli, T., Garza, P., Pulvirenti, F., Venturini, L.: Frequent itemsets mining for big data: a comparative analysis. Big Data Res. **2017**(9), 67–83 (2017)
16. Farhan, N., Habib, A., Ali, A.: A study and performance comparison of Mapreduce and apache spark on twitter data on Hadoop cluster. Int. J. Inf. Technol. Comput. Sci. (IJITCS) **10**(7), 61–70 (2018)
17. Lenka, R.K., Barik, R.K., Panigrahi, S., Panda, S.S.: An improved hybrid distributed collaborative filtering model for recommender engine using apache spark. Int. J. Intell. Syst. Appl. (IJISA) **10**(7), 74–81 (2018)
18. Daniel, J.: Data Science with Python and Dask. Manning Publications (2019)
19. Henriques, J., Caldeira, F., Cruz, T., Simões, P.: Combining K-Means and XGBoost models for anomaly detection using log datasets. Electronics **9**(7), 1164 (2020)
20. Flores, J.J., et al.: Parallel mining of frequent patterns for school records analytics at the Universidad Michoacana. In: 2017 IEEE International Autumn Meeting on Power, Electronics and Computing (ROPEC), pp. 1–6 (2017)

21. Madyatmadja, E.D., Sembiring, D.J.M., Angin, S.M.B.P., Ferdy, D., Andry, J.F.: Big data in educational institutions using RapidMiner to predict learning effectiveness. J. Comput. Sci. **17**(4), 403–413 (2021)
22. Scheffe, H.: The Analysis of Variance. Wiley, New York (1963)
23. Kristiani, E., Yang, C.-T., Wang, Y.T., Huang, C.-Y.: Implementation of an edge computing architecture using openstack and Kubernetes. In: Kim, K.J., Baek, N. (eds.) ICISA 2018. LNEE, vol. 514, pp. 675–685. Springer, Singapore (2019). https://doi.org/10.1007/978-981-13-1056-0_66

Software System for Processing and Visualization of Big Data Arrays

Fedorova Nataliia, Havrylko Yevgen, Kovalchuk Artem, Husyeva Iryna[✉],
Zhurakovskiy Bohdan, and Zeniv Iryna

National Technical University of Ukraine "Igor Sikorsky Kyiv
Polytechnic Institute", Kyiv 03056, Ukraine
i.husyeva@kpi.ua

Abstract. The article presents the description of the software system for collecting data from sensors of the Internet of Things (IoT), processing, analysing using models and algorithms of machine learning and visualization the results close to "real-time". The system processes big data and is stress-resistant to loads. The existing methods of data processing were improved, namely methods for data cleaning, analysis and visualization. Pre-processing methods were used to achieve the best result of the analysis for more accurate visualization. A Long Short Term Memory (LSTM) Autoencoder model was used to predict and detect anomalies. The efficiency of the software system was increased by optimization of the development process. Proposed software system ensures the integration of data from IoT sensors into the Enterprise Resource Planning System (ERP). Test results from different environments are presented. The practical significance of the research lies in the ability to process, analyze, visualize and integrate data from IoT devices into the ERP system.

Keywords: Big Data · Internet of Things · ERP-system · Sensors · IoT devices · Machine learning · Apache Kafka · Docker

1 Introduction

The development of computer technology today rises a problem of processing, storing, analysing and visualizing large amounts of data. Therefore, there is a need for systems that will collect data from various sources of information, efficiently process, store and display it in real time. Big data processing and analysis are used in diverse areas. ERP systems enable efficient management of processing, analysis and visualization of the results. It is also necessary to customize the processing of data for efficient decision making.

The close integration of Big Data, IoT and ERP systems plays an important role in building successful business development strategies. The aim of this article is to consider a system that simplifies the process of data collection, processing, analysis and visualization, namely a real-time system which collects data from IoT sensors, processes collected data, analyzes it using machine learning algorithms and visualizes the results. The system also has the ability to integrate processed data with third-party ERP systems, which currently is an urgent task.

© The Author(s), under exclusive license to Springer Nature Switzerland AG 2022
Z. Hu et al. (Eds.): ICCSEEA 2022, LNDECT 134, pp. 324–336, 2022.
https://doi.org/10.1007/978-3-031-04812-8_28

2   Analysis of Existing Solutions for Collecting, Processing, Analyzing and Visualizing Data from Iot Devices

Among the existing solutions for data analysis and visualization are the following platforms.

Google Cloud IoT. Google has launched its IoT platform based on its end-to-end Google cloud platform. It is currently one of the world's leading platforms for the IoT. Google Cloud IoT is an integration of various services that add value to connected solutions. The service allows to register devices, monitor and configure devices, capture and process device data.

Cisco IoT Cloud Connect. This is a cloud software package for mobile devices. This IoT solution is used for mobile operators and fully optimizes the network. Cisco offers IoT solutions for security, networking, and data management [1, 2].

Salesforce IoT Cloud. Salesforce specializes in customer relation management and skillfully expands this segment with IoT solutions. The Salesforce IoT Cloud platform collects valuable information from connected devices to provide a personalized experience and build stronger relationships with customers. It works in tandem with Salesforce CRM: data from connected assets is delivered directly to the CRM system, where context-based actions begin instantly.

One of the leaders in the market, Amazon AWS IoT Core, allows to connect devices to AWS cloud services without having to manage servers. The platform provides reliability and security for managing millions of devices.

With Microsoft Azure IoT Hub it is possible to build scalable and secure border-to-cloud solutions. Flexible applications can be developed using ready-to-use tools, templates and services.

Oracle IoT. Oracle's Cloud Internet Service is a managed platform as a service (PaaS) for connecting devices to the cloud.

Particle platform offers an IoT edge-to-cloud platform for global device connectivity and management, as well as hardware solutions including development kits, production modules and asset tracking devices.

Raspberry Pi is a platform used in popular IoT devices. Comes with pre-installed IoT software such as Chromium, Mathematica and Minecraft Pi Edition. The Raspberry Pi user interface is similar to macOS, Windows, Ubuntu Linux and more. This makes it extremely useful for experimental, educational and other purposes, ensuring continuous development.

ThingWorx platform offers growth management solutions with low application development costs and application development time. It has flexible solutions for the distribution of the complete program design, runtime and consists of an intelligent environment. The platform is popular due to the fast process of application development and the spread of various IoT solutions. It also offers flexibility and scalability.

IRI Voracity. A very fast platform for data discovery, management, integration, migration and analytics, which has full power for transformation, and also transmits analytics device data via Kafka or MQTT. Voracity has a data manipulation mechanism for fast aggregation at the border, which has a full-flow Eclipse IDE for metadata management and analytics.

Altair SmartWorks. IoT platform which offers the platform as a service. Altair Smart-Works helps to connect devices, collect data, manage devices and data, and create and run applications. It offers a variety of functionalities such as listening, rules, device management, user reminders, triggers and data export [2].

3   Task Statement

The analysis of existing IoT platforms showed that each of them has a number of advantages and disadvantages. Therefore, the article considers the problem of system development using the latest tools and technologies available, which will provide an opportunity to maximize the potential of the system and increase its efficiency together with close integration of Big Data, IoT and ERP systems. It is also necessary to improve the system in terms of working with methods for cleaning, analysis and visualization of data. When preparing data from IoT sensors in real time, it is proposed to use pre-processing methods to achieve maximum results in the analysis and a more accurate visualization. As well as the use of the LSTM Autoencoder model to predict and detect anomalies.

The proposed system can be used to solve problems in various fields, as well as be easy to configure, deploy, scale and have the necessary basic functionality that can be used by customers to solve their tasks.

4   Basic Methods for Data Processing and Visualization

Data processing means not only transformation and enriching data, it is more complex and has its issues at each stage. The data can contain inconsistencies and missing records. Data cleaning is used to handle such situations. This involves processing missing data, noise and uncertainties. Basic methods for data processing and visualization are following [3].

Regression. This is a method of predicting the target variable by establishing the best linear relationship between the dependent and independent variable.

Data can be made smooth by applying a regression function. Regression can be linear (with one independent variable) or multiple (with several independent variables). Figure 1 shows an example of regression [4].

Clustering. Cluster analysis is a method of machine learning without a teacher that divides/groups data points into clusters or groups, so that all data points in one cluster/group have similar properties. There are four main categories for cluster analysis: distribution methods (K-means) [5], hierarchical methods (BIRCH), density-based methods (DBSCAN) [6], and grid-based methods.

Density-based spatial clustering of applications with noise (DBSCAN) is a method that identifies different clusters in data based on the idea that the cluster is a group of high data point density, separated from other clusters by areas with low data point density. The basic idea is to find very dense regions and consider them as one cluster. It can easily detect clusters of various shapes and sizes from large amounts of data that contain noise and emissions. Figure 2 shows an example of a DBSCAN.

There are three different types of data points in this method. Primary data point: a data point that has at least "minPts" at a distance of "ε". Data boundary point: a data

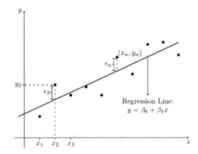

Fig. 1. Regression example

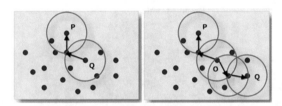

Fig. 2. DBSCAN example

point that is at a distance "ε" from the main data point, but is not the main point. Noise point: a data point that is neither a base nor a boundary. Figure 3 shows an example of noise detection.

The main types of data visualization are:

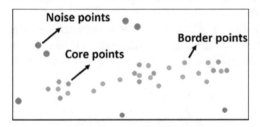

Fig. 3. Noise detection example

1. Line chart. The simplest method - a linear graph - is used to plot the relationship or dependence of one variable on another. Figure 4 shows an example of a line graph.
2. Bar Chart. Used to compare quantities of different categories or groups. Category values are displayed using bars, and they can be adjusted using vertical or horizontal bars, with the length or height of each column representing the value. Figure 5 shows an example of a Bar Chart.

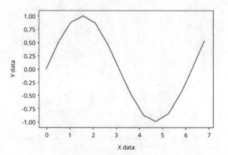

Fig. 4. Linear graph

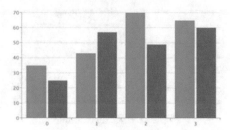

Fig. 5. Bar chart

3. Pie/Donut Charts. They are used to compare parts of the whole and are most effective when there are limited components and when text and percentages are included to describe the content. However, they can be difficult to interpret because it is hard for the human eye to estimate areas and compare angles. Figure 6 shows an example of Pie/Donut Charts.

4. Histograms. Histogram, which represents the distribution of a continuous variable over a period of time, is one of the most commonly used data visualization techniques in machine learning. Figure 7 shows an example of a histogram.

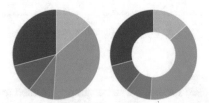

Fig. 6. Pie/Donut charts

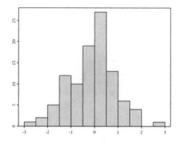

Fig. 7. Histogram

5   Description of Software Implementation

5.1   Description of Precedents and Processes in the System

The suggested system provides such processes as obtaining, processing, analysing data and visualization in a user-friendly way. The data comes from various IoT sensors, the system converts it into the required format, processes the data (removes duplicates, noise, replaces empty data), analyses and stores it in the data warehouse. Afterwards the user can perform analysis and visualization of data [7].

It is suggested to design the system according to the Kappa architecture. Java and Python languages are used to implement the necessary services. Tensorflow and Keras are used to construct analytical models. Apache Kafka broker is used to interact with IoT sensors, which allows to process and transmit events in a continuous stream, Apache Pinot - as a data warehouse, allows to efficiently allocate resources in the cluster and execute a huge number of requests almost without delay. The choice of technologies is justified based on the speed of development, the architecture of applications and the range of solutions.

The suggested system is able to receive, process, analyse and display big data in real-time. The system has the ability to get the input data from IoT devices using Apache Kafka [8]. Based on the types of sensors, the data is sent to various topics. A cluster based on Kafka nodes in Zookeeper nodes allows the system to withstand heavy loads, be stress-resistant and remain active until the last node fails. The administrator can monitor the status of the cluster online. It is achieved by collecting metrics from the cluster at regular intervals and displaying them using the Grafana framework.

The directory administrator precedence diagram is shown in Fig. 8.

The administrator is able to monitor the resource utilization for further adjustment in accordance with the obtained metrics.

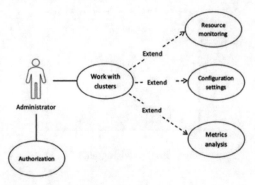

Fig. 8. System administrator precedence diagram

The system consists of several levels that are responsible for data processing. Diagram of precedence of the system is shown at Fig. 9 [9].

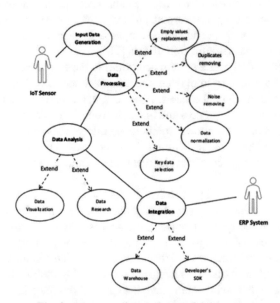

Fig. 9. Diagram of precedence of the system

The data comes from IoT sensors, which is immediately processed in a continuous stream of data using Kafka Streams: duplicates and noise are removed, empty values are replaced, the data is normalized and converted into a usable form for analysis. After this process, the data is stored in a highly accessible Apache Pinot repository, which allows to perform queries for analytics and display the data with low latency. It is also easy to set the data format of the data development toolkit for different developers which is required to integrate with the ERP system.

5.2 Description of the Conceptual Database

The column-oriented Apache Pinot database was used as a data warehouse [10]. The user can create a schema and table in the repository using a JSON configuration file. Each table in Pinot is associated with a schema. The scheme determines which fields are present in the table together with the data types [10].

The schema is stored in Zookeeper along with the table configuration. The scheme also determines to which category the column belongs. The columns in the Pinot table can be divided into three categories: - Dimension, - Metric, - DateTime.

This column represents the time columns in the data. There can be several time columns in a table, but only one of them can be considered as the main one. The main time column is the column which is present in the segment configuration and is used by Pinot to maintain the time limit between offline and real-time data in the hybrid table and to manage storage. The main time column is required if the push type for the table is APPEND, and optional - if the push type is REFRESH [10, 11].

5.3 Description of System Architecture

The server part of the system is presented in the form of microservices that can be scaled according to the load. Each microservice uses different language and framework according to the task. Data exchange and stream processing is developed by means of Apache Kafka. Analytical models are built using Python, Tensorflow and Keras. Visualization is done using libraries available in Python. Java is used to process the logic of data exchange, as it has native support and interaction with Apache Kafka. Prometheus and Grafana are used to collect and store metrics from different modules for their interactive display. All services are deployed in Docker containers [12, 13].

Traditionally, most machine learning (ML) models use samples as input characteristics, but there is no time measurement in the data. Time series prediction models are models that are able to predict future values based on previously observed values. Time series prediction is widely used for non-stationary data [11, 14].

Long Short Term Memory (LSTM) can be used as analytical models. Before deep learning neural networks became popular, in particular, recurrent neural networks, there were a number of classical analytical methods to predict time series - AR, MA, ARMA, ARIMA [15], SARIMA, etc. LSTM is a popular variant of RNN that solves problems in conventional RNNs, such as the "gradient disappearance problem" in very deep RNNs, which interferes with the learning process at entry levels when error gradients are transmitted by time-propagation (BPTT) in RNNs with many hidden layers [16].

Light Gradient Boosting Machine (LightGBM) is another popular machine learning algorithm for solving problems (forecasting, detecting anomalies). LightGBM uses a tree-based learning algorithm which grows vertically, while in other algorithms the tree grows horizontally.

6 Experimental Results

6.1 System Requirements and Additional Software

There are system requirements imposed on the characteristics of the hardware when processing large amounts of data. Minimum requirements: Intel Core i5 processor, 16 GB of RAM, 256 GB of free hard disk space. Additional software: Docker and docker-compose to run the containers.

6.2   The Results of the Program

The operation of the Apache Pinot data warehouse is shown in Figs. 10, 11, 12, 13, 14, 15 and 16.

Fig. 10. Page of tables, diagrams

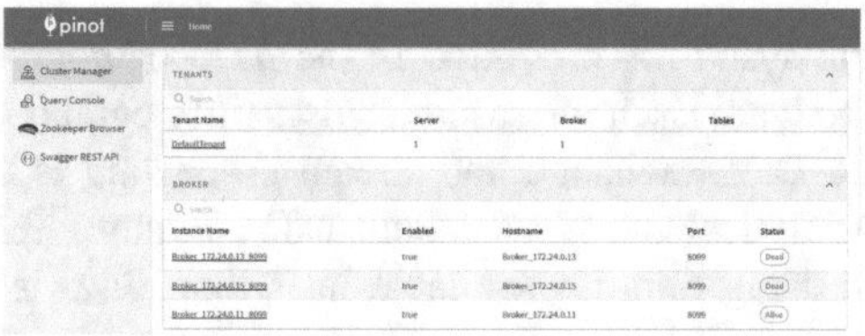

Fig. 11. Apache Pinot Broker information page

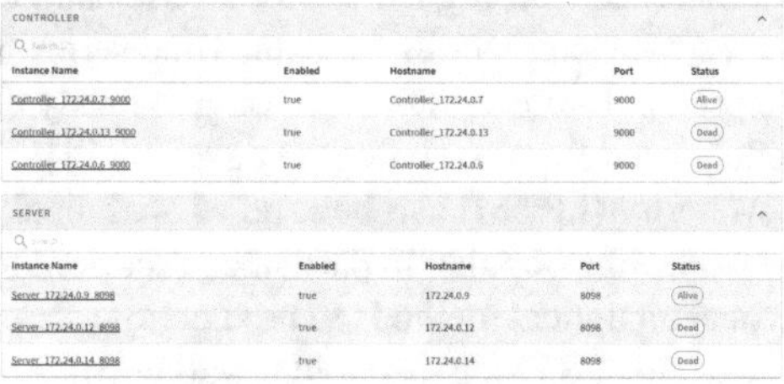

Fig. 12. Page with information about the controller, Apache Pinot server

It can be concluded that the repository works using only one node.

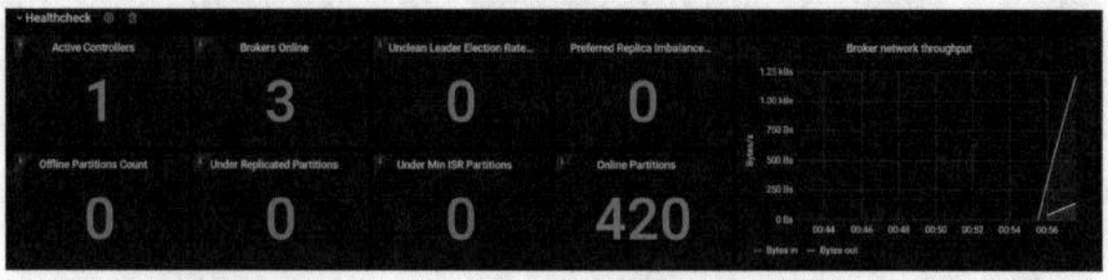

Fig. 13. Producer's Apache Kafka cluster metric information page using Prometheus, Grafana

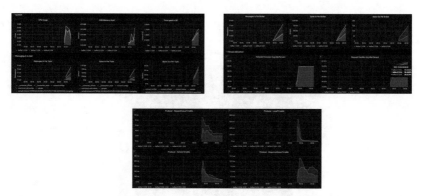

Fig. 14. Consumer's Apache Kafka cluster metric information page using Prometheus, Grafana

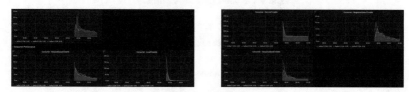

Fig. 15. Consumer Apache Kafka metrics information page using Prometheus, Grafana

Fig. 16. Page with information about Apache Zookeeper metrics using Prometheus, Grafana

The figures show that the cluster has started working and is processing the incoming events.

Figure 17 shows the correlation matrix. The correlation matrix is a table that represents the values of correlation coefficients for different variables. It shows the numerical value of the correlation coefficient for all possible combinations of variables. It is mainly used when it is necessary to clarify the relation between more than two variables.

Figure 18 and Fig. 19 show comparing the input values with the generated (predicted) and the detected anomalies.

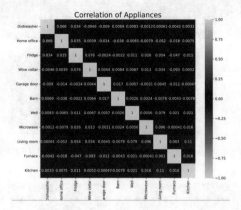

Fig. 17. Correlation matrix

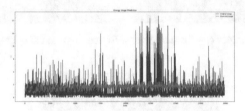

Fig. 18. Comparison of input values and generated (predicted)

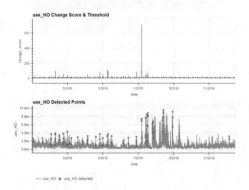

Fig. 19. Detected anomalies

7 Summary and Conclusion

1. A software system for processing and visualization of big data arrays in real time using data from IoT sensors is proposed. The methods and approaches that have

been studied have become the basis for further improvement, namely data cleaning, analyzing and visualizing.

2. During data preparation, pre-processing methods were used to achieve the best results in the analysis and a more accurate visualization. The LSTM Autoencoder model was used to predict and detect anomalies, and the results were described and analyzed.

3. The system is developed using the latest tools and technologies that allow to maximize the potential of the system. The efficiency of the software system was increased by optimization of the development process.

4. The application of the developed system is not limited to a particular industry, but is inherent in the use of different areas of the market. The system is easy to set up and deploy, easy to scale and has the necessary basic functionality that can be used by customers to solve their tasks.

5. Based on a subject area analysis, including requirements analysis, technology selection, system architecture, program code organization, the software system has been created to ensure the integration of data from IoT sensors into the ERP system.

6. The practical application of the results lies in the ability to process, analyse, visualize and integrate data from IoT devices into the ERP system. This software solution is easy to maintain, scale and refine.

References

1. Comparing top IoT development platforms [Electronic resource] (2020). https://www.dig iteum.com/top-iot-development-platforms/

2. Top IoT development tools and platforms with comparison [Electronic resource] (2021). https://www.intuz.com/blog/top-iot-development-platforms-and-tools

3. Ways to handle missing values in machine learning/Satyam Kumar [Electronic resource] (2020). https://towardsdatascience.com/7-ways-to-handle-missing-values-in-mac hine-learning-1a6326adf79e

4. The six types of data analysis/Benedict Neo [Electronic resource] (2020). https://towardsda tascience.com/the-six-types-of-data-analysis-75517ba7ea61

5. Laskhmaiah, K., Krishna, S.M., Reddy, B.E.: An Optimized K-means with density and distance-based clustering algorithm for multidimensional spatial databases. Int. J. Comput. Network Inf. Secur. (IJCNIS) **13**(6), 70–82 (2021). https://doi.org/10.5815/ijcnis.2021.06.06

6. Fahim, A.: A clustering algorithm based on local density of points. Int. J. Mod. Educ. Comput. Sci. (IJMECS) **9**(12), 9–16 (2017). https://doi.org/10.5815/ijmecs.2017.12.02

7. Fedorova, N.V., Bandurka, O.I., Khomenko, O.M.: Methods of processing and visualization in the ERP system of large data sets/modern engineering and innovative Technologies. Deutschland (2021). https://www.moderntechno.de/index.php/meit

8. Samizadeh, I.: A brief introduction to two data processing architectures—Lambda and Kappa for Big Data [Electronic resource] (2018). https://towardsdatascience.com/a-brief-introduct ion-to-two-data-processing-architectures-lambda-and-kappa-for-big-data-4f35c28005bb

9. Patel, H.: Apache Kafka Tutorial - Kafka For Beginners [Electronic resource] (2018). https://medium.com/@patelharshali136/apache-kafka-tutorial-kafka-for-beginners-a58140cef84f

10. Apache Pinot docs. Architecture [Electronic resource] (Cloud data storage, 2022). https://docs.pinot.apache.org/basics/architecture

11. Apache Pinot docs. Table [Electronic resource] (Cloud data storage, 2022). https://docs.pinot.apache.org/configuration-reference/table

12. Loobuyck, U.: Scikit-learn, TensorFlow, PyTorch, Keras… but where to begin? [Electronic resource] (2020). https://towardsdatascience.com/scikit-learn-tensorflow-pytorch-keras-but-where-to-begin-9b499e2547d0
13. Python Documentation. What is Python? Executive Summary [Electronic resource] (2022). https://www.python.org/doc/essays/blurb/
14. Loukas, S.: Time-Series Forecasting: Predicting Stock Prices Using An LSTM Model [Electronic resource] (2020). https://towardsdatascience.com/lstm-time-series-forecasting-predicting-stock-prices-using-an-lstm-model-6223e9644a2f
15. Jain, G., Mallick, B.: A study of time series models ARIMA and ETS. Int. J. Mod. Educ. Comput. Sci. (IJMECS) **9**(4), 57–63 (2017). https://doi.org/10.5815/ijmecs.2017.04.07
16. Suryanarayanan, R.: Time Series Forecast Using Deep Learning [Electronic resource] (2021). https://medium.com/geekculture/time-series-forecast-using-deep-learning-adef5753ec85

GEOCLUS: A Fuzzy-Based Learning Algorithm for Clustering Expression Datasets

Zhengbing Hu[1] ⓘ, Esha Kashyap[2] ⓘ, and Oleksii K. Tyshchenko[3](✉) ⓘ

[1] National Aviation University, Liubomyra Huzara Ave 1, Kyiv 03058, Ukraine
[2] Department of Mathematics, Ramanujan School of Mathematical Sciences, Pondicherry University, Kalapet, India
[3] Institute for Research and Applications of Fuzzy Modeling, CE IT4Innovations, University of Ostrava, 30. dubna 22, 701 03 Ostrava, Czech Republic
lehatish@gmail.com

Abstract. Microarray experiments monitor the expression of genes over a set of samples or experimental conditions. Clustering approaches focused on this genomic information determine sub-population of patients with analogous prognostic characteristics. These datasets have an intricate high-dimensional data structure associated with noise and imprecise information. The motive of this research is to develop an efficient clustering algorithm for cancer patient stratification. This article proposes a fuzzy-based clustering algorithm, Geodesic Fuzzy Clustering with Local Information (GEOCLUS). Our approach embeds a microarray dataset on a Riemannian manifold of constant curvature while retaining trends and patterns. GEOCLUS is built on the Fuzzy C-Medoids algorithm and factors into the local non-linear interactions between the dataset's features. Four cancer expression datasets assess the robustness of the proposed algorithm. GEOCLUS searches for treatment groups; it partitions cancer samples with distinct histopathological characteristics. It uses geodesics as distance measure. The performance of the proposed approach is compared with the generic K-Means and Fuzzy C-Means (FCM) algorithms. The experimental findings suggest that the proposed approach be a better classifier for clustering cancer gene expression datasets.

Keywords: Geodesic fuzzy clustering · Gene expression · Distance measure · Feature vector · Riemannian manifold

1 Introduction

DNA microarray experiments, like cDNA and oligonucleotide microarrays, monitor the expression levels of many genes together. These experiments further our understanding of the fundamental biological processes in an organism. In cancer patients, genetic alterations result in abnormal gene functioning, that furthers gene mutations, resulting in tumorigenesis. Microarray datasets capture these genetic variations. Clustering algorithms meliorate the understanding of gene functions, narrow down co-expressed genes, differentiate cellular processes and assist in patient stratification.

Z. Hu et al. (Eds.): ICCSEEA 2022, LNDECT 134, pp. 337–349, 2022.
https://doi.org/10.1007/978-3-031-04812-8_29

Clustering algorithms discover hidden patterns and natural grouping of homogeneous structures and substructures of biological significance from an expression dataset. Clustering an expression dataset meliorates the understanding of intricate biological networks and assists in assimilating the fundamental biological processes in an organism. Introductory works for clustering gene expression data were borrowed methods that were created for some other domain. These include algorithms like Hierarchical Clustering, K-Means, Fuzzy C-Means, Self-Organizing Maps, and Genetic Algorithms [1–5]. These algorithms are sensitive to noise in a dataset and tend to converge to the local minima, dependent on the parameters' choice and initial cluster centroids [6–12].

Gene expression datasets have an intricate high-dimensional data structure, associated with imprecise information [13]. It contains noise. One of the dominating challenges of clustering gene expression data is the 'big p small n' problem. A scenario is characterized by a prominent attribute or feature size in comparison with the sample size. The expression profiles of the prognostic clusters in a gene expression dataset are similar; the boundaries between the clusters are vague. Generic clustering algorithms fail to accommodate the domain-specific characteristics of the expression dataset. A clustering algorithm must be tailored to the domain-specific requirements of the expression dataset, articulate in handling redundant cluster structures and humor unstructured patterns or noise to remain unclustered.

The fundamental motive of this research is to develop an efficient clustering algorithm for cancer patient stratification. This paper presents a robust fuzzy-based clustering algorithm, Geodesic Fuzzy Clustering with Local Information (GEOCLUS). The algorithm searches for treatment groups in the cancer datasets. It partitions cancer samples with distinct histopathological characteristics. Multiple genes impact a diseased state through non-linear interactions; regardless, a single gene contributes but a little to the diseased state. The correlation among expressions is also well-documented in the literature. These two observations nurture the idea that gene expressions can be transformed to a low-dimensional space that captures the variations in the data structure [14]. Furthermore, gene expressions are high-dimensional directional data, and prior research suggests a spherical space or a positive curvature space be a more potent choice for assessing directional resemblance [15].

Our approach embeds the microarray data onto the Riemannian manifold with the Fisher-Rao information metric. Manifolds with the Riemannian metrics maintain intrinsic characteristics of the original space. Therein the data is mapped to a hypersphere, furthering straightforward intuitions. GEOCLUS factors into the local non-linear interactions between the features. It is established on the Fuzzy C-Medoids algorithm. In C-Medoids algorithms, a medoid is used as a cluster representative, and a usual preference is the cluster mean [16]. GEOCLUS favors the geometric median as a medoid, a more robust central representative for datasets with noise. In the proposed approach, a distance between feature vectors is measured by geodesics. Geodesics on the hypersphere correspond to line segments in $\mathbb{R}^n$.

The remainder of this paper is composed as follows. Section 2 introduces the recent related works published within the scope of research. Section 3 provides a theoretical formulation of the proposed approach. Section 4 presents the experimental findings. The robustness of the proposed approach is assessed. Section 5 concludes the research.

2   Related Works

Traditional clustering approaches are linear algorithms. These approaches are not built to deal with the non-linear interactions among the features of an expression dataset. A suitable metric is required to obtain the true partitioning of these datasets. Learning algorithms' search for resemblance in the shape of expression patterns. In literature, metrics based on angular separation were the most competent in assessing directional similarities.

In [17], the authors provided a framework for efficient classification of directional data, the Spherical K-Means algorithm. In this framework, the Euclidean metric in the generic K-Means algorithm was substituted by the cosine metric. The cosine distance measure calibrates the angular separation between feature vectors. It is independent of the magnitude of the feature vectors. The findings in [17] were promising but were restricted to sparse high-dimensional text data. The authors in [15], developed two model-based learning algorithms: soft-movMF and hard-movMF to classify the directional data. These algorithms were the Expectation Maximization (EM) algorithms on a mixture of von Mises-Fisher (vMF) distribution. soft-movMF concentrated on the soft assignment of the feature vectors while hard-movMF focused on the hard assignment. In [18], soft-movMF and hard-movMF partitioned the directional data on a hypersphere. L_2 normalized the high-dimensional directional data to have characteristics that approximate the modeling assumptions of a vMF mixture model. Thereupon, both algorithms demonstrated the robust performance gains compared to generic algorithms. The experimental findings in articles, [17] and [15] focused on text and expression datasets.

The authors in [19] proposed a modification of the Similarity-Based Clustering Method (SCM) focused on the directional data. SCM algorithms are insensitive to initialization and hence are advantageous to the EM algorithms. The SCM algorithm is limited to the Euclidean space. The modified variation in [19], clusters directional datasets on a unit hypersphere without initialization. The efficiency of this theoretical fabrication was validated by multiple simulated and real-world datasets. A fuzzy-based clustering algorithm, Hyper-Spherical Fuzzy C-Means (HFCM) was developed in [20]. HFCM is a variation of the Fuzzy C-Means (FCM) algorithm. It used the cosine metric for document clustering. HFCM was built for static datasets. The authors in [21] proposed a refined version of HFCM, the Online Hyperspherical Fuzzy C-Means (OnHFCM) algorithm, to handle non-static datasets. OnHFCM has two variations: OnHFCM-m and OnHFCM-e. OnHFCM-m is a classic variation of the fuzzy clustering algorithm, a fuzzification parameter m is used, while OnHFCM-e is a regularization algorithm. Both variations utilize the cosine metric to assess the likeness among the feature vectors. In [22], another modified version of HFCM was proposed, the Hierarchical Hyper-Spherical Fuzzy C-Means (H^2FCM) algorithm for text data. It grouped data while establishing a hierarchy among clusters. The Hyperspherical Possibilistic Fuzzy C-Means (HPFCM) method was proposed in [23]. This algorithm partitioned the directional data by the Possibilistic Fuzzy C-Means (PFCM) algorithm on a unit hypersphere with the cosine measure.

Geodesic-based algorithms have gained recognition in recent times. A geodesic distance reflects an inherent geometric structure of data [24, 25]. The Geodesic K-Means [24] algorithm altered the Euclidean metric of the generic K-Means algorithm by the geodesic distance. The algorithm was passed through three artificial datasets and the benchmark Iris data. It proved to be efficient in partitioning datasets with non-linear cluster structures. In [16], the authors proposed a fuzzy-based framework with the geodesic distance. Two versions were proposed. In the first approach, Isomap embedded a dataset on a low-dimensional manifold and then partitioned the dataset by FCM. Isomaps are efficient in discovering non-linear smooth manifolds. Clustering and Isomap algorithms have different objectives, while the clustering approaches search for distinct substructures in the dataset, Isomap hunts for connected graphs. The diametric nature of these algorithms routed the second version, in which the dataset was partitioned by Fuzzy C-Medoids using the geodesic distance. The idea of a geodesic kernel was introduced in [26]. The geodesic RBF kernel recognized the underlining manifold structure, and captured the non-linear interactions among the feature vectors. The authors in [26] presented a soft clustering algorithm that grouped non-convex substructures in the dataset with competence.

3   The Developed Method

The section presents the theoretical foundation of the proposed approach. The proposed approach is built focused on microarray datasets. The proposed approach generates a low-dimensional embedded representation of the microarray data preserving angular resemblance. Clustering occurs in this ambient space. This section is organized as follows. Section 3.1 describes the mapping of microarray datasets on the positive curvature space H^+. Section 3.2 defines geodesics on the hypersphere. Section 3.3 defines a local information ω_l. ω_l helps locate outliers in a dataset. Section 3.4 provides an objective for the proposed approach.

3.1   Mapping the Expression Dataset on the Riemannian Manifold

DNA microarray experiments record the expression of genes over a set of experimental conditions. A matrix can represent this high-dimensional expression dataset. Considering a gene expression matrix.

$$G = [x_{lk}]_{n \times m} = \begin{bmatrix} x_{11} & x_{12} & x_{13} & \cdots & x_{1m} \\ x_{21} & x_{22} & x_{23} & \cdots & x_{2m} \\ \vdots & \vdots & \vdots & \vdots & \vdots \\ x_{n1} & x_{n2} & x_{n3} & \cdots & x_{nm} \end{bmatrix},$$

the row vectors in the matrix $G = [x_{lk}]_{n \times m}$ represent the samples taken under consideration. The l th sample, s_l is a vector of length m. It is a $m-$ tuple of the form: $s_l = (x_{l1}, x_{l2}, x_{l3} \ldots, x_{lm})$ where x_{lk} records the expression of the k th gene on the l th sample.

Each sample vector s_l is transmuted to a probability vector s_l^* where each expression level x_{lk} is mapped to $x_{lk}^* = \frac{x_{lk}}{\sum_{h=1}^{m} x_{lh}} \cdot x_{lk}^*$ represents the maximum likelihood estimation (MLE) of the expression's probability for the k th gene on the l th sample, $\sum_{k=1}^{m} x_{lk}^* = 1$. This maps each sample to a point on a $(m-1)-$ dimensional probability simplex $P(G)$.

$P(G)$ is a set of the probability functions $x_{lk}^* : G \to \mathbb{R}_{\geq 0}$, $\sum_{k=1}^{m} x_{lk}^* = 1$. It is a mathematical space where each vector x_{lk}^* identifies a probability distribution among mutually exclusive events, finite in a number. Here, $P(G) \subset \mathbb{R}^{m-1}$.

$$P(G) = \left\{ x_{lk}^* \in \mathbb{R}^G \,\middle|\, \sum_{k=1}^{m} x_{lk}^* = 1, \, x_{lk}^* \geq 0 \right\}.$$

Further, the relative topological interior $P^o(G)$ corresponding to $P(G)$

$$P^o(G) = \left\{ x_{lk}^* \in \mathbb{R}^G \,\middle|\middle|\, \sum_{k=1}^{m} x_{lk}^* = 1, \, x_{lk}^* > 0 \right\}$$

generates a differentiable Riemannian manifold M when the Fisher-Rao information metric is considered.

Let H be a $(m-1)$-dimensional hypersphere centered at the origin and of radius 2. Let $H^+(\subset H)$ be the positive orthant. A homorphism $\gamma : M \to H^+$, $\gamma(s_l^*) = a_l$ with $a_l = \left(2\sqrt{x_{l1}^*}, 2\sqrt{x_{l2}^*}, \ldots, 2\sqrt{x_{lm}^*} \right)$ maps the probability vector s_l^* corresponding to the l th sample from the manifold M onto a point u_l on the positive orthant H^+ with $\sum_{k=1}^{m} \left\{ 2\sqrt{x_{lk}^*} \right\}^2 = 4$, $x_{lk}^* > 0$.

The corresponding tangent space at $a_l \in H^+$ is generated as

$$\Delta_\gamma : T_{x_{lk}^*} M \to T_{a_l} H^+, \ \Delta_\gamma(a_l) = \sigma_l$$

with $\sigma_l = \left(\frac{a_{l1}}{\sqrt{x_{l1}^*}}, \frac{a_{l2}}{\sqrt{x_{l2}^*}}, \ldots, \frac{a_{lm}}{\sqrt{x_{lm}^*}} \right)$. γ preserves the Riemannian metrics. Thus, the sample vectors from the gene expression matrix G are embedded to a constant curvature space H^+. Each l th sample vector s_l from the matrix G is mapped to a point a_l on the positive orthant of a $(m-1)$-dimensional hypersphere with coordinates $(a_{l1}, a_{l2}, \ldots, a_{lm})$ where

$$a_{lk} = 2\sqrt{\frac{x_{lk}}{\sum_{h=1}^{m} x_{lh}}}.$$

3.2   Geodesics on the Hypersphere

Geodesics in the Riemannian manifolds are analogous to line segments in $\mathbb{R}^n$. Geodesics are local length minimizing curves. Geodesics on the hypersphere correspond to a length of the shortest curve between two vectors. On the hypersphere, the geodesic distance d_g between two feature vectors $a = (a_1, a_2, \ldots, a_n)$ and $b = (b_1, b_2, \ldots, b_n)$ is calculated as follows:

$$d_g(a, b) = r \ \arccos\left(\frac{a \cdot b}{r^2} \right) = r \ \arccos\left(\frac{\sum_{l=1}^{n} a_l \cdot b_l}{r^2} \right). \tag{1}$$

where r is a radius of the hypersphere.

3.3 Local Information ϖ_l

Before the clustering process, a local information ϖ_l is associated with each feature vector x_l. ϖ_l is determined as follows:

$$\varpi_l = \sum_{h=1}^{n} X\left(d_g(x_l, x_h) - t_d\right) \tag{2}$$

where

$$X(u) = \begin{cases} 1, & u < 0 \\ 0, & otherwise \end{cases},$$

$d_g(x_l, x_h)$ measures a distance between the feature vectors, x_l and x_h in the ambient space calculated according to (1), and t_d is a threshold distance. The threshold distance t_d is based on a standard deviation of the attributes. t_d is defined as follows:

$$t_d = \eta c \sqrt{\sum_{k=1}^{m} \mu_k \sigma_k} \tag{3}$$

where μ_k is the mean expression across the m attributes; σ_k measures the dispersion of features about the mean expression. $c = {}^{m-1}\!/_{2m^2}$ is a constant, $\eta \in (0, 1]$ is a user defined parameter.

3.4 Geodesic Fuzzy Clustering with Local Information (GEOCLUS)

Geodesic Fuzzy Clustering with Local Information (GEOCLUS) is based on the minimization of the following functional:

$$J = \sum_{l=1}^{n} \sum_{k=1}^{c} u_{lk}^m d_g^2(x_l, v_k) \tag{4}$$

where $u_{lk} \in U$ stands for membership functions. The set U is defined as

$$U = \left\{ u_{lk} : u_{lk} \in [0, 1], \sum_{k=1}^{c} u_{lk} = 1, \sum_{l=1}^{n} u_{lk} > 0, \forall l, k \right\}. \tag{5}$$

$\{v_k\}_{k=1}^{c}$ are the cluster centroids. $m \in [1, \infty)$. m controls the fuzziness of the membership assignments. Distance is measured in geodesics. $\overline{d}_g$ captures local non-linear interactions among the features. $\overline{d}_g$ is defined as follows:

$$\overline{d}_g^2(x_l, v_k) = \varpi_l * d_g^2(x_l, v_k).$$

Minimization of the functional in (4) results in the following memberships:

$$u_{lk} = \frac{1}{\sum_{h=1}^{c} \dfrac{\overline{d}_g^2(x_l, v_k)^{\frac{2}{m-1}}}{\overline{d}_g^2(x_l, v_h)}}, \quad l = 1, 2, \ldots, n, \ k = 1, 2, \ldots, c. \tag{6}$$

The cluster centroids are updated by the geometric median of the cluster structure.

4   Experiments

This section presents a series of experiments to investigate the performance of the proposed algorithm GEOCLUS. The introduced fuzzy-based clustering algorithm is presented to assist in cancer patient stratification. Principal Component Analysis (PCA) is performed before the clustering process. PCA projects data on a low-dimensional space while maintaining the intrinsic data characteristics. The performance of the proposed algorithm is assessed with multiple expression datasets. The proposed scheme is implemented in Python.

4.1   Data

The experimental investigation considers four cancer datasets registered at the GEO database of NCBI as GSE129617 [27], GSE2034 [28], GSE15484 [29], and GSE50161 [30]. Each dataset is associated with prognostic clinical information. Table 1 records the data statistics.

GSE2034 is an expression dataset of 286 lymph-node negative breast cancer patients, with 180 relapse-free patients and 106 patients with developed distant metastasis.

GSE129617 records the expression signatures of 25 Ovarian Clear Cell Carcinoma (OCCC) patients with 18 EpiCC (the early-stage OCCC having promising prognosis) and 7 MesCC samples (the advanced stage OCCC and has a poor prognostic histology).

GSE15484 records the expression changes in human prostate cancer (PC) patients. It contains an expression pattern of 65 prostate cancer patients with 13 benign tissues, 25 tumor tissues with the Gleason score of 6, and 27 tumor tissues with the Gleason score of 8. The Gleason score is a standard grading used to assess the aggressiveness of prostate cancers. It ranges between 6–10. Tumor tissues with a Gleason score of 10 are the most aggressive ones.

GSE50161 contains an expression profile of 130 brain cancer patients with 15 benign tissues and four neuroepithelial tumor tissue groups: ependymoma (46), glioblastoma (34), medulloblastoma (22), and pilocytic astrocytoma (13). Each prognostic group is associated with unique histopathological characteristics and patient survival.

Table 1. Data statistics

GEO accession	Tissue	Dimension	Clusters	Data distribution
GSE2034	Breast	286×22283	2	106:180
GSE129617	Ovary	25×43287	2	7:18
GSE15484	Prostate	65×12282	3	13:25:27
GSE50161	Brain	130×21050	5	13:15:22:34:46

4.2   Partitioning the Expression Dataset

The experimental design handles high-dimensional microarray datasets. These datasets are associated with redundant features and experimental noise. The data is preprocessed before clustering. The partitioning process starts with feature transformation, Principal Component Analysis (PCA) is used for this reason. PCA projects the data on a low-dimensional space while maintaining the intrinsic data characteristics. It filters out noise. Here, the microarray datasets are projected on a three-dimensional subspace.

In Figs. 1, 2 and 3 , an illustration of the clustering process on the prostate cancer dataset GSE15484 for the proposed approach is presented. GSE15484 is high-dimensional data with three distinct classes. It has dimensions of 65×12282. PCA projects the data on a low-dimensional space with three principal components. Figure 1 captures this projection for the data GSE15484. PCA substitutes original features with principal components. The principal components are orthogonal with decreasing variance. The principal components of GSE15484 have a variance of 0.3362, 0.1285, and 0.0684. Herein, the data is embedded in the Riemannian manifold H^+. The manifold is of positive curvature. H^+ is a positive orthant of the hypersphere of radius two. Figure 2 illustrates the low-dimensional embedding of the prostate cancer data GSE15484 on H^+. GSE15484 is mapped on the surface of a three-dimensional hypersphere of radius two. The clustering process occurs on this three-dimensional subspace. In the proposed approach, first, the threshold distance t_d is calculated. Table 2 records the user-defined parameter η and the threshold distance t_d. The local information ϖ_l is then calculated. It assigns higher weights to outliers in the dataset. The proposed approach clusters the dataset with the iterative minimization of the functional (4). Figure 3 captures the clustering outcome of GSE15484 for the proposed approach. The proposed approach generates the non-overlapping distinct partition of the data. In GEOCLUS, a fuzzy parameter m is chosen to be two across all the datasets.

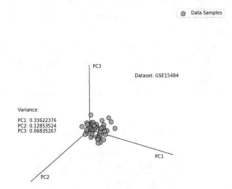

Fig. 1. GSE15484: PCA projects the microarray dataset on a three-dimensional subspace.

4.3 Performance Evaluation

The primary motivation of this research is to develop a robust scheme to cluster different expression datasets into their prognostic groups. K-Means and Fuzzy C-Means (FCM) are used for comparative investigation. The proposed plan obtains comparable results to these generic approaches. Performance is evaluated with an external criterion, classification rate. Classification rate measures the performance of datasets with a known data

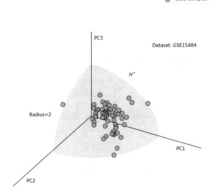

Fig. 2. GSE15484: Mapping on a positive curvature space.

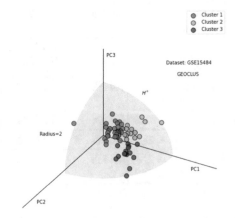

Fig. 3. GSE15484: Data partitioning with GEOCLUS.

Table 2. The threshold distance t_d

GEO accession	η	t_d
GSE2034	0.000002	0.7778
GSE129617	0.0003	1.0859
GSE15484	0.0008	1.6141
GSE50161	0.00002	1.9746

structure (classes). The classification rate calculates the accuracy of prediction. Besides that, some additional performance metrics such as Silhouette Index (S(C)) [31], Davies-Bouldin (DB) Index [32], and Calinski-Harabasz (CH) Index [33] are used. These three performance metrics are internal measures that depend on information from the feature vectors in the dataset. Optimal partitioning occurs at the minimum of DB and at the maximum of S(C) and CH.

Table 3. Clustering performance

Data	Algorithm	Parameters	Classification rate (↑)	S(C) (↑)	DB (↓)	CH (↑)
GSE2034	K-Means	$c = 2$	0.5140	0.2560	1.5217	95.7954
	FCM	$c = 2, m = 2$	0.4476	0.2482	1.5568	92.7159
	GEOCLUS	$c = 2, m = 2, \eta = 0.000002$	**0.5944**	**0.3334**	**1.2459**	**135.9781**
GSE129617	K-Means	$c = 2$	0.6400	0.3370	1.2083	9.6334
	FCM	$c = 2, m = 2$	0.6400	0.2921	1.4525	9.5555
	GEOCLUS	$c = 2, m = 2, \eta = 0.0003$	**0.6400**	**0.4743**	**1.0421**	**12.4438**
GSE15484	K-Means	$c = 3$	0.4000	0.4193	0.9314	41.0276
	FCM	$c = 3, m = 2$	0.4154	0.4090	0.9318	40.0387
	GEOCLUS	$c = 3, m = 2, \eta = 0.00008$	**0.4462**	**0.4198**	**0.8778**	**46.3458**
GSE50161	K-Means	$c = 5$	0.4308	0.5059	0.7491	185.0006
	FCM	$c = 5, m = 2$	0.4385	**0.5085**	**0.7173**	183.8097
	GEOCLUS	$c = 5, m = 2, \eta = 0.00002$	**0.6308**	0.5016	0.8056	**218.5749**

An overview of the clustering performance for the compared algorithms in terms of the classification rate is given in Table 3. K-Means and Fuzzy C-Means (FCM) are used for comparative investigation. The findings reveal that the performance enhances significantly by GEOCLUS on the datasets listed in Table 1. The classification rate for GSE2034 is observed to have gone 32.80% higher when GEOCLUS is used rather than the generic FCM algorithm. Further, a 46.43% increase in the classification rate is noted for the GSE50161 dataset, and a 11.55% enhancement in the classification rate was sighted for GSE15484 when GEOCLUS is used over the K-Means algorithm. There is no performance gain in terms of the classification rate for the OCCC dataset GSE129617. Each algorithm partitions GSE129617 with the classification rate of 0.64. It is observed that GEOCLUS has the optimal S(C) and DB scores for all the datasets except for GSE50161. The fact that GEOCLUS has the highest CH score for all the datasets indicates cohesive well-separated cluster structures. In Table 3, the highest performance of a measure is marked in bold.

5 Conclusion

This paper presented a robust fuzzy-based clustering algorithm (GEOCLUS) for cancer patient stratification. The proposed approach is focused on microarray datasets. It is based on the Fuzzy C-Medoids algorithm. The clustering process occurs in a constant curvature space. The performance of the proposed approach is compared with the generic K-Means and FCM algorithm in terms of classification rate on the microarray dataset. Three internal measures are also used to assess the cluster structures obtained while partitioning the data. The experimental findings indicate that GEOCLUS can be a plausible tool for clustering expression datasets. The proposed approach partitions the data into distinct groups for further analysis.

Microarray datasets are associated with non-informative features. The expression data are pre-processed; each partitioning process starts with feature transformation. We use the generic PCA algorithm for this purpose. The future scope of this research would be to develop and operate a data-centric feature selection technique in conjunction with the proposed approach.

Acknowledgement. The research by Oleksii Tyshchenko is fulfilled within the project CZ.01.1.02/0.0/0.0/17_147/0020575.

The manuscript is supported by the International Center of Informatics and Computer Science (ICICS).

References

1. Di Gesu, V., et al.: GenClust: a genetic algorithm for clustering gene expression data. BMC Bioinf. **6**(1), 1–11 (2005)
2. Eisen, M.B., et al.: Cluster analysis and display of genome-wide expression patterns. Proc. Natl. Acad. Sci. **95**(25), 14863–14868 (1998)
3. Gasch, A.P., Eisen, M.B.: Exploring the conditional coregulation of yeast gene expression through fuzzy k-means clustering. Genome Biol. **3**(11), 1–22 (2002)
4. Hruschka, E.R., Campello, R., De Castro, L.N.: Evolving clusters in gene-expression data. Inf. Sci. **176**(13), 1898–1927 (2006)
5. Yang, Y., Chen, J.X., Kim, W.: Gene expression clustering and 3d visualization. Comput. Sci. Eng. **5**(5), 37–43 (2003)
6. Celebi, M.E., Kingravi, H.A., Vela, P.A.: A comparative study of efficient initialization methods for the k-means clustering algorithm. Expert Syst. Appl. **40**(1), 200–210 (2013)
7. Laskhmaiah, K., Murali Krishna, S., Eswara Reddy, B.: An optimized k-means with density and distance-based clustering algorithm for multidimensional spatial databases. Int. J. Comput. Netw. Inf. Secur. (IJCNIS) **13**(6), 70–82 (2021). https://doi.org/10.5815/ijcnis.2021.06.06
8. Alsmadi, M.K.: A hybrid firefly algorithm with fuzzy-C mean algorithm for MRI brain segmentation. Am. J. Appl. Sci. **11**(9), 1676–1691 (2014)
9. Kundu, M., Nashiry, A., Kumar Dipongkor, A., Sarmin Sumi, S., Hossain, A.: An optimized machine learning approach for predicting parkinson's disease. Int. J. Mod. Educ. Comput. Sci. (IJMECS) **13**(4), 68–74 (2021). https://doi.org/10.5815/ijmecs.2021.04.06
10. Pal, N.R., et al.: A possibilistic fuzzy c-means clustering algorithm. IEEE Trans. Fuzzy Systems **13**(4), 517–530 (2005)

11. Iqbal, A., Aftab, S.: A classification framework for software defect prediction using multifilter feature selection technique and MLP. Int. J. Mod. Educ. Comput. Sci. (IJMECS) **12**(1), 18–25 (2020). https://doi.org/10.5815/ijmecs.2020.01.03

12. Xu, D., Tian, Y.: A comprehensive survey of clustering algorithms. Ann. Data Sci. **2**(2), 165–193 (2015)

13. Bandyopadhyay, S., Mukhopadhyay, A., Maulik, U.: An improved algorithm for clustering gene expression data. Bioinformatics **23**(21), 2859–2865 (2007)

14. Zhou, Y., Sharpee, T.O.: Hyperbolic geometry of gene expression. Iscience **24**(3), 102225 (2021)

15. Banerjee, A., et al.: Generative model-based clustering of directional data. In: Proceedings of the Ninth ACM SIGKDD International Conference on Knowledge Discovery and Data Mining, pp. 19–28 (2003)

16. Feil, B., Abonyi, J.: Geodesic distance based fuzzy clustering. In: Soft Computing in Industrial Applications, pp. 50–59 (2007)

17. Dhillon, I.S., Modha, D.S.: Concept decompositions for large sparse text data using clustering. Mach. Learn. **42**(1), 143–175 (2001)

18. Banerjee, A., et al.: Clustering on the unit hypersphere using von mises-fisher distributions. J. Mach. Learn. Res. **6**(9), 1345–1382 (2005)

19. Yang, M.-S., Chang-Chien, S.-J., Hung, W.-L.: An unsupervised clustering algorithm for data on the unit hypersphere. Appl. Soft Comput. **42**, 290–313 (2016)

20. Mendes Rodrigues, M.E.S., Sacks, L.: Dynamic knowledge representation for e-learning applications. In: Enhancing the Power of the Internet, pp. 259–282 (2004)

21. Mei, J.-P., Wang, Y.: Hyperspherical fuzzy clustering for online document categorization. In: 2016 IEEE International Conference on Fuzzy Systems (FUZZ-IEEE), pp. 1487–1493 (2016)

22. Mendes Rodrigues, M.E.S., Sacks, L.: A scalable hierarchical fuzzy clustering algorithm for text mining. In: Proceedings of the 5th International Conference on Recent Advances in Soft Computing, pp. 269–274 (2004)

23. Yan, Y., Chen, L.: Hyperspherical possibilistic fuzzy c-means for high-dimensional data clustering. In: 2009 7th IEEE International Conference on Information, Communications and Signal Processing (ICICS), pp. 1–5 (2009)

24. Asgharbeygi, N., Maleki, A.: Geodesic k-means clustering. In: 2008 19th IEEE International Conference on Pattern Recognition, pp. 1–4 (2008)

25. Yang, T., et al.: Geodesic clustering in deep generative models. arXiv preprint arXiv:1809.04747 (2018)

26. Kim, J., Shim, K.-H., Choi, S.: Soft geodesic kernel k-means. In: 2007 IEEE International Conference on Acoustics, Speech and Signal Processing (ICASSP'07) 2, pp. II–429 (2007)

27. Tan, T.Z., et al.: Analysis of gene expression signatures identifies prognostic and functionally distinct ovarian clear cell carcinoma subtypes. EBioMedicine **50**, 203–210 (2019)

28. Wang, Y., et al.: Gene-expression profiles to predict distant metastasis of lymph-node-negative primary breast cancer. The Lancet **365**(9460), 671–679 (2005)

29. Pressinotti, N.C., et al.: Differential expression of apoptotic genes PDIA3 and MAP3K5 distinguishes between low-and high-risk prostate cancer. Mol. Cancer **8**(1), 1–12 (2009)

30. Griesinger, A.M., et al.: Characterization of distinct immunophenotypes across pediatric brain tumor types. J. Immunol. **191**(9), 4880–4888 (2013)

31. Starczewski, A., Krzyżak, A.: Performance evaluation of the silhouette index. In: Rutkowski, L., Korytkowski, M., Scherer, R., Tadeusiewicz, R., Zadeh, L.A., Zurada, J.M. (eds.) ICAISC 2015. LNCS (LNAI), vol. 9120, pp. 49–58. Springer, Cham (2015). https://doi.org/10.1007/978-3-319-19369-4_5

32. Davies, D.L., Bouldin, D.W.: A cluster separation measure. IEEE Trans. Pattern Anal. Mach. Intell. **2**, 224–227 (1979)
33. Calinski, T., Harabasz, J.: A dendrite method for cluster analysis. Commun. Stat.-Theory Methods **3**(1), 1–27 (1974)

An Efficient Storage Architecture Based on Blockchain and Distributed Database for Public Security Big Data

Duoyue Liao, Xinhua Dong, Zhigang Xu$^{(\boxtimes)}$, Hongmu Han, Zhongzhen Yan, Qing Sun, and Qi Li

Institute of Public Security Big Data, School of Computer Science, Hubei University of Technology, Wuhan 430068, China
386160844@qq.com

Abstract. For the massive multi-source and heterogeneous public security big data, a distributed, secure and large-scale storage platform is urgently needed. As one of the most popular distributed ledger technologies, blockchain satisfies the needs of distribution and security, but there are still deficiencies in the storage of massive data. This paper provides an in-depth analysis of the advantages and disadvantages of blockchain and distributed databases in various aspects, and combines the characteristics of distributed databases to improve the performance of blockchain in terms of storage and query. HBase is proposed to extend the data storage layer of Hyperledger Fabric. The hash of the bulk data is first calculated and the complete data is stored into HBase by calling chaincode to store only the hash onto the chain. A hash-based rowkey construction has also been designed in an attempt to improve the efficiency of reading data in the blockchain. Finally, this architecture is applied to realize the storage of massive public security big data and the results demonstrate an ideal case of a 6% increase in throughput, 5% reduction in block size, and 13% reduction in transaction latency.

Keywords: Blockchain · Distributed database · Fabric · HBase · Performance optimization

1 Introduction

With the rapid growth of big data in public security, a secure, decentralized data storage platform is needed to speed up the process of research and crime-solving. Due to its decentralized, tamper-proof, traceable, and data sharing features, the blockchain [1–4] system is very suitable for solving data storage and sharing in situations such as trust crises and data silos, as well as ensuring data security. However, blockchain still has some drawbacks as a storage medium compared to traditional databases. For example, high storage pressure, low throughput and low query efficiency.

The interplay and rapid development of blockchain and databases are witnessed by all [5–8]. To expand blockchain technology to a wide range of fields, it is necessary to draw on the mature technology accumulated by distributed databases over the years; while distributed databases, to improve data security and avoid over-centralization, should also refer to the technical concept of blockchain for theoretical and technical innovation. Therefore, the combination of blockchain and database can be considered, and the advantages of the database can be used to improve the performance of blockchain [9–11]. FalconDB [12] utilizes the blockchain platform to achieve a low storage cost and efficient database collaboration system and provides certain incentives to prevent server nodes from being evil. However, it wastes too much of the blockchain's storage resources, lacks performance in various aspects such as throughput, and is not sufficiently complete in terms of security. BigchainDB [13] is a decentralized distributed database that incorporates many blockchain features and uses MongoDB as the backend database. But it is essentially a distributed database and the blockchain is only part of the system used to maintain security, with data query functions left entirely to external databases.

Therefore, to solve the above-mentioned problems, this paper fuses Fabric with an external distributed database HBase [14] to provide blockchain with massive data storage function under a clustered distributed file system based on Hadoop, which solves the storage overhead that is difficult to be alleviated by blockchain and further optimizes the performance of Fabric by using HBase. Specifically, this paper does the following work:

First, optimize the storage performance of blockchain. Adopt HBase distributed database of Hadoop platform as the off-chain big data storage layer of Fabric blockchain. The huge amount of complete data is stored in HBase, and the data is packaged in bulk to calculate the hash value, and only the hash value is uploaded to the chain. This method not only makes the blocks lighter, but also relieves the storage load pressure of the blockchain, and provides high throughput for the blockchain. Next, optimize the query performance of blockchain. Using the high concurrent read performance of external distributed database HBase, we design the rowkey arrangement method through the hash field on the chain to achieve efficient querying of data records.

The rest of the paper is organized as follows. Section 2 mainly introduces the basic concepts in Fabric transaction flow and focuses on previous work that is closely related to our work, including work on blockchains, distributed databases, and blockchain performance optimization. Section 3 describes the defined data storage algorithm and system architecture in detail. Section 4 presents experimental tests of the methods in the previous section, compares them with traditional methods, and evaluates a range of performance metrics. Section 5 summarizes the article and the areas to be improved or a prospect of the future.

2   Background and Related Work

2.1   Blockchain

In 2008, Satoshi Nakamoto, the father of Bitcoin, published a paper called "Bitcoin: A Peer-to-Peer Electronic Cash System" [1], describing a "Bitcoin electronic currency and its algorithm. Since then, there has been a blockchain boom. Other blockchain application platforms have been created following the traditional Bitcoin blockchain, such as Ethereum [2]. They belong to the same class of blockchains like Bitcoin, called public chains, also called permissionless blockchains. Hyperledger Fabric [3] is a federated-chain-based, open-source, enterprise-grade permissioned distributed ledger technology with superior performance, better mechanisms, and elimination of resource waste compared to past blockchain platforms such as Bitcoin and Ethereum, and introduces a polycentric federated chain that can be applied to more innovative enterprise cases. Fabric is a permission blockchain that limits the number of nodes that can join a blockchain network, and nodes need to be approved before they can join the blockchain.

2.2   Fabric Transaction Flow

In pursuit of high performance, this paper uses Hyperledger Fabric, one of the most popular licensed blockchain platforms developed by the Linux Foundation, which has a modular design, allowing for high throughput through a trust model and pluggable components.

As shown in Fig. 1, Unlike other blockchain network execution transaction models, Fabric uses a new transaction architecture called execute-order-validate [15], which is a simulated execution-commit and validates the model. It consists of a 3-step transaction flow:

1) endorsement phase—simulating transactions on selected peers and signing them;
2) ordering phase—ordering transactions by consensus protocol;
3) validation phase—submitting to the general ledger after validation.

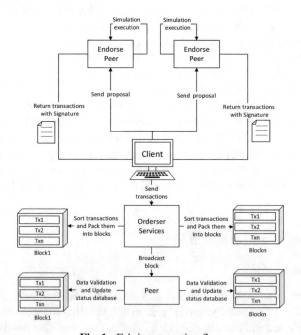

Fig. 1. Fabric transaction flow

2.3 Distributed Database

Distributed databases today, whether they are SQL relational or NoSQL non-relational, are inherently centralized. Centralized databases address the pain points of speed and bandwidth, allowing for adding, deleting, reading, and writing at a high rate. On the other hand, a distributed database is a database in which nodes in different geographical locations store their copies, forming a complete, logically unified, and physically dispersed database through network connections. A distributed database is logically a unified whole, and physically stored separately on different physical nodes. However, this so-called distributed database is only distributed storage in the sense of server clusters, with individual nodes distributed in different geographical locations, but it is still uniformly and centrally managed. The distributed database used in this paper is Apache Hbase, similar to Google's Bigtable [16], which is an open-source, distributed, scalable, versioned, non-relational, column family-oriented NoSQL database. It supports random, real-time read and write access to big data, and provides a fault-tolerant way to store large amounts of sparse data.

2.4 Blockchain Performance Optimization

This article focuses on the performance optimization of blockchain data storage and query. Many researchers and scholars have tried some solutions to the following problems of blockchain at present:

1) Storage space overhead

   When the scale of blockchain keeps increasing, the storage volume and read/write frequency required by the whole network will also grow exponentially, and it may be difficult for individual nodes to bear the storage pressure of blocks. BlockchainDB [9] and FalconDB [10] introduce database features on top of blockchain systems, allowing mutually untrusted multiple parties to jointly participate in maintaining a verifiable database.

2) Insufficient throughput

   Due to the decentralized and tamper-proof nature of blockchain centralization, tamper-proof and other characteristics, numerous nodes are required to participate in maintenance, and various consensus mechanisms and smart contracts are introduced for transaction processing to ensure the security and stability of the blockchain so that the throughput of the vast majority of blockchain systems is less than that of traditional databases. The NUS R&D team has developed the FabricSharp [17] project, which uses ForkBase [18], a distributed ledger storage engine deeply optimized for blockchain data features, to greatly increase the throughput of blockchain data storage. Ankur Sharma et al. [19] propose transaction reordering and early interruption of transactions for the transaction flow in Fabric to improve the throughput of Hyperledger Fabric. Lu Xu et al. [20] improve the read speed of data in Fabric by caching and increasing the cache log to ensure the consistency of data.

3) Query performance bottleneck

   The underlying storage system of most blockchains uses levelDB [21] for storing information such as state of the world, block index, etc. levelDB is stored as key-value form, which has high random write and sequential read/write performance, but the performance of random read is very general, which is suitable for application in scenarios where the number of queries is small and writes are very frequent. Yang L et al. [22] proposed EtherQL system, which is a query layer designed in the outer layer of the blockchain. the main idea is to copy the blockchain data into an external database and design the query layer with the functional interface provided by the external database. This method does not need to modify the original blockchain system, and it is a simple and feasible solution to implement the query layer directly on top of the blockchain system.

3   Storage Architecture Based on Fabric and HBase

3.1   Data Storage Algorithm

The Fabric has many transaction message queues, and conducting transactions takes a lot of time through the transaction flow. Usually, it is only suitable to store them by row, constructing JSON data structure for each row of data and writing them to the state database in the form of key-value when using Fabric to store datasets. However, when it comes to datasets with a large number of fields and columns, the performance of using Fabric to store the complete data will be very pessimistic. Therefore, it is considered to store only the hash values [23] of the data into the state database and store the complete data into the under-chain database. The following algorithm illustrates the specific steps of the data writing method in this paper.

Algorithm 1: Write Data

    Data: datasetFile

    Result: Is the data write valid (True or
           False)

1   initialization;

2   **while** *The data set is entered* **do**

3   String Hash;

4   Hash = getSHA256Str(datasetFile);

5   **if** *!Hash.isEmpty()* **then**

6   WriteDataHashIntoFabric(Hash);
             putDateSetIntoHbase(datasetFile);
             return true;

7   **else**

8   return false;

9   **end**

10 **end**

3.2   Data Storage Structure

The block is composed of the block header, block body, and block metadata. The information such as timestamp, block hash is stored in the block header, while the transaction information is stored in the block body. The data hash is generated by the SHA256 algorithm, which is stored as a transaction on the chain, and then the timestamp and hash are used to construct the row key of HBase, and the complete data is stored in HBase by column. In this way, the hash value obtained on the chain can be accurate to a row of data in the distributed database under the chain. Figure 2 shows our designed Fabric blockchain and the data storage structure in HBase and its correspondence.

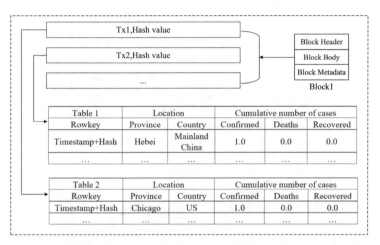

Fig. 2. On-chain and off-chain data storage structure

3.3   System Architecture

In this paper, a storage layer is introduced on top of Fabric, and HBase is used as an external database to improve the throughput for Fabric. As shown in Fig. 3, the Fabric blockchain contains all the peer nodes and sorting nodes in the channel, and the orderer nodes form the orderer service through a consensus mechanism. A channel corresponds to a ledger, and all nodes in the channel will jointly maintain a ledger.HBase relies on the Datanode cluster in HDFS as the underlying data storage service and uses Zookeeper to do the high availability of HMaster, monitoring of RegionServer, entry of metadata, and maintenance of cluster configuration. Data storage structures have been designed in Fabric blockchain and HBase respectively.

The client in the channel calls the blockchain storage layer, first accesses zookeeper, gets the corresponding RegionServer address in HBase, and then initiates a written request to the RegionServer, which accepts the data and writes it to memory. When the size of MemStore reaches a certain value, it is flushed to StoreFile and stored to HDFS.

After that, the client initiates a transaction proposal, which is simulated, endorsed, and signed by the peer node and then sent to the orderer service, where the orderer nodes sort, package, and generate a block through, which is then broadcast to the peer node, linking the block to the tail of the entire Fabric blockchain.

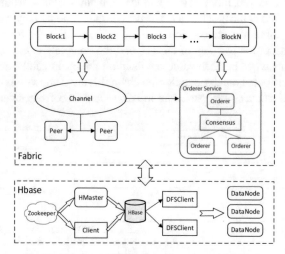

Fig. 3. Fabric and HBase based system architecture

4   Experiment and Performance Evaluation

This section introduces the experimental environment, the experimental data set, and the comparative experiments conducted on metrics such as throughput, block size, and transaction latency.

4.1   Experimental Environment

The cluster hardware configuration for this experiment was an eight-core, sixteen-thread AMD Ryzen 7 4800U 1.80 GHz CPU with 16 GB of RAM, a virtual machine using Vmware WorkStation Pro 15.5, a virtual machine system with Linux CentOS7, Fabric version 2.2, Golang version v1.14.2 Linux/amd64, and Docker version was v20.10.7. In our experiments, we deployed a five-node Hadoop cluster, a single orderer node, a dual-organization dual-peer node, three HMaster nodes, three zookeeper nodes, and two HregionServer. Smart contracts were configured to be identified by a single peer node. Either of the two peers can act as an endorsing peer to spread the workload.

4.2   Experimental Dataset

The experimental dataset used was daily level information on the number of Covid 2019 affected cases across the globe, with more than 300,000 rows of data, containing fields such as observation date, province, country, last update date time in UTC, the cumulative number of confirmed cases, deaths cases and recovered cases. Test results were generated using the Caliper tool, a blockchain performance benchmarking tool and allows user-defined use cases to test various aspects of the blockchain performance metrics.

4.3 Performance Evaluation

The performance metrics evaluated in this paper are throughput, block size, and transaction latency, with the following definitions for the sake of uniformity.

Throughput, the number of transactions per second, is one of the most important indicators of system performance. Block size, which counts the space occupied by a block ledger in a node in the blockchain. Transaction latency is the average time for each complete transaction flow (all three phases of the E-O-V model), including failed transactions and successful transactions.

4.4 Throughput

1) Write data throughput experiment

In order to demonstrate the advantages of the HBase off-chain storage method proposed in this paper, two methods were used for data storage in the experiments: one was the traditional ledger data storage method, where each row of data of the dataset was stored in the blockchain ledger and the process of storing each row of data was regarded as a transaction in the blockchain. The other was to calculate the hash of the bulk data of the dataset, store the hash in the blockchain ledger, and store the specific dataset in HBase, the process of storing each hash was also considered as a transaction in the blockchain.

The throughput of the two blockchain data storage methods was tested with a fixed number of 50 to 400 transactions by setting Caliper's workload module to one test worker and a transaction sending rate of 100 TPS per round.

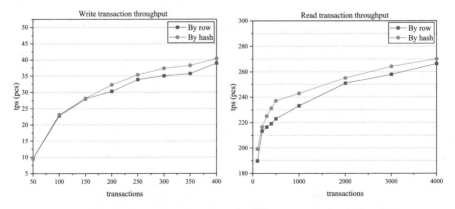

Fig. 4. Transaction throughput by different storage methods

As shown in Fig. 4, the experimental results show that the throughput of both storage methods improves as the total number of transactions tested increases, and data storage by hash value always has higher throughput than per-row data storage, and this method can store more data in the off-chain database when the same number of transactions are performed.

2) Read data throughput experiment

The results are shown in Fig. 4. For the data stored in both ways in the blockchain ledger, the throughput of randomly reading 100 to 4000 data, i.e., 100 to 4000 transactions under each of them, was tested. As shown in the figure, the experimental results show that the data throughput is higher for random reads using off-chain storage. Also, as shown in Table 1, we counted the time taken to read the data and found that the method reduced the time to read the data even more significantly at low transaction volumes.

Table 1. The time taken by different storage methods

Transactions		400	1000	2000	3000	4000	5000
Time (s)	By row	4.356	4.583	8.741	11.797	15.025	18.396
	By hash	1.916	4.363	8.051	11.664	15.023	18.351

At the same time, a rowkey based on hash field construction was also designed for HBase, an off-chain database, on which random data was read, and the experimental results were compared with random data read in Fabric using the traditional method. In the experiments, random reads were performed in Fabric and HBase for 500 to 6000 data items. As shown in Fig. 5, the experimental results show that the throughput of random reads using the off-chain database HBase was much higher than that of Fabric,

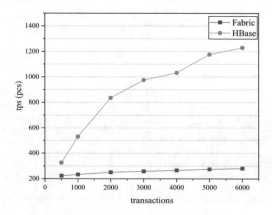

Fig. 5. On-chain and off-chain read throughput

which proves the excellent query efficiency of using the off-chain database and shows that the combination of Fabric and HBase can bring superior performance.

4.5 Block Size

To test the optimization of block storage load of this method, the volume sizes of all blocks in the resulting blockchain ledger were counted for the cases of 1000 to 5000 transaction volumes and compared with the traditional method of Fabric data storage, and the results are shown in Fig. 6. The experimental results show that as the transaction volume increases, the block size also increases. For the same number of transactions, the block storage load of this method was significantly reduced, and more data can be stored in the off-chain database.

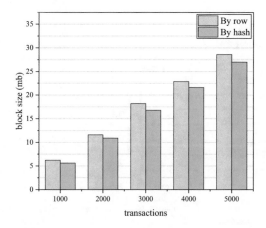

Fig. 6. Block size by different storage methods

4.6 Latency

Caliper generates latency figures in addition to measuring the throughput of transactions. We compare the average latency of write transactions with the maximum latency of read transactions. Since the throughput of reading transactions is much greater than that of writing transactions, both have lower average latency, so only the maximum transaction latency of the two was compared here.

Table 2 shows the average latency for 100 to 400 write transactions and the maximum latency for 500 to 5000 read transactions using the storage method defined in this paper compared to the traditional storage structure. Fig. 7 shows the trend of the two types of transaction latency.

Experiments show that the write average transaction latency and read maximum transaction latency is significantly improved by our approach. It can be seen that as the number of read transactions increases, the maximum latency for reading data stored by rows is about 70 ms, while the maximum latency for reading data stored with hashes for on-chain and off-chain collaboration is stable at about 40 ms. Moreover, for our approach, the average latency of write transactions is always lower.

Table 2. Latency by different storage methods

Transactions		100	150	200	250	300	350
Avg latency (ms)	By row	480	420	350	330	370	340
	By hash	450	410	330	320	320	300
Transactions		500	1000	2000	3000	4000	5000
Max latency (ms)	By row	40	30	30	60	60	70
	By hash	30	30	30	40	40	30

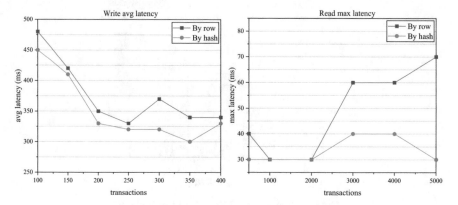

Fig. 7. Transaction latency by different storage methods

5 Conclusion

This paper proposes efficient storage architecture based on Hyperledger Fabric and HBase. A data writing algorithm was designed to perform bulk packing to calculate hash values, then store the hash values on the chain and construct the low position of rowkey with the calculated hash values to store the complete data in the under-chain database— HBase. By using the dataset for testing experiments, the results show that this method can improve the read and write throughput of data in the Fabric blockchain, effectively reduce the storage load of the blockchain and reduce the latency of transactions compared to the traditional method of storing data in the blockchain, and achieve a storage architecture that meets the needs of public security big data. At the same time, research has shown that blockchain performance can be effectively improved by incorporating the benefits of distributed database technology. This architecture can also be extended to finance, transport, healthcare [24] and other areas. The subsequent research direction is to study the scaling mechanism of the underlying blockchain storage to further ensure the security of data.

Due to machine performance and time constraints, the network nodes in this paper were relatively small, making it difficult to realize the extreme performance advantages of blockchain, but we will then work to address this issue and further optimize our model.

Acknowledgment. This work is supported by the National Natural Science Foundation of China under Grant No.61772180, the Key-Area Research and Development Program of Guangdong Province 2020B1111420002, and the Innovation Fund of Hubei University of Technology BSQD2016019. We sincerely thank the anonymous reviewers for their very comprehensive and constructive comments.

References

1. Nakamoto, S.: Bitcoin: a peer-to-peer electronic cash system. Decentralized Bus. Rev., 21260 (2008)
2. Buterin, V.: A next-generation smart contract and decentralized application platform. White paper **3**(37) (2014)
3. Androulaki, E., Barger, A., Bortnikov, V., et al.: Hyperledger fabric: a distributed operating system for permissioned blockchains. In: Proceedings of the Thirteenth EuroSys Conference, pp. 1–15 (2018)
4. Anwar, S., Anayat, S., Butt, S., et al.: Generation analysis of blockchain technology: bitcoin and ethereum. Int. J. Inf. Eng. Electron. Bus. **12**(4), 30–39 (2020)
5. Huang, H., Kong, W., Zhou, S., et al.: A survey of state-of-the-art on blockchains: Theories, modelings, and tools. ACM Comput. Surv. (CSUR) **54**(2), 1–42 (2021)
6. István, Z., Sorniotti, A., Vukolić, M.: Streamchain: do blockchains need blocks?. In: Proceedings of the 2nd Workshop on Scalable and Resilient Infrastructures for Distributed Ledgers, pp. 1–6 (2018)
7. Mohan, C.: Blockchains and databases: a new era in distributed computing. In: 2018 IEEE 34th International Conference on Data Engineering (ICDE), pp. 1739–174. IEEE (2018)
8. Raikwar, M., Gligoroski, D., Velinov, G.: Trends in development of databases and blockchain. In: Chacko, J.A., Mayer, R., Jacobsen, H.A. (eds.) 2020 Seventh International Conference on Software Defined Systems (SDS), pp. 177–182. IEEE (2020)
9. Sharma, A., Schuhknecht, F.M., Agrawal, D., et al.: How to databasify a blockchain: the case of hyperledger fabric. arXiv preprint arXiv:1810.13177 (2018)
10. Ruan, P., Dinh, T.T.A., Loghin, D., et al.: Blockchains vs. distributed databases: dichotomy and fusion. In: Proceedings of the 2021 International Conference on Management of Data, pp. 1504–1517 (2021)
11. Ruan, P., Anh Dinh, T.T., Lin, Q., et al.: Revealing every story of data in blockchain systems. ACM Sigmod Record **49**(1), 70–77 (2020)
12. Peng, Y., Du, M., Li, F., et al.: FalconDB: Blockchain-based collaborative database. In: Proceedings of the 2020 ACM SIGMOD International Conference on Management of Data, pp. 637–652 (2020)
13. El-Hindi, M., Binnig, C., Arasu, A., et al.: BlockchainDB: a shared database on blockchains. Proc. VLDB Endowment **12**(11), 1597–1609 (2019)
14. Vora, M.N.: Hadoop-HBase for large-scale data. In: Proceedings of 2011 International Conference on Computer Science and Network Technology, vol. 1, pp. 601–605. IEEE (2011)

15. Why do my blockchain transactions fail? a study of hyperledger fabric. In: Proceedings of the 2021 International Conference on Management of Data, pp. 221–234 (2021)
16. Chang, F., Dean, J., Ghemawat, S., et al.: Bigtable: a distributed storage system for structured data. ACM Trans. Comput. Syst. (TOCS) **26**(2), 1–26 (2008)
17. Ruan, P., Loghin, D., Ta, Q.T., et al.: A transactional perspective on execute-order-validate blockchains. In: Proceedings of the 2020 ACM SIGMOD International Conference on Management of Data, pp. 543–557 (2020)
18. Wang, S., Dinh, T.T.A., Lin, Q., et al.: Forkbase: an efficient storage engine for blockchain and forkable applications. Proc. VLDB Endowment **11**(10), 1137–1150 (2018)
19. Sharma, A., Schuhknecht, F.M., Agrawal, D., et al.: Blurring the lines between blockchains and database systems: the case of hyperledger fabric. In: Proceedings of the 2019 International Conference on Management of Data, pp. 105–122 (2019)
20. Xu, L., Chen, W., Li, Z., et al.: Solutions for concurrency conflict problem on Hyperledger Fabric. World Wide Web **24**(1), 463–482 (2021)
21. Dent, A.: Getting started with LevelDB. Packt Publishing Ltd. (2013)
22. Li, Y., Zheng, K., Yan, Y., Liu, Q., Zhou, X.: EtherQL: a query layer for blockchain system. In: Candan, S., Chen, L., Pedersen, T.B., Chang, L., Hua, W. (eds.) Database Systems for Advanced Applications, pp. 556–567. Springer International Publishing, Cham (2017). https://doi.org/10.1007/978-3-319-55699-4_34
23. Kuznetsov, A., Oleshko, I., Tymchenko, V., et al.: Performance analysis of cryptographic hash functions suitable for use in blockchain. Int. J. Comput. Netw. Inf. Secur. **13**(2), 1–15 (2021)
24. Hapiffah, S., Sinaga, A.: Analysis of blokchain technology recommendations to be applied to medical record data storage applications in Indonesia. Int. J. Inf. Eng. Electron. Bus. **12**(6), 13–27 (2020)

Generalization of the Formal Method for Determining the State of Processors of a Multiprocessor System Under Testing

Alexei M. Romankevich, Kostiantyn V. Morozov$^{(\boxtimes)}$, and Vitaliy A. Romankevich

National Technical University of Ukraine "Igor Sikorsky Kyiv
Polytechnic Institute", Kyiv 03056, Ukraine
mcng@ukr.net

Abstract. The paper proposes a generalization of the formal procedure for establishing the state of processors of an M-diagnosed system based on the results of performing a given set of mutual test checks of processors, in accordance with the Preparata-Metze-Chien model. An approach to solving the problem of a possible assessment of the sufficiency of a set of test checks to unambiguously determine the state of the system is also proposed. In addition, as a result of special transformations, the process of establishing the state of the system processors based on the results of the tests carried out can be greatly simplified. Thus, a significant part of the calculations aimed at performing self-diagnostics of the system is transferred from the operational stage to the development stage.

Keywords: Multiprocessor systems · Mutual testing of processors ·
PMC-model · M-diagnosable systems

1 Introduction

Modern control systems for complex objects often have increased requirements for both performance and reliability [1–11]. Building them on the basis of fault-tolerant multiprocessor systems (FTMS) [12–15] allows solving both of the abovementioned problems [16–20]. One of the goals that appears while constructing FTMS is the developing of their self-testing process [21–24], i.e., determination of states (operational or faulty) of each processor of the system during its functioning. Information about the state of each processor of the system allows to optimally organize the system reconfiguration to maintain its performance during operation [25–29].

There are two main approaches for testing FTMS processors: implementation of testing via a dedicated test device and cross-processor testing approach [30–32].

An advantage of the first approach is the relative ease of implementation, however, there is a disadvantage of the need of insertion an additional node into the system, which can also fail, and for this particular reason, appears to be a bottleneck in terms of reliability. This approach can be used, in particular, for testing the system at the stage of production or service.

Z. Hu et al. (Eds.): ICCSEEA 2022, LNDECT 134, pp. 363–375, 2022.
https://doi.org/10.1007/978-3-031-04812-8_31

The second approach is based on mutual test checks of system processors [33]. Its organization is often much more complicated; however, this approach allows to avoid the insertion of additional nodes, which is more preferable at the stage of system operation. It is to this particular direction that the work is devoted.

Testing the processor can be implemented by running a certain set of tests and comparing the results with the benchmark: if the test results do not match, the processor is considered faulty. The work assumes that mutual testing of the system processors is carried out in accordance with the Preparata-Metze-Chien (PMC) model [34], i.e. as a result of testing a certain processor with a serviceable processor, the value 0 or 1 is obtained, depending on the state of the tested processor (serviceable or defective, respectively); if the testing processor is faulty, then the test result can be either 0 or 1, regardless of the state of the tested processor.

The capabilities of the system in terms of diagnostics (i.e., carrying out mutual test checks) can be reflected using a digraph, where each of the vertices corresponds to a certain processor of the system, and the existence of an arc from vertex a to vertex b – the possibility of performing a test check of the processor corresponding to vertex a by the processor corresponding to the vertex b.

In the case of implementation of processors testing on the principle of each-with-each (i.e., if the system corresponds to a complete graph), an algorithm was proposed [35], which allows determining the state of the system based on the execution of no more than $N + 2p$ test checks, where N is the number of system's processors, and p is the number of actually faulty processors. In [34] it was proved that the state of such a system can always be determined if no more than $[(N - 1) / 2]$ processors fail in it, and in [36] it was shown that this is practically always possible for more failures. However, real structures do not necessarily correspond to a complete graph.

In [37], a formal method was proposed that allows one to establish the states (operational or faulty) of multiprocessor system's processors based on the results of their mutual test checks. In this case, an arbitrary topology of the graph is allowed, reflecting the capabilities of the system in terms of diagnostics.

According to the method proposed in [37], for each of the processors of the M-diagnosed system [38] a boolean variable x_i (where i is the processor number) is associated, which takes the value 1 if it is operational and 0 if it fails. Obviously, any state of the system will uniquely correspond to some elementary conjunction of variables x_i (constituent of one), and it is for a real state that it will have a value equal to one.

Let's also note that it is the M-diagnosed system that is considered, i.e. such a system, the state of which can be established by executing mutual test checks of processors only in case if the number of their failures does not exceed the value M [39, 40].

According to [37], each of the test experiments is associated with a specially obtained expression. Let some test experiment rest on testing the j-th processor by the i-th processor, as a result of which the value r_{ij} is obtained. Then the corresponding expression R_{ij} will have the form:

$$R_{ij}(x_i, x_j) \triangleq \begin{cases} x_i x_j \vee \bar{x}_i \equiv x_j \vee \bar{x}_i, \text{ when } r_{ij} = 0 \\ x_i \bar{x}_j \vee \bar{x}_i \equiv \bar{x}_j \vee \bar{x}_i, \text{ when } r_{ij} = 1 \end{cases} \tag{1}$$

It was shown in [37] that with such a construction the equality $R_{ij}(x_i, x_j) = 1$ will be valid.

Further, according to the results of K tests carried out, the expression V_K, is formed:

$$V_k \triangleq \bigwedge_{R \in S_K},\tag{2}$$

where S_K is a set of expressions corresponding to the results of the tests executed. Let's note that the equality $V_K = 1$ may exist in this context.

Further, expression (2) is transformed into a perfect disjunctive normal form (PDNF, a disjunction of elementary conjunctions) with the simultaneous exclusion of all conjunctions containing more than M inversions:

$$V_K = \bigvee_{l=1}^{L} C_l,\tag{3}$$

where each of C_l is a constituent of one, and L is their number. In [37] it is shown that if in the system no more than M processors have actually failed, the transformation described above will not lead to the exclusion of constituent of one which corresponds to the real state of the system.

On the other hand, remember that $V_K = 1$. Therefore, the equality (4) takes place:

$$\bigvee_{l=1}^{L} C_l = 1.\tag{4}$$

If, as a result, the only single constituent of one ($L = 1$) was obtained, then it obviously corresponds to the real state of the system (as already mentioned, it is it that has a value equal to 1): the variables included in it without inversions correspond to operational processors, and all the rest are faulty.

If several constituents are obtained ($L > 1$), then the state cannot be unambiguously determined. However, the constituent corresponding to the real state will exist among them. In order to determine the state of the system, k additional tests can be carried out, the results of which will correspond to the set S_k of expressions of the form R_{ij}. Next, the expression is constructed and similarly converted to PDNF:

$$V_{K+k} \triangleq V_K \wedge \bigwedge_{R \in S_k} R.\tag{5}$$

If all the allowable tests have already been carried out, then the system is obviously not M-diagnosable.

It should be noted that the above is true only if no more than M processors have actually failed in the system.

If no constituents have been received ($L = 0$), then this corresponds to a situation when more than M processors have failed in the system. However, the opposite is generally not true: in this case, depending on the combination of test results, any of the three situations is possible.

2 Generalization of the Formal Method

The abovementioned approach allows determining the state of the system in accordance with the results of already executed test checks. However, each of these checks can lead to significant resource overheads for both the testing and the tested processors. Therefore, it appears also of interest ability of analyzing the test set even before carrying out test checks.

On the basis of such an analysis, in particular, the selection of the optimal test set is carried out. It can be especially useful for choosing a set of additional test checks, in case, if, according to the results of previous tests, an unambiguous result was not obtained. The computational complexity of the analysis can be significantly lower than the complexity of executing additional test checks.

To implement the possibility of such an analysis, let's write the expressions corresponding to test checks in a different form in a relation to [37]. Let the test experiment consist in testing the j-th processor by the i-th processor, as a result of which the value r_{ij} is obtained. Then the corresponding expression will have the form:

$$T_{ij} \triangleq (r_{ij} \oplus x_j) \bigvee \bar{x}_i \equiv r_{ij}\bar{x}_j \vee \bar{r}_{ij}x_j \bigvee \bar{x}_i \equiv r_{ij}x_i\bar{x}_j \vee \bar{r}_{ij}x_ix_j \bigvee \bar{x}_i = 1 \qquad (6)$$

Note that with such a construction, substituting the corresponding value of r_{ij} into (6), we obtain expression (1), i.e. $T_{ij} \equiv R_{ij}$.

Let K test checks were carried out, as a result of which, in accordance with [37], the expression V_K was obtained. At first, before carrying out checks, consider that $V_K \triangleq V_0 \triangleq 1$. Let's also consider a set of additional tests (k pieces), which correspond to the set U_k of expressions of the form T_{ij}. Let's build an expression:

$$V_{K+k} \triangleq V_K \wedge \bigwedge\nolimits_{T \in U_k} T. \qquad (7)$$

We transform it to the following form, excluding all conjunctions containing more than M inversions of the variables x_q:

$$V_{K+k} = \bigvee\nolimits_{l=1}^{L} P_l C_l, \qquad (8)$$

where C_l is an elementary conjunction containing all variables of the form x_q, and P_l is an expression consisting of variables of the form r_{ij}. In what follows, expressions of the form C_l will be called constituents, and P_l – coefficients.

Let's recall that $V_K = 1$ and $T_{ij} = 1$ are true, therefore, it is also true that:

$$\bigvee\nolimits_{l=1}^{L} P_l C_l = 1. \qquad (9)$$

Statement 1. If for any two different coefficients (let's denote them P_a and P_b) the identity $P_a \wedge P_b \equiv 0$ is valid, then the considered set of test checks is sufficient to unambiguously determine the state of the system (in case of failure of no more than M processors).

Proof. Suppose it is not. I.e. as a result of the tests and the corresponding transformations, more than one constituent remains in the expression. Suppose these are conjunctions C_a and C_b, $a \neq b$. Therefore, for a given combination of test results values (r_{ij}), the values of the coefficients P_a and P_b are equal to 1. In this case, $P_a \wedge P_b = 1$, which contradicts the original statement. ∎

Statement 2. If the condition of statement 1 is satisfied, and for some combination of test results one of the coefficients P_h acquires a nonzero value, then the state of the system corresponds to the constituent C_h.

Proof. Due to the fulfillment of the condition of statement 1, none of the coefficients, except for P_h, can have a nonzero value. Therefore, expression (8) takes the form.

$$\bigvee_{l=1}^{L} P_l C_l \equiv P_h C_h \equiv C_h. \tag{10}$$

On the other hand, according to (9)

$$C_h = 1. \tag{11}$$

Thus, as a result of the transformations, an equation was obtained containing exactly one constituent C_h, which, as already mentioned, uniquely corresponds to the actual state of the system. ∎

By expressing the coefficient P_l in disjunctive normal form (DNF), one can easily establish combinations of test results that lead to its non-zero value. On the other hand, a certain constituent C_l, which characterizes a certain state of the system, corresponds to each coefficient P_l. Further, at the stage of testing the system, the real values of test checks can be easily compared with one of such combinations, which allows one to determine the real state.

Thus, the implementation of complex transformations can be transferred from the stage of self-testing of the system (where the additional computational load is obviously extremely undesirable) to the stage of system development (where it is enough to perform the transformations once and, in addition, often disproportionately larger computing resources are available).

3 Examples

Let's move on to some examples. To reduce the length of the article, fairly simple examples are provided, which, however, fully demonstrate the application of the solutions proposed in the work.

Example 1. Consider a 1-diagnosable system consisting of 4 processors, which we denote by a, b, c and d. A graph representing its capabilities in terms of diagnostics (coupling topology) is shown in Fig. 1. In accordance with the graph, the following test checks are available: $a \rightarrow b$, $b \rightarrow c$, $c \rightarrow d$ and $d \rightarrow a$ (here the testing processor is specified on the left of the arrow, and the one being tested is on the right).

Let's construct and transform the expression W, based on the expressions corresponding to these test checks according to (7), (8):

$$
\begin{aligned}
W &\triangleq T_{ab}T_{bc}T_{cd}T_{da} = \left(r_{ab}\bar{a}\bar{b} \vee \bar{r}_{ab}ab \vee \bar{a}\right)\left(r_{bc}\bar{b}\bar{c} \vee \bar{r}_{bc}bc \vee \bar{b}\right) \\
&\wedge \left(r_{cd}\bar{c}\bar{d} \vee \bar{r}_{cd}cd \vee \bar{c}\right)\left(r_{da}\bar{a}\bar{d} \vee \bar{r}_{da}ad \vee \bar{d}\right) \\
&= \left(r_{ab}\bar{a}\bar{b} \vee \bar{r}_{ab}r_{bc}ab\bar{c} \vee \bar{r}_{ab}\bar{r}_{bc}abc \vee r_{bc}\bar{a}\bar{b}\bar{c} \vee \bar{r}_{bc}\bar{a}bc \vee \bar{a}\bar{b}\right) \\
&\wedge \left(r_{cd}\bar{c}\bar{d} \vee \bar{r}_{cd}r_{da}\bar{a}cd \vee \bar{r}_{cd}\bar{r}_{da}acd \vee r_{da}\bar{a}\bar{c}\bar{d} \vee \bar{r}_{da}a\bar{c}d \vee \bar{c}\bar{d}\right) \\
&= r_{ab}r_{cd}\bar{a}\bar{b}\bar{c}\bar{d} \vee r_{ab}\bar{r}_{bc}r_{cd}abcd \vee \bar{r}_{bc}r_{cd}\bar{a}\bar{b}\bar{c}\bar{d} \vee \bar{r}_{bc}\bar{r}_{cd}r_{da}\bar{a}bcd \\
&\vee r_{ab}\bar{r}_{cd}\bar{r}_{da}\bar{a}\bar{b}cd \vee r_{ab}\bar{r}_{bc}\bar{r}_{cd}\bar{r}_{da}abcd \vee r_{ab}\bar{r}_{da}\bar{a}\bar{b}\bar{c}\bar{d} \vee \bar{r}_{ab}r_{bc}\bar{r}_{da}ab\bar{c}\bar{d} \\
&= \bar{r}_{ab}\bar{r}_{bc}\bar{r}_{cd}\bar{r}_{da}abcd \vee \bar{r}_{bc}\bar{r}_{cd}r_{da}\bar{a}bcd \vee r_{ab}\bar{r}_{cd}\bar{r}_{da}a\bar{b}\bar{c}d \vee \bar{r}_{ab}r_{bc}\bar{r}_{da}ab\bar{c}\bar{d} \\
&\vee \bar{r}_{ab}\bar{r}_{bc}r_{cd}abc\bar{d}
\end{aligned}
$$

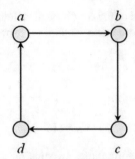

Fig. 1. Coupling topology of system (Example 1)

In this case, the expression has five members. Let's denote the coefficients $P_1 = \bar{r}_{ab}\bar{r}_{bc}\bar{r}_{cd}\bar{r}_{da}$, $P_2 = \bar{r}_{bc}\bar{r}_{cd}r_{da}$, $P_3 = r_{ab}\bar{r}_{cd}\bar{r}_{da}$, $P_4 = \bar{r}_{ab}r_{bc}\bar{r}_{da}$ and $P_5 = \bar{r}_{ab}\bar{r}_{bc}r_{cd}$, as well as constituents $C_1 = abcd$, $C_2 = \bar{a}bcd$, $C_3 = a\bar{b}cd$, $C_4 = ab\bar{c}d$ and $C_5 = abc\bar{d}$. Note that C_1 corresponds to a completely operational state of the system (all processors are serviceable), and C_2, C_3, C_4 and C_5 correspond to states in which one of the processors is faulty: a, b, c and d, respectively.

Note that for any $a \neq b$, $(a, b = 1, 2, ..., 5)$, regardless of the r_{ij} values, $P_a \wedge P_b = 0$ is true. Consequently, in accordance with Statement 1, this set of tests is sufficient to unambiguously determine the state of the system (i.e., the system is indeed 1-diagnosable). Using the expressions for the P_l coefficients, let's construct a table of correspondence of the test results to the state of the system (Table 1). In the table, the symbol "*" denotes an arbitrary value of the test result (i.e., in fact, the independence of the set state of the system from its value is shown). So, for example, if the checks $a \to b$ and $b \to c$ gave the result 0, and the check $c \to d$ gave 1, then the check $d \to a$ doesn't need to be executed. This situation corresponds to a system state in which all processors are serviceable except for d. In addition, to determine the state of the system, it is enough just to find the corresponding row in the table, without the need to construct and transform a special boolean expression (as was suggested in [37]).

Table 1. Correspondence of test results to the state of the system (Example 1)

Test result				System state vector			
$a \to b$	$b \to c$	$c \to d$	$d \to a$	a	b	c	d
0	0	0	0	1	1	1	1
*	0	0	1	0	1	1	1
1	*	0	0	1	0	1	1
0	1	*	0	1	1	0	1
0	0	1	*	1	1	1	0

Example 2. Consider a 2-diagnosable system consisting of 8 processors denoted as a, b, c, d, e, f, g and h. The graph representing the capabilities of this system in terms of diagnostics (coupling topology) is shown in Fig. 2. Suppose that 2 processors in the system have failed: b and d. A number of test checks were also carried out: $a \to b$, $c \to d$, $e \to f$ and $g \to h$, the results of which were respectively $r_{ab} = 1$, $r_{cd} = 1$, $r_{ef} = 0$ and $r_{gh} = 0$.

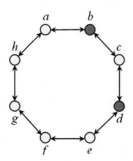

Fig. 2. Coupling topology of system (Example 2)

In accordance with [37], a boolean expression was constructed and transformed:

$$V_4 \triangleq \left(a\overline{b} \vee \overline{a} \right)\left(c\overline{d} \vee \overline{c} \right)(ef \vee \overline{e})\left(gh \vee \overline{h} \right)$$
$$= \left(a\overline{b}c\overline{d} \vee a\overline{b}\overline{c} \vee \overline{a}c\overline{d} \vee \overline{a}\overline{c} \right)\left(efgh \vee ef\overline{h} \vee \overline{e}gh \vee \overline{e}\overline{h} \right)$$
$$= a\overline{b}c\overline{d}efgh \vee a\overline{b}\overline{c}efgh \vee \overline{a}c\overline{d}efgh \vee \overline{a}\overline{c}efgh$$
$$\vee a\overline{b}c\overline{d}ef\overline{h} \vee a\overline{b}\overline{c}ef\overline{h} \vee \overline{a}c\overline{d}ef\overline{h} \vee \overline{a}\overline{c}ef\overline{h}$$
$$\vee a\overline{b}c\overline{d}\overline{e}gh \vee a\overline{b}\overline{c}\overline{e}gh \vee \overline{a}c\overline{d}\overline{e}gh \vee \overline{a}\overline{c}\overline{e}gh$$
$$\vee a\overline{b}c\overline{d}\overline{e}\overline{h} \vee a\overline{b}\overline{c}\overline{e}\overline{h} \vee \overline{a}c\overline{d}\overline{e}\overline{h} \vee \overline{a}\overline{c}\overline{e}\overline{h} = a\overline{b}c\overline{d}efgh$$
$$\vee a\overline{b}\overline{c}defgh \vee \overline{a}bc\overline{d}efgh \vee \overline{a}b\overline{c}defgh$$
$$\vee \overline{a}\overline{b}\overline{c}defgh \vee \overline{a}\overline{b}\overline{c}defgh \vee \overline{a}b\overline{c}defgh \vee \overline{a}b\overline{c}defgh$$
$$= a\overline{b}c\overline{d}efgh \vee a\overline{b}\overline{c}defgh \vee \overline{a}bc\overline{d}efgh \vee \overline{a}b\overline{c}defgh$$

As a result of the transformation, the PDNF was obtained, consisting of 4 conjuncts, which means the need for additional tests. Also let's denote $C_1 = a\overline{b}c\overline{d}efgh$, $C_2 = a\overline{b}\overline{c}defgh$, $C_3 = \overline{a}bc\overline{d}efgh$ and $C_4 = \overline{a}b\overline{c}defgh$.

As an example, consider two possible sets of additional tests, each containing two checks: $(f \to g, c \to b)$ and $(h \to a, e \to d)$. For each of the sets, we construct and transform boolean expressions, denoting them $W^{(1)}$ and $W^{(2)}$, respectively:

$$W^{(1)} \triangleq T_{fg}T_{cb} = \left(r_{fg}f\overline{g} \vee \overline{r}_{fg}fg \vee \overline{f} \right)\left(r_{cb}\overline{b}c \vee \overline{r}_{cb}bc \vee \overline{c} \right)$$
$$= r_{cb}r_{fg}\overline{b}cf\overline{g} \vee r_{cb}\overline{r}_{fg}\overline{b}cfg \vee r_{cb}\overline{b}c\overline{f} \vee \overline{r}_{cb}r_{fg}bcf\overline{g} \vee \overline{r}_{cb}\overline{r}_{fg}bcfg$$
$$\vee \overline{r}_{cb}bc\overline{f} \vee r_{fg}\overline{c}f\overline{g} \vee \overline{r}_{fg}\overline{c}fg \vee \overline{c}\overline{f}$$
$$W^{(2)} \triangleq T_{ha}T_{ed} = \left(r_{ha}\overline{a}h \vee \overline{r}_{ha}ah \vee \overline{h} \right)\left(r_{ed}\overline{d}e \vee \overline{r}_{ed}de \vee \overline{e} \right)$$
$$= r_{ha}r_{ed}\overline{a}\overline{d}eh \vee \overline{r}_{ha}r_{ed}a\overline{d}eh \vee r_{ed}\overline{d}e\overline{h} \vee r_{ha}\overline{r}_{ed}\overline{a}deh \vee \overline{r}_{ha}\overline{r}_{ed}adeh$$
$$\vee \overline{r}_{ed}de\overline{h} \vee r_{ha}\overline{a}e\overline{h} \vee \overline{r}_{ha}a e\overline{h} \vee e\overline{h}$$

In the case of the first set of test checks, in accordance with (7) and (8), the expression can be constructed

$$V_6^{(1)} \triangleq V_4 W^{(1)} = \left(\overline{abcd}efgh \vee a\overline{bc}defgh \vee \overline{abc}defgh \vee \overline{abc}defgh \right)$$
$$\wedge \left(r_{cb}r_{fg}\overline{bcf}\,\overline{g} \vee r_{cb}\overline{r}_{fg}\overline{bc}fg \vee r_{cb}\overline{bcf} \vee \overline{r}_{cb}r_{fg}bcf\,\overline{g} \vee \overline{r}_{cb}\overline{r}_{fg}bcfg \vee \overline{r}_{cb}bc\overline{f} \vee \vee r_{fg}\overline{cf}\,\overline{g} \vee \overline{r}_{fg}\overline{c}fg \vee \overline{cf} \right)$$
$$= r_{cb}\overline{r}_{fg}\overline{abcd}efgh \vee \overline{r}_{fg}a\overline{bc}defgh \vee \overline{r}_{cb}\overline{r}_{fg}\overline{abc}defgh \vee \overline{r}_{fg}\overline{abc}defgh$$
$$= \overline{r}_{fg}\left(r_{cb}\overline{abcd}efgh \vee \overline{r}_{cb}\overline{abc}defgh \vee a\overline{bc}defgh \vee \overline{abc}defgh \right)$$

Let's denote $P_1^{(1)} = r_{cb}\overline{r}_{fg}$, $P_2^{(1)} = \overline{r}_{fg}$, $P_3^{(1)} = \overline{r}_{cb}\overline{r}_{fg}$ and $P_4^{(1)} = \overline{r}_{fg}$. It can be noted that for this set of coefficients, the condition of Statement 1 is not met. So, for example, $P_2^{(1)} \wedge P_4^{(1)} = \overline{r}_{fg}$. Consequently, the proposed test suite may not be enough to unambiguously determine the state of the system.

In the case of the second set of test checks, in accordance with (7) and (8), the expression can be constructed

$$V_6^{(2)} \triangleq V_4 W^{(2)} = \left(a\overline{bc}defgh \vee a\overline{bc}defgh \vee \overline{abc}defgh \vee \overline{abc}defgh \right)$$
$$\wedge \left(r_{ha}r_{ed}\overline{ad}eh \vee \overline{r}_{ha}r_{ed}a\overline{d}eh \vee r_{ed}\overline{d}eh \vee r_{ha}\overline{r}_{ed}\overline{ad}eh \vee \overline{r}_{ha}\overline{r}_{ed}adeh \vee \overline{r}_{ed}deh \vee r_{ha}\overline{ae}h \vee \overline{r}_{ha}a\overline{e}h \vee \overline{eh} \right)$$
$$= \overline{r}_{ha}r_{ed}a\overline{bc}defgh \vee \overline{r}_{ha}\overline{r}_{ed}a\overline{bc}defgh \vee r_{ha}r_{ed}\overline{abc}defgh \vee r_{ha}\overline{r}_{ed}\overline{abc}defgh$$

Let's set $P_1^{(2)} = \overline{r}_{ha}r_{ed}$, $P_2^{(2)} = \overline{r}_{ha}\overline{r}_{ed}$, $P_3^{(2)} = r_{ha}r_{ed}$ and $P_4^{(2)} = r_{ha}\overline{r}_{ed}$. Note that, for a given set of coefficients, the condition of Statement 1 is satisfied, therefore, it is sufficient to uniquely determine the state of the system. Thus, the choice of the second test case is more preferable for this particular situation.

One can also build a table of correspondence of the results of test checks to the state of the system (Table 2).

Table 2. Correspondence of test results to the state of the system (Example 2)

Test result		System state vector							
$h \to a$	$e \to d$	a	b	c	d	e	f	g	h
0	0	1	0	0	1	1	1	1	1
0	1	1	0	1	0	1	1	1	1
1	0	0	1	0	1	1	1	1	1
1	1	0	1	1	0	1	1	1	1

It should be noted that the example under consideration is degenerate. The proposed set of additional tests will be sufficient only for the case of the considered structure and a combination of the results of the initial test checks. In other cases, the set of additional tests may differ (including their number).

Example 3. Consider a 1-diagnosable system consisting of 5 processors denoted as a, b, c, d and e. The graph representing the capabilities of this system in terms of diagnostics

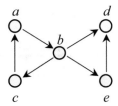

Fig. 3. Coupling topology of system (Example 3)

(coupling topology) is shown in Fig. 3. Suppose that processors a, b and c are serviceable. Three control checks were carried out: $a \rightarrow b$, $b \rightarrow c$ and $c \rightarrow a$, the results of which were respectively $r_{ab} = 0$, $r_{bc} = 0$ and $r_{ca} = 0$.

In accordance with [37], a boolean expression was constructed and transformed:

$$V_3 = \left(ab \vee \bar{a}\right)\left(bc \vee \bar{b}\right)\left(ca \vee \bar{c}\right) = \left(abc \vee \bar{a}bc \vee \bar{a}\bar{b}\right)\left(ca \vee \bar{c}\right) = abc$$
$$= abcde \vee abc\bar{d}e \vee abcd\bar{e} \vee abc\bar{d}\bar{e}$$

As a result of the transformation, the PDNF was obtained, consisting of 3 conjuncts. According to it the states of processors a, b and c are determined as serviceable. However, to determine states of processors d and e additional tests should be carried out.

There are three test checks available: $b \rightarrow d$, $b \rightarrow e$ and $e \rightarrow d$. Let's construct and transform a boolean expression, denoting it W:

$$W = T_{bd}T_{be}T_{ed} = \left(r_{bd}b\bar{d} \vee \bar{r}_{bd}bd \vee \bar{b}\right)\left(r_{be}b\bar{e} \vee \bar{r}_{be}be \vee \bar{b}\right)\left(r_{ed}e\bar{d} \vee \bar{r}_{ed}ed \vee \bar{e}\right) =$$
$$= \left(r_{bd}r_{be}b\bar{d}\bar{e} \vee r_{bd}\bar{r}_{be}b\bar{d}e \vee \bar{r}_{bd}r_{be}bd\bar{e} \vee \bar{r}_{bd}\bar{r}_{be}bde \vee \bar{b}\right)\left(r_{ed}e\bar{d} \vee \bar{r}_{ed}ed \vee \bar{e}\right) =$$
$$= \cancel{r_{bd}r_{be}b\bar{d}\bar{e}} \vee r_{bd}\bar{r}_{be}r_{ed}b\bar{d}e \vee \bar{r}_{bd}r_{be}bd\bar{e} \vee \bar{r}_{bd}\bar{r}_{be}\bar{r}_{ed}bde \vee \cancel{r_{ed}b\bar{e}d} \vee \bar{r}_{ed}b\bar{e}d \vee \cancel{\bar{b}\bar{e}} =$$
$$= r_{bd}\bar{r}_{be}r_{ed}b\bar{d}e \vee \bar{r}_{bd}r_{be}bd\bar{e} \vee \bar{r}_{bd}\bar{r}_{be}\bar{r}_{ed}bde \vee \bar{r}_{ed}b\bar{e}d$$

In accordance with (7) and (8), the following expression can be constructed:

$$V_6 = V_3 W = abc\left(r_{bd}\bar{r}_{be}r_{ed}\bar{d}e \vee \bar{r}_{bd}r_{be}bd \vee \bar{r}_{bd}\bar{r}_{be}\bar{r}_{ed}bde \vee \bar{r}_{ed}\bar{b}ed\right)$$
$$= r_{bd}\bar{r}_{be}r_{ed}abc\bar{d}e \vee \bar{r}_{bd}r_{be}abcd\bar{e} \vee \bar{r}_{bd}\bar{r}_{be}\bar{r}_{ed}abcde$$

Let's denote coefficients $P_1 = r_{bd}\bar{r}_{be}r_{ed}$, $P_2 = \bar{r}_{bd}r_{be}$, $P_3 = \bar{r}_{bd}\bar{r}_{be}\bar{r}_{ed}$ and constituents $C_1 = abc\bar{d}e$, $C_1 = abcd\bar{e}$, $C_1 = abcde$. Note that, for a given set of coefficients, the condition of Statement 1 is satisfied, therefore, it is sufficient to uniquely determine the state of the system. Let's also build a table of correspondence of the results of test checks to the state of the system (Table 3).

In Table 3, the values of test checks results can be used as a unique key to obtain the corresponding state of system's processors. It can be noticed that considered test checks set is redundant and exclusion of test $e \rightarrow d$ (3-rd column of the table) does not lead to violation of the unique key condition. Let's show that tests set can indeed be reduced by means of exclusion of this test check.

Table 3. Correspondence of test results to the state of the system (Example 2)

Test result			System state vector				
$b \to d$	$b \to e$	$e \to d$	a	b	c	d	E
1	0	1	1	1	1	0	1
0	1	*	1	1	1	1	0
0	0	0	1	1	1	1	1

This time, we construct and transform a boolean expression for case of test checks $b \to d$ and $b \to e$, denoting it W':

$$W' = T_{bd}T_{be} = \left(r_{bd}b\overline{d} \vee \overline{r}_{bd}bd \vee \overline{b} \right)\left(r_{be}b\overline{e} \vee \overline{r}_{be}be \vee \overline{b} \right)$$
$$= r_{bd}r_{be}b\overline{de} \vee r_{bd}\overline{r}_{be}b\overline{d}e \vee \overline{r}_{bd}r_{be}bd\overline{e} \vee \overline{r}_{bd}\overline{r}_{be}bde \vee \overline{b}$$

In accordance with (7) and (8), the following expression can be constructed:

$$V_6' = V_3 W' = abc\left(r_{bd}r_{be}b\overline{de} \vee r_{bd}\overline{r}_{be}b\overline{d}e \vee \overline{r}_{bd}r_{be}bd\overline{e} \vee \overline{r}_{bd}\overline{r}_{be}bde \vee \overline{b} \right)$$
$$= r_{bd}\overline{r}_{be}abc\overline{d}e \vee \overline{r}_{bd}r_{be}abcd\overline{e} \vee \overline{r}_{bd}\overline{r}_{be}abcde$$

Let's denote coefficients $P_1' = r_{bd}\overline{r}_{be}$, $P_2' = \overline{r}_{bd}r_{be}$, $P_3' = \overline{r}_{bd}\overline{r}_{be}$. Note that, for a given set of coefficients, the condition of Statement 1 is also satisfied, therefore, it is sufficient to uniquely determine the state of the system. Let's build a table of correspondence of the results of test checks to the state of the system (Table 4).

Table 4. Correspondence of reduced tests set results to the state of the system (Example 3)

Test result		System state vector				
$b \to d$	$b \to e$	a	b	c	d	E
1	0	1	1	1	0	1
0	1	1	1	1	1	0
0	0	1	1	1	1	1

As in the previous case, example under consideration is degenerate. The proposed set of additional tests will be sufficient only for the case of the considered structure and a combination of the results of the initial test checks.

4   Conclusion

In this work, a generalization of the formal method for determining the state of processors of a multiprocessor system during testing was proposed, which allows to analyze a combination of test checks before executing tests. In accordance with the proposed synthesis, for each admissible state of the system, a certain coefficient can be determined, represented by a boolean expression, consisting of the values of the test results. Criteria are formulated that make it possible, on the basis of the analysis of these coefficients, to assess the sufficiency of a combination of test checks to unambiguously determine the state of the system. In addition, on the basis of the proposed approach, the search for an optimal set of test checks can be implemented.

One of the advantages of the approach proposed in the work is that the implementation of complex transformations can be transferred from the stage of self-testing of the system to the stage of system development.

References

1. Drozd, A., et al.: Green experiments with FPGA. In: Kharchenko, V., Kondratenko, Y., Kacprzyk, J. (eds.) Green IT Engineering: Components, Networks and Systems Implementation. SSDC, vol. 105, pp. 219–239. Springer, Cham (2017). https://doi.org/10.1007/978-3-319-55595-9_11
2. Avižienis, J., Laprie, B., Randell, C.: Dependability and its threats: a taxonomy. In: Building the Information Society, pp. 91–120 (2004)
3. Kuo, W., Zuo, M.J.: Optimal Reliability Modeling: Principles and Applications. John Wiley & Sons Inc., New Jersey, USA (2003)
4. Kaswan, K.S., Choudhary, S., Sharma, K.: Software reliability modeling using soft computing techniques: critical review. IJITCS 7(7), 90–101 (2015)
5. Wason, R., Soni, A.K., Rafiq, M.Q.: Estimating software reliability by monitoring software execution through OpCode. IJITCS 7(9), 23–30 (2015)
6. Thomas, M., Rad, B.: Reliability evaluation metrics for internet of things, car tracking system: a review. Int. J. Inf. Technol. Comput. Sci. 9(2), 1–10 (2017). https://doi.org/10.5815/ijitcs.2017.02.01
7. Ushakov, I. (ed.): Reliability of Technical Systems: Handbook. Radio i Sviaz, Moskov (1985). (in Russian)
8. Mikhaylov, D., Zhukov, I., Starikovskiy, A., Zuykov, A., Tolstaya, A., Fomin, M.: Method and system for protection of automated control systems for "smart buildings." Int. J. Comput. Netw. Inf. Secur. 5(9), 1–8 (2013). https://doi.org/10.5815/ijcnis.2013.09.01
9. Yalcinkaya, E., Maffei, A., Onori, M.: Application of attribute based access control model for industrial control systems. Int. J. Comput. Netw. Inf. Secur. 9(2), 12–21 (2017). https://doi.org/10.5815/ijcnis.2017.02.02
10. Gokhale, S., Dalvi, A., Siddavatam, I.: Industrial control systems honeypot: a formal analysis of conpot. Int. J. Comput. Netw. Inf. Secur. 12(6), 44–56 (2021). https://doi.org/10.5815/ijcnis.2020.06.04
11. Keshtgar, S.A., Arasteh, B.B.: Enhancing software reliability against soft-error using minimum redundancy on critical data. Int. J. Comput. Netw. Inf. Secur. (IJCNIS) 9(5), 21–30 (2017)
12. Avizienis, A.: Fault-tolerance: the survival attribute of digital systems. Proc. IEEE 66(10), 1109–1126 (1978)
13. Arfat, Y., Eassa, F.E.: A survey on fault tolerant multi agent system. Int. J. Inf. Technol. Comput. Sci. 9, 39–48 (2016)
14. Romankevich, A., Feseniuk, A., Maidaniuk, I., Romankevich, V.: Fault-tolerant multiprocessor systems reliability estimation using statistical experiments with GLmodels. In: Advances in Intelligent Systems and Computing, vol. 754, pp. 186–193 (2019)

15. Romankevich, A., Maidaniuk, I., Feseniuk, A., Romankevich, V.: Complexity estimation of GL-models for calculation FTMS reliability. In: Hu, Z., Petoukhov, S., Dychka, I., He, M. (eds.) ICCSEEA 2019. AISC, vol. 938, pp. 369–377. Springer, Cham (2020). https://doi.org/10.1007/978-3-030-16621-2_34

16. Romankevich, A., Feseniuk, A., Romankevich, V., Sapsai, T.: About a fault-tolerant multiprocessor control system in a pre-dangerous state. In: 2018 IEEE 9th International Conference on Dependable Systems, Services and Technologies (DESSERT), pp. 207–211 (2018)

17. Romankevich, A.M., Morozov, K.V., Romankevich, V.A.: Graph-logic models of hierarchical fault-tolerant multiprocessor systems. IJCSNS Int. J. Comput. Sci. Netw. Secur. **19**(7), 151–156 (2019)

18. Belarbi, M.: Formal and informal modeling of fault tolerant noc architectures. Int. J. Intell. Syst. Appl. **12**, 32–42 (2015)

19. Wang, X., Li, S., Liu, F., Fan, X.: Reliability analysis of combat architecture model based on complex network. IJEM **2**(2), 15–22 (2012)

20. Rahdari, D., Rahmani, A.M., Aboutaleby, N., Karambasti, A.S.: A distributed fault tolerance global coordinator election algorithm in unreliable high traffic distributed systems. IJITCS **7**(3), 1–11 (2015). https://doi.org/10.5815/ijitcs.2015.03.01

21. Romankevich, V.A.: Self-testing of multiprocessor systems with regular diagnostic connections. Autom Remote Control. **78**(2), 289–299 (2017)

22. Drozd, J., Drozd, A., Al-dhabi, M.: A resource approach to on-line testing of computing circuits. In: Proceedings of IEEE East-West Design & Test Symposium. Batumi, Georgia, pp. 276–281 (2015)

23. Mikeladze, M.A.: Development of basic self-diagnosis models for complex engineering systems. Autom. Remote Control **56**(5), 611–623 (1995)

24. Grishin, V.Y., Lobanov, A.V., Sirenko, V.G.: Distributed system diagnosis of byzantine failures in partially connected multicomputer systems. Autom. Remote Control **66**(2), 304–312 (2005)

25. Nazemi, E., Talebi, T., Elyasi, H.: Self-healing mechanism for reliable architecture with focus on failure detection. Int. J. Inf. Eng. Electron. Bus. **7**(3), 32–38 (2015)

26. Sinha, B., Singh, A., Saini, P.: A failure detector for crash recovery systems in cloud. Int. J. Inf. Technol. Comput. Sci. **11**(7), 9–16 (2019)

27. Vedeshenkov, V.A.: Organization of diagnostics of digital systems with the structure of a symmetric bipartite graph. Control Sci. **6**, 59–67 (2009)

28. Drozd, A., et al.: Checkable FPGA design: energy consumption, throughput and trustworthiness. In: Kharchenko, V., Kondratenko, Y., Kacprzyk, J. (eds.) Green IT Engineering: Social, Business and Industrial Applications, pp. 73–94. Springer International Publishing, Cham (2019). https://doi.org/10.1007/978-3-030-00253-4_4

29. Karavaj, M.F., Podlazov, V.S.: Extended generalized hypercube as a fault-tolerant system network for multiprocessor systems. LSS **45**, 344–371 (2013)

30. Drozd J., Drozd A., Antoshchuk S., Kharchenko V. Natural development of the resources in design and testing of the computer systems and their components. In: 7th IEEE International Conference on Intelligent Data Acquisition and Advanced Computing Systems: Technology and Applications, Berlin, Germany, pp. 233–237 (2013)

31. Drozd, A., Drozd, M., Antonyuk, V.: Features of hidden fault detection in pipeline components of safety-related system. CEUR Workshop Proc. **1356**, 476–485 (2015)

32. Drozd, A., Drozd, J., Antoshchuk, S., Nikul, V., Al-dhabi, M.: Objects and methods of on-line testing: main requirements and perspectives of development. In: Proceedings of IEEE East-West Design & Test Symposium, Yerevan, Armenia, pp. 72–76 (2016)

33. Parkhomenko, P.P.: Checking multiprocessor computer systems for serviceability by analyzing their syndrome graphs. Avtom. Telemekh. **5**, 126–135 (1999)

34. Preparata, F., Metze, G., Chien, R.: On the connection assignment problem of diagnosable systems. IEEE Trans. Electron. Comput. **ES-16**(6), 848–854 (1967)

35. Belyavskii, V.E., Valuiskii, V.N., Romankevich, A.M., Romankevich, V.A.: Self-diagnosable multimodular systems: some estimates of testing. Autom. Remote Control **60**(8), 1179–1183 (1999)
36. Romankevich, A.M., Romankevich, V.A.: Diagnosis of multiprocessor systems under failure of more than half processors. Autom. Remote Control **78**(9), 1614–1618 (2017)
37. Romankevich, A.M., Morozov, K.V., Romankevich, V.A.: A formal method for determining the state of processors in a multiprocessor system under testing. Autom. Remote Control **82**(3), 460–467 (2021)
38. Hakimi, S.L., Amin, A.T.: Characterization of connection assignment of diagnosable systems. IEEE Trans. Comput. **C–23**(1), 86–88 (1974)
39. Dimitriev, Y.K.: On t-diagnosability of multicore systems with symmetric circulant structure. Autom Remote Control. **74**(1), 105–112 (2013)
40. Dimitriev, Y.K.: Necessary and sufficient conditions for t-diagnosability of multiprocessor computer systems for various models of nonreliable testing established using the system graph-theoretical model. Autom. Remote Control **76**(7), 1260–1270 (2015)

Risk Assessment of Cold Chain Logistics Chain Breaking Based on BP Neural Network

Jun Yuan[1,2], Peilin Zhang[1(✉)], Bingbing Li[2], and Bing Tang[2]

[1] Transportation Planning and Management, Wuhan University of Technology, Wuhan 430063, China
plzhanghq@126.com

[2] School of Logistics, Wuhan Technology and Business University, Wuhan 430065, China

Abstract. Chain breaking will result in huge losses for both the practitioners of the supply chain and the consumers of the supply chain. Once the cold chain logistics is broken, the employees will face huge economic and reputation losses; Consumers may suffer physical and psychological harm directly due to the spoilage of products caused by chain breaking. There is no doubt that how to solve the problem of cold chain breaking has become an important issue in the cold chain logistics industry. Based on the analysis of the main causes of chain break in cold chain logistics, this paper constructs a breaking risk assessment index system of the cold chain logistics, and uses MATLAB software to build an evaluation model using BP neural network to evaluate the possible chain break risks in the example of cold chain logistics enterprises. The risk assessment model of cold chain logistics based on BP neural network can provide some reference for cold chain logistics enterprises to preestimate the risk of chain breakage and formulate risk prevention measures.

Keywords: Chain breaking · Cold chain logistics · BP neural network

1 Introduction

Cold chain logistics is a branch of logistics industry with large basic investment and high operation cost. Due to the particularity of its service object, it is necessary to ensure that the goods are always in a low temperature environment in the whole logistics process. Due to various external reasons and the employees' own non-standard behavior, the goods are exposed to normal temperature or high temperature environment in the logistics process, which results in spoilage. This consequence is called "chain breaking" [1].

Chain breaking will bring huge losses to both the practitioners of the supply chain and the consumers of the supply chain. Once the cold chain logistics is broken, the employees will face huge economic and reputation losses [2]; Consumers may be harmed physically and psychologically directly due to the spoilage of products caused by chain breaking. There is no doubt that how to solve the problem of cold chain breaking has become an important issue in the cold chain logistics industry.

Z. Hu et al. (Eds.): ICCSEEA 2022, LNDECT 134, pp. 376–386, 2022.
https://doi.org/10.1007/978-3-031-04812-8_32

According to the Research Report on China's fresh supply chain production, China's annual agriculture product consumption is over 300 million tons, increasing year by year. However, as the supply side of the fresh supply chain, agriculture has many links, long chain and large consumption. The annual economic loss caused by the broken chain of cold chain logistics can reach more than 100 billion [3].

The phenomenon of chain breaking itself can be regarded as a possible risk in the operation of cold chain logistics. Due to the influence of different factors, its occurrence is uncertain. Therefore, how to preestimate the possible chain breaking risks in the cold chain enterprises and find out the most important risk factors that may lead to broken chain prevention, has become a key issue that cold chain enterprises and even the whole cold chain industry must pay attention to.

Based on the analysis of the main causes of chain break in cold chain logistics, this paper constructs a breaking risk assessment index system of the cold chain logistics, and uses MATLAB software to build an evaluation model using BP neural network to evaluate the possible chain break risks in the example of cold chain logistics enterprises. The risk assessment model of cold chain logistics based on BP neural network can provide some reference for cold chain logistics enterprises to preestimate the risk of chain breakage and formulate risk prevention measures.

2 Construction of the Breaking Risk Assessment Index System of Cold Chain Logistics

2.1 Analysis on Risk Factors of Chain Break in Cold Chain Logistics

At present, the main cold chain logistics business in China is still mainly meat and agricultural products cold chain logistics, among which meat is still the main source of cold chain logistics supply, while the cold chain of agricultural products represented by fruits are the fastest growing business. Therefore, this paper takes meat and agricultural products, the two most representative products, as the main objects of cold chain logistics risk factor analysis [4].

The main processes of cold chain logistics of the above two types of products are as Fig. 1.

According to the basic definition of chain break, in the process of cold chain logistics, if the goods are not in the low temperature environment in any link, it may lead to chain break. By analyzing the cases of chain breaking caused by various reasons in the process of product processing, storage, transportation and sales in cold chain logistics, the risk factors leading to chain breaking can be divided into six categories: risks caused by employees, risks caused by infrastructure equipment, risks caused by node connection, risks caused by enterprise operation, risks caused by relevant technical level of cold chain, and other risks [5].

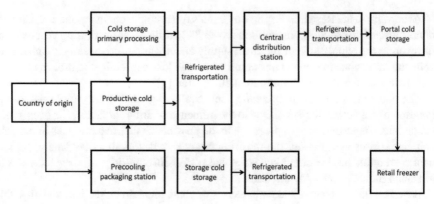

Fig. 1. The main processes of cold chain logistics

2.2 The Breaking Risk Assessment Index System of Cold Chain Logistics

According to the above analysis on the influencing factors of cold chain logistics chain breaking risk, the index system of cold chain logistics chain breaking risk assessment is divided into six categories, with the specific titles as follows:

Risks caused by enterprise operation includes: Risk1 Enterprise's cold chain safety production awareness; Risk2 Standardization of enterprise cold chain operation; Risk3 Integrity rate of enterprise cold chain production line; Risk4 Stability of enterprise capital chain.

Risks caused by employees includes: Risk5 Training rate of cold chain basic ability of cold chain employees; Risk6 Cold chain awareness of cold chain practitioners.

Risks caused by infrastructure equipment includes: Risk7 Mechanization level of cold chain processing equipment; Risk8 Automation level of cold chain warehouse; Risk9 Level of cold chain logistics transportation tools.

Risks caused by node connection includes: Risk10 Penetration of temperature detection during node connection; Risk11 Joint operation efficiency of node enterprises; Risk12 Node enterprise information sharing rate; Risk13 Product loss rate during node connection.

Risks caused by cold chain technology level includes: Risk14 Qualification rate of whole process temperature control; Risk15 Loading and unloading automation level; Risk16 Application degree of cold chain logistics information technology.

Other risks includes: Risk17 Stability of regional power supply; Risk18 Regional climate stability; Risk19 Supervision and protection of policies and regulations on the industry [6].

The breaking risk assessment index system of cold chain logistics is as Table 1:

Table 1. The breaking risk assessment index system of cold chain logistics

Risks caused by enterprise operation	Risk1 Enterprise's cold chain safety production awareness Risk2 Standardization of enterprise cold chain operation Risk3 Integrity rate of enterprise cold chain production line Risk4 Stability of enterprise capital chain
Risks caused by employees	Risk5 Training rate of cold chain basic ability of cold chain employees Risk6 Cold chain awareness of cold chain practitioners
Risks caused by infrastructure equipment	Risk7 Mechanization level of cold chain processing equipment Risk8 Automation level of cold chain warehouse Risk9 Level of cold chain logistics transportation tools
Risks caused by node connection	Risk10 Penetration of temperature detection during node connection Risk11 Joint operation efficiency of node enterprises Risk12 Node enterprise information sharing rate Risk13 Product loss rate during node connection
Risks caused by cold chain technology level	Risk14 Qualification rate of whole process temperature control Risk15 Loading and unloading automation level Risk16 Application degree of cold chain logistics information technology
Other risks	Risk17 Stability of regional power supply Risk18 Regional climate stability Risk19 Supervision and protection of policies and regulations on the industry

3   Construction of Risk Assessment Model of Cold Chain Logistics Based on BP Neural Network

3.1   Design of BP Neural Network Structure

As the most widely used neural network model, BP neural network has the advantages of automatically optimizing the weight and modifying the error to the allowable range in the function of risk assessment. Therefore, this paper uses BP neural network to build an evaluation model of cold chain logistics chain breaking risk [7].

Before establishing BP neural network model, the structure of BP neural network should be determined first. In this paper, the structure design of BP neural network is as follows:

3.1.1 Design of Input Layer

According to the cold chain logistics chain breaking risk assessment index system established earlier, there are 19 main factors affecting cold chain logistics chain breaking, so the number of input layer nodes designed in this paper is $m = 19$.

3.1.2 Design of Output Layer

The risk of enterprise cold chain logistics chain breaking can be roughly divided into three levels: low chain breaking risk, chain breaking risk and high chain breaking risk. The final output is the comprehensive rating data of enterprise cold chain logistics chain breaking risk. Therefore, the output layer designed in this paper is a single neuron, that is, the number of nodes in the output layer is $n = 1$.

3.1.3 Design of Hidden Layer

The number of hidden layers designed in this paper is one layer. In order to reduce the error of output results, the number of neurons in hidden layer can be increased appropriately. The common calculation formula for the number of hidden layer stages is:

$$
\begin{aligned}
l &= \sqrt{m+n} + \alpha \\
l &= \sqrt{mn} \\
l &= \log 2^n
\end{aligned}
\tag{1}
$$

The number of input layer nodes designed by BP neural network is $m = 19$; The number of output nodes is $n = 1$ substituted into the number of hidden layer nodes calculated by the three formulas, and the BP neural network is trained respectively. The number of nodes with the smallest measured final error is 19, so the number of hidden layer nodes finally set in this paper is $l = 19$ [8].

To sum up, this paper uses the BP neural network toolbox of MATLAB to generate the BP neural network of 19X19X1.

3.2 Design of BP Neural Network Training Algorithm

In this paper, the algorithm design of BP neural network training is as follows:

3.2.1 Initialize Weights and Thresholds

Set the number of BP neural network input layer nodes $m = 19$, output layer nodes $n = 1$ and hidden layer nodes $l = 19$.

The connection weight from the input layer data x_i to the hidden layer data y_j is W_{ij}, and the connection weight from the hidden layer data y_j to the output layer data z_k is V_{jk}. The hidden layer unit threshold is θ_j, and the output layer unit threshold is θ_k.

The smaller value between $(0,1)$ is randomly assigned as the connection weight from the input layer to the hidden layer, the connection weight from the hidden layer to the output layer, the hidden layer unit threshold and the output layer unit threshold of BP neural network to initialize the weight and threshold [9].

3.2.2 Input of Training Samples

Input training samples, which include actual input value and output expected value.

If The number of input data is m, Then the actual input value vector is [10]

$$X_i = (x_1, x_2, \cdots, x_m) \tag{2}$$

If the number of output data is n, Then the Expected output value vector is

$$\hat{Z}_i = \left(\hat{z}_1, \hat{z}_2, \cdots, \hat{z}_n\right) \tag{3}$$

The actual input values are input into the input layer nodes in turn to obtain the data of the hidden layer and the output layer. The calculation formulas are as follows [11]:

$$y_j = f\left(\sum_{i=1}^{m} W_{ij}X_i - \theta_j\right) (j = 1, 2, \cdots, l) \tag{4}$$

$$z_k = f\left(\sum_{j=1}^{l} V_{jk}Y_j - \theta_k\right) (k = 1, 2, \cdots, n) \tag{5}$$

3.2.3 Error Inspection and Correction

Calculate the error between the expected output value and the actual output value, and the calculation formula is

$$\delta_k = z_k(1 - z_k)\left(\hat{z}_k - z_k\right) \tag{6}$$

According to the connection weight between the hidden layer and the output layer, the error of the hidden layer value is inversely calculated, and the calculation formula is

$$\delta_j = y_j(1 - y_j)\sum_{k=1}^{n} V_{jk}\delta_k \tag{7}$$

The connection weight from the input layer to the hidden layer W_{ij}, the connection weight from the hidden layer to the output layer V_{jk}, the hidden layer unit threshold θ_j and the output layer unit threshold θ_k are corrected, and the correction formula is

$$W_{ij}(t + 1) = W_{ij}(t) + \eta \delta_j x_i \tag{8}$$

$$\theta_j(t + 1) = \theta(t) + \eta \delta_j \tag{9}$$

$$V_{jk}(t + 1) = V_{jk}(t) + \eta \delta_k y_j \tag{10}$$

$$\theta_k(t + 1) = \theta_k(t) + \eta \delta_k \tag{11}$$

Replace the original connection weight and threshold with the modified connection weight and threshold, repeatedly input the training samples for training, and repeat the iteration until the error is less than the ideal value [12].

4 Test of Risk Assessment Model of Cold Chain Logistics Based on BP Neural Network

4.1 Data Source and Data Processing of Inspection Object

In this paper, the model test selects the data of 10 cold chain logistics enterprises in a certain place as the training sample. The original data of the enterprises are shown in the Table 2:

Table 2. Original data of test case

Index	1	2	3	4	5	6	7	8	9	10	11	12	13	14	15	16	17	18	19
CASE 1	5	5	0.90	5	0.97	5	5	4	4	0.93	4	5	−0.04	0.94	4	4	4	4	5
CASE 2	3	4	0.80	4	0.88	4	3	4	3	0.87	4	4	−0.03	0.84	3	4	4	4	5
CASE 3	4	3	0.75	4	0.92	5	4	3	4	0.85	4	3	−0.07	0.78	3	4	4	4	5
CASE 4	4	3	0.75	3	0.83	3	3	4	3	0.79	3	2	−0.04	0.74	2	3	4	4	5
CASE 5	3	2	0.66	4	0.85	3	4	2	2	0.74	3	4	−0.05	0.50	2	4	4	4	5
CASE 6	4	3	0.78	3	0.85	4	4	4	3	0.91	3	4	−0.04	0.82	4	4	4	4	5
CASE 7	5	4	0.85	5	0.89	4	3	5	4	0.89	3	5	−0.04	0.74	4	3	4	4	5
CASE 8	2	2	0.72	3	0.90	3	4	4	3	0.82	4	4	−0.07	0.64	3	2	4	4	5
CASE 9	3	4	0.80	4	0.83	4	4	2	4	0.85	3	3	−0.06	0.81	3	4	4	4	5
CASE 10	4	3	0.90	4	0.92	4	4	4	4	0.82	4	4	−0.04	0.83	4	3	4	4	5

In this index system, fourteen indicators are qualitative indicators, and the value range is rounded to (1,5). The higher the score, the smaller the negative impact on the chain breaking risk of cold chain logistics enterprises; five indicators are quantitative indicators, and the actual values are directly selected [13].

In order to facilitate the data input of BP neural network model, all are standardized. In this paper, the maximum and minimum value method is used to process the data, and the specific calculation formula is as follows:

$$x'_i = \frac{x_i - min(x)}{max(x) - min(x)} \tag{12}$$

The risk rating of cold chain logistics enterprises with chain breakage is divided into three levels. The high chain breakage risk is assigned 1, the medium chain breakage risk is assigned 0.5 and the low chain breakage risk is assigned 0. An expected value is given to the risk of 10 enterprises through expert evaluation method, The enterprise input obtained after standardization is shown in the Table 3:

Table 3. Standardized data of test case

Index	1	2	3	4	5	6	7	8	9	10	11	12	13	14	15	16	17	18	19	V
CASE 1	1	1	1	1	1	1	1	0.67	1	1	1	1	0.75	1	1	1	1	1	1	1
CASE 2	0.33	0.67	0.58	0.5	0.36	0.5	0	1	0.5	0.68	1	0.67	1	0.77	0.5	1	1	1	1	0.5
CASE 3	0.67	0.33	0.38	0.5	0.64	1	0.5	0.33	1	0.58	1	0.33	0	0.64	0.5	1	1	1	1	0.5
CASE 4	0.67	0.33	0.38	0	0	0	0	0.67	0.5	0.26	0	0	0.75	0.55	0	0.5	1	1	1	0.5
CASE 5	0.33	0	0	0.5	0.14	0	0.5	0	0	0	0	0.67	0.5	0	0	1	1	1	1	0
CASE 6	0.67	0.33	0.5	0	0.14	0.5	0.5	0.67	0.5	0.89	0	0.67	0.75	0.73	1	1	1	1	1	0.5
CASE 7	1	0.67	0.79	1	0.42	0.5	0	1	1	0.8	0	1	0.75	0.55	1	0.5	1	1	1	0.5
CASE 8	0	0	0.25	0	0.5	0	0.5	0.67	0.5	0.42	1	0.67	0	0.32	0.5	0	1	1	1	0
CASE 9	0.33	0.67	0.58	0.5	0	0.5	0.5	0	1	0.58	0	0.33	0.25	0.7	0.5	1	1	1	1	0.5
CASE 10	0.67	0.33	1	0.5	0.64	0.5	0.5	0.67	1	0.42	1	0.67	0.75	0.75	1	0.5	1	1	1	1

4.2 Model Checking Based on MATLAB Software

The BP neural network of 19X19X1 is generated by using the BP neural network toolbox of MATLAB. The standardized enterprise data are input, and the input data are the top 9 enterprises.

Model setting parameters: the hidden layer excitation function is Tansig, the output layer excitation function is purelin, the training function is trainlm, the training times are 1000, the target error is 0.00000 1 and the learning rate is 0.0001 [14].

The data after training are shown in the Table 4.

Table 4. Training case data

Case	1	2	3	4	5	6	7	8	9
Raw data output	1.000	0.500	0.500	0.500	0.000	0.500	0.500	0.000	1.000
Fitting data output	1.000	0.484	0.513	0.541	0.000	0.477	0.522	0.043	0.717

The fitting curve between the output data and the original data is shown in the Fig. 2:

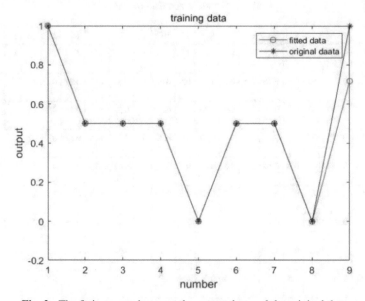

Fig. 2. The fitting curve between the output data and the original data

Accordingly, the risk assessment values of the top 9 cold chain logistics enterprises can be obtained. Among them, the first enterprise has the lowest cold chain logistics chain breaking risk, and the fifth enterprise has the highest cold chain logistics chain breaking risk [15].

The data of the 10th enterprise is used for verification, and the results are shown in the Table 5:

Table 5. Testing case data

Raw data output	0.5
Fitting data output	0.5047

The fitting curve between the output data and the original data is shown in the Fig. 3:

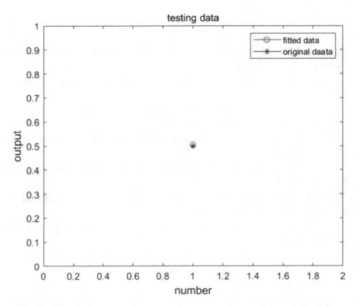

Fig. 3. The fitting curve between the output data and the original data

From the above test and calculation results, it can be seen that the fitting result of the BP neural network model is good, and the model has a good ability of chain breaking risk assessment of cold chain logistics enterprises [16].

5 Conclusion

In this paper, we establish the risk assessment model of cold chain logistics enterprises based on BP neural network, give the specific method of data assessment using MATLAB software. 10 cold chain logistics enterprises are used to test the model, which proves that the model has relatively good assessment ability. However, due to the imperfections in the selection of evaluation indicators and data research, there is still room for improvement. But the idea of the model itself has certain feasibility, which can be used for cold chain logistics enterprises to evaluate their own chain breaking risk.

References

1. Han, Y.: Research on the development status and countermeasures of cold chain logistics in China. Logist. Eng. Manag. **43**(10) (2021). (in Chinese)
2. Net, M., Trias, E., Ruiz, A., et al.: Cold chain maintaining in food trade. Food Control (2) (2004)
3. Airy consulting. China fresh supply chain Market Research Report 2020 9 (2020). (in Chinese)
4. White, P.N., Kitinoja, L.P.: Use of cold chains for reducing food losses in developing countries. Population **6**(13), 1–16 (2013)
5. Liu, Y., Bai, X.: On early warning index system of supply chain risk. Logist. Technol. **25**(10), 55–57 (2006). (in Chinese)
6. Crouhy, M., Galai, D., Robert, M.: A comparative analysis of current credit risk modelsr. J. Bank. Finan. **24**, 59–117 (2002)
7. Liu, S., Jiang, L.: Construction and application of flower cold chain logistics service quality evaluation model based on FAHP. Rail. Freight Transp. **34**(07), 17–21 (2016). (in Chinese)
8. Vincenzo, P., Michele, A.: An artificial neural network approach for credit risk management. J. Intell. Learn. Syst. Appl. **3**, 58–63 (2011)
9. Shi, D.: A review of enterprise supply chain risk management. J. Syst. Sci. Syst. Eng. **13**(2), 19–44 (2004)
10. Ojala, M., Hallikas, J.: Investment decision-making in supplier networks: Management of risk. Int. J. Prod. Econ. **104**(1), P201-213 (2006)
11. Han, L.: The Theory, Design and Application of the Artificial Neural Network. Chemical Industry Press, Beijing (2004). (in Chinese)
12. Su, G., Deng, F.: On the improvement of BP neural network based on MATLAB language Algorithm. Sci. Technol. Bull. **19**(2), 130–135 (2003). (in Chinese)
13. Liu, X., Yang, M.: Simultaneous curve registration and clustering for functional data. Comput. Stat. Data Anal. **53**(4), 1361–1376 (2019)
14. Rao, S., Tomar, R.: A new MATLAB based microstrip filter design tool. Int. J. Wirel. Microwave Technol. **7**(5), 49–70 (2017)
15. Fenghao, L., Sun Yun, S., Jun, S.W.: A node localization algorithm based on woa-bp optimization. Int. J. Wirel. Microwave Technol. **11**(3), 30–39 (2021)
16. Zhou, D.: Optimization modeling for GM(1,1) model based on BP neural network. Int. J. Comput. Netw. Inf. **4**(1), 24 (2012)

Construction of Logistics Distribution Information Sharing Platform Based on Internet of Things Technology

Liwei Li and Zhong Zheng[✉]

Nanning University, Guangxi 530200, China
523201940@qq.com

Abstract. Focusing on the new logistics management and operation mode under the concept of sharing economy, and taking this as the starting point and purpose of the research, in order to realize the effective control of logistics distribution cost, this paper constructs a logistics distribution information sharing platform based on Internet of things technology. Based on the current situation of sub sharing platform architecture connection, this paper analyzes the functional requirements of logistics distribution for supplier role management, demand role management, supply information release and search and demand information release and search, and determines the business logic form. By improving the three IOT cloud modes of single center multi terminal, multi center multi terminal, information and application layering, Study the practical application value of distribution and transportation contract, analyze the logistics distribution demand based on Internet of things technology, and complete the construction of logistics distribution information sharing platform based on Internet of things technology. Experiments show that, compared with the blockchain sharing system, the sharing platform supported by Internet of things technology can accurately record the actual transmission behavior of logistics distribution information, and can effectively control the logistics distribution cost while improving the logistics management and operation mode.

Keywords: Internet of Things technology · Logistics distribution · Business logic · Network cloud model · Transportation contract

1 Introduction

Internet of things is a new information bearing structure based on traditional telecommunication network and Internet system. It can establish interconnection mapping relationship between all independent addressing hosts and physical objects, which is also called "universal connection application Internet". Generally, it is available to be understood as an expanded and extended network derived from the Internet foundation, which can combine various types of information sensing devices with network hosts one by one, so as to form a huge and complete network system [1–3]. Due to the openness and inclusiveness of Internet space, people or hosts at anytime and anywhere can realize information exchange and transmission with the help of Internet of things (or IoT as abbreviated)

© The Author(s), under exclusive license to Springer Nature Switzerland AG 2022
Z. Hu et al. (Eds.): ICCSEEA 2022, LNDECT 134, pp. 387–397, 2022.
https://doi.org/10.1007/978-3-031-04812-8_33

system [4]. Most of the information transmitted by the IoT belongs to perception and identification data. The so-called information perception refers to the ability of the IoT host to always be sensitive to the change mode and attribute state of things; Information recognition means that the host of the IoT is capable to display the perceived data state in a special form [5].

With the continuous development of the concept of sharing economy, the existing management and operation mode of logistics distribution cannot fully meet the needs of practical application, and some logistics goods will even lag behind transportation [6, 7]. In order to avoid the above situation, the traditional blockchain sharing system adopts the most basic form of distributed storage, queries the basic information of distribution users through the Ethernet platform, and then uses the blockchain host to write logistics and transportation orders in real time [8]. However, this type of application structure cannot effectively control the logistics distribution cost, which easily leads to the inaccurate information sharing behavior recorded by the host structure [9]. To solve this problem, a logistics distribution information sharing platform based on IoT technology is designed. While dealing with the functional requirements and business logic functions of logistics distribution information, the parameters of IoT are accurately defined, and then the necessary distribution and transportation contracts are set according to the actual connection requirements of IoT cloud mode [10].

2   Design of Logistics Distribution Information Sharing Platform

The logistics distribution information sharing platform takes the architecture system as the foundation. While analyzing the functional requirements and business logic, it improves the execution and application capabilities of relevant equipment structures [11].

2.1   Infrastructure Platform Architecture

The infrastructure of logistics distribution information sharing platform is composed of user layer and information storage layer. The former describes the distribution subjects participating in logistics information sharing services, and the latter carries the information storage host in the IoT environment. The user layer includes multiple sharing subjects and IoT service objects, which can connect multiple objects such as goods source, logistics enterprise, warehouse source and vehicle source with the help of logistics distribution network, so as to form a shared architecture with distribution object as the core [12]. Generally, logistics enterprises can deliver goods from the target location to the actual location only with the support of the demander and supplier objects. In the logistics distribution information sharing platform, the roles of the participating objects will not be limited, and their participating identities can be converted to each other [13]. In the IoT environment, due to the existence of data block, digital signature and other structures, the database host can schedule and update the logistics distribution information stored therein in real time, and with the improvement of data storage security level, most shared objects can spontaneously participate in the management process of logistics distribution information [14–16].

The infrastructure of logistics distribution information sharing platform is shown in Fig. 1.

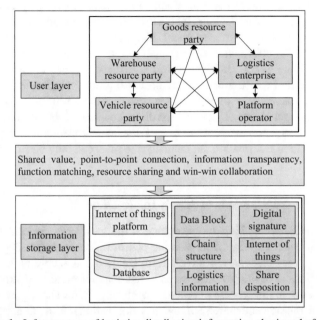

Fig. 1. Infrastructure of logistics distribution information sharing platform

2.2 Functional Requirements Analysis

The functional requirements analysis of logistics distribution information sharing platform consists of four parts: supplier role management, demand role management, supply information release and search, and demand information release and search.

2.3 Supplier Role Management

Supplier role management refers to the IoT services provided to the service providers participating in the information sharing platform, including personal information processing, user registration, login audit and other operation links [17]. Logistics distribution objects can be registered using the IoT platform, and during the registration process, they can set constraints matching with the supplier according to personal needs, such as warehouse source information, vehicle source information, etc. when the registration conditions are reviewed by the IoT host, the information sharing platform will provide corresponding logistics distribution services for the object. The user name and password used by the user object when registering personal information can be continuously improved in the subsequent application process.

2.4 Demand Side Role Management

Generally, it is similar to the functional requirements of the supplier role management, but the demander role pays more attention to the management of personal logistics information and goods distribution, such as personal information management, login and registration information management. Compared with the supplier entity of the logistics distribution information sharing platform, the demand side role management does not require the direct participation of the IoT host. After the user completes the personal information registration, the IoT host will issue execution instructions to the database and extract a large number of logistics distribution information that can be selected by the demand side object, and the direct cooperation of the audit entity is not required in the whole process. in order to meet the actual transmission requirements of logistics distribution information, all demand side roles can only be stored in the same client host.

2.5 Publishing and Searching of Supply Information

Supply information publishing and searching is the most basic executive function of the logistics distribution information sharing platform. The logistics supplier can publish all idle logistics resources to the platform environment according to the standard configuration format of the platform, so as to facilitate information sharing with other registered objects. In the process of practical application, This sharing mode can better ensure the transmission authenticity of logistics distribution information, and all sharing objects participating in logistics distribution services need to upload detailed information vouchers at the time of registration, and take it as the only object authentication resource. In the process of reviewing registered objects, the release of supply information and search instructions are usually carried out synchronously. The IoT host provides a release and application platform for logistics distribution information, and the logistics distribution information to be shared can also be directly transmitted to the client host for direct search of registered objects [18].

2.6 Demand Information Release and Search

In the logistics distribution information sharing platform, the demander can search the online logistics distribution information according to his actual demand for goods, such as storage resource search and vehicle resource search. On the premise of knowing the intelligent matching function of the platform, the demander user can arrange the logistics transportation route according to the actual distribution target of goods, Moreover, due to the existence of multiple monitoring equipment elements such as IoT host, during the whole transportation process, the demander can accurately grasp the actual transportation situation of distributed goods, and can feed back these information to multiple operators such as warehouse source party, goods source party and logistics enterprise with the help of sharing platform, so as to build an adequate and stable logistics supply system [19].

2.7 Business Logic Analysis

Logistics distribution information sharing platform is an asymmetric data query and transmission platform based on IoT technology. According to different business processes, it can be divided into five logical routes: demander, supplier, order contract, payment contract and shared service. For the logistics demander and the logistics supplier, the access of any service object must first go through the registration and login process. The former can independently select relevant storage and application resources, and establish the shared connection relationship between the demander object and the IoT users after completing the order payment; The login behavior of the latter requires the authentication and audit of the IoT host. After generating the logistics distribution order, the distribution mechanism will share the existing logistics information and finally help enterprises choose the most suitable logistics transportation unit. There is an interconnection influence relationship among the three business behaviors of order contract, payment contract and shared service. After the order confirmation processing is completed, the relevant enterprise units will pay the transportation expenses to the logistics platform, and can record the sharing behavior in a unified contract account. The business logic framework of logistics distribution information sharing platform is analyzed as shown in Fig. 2.

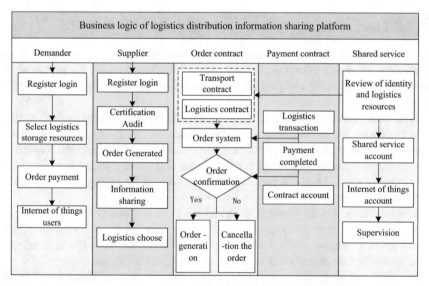

Fig. 2. Business logic analysis of logistics distribution information sharing platform

3 Logistics Distribution Demand Analysis Based on IoT Technology

With the support of shared platform application equipment architecture, logistics distribution demand analysis based on Internet of things technology is realized according to the processing flow of IoT definition, cloud model establishment and distribution and transportation contract connection.

3.1   Basic Definition of IoT

Most logistics distribution enterprises believe that the IoT system should have several application characteristics, such as wide network, rapid expansion, resource sharing, on-demand service and so on. The basic principle of the IoT is to apply a variety of virtualization processing technologies to the distributed computer at the same time, so that all connected objects participating in the distribution network can find their actual logistics distribution information in a short time.

Generally, the IoT cannot exist independently as a software execution environment, but needs to change the actual action form of internal processing elements of the host on the premise of meeting the needs of logistics distribution information sharing. On the one hand, it can effectively store various types of transmission information, and on the other hand, it can place relevant sharing objects in a proper service location. The reason why IoT devices can be widely used is inseparable from the sharing demand of logistics distribution information. The former provides a relatively stable service platform for the latter, and always maintains a relatively open and inclusive state for the logistics information input. It can not only run multiple types of transmission information at the same time, but also on the basis of no external platform, Realize the effective maintenance of the balance ability of shared service mode.

3.2   IoT Cloud Model

The cloud mode of the IoT is equivalent to the five features and limbs of the logistics distribution information sharing platform. It can effectively classify the supply objects and service objects in the collaborative work environment. Generally, it can be divided into the following three modes:

Single center multi terminal mode: compared with other IoT cloud modes, the single center multi terminal mode has the smallest distribution range in the logistics distribution information sharing platform. It requires the IoT terminal to take the cloud center as the core information processing structure. Before large-scale storage of logistics distribution information, it must obtain the permission instructions of the supply object and service object for storage behavior.

Multi center and multi terminal mode: it includes two types of information storage spaces: Private Cloud and public cloud. The former mainly takes the logistics distribution information without sharing ability as the storage object, while the latter takes the logistics distribution information with sharing ability as the storage object.

Information and application layering mode: for logistics distribution information sharing objects under wide area conditions, the shared information can be fed back to IOT hosts at all levels with the help of transmission channels on the premise of ensuring safety and demand capacity. The specific contents of the cloud mode of the IoT are shown in Table 1.

3.3   Distribution and Transportation Contract

Distribution and transportation contract is also called the IoT standard agreement of logistics distribution information sharing platform. For storage contract and transportation

Table 1. List of cloud modes of IoT

Cloud mode category	Advantage	Inferiority
Single center multi terminal mode	Sharing and storage of logistics distribution information with the permission of supply object and service object	Limited distribution
Multi center multi terminal mode	Shared and non-shared logistics distribution information stored separately	Complex action ability
Information and application layering model	It can establish a shared connection with the IoT host	For wide area distribution objects

contract, this application protocol has strong consistency in business function maintenance and participant coordination, and can evaluate and plan physical parameters such
as vehicle source, goods source and order information at the same time, Thus, the sharing
party can receive the data parameters related to logistics distribution information in a
short time. Under the function of the IOT host, the distribution and transportation contract can directly restrict the debugging ability of the information query mechanism of
the shared platform. Due to the richness of protocol types, the IOT host can feed back the
data information to the natural transmission environment when the logistics distribution
destination is unknown. On the one hand, it is conducive to the direct access of different cloud model systems, On the other hand, it can also better realize the hierarchical
storage of logistics distribution information. Suppose e_1 and e_2 respectively represent
two different logistics distribution information sharing eigenvalues, and the connection
capacity of distribution transportation contract can be defined as:

$$T = \frac{|\bar{p}|}{f(e_1 + e_2)} \tag{1}$$

where, $\bar{p}$ represents the average amount of logistics distribution information transmission that can be borne by the Internet host in unit time, and f represents the sharing
coefficient. So far, we have completed the definition of various IoT application permissions, combined supply objects and service objects, and realized the smooth application
of logistics distribution information sharing platform.

4   Analysis of Experimental Results

Select a logistics enterprise with relatively stable distribution capacity as the experimental object, make distributors and dealers at all levels purchase their required logistics
products on demand, and summarize the actual distribution information into the order
processing system. The transportation behavior of the experimental group is matched
with the logistics distribution information sharing platform based on IoT technology,
and the transportation behavior of the control group is matched with the blockchain

sharing system. Considering the integrity of the experiment, the logistics transportation behavior of the experimental group and the control group is involved in the whole processing process from the delivery of goods to the purchase of consumers. The logistics distribution network is shown in Fig. 3.

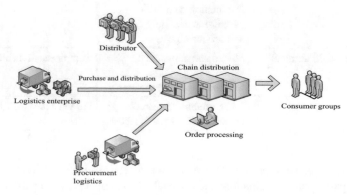

Fig. 3. Logistics distribution network

Table 2 records the physical values between the actual distribution flow of the experimental group and the control group and the distribution flow recorded by the shared host.

Table 2. Value of distribution flow

Delivery flow of experimental group (pieces)		Distribution flow of control group (piece)	
Actual value	Record value	Actual value	Record value
1089	1091	1089	1102
1132	1132	1132	1130
2504	2504	2504	2505
2361	2361	2361	2361
1870	1871	1870	1872
1965	1965	1965	1967
2143	2143	2143	2140
1796	1796	1796	1795
1533	1533	1533	1531
3917	3917	3917	3915

It can be seen from Table 2 that the recorded values of the experimental group are inconsistent with the actual values only twice, the difference level is relatively small, and the average value of the overall error is less than 1%. The recorded values of the control group were completely consistent with the actual values only once, and the results of the remaining nine records maintained certain differences with the actual values, and the overall mean error was much higher than that of the experimental group.

Figure 4 shows the transmission accuracy of logistics distribution information in the experimental group and the control group over a period of time.

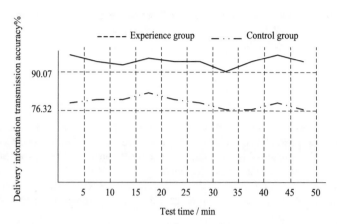

Fig. 4. Accuracy of logistics distribution information transmission

As can be seen from Fig. 4, the transmission accuracy of logistics flow distribution information in the experimental group and the control group showed a slight fluctuation, but the average level of the experimental group was significantly higher than that of the experimental group, and the minimum recorded value was 90.07%, while the minimum value of the control group was as low as 76.32%.

Combined with the above experimental research results, the logistics distribution information sharing platform based on IoT technology can accurately record the value of distribution logistics volume and realize the accurate transmission of distribution information, which is more in line with the actual application requirements.

5   Conclusion

The traditional blockchain sharing system improves the storage form, but this type of application structure cannot effectively control the logistics distribution cost, and the recorded information sharing behavior is not very accurate. To solve this problem, with the support of Internet of things technology, a logistics distribution information sharing platform based on Internet of things technology is proposed, which solves the problems existing in logistics distribution from many aspects.

The paper studies the new logistics distribution management and operation mode based on the Internet of things, and continuously improves the cloud mode of the Internet of things by means of functional requirements and business logic settings.

Evaluate and plan the physical parameters such as vehicle source, goods source and order information, and clarify the distribution and transportation contract.

By comparing with the distribution information transmission accuracy of the blockchain sharing system, it is verified that in terms of logistics distribution information sharing, the accuracy of the Internet of things sharing system has been improved compared with the more advanced blockchain sharing system, and the cost of logistics distribution has been effectively controlled.

Acknowledgment. This project is supported by Guangxi young and middle-aged teacher's basic ability enhancement project (2021KY1807) and Guangxi young and middle-aged teacher's basic ability enhancement project (2018KY0744).

References

1. Um, T.-W., Lee, E., Lee, G.M., Yoon, Y.: Design and implementation of a trust information management platform for social internet of things environments. Sensors **19**(21), 4707 (2019)
2. Hasegawa, Y., Yamamoto, H.: Reliable IoT data management platform based on real-world cooperation through blockchain. IEEE Consum. Electron. Mag. **10**, 82–92 (2020)
3. Koot, M., Mes, M.R., Iacob, M.E.: A systematic literature review of supply chain decision making supported by the internet of things and big data analytics. Comput. Ind. Eng. **154**, 107076 (2021)
4. Yang, T., Shi, Y., Chao, L.: Application of Internet of things technology in financial leasing industry – Based on the construction idea of closed-loop real estate management platform. China's Circ. Econ. **33**(3), 112–120 (2019). (in Chinese)
5. Ma, X., Zhang, X., Peng, R.: Research on reliability of intelligent management of science and technology infrastructure based on Internet of things. Control Decis. Mak. **34**(5), 1116–1120 (2019). (in Chinese)
6. Cagliano, A.C., Marco, A.D., Mangano, G., et al.: Levers of logistics service providers' efficiency in urban distribution. Oper. Manag. Res. **10**, 104–117 (2017)
7. Prasad, T.V.S.R.K., Veeraiah, T., Kiran, Y., Srinivas, K., Srinivas, C.: Decentralized production-distribution planning in a supply chain: computer experiments. Mater. Today Proc. **18**, A1–A11 (2019)
8. Atlam, H., Alenezi, A., Alassafi, M., Wills, G.: Blockchain with internet of things: benefits, challenges, and future directions. Int. J. Intell. Syst. Appl. **10**(6), 40–48 (2018). https://doi.org/10.5815/ijisa.2018.06.05
9. Lusiantoro, L., Yates, N., Mena, C., et al.: A refined framework of information sharing in perishable product supply chains. Int. J. Phys. Distrib. Logist. Manag. **48**(3), 254–283 (2018)
10. Malik, B.H., Zainab, Z., Mushtaq, H., et al.: Investigating technologies in decision based internet of things, internet of everythings and cloud computing for smart city. Int. J. Adv. Comput. Sci. Appl. (IJACSA) **10**(1), 56–61 (2019)
11. Wang, L., Pang, X., Zhu, Y., et al.: Research and application of secondary equipment operation and maintenance technology based on Internet of things and mobile Internet. China Power **52**(3), 182–189 (2019). (in Chinese)
12. Zhang, J., Yun, L., Wang, Y., et al.: Design of intelligent monitoring and early warning system for storage banana maturity based on Internet of things. J. Anhui Agric. Univ. **46**(4), 193–198 (2019). (in Chinese)
13. Lu, J.: Distribution vehicle scheduling of Internet of things based on perturbation shrinkage particle swarm optimization. Highw. Transp. Technol. **37**(4), 114–120 (2020). (in Chinese)
14. Ge, L., Ji, X., Jiang, T., et al.: Internet of things information sharing security mechanism based on blockchain technology. Comput. Appl. **39**(2), 458–463 (2019). (in Chinese)

15. Zhou, Q., Deng, Z., Zou, P., et al.: Block chain based DDoS attack method for protecting Internet of things devices. J. Appl. Sci. **37**(2), 213–223 (2019). (in Chinese)
16. Terrada, L., Bakkoury, J., El Khaili, M., Khiat, A.: Collaborative and communicative logistics flows management using the internet of things. Adv. Intell. Syst. Comput. **756**, 216–224 (2019)
17. Atta, N.: Internet of things, big data and information platforms for advanced information management within FM processes. In: Atta, N. (ed.) Internet of Things for Facility Management: Strategies of Service Optimization and Innovation, pp. 41–49. Springer International Publishing, Cham (2021). https://doi.org/10.1007/978-3-030-62594-8_4
18. Ahmed, H., Nasr, A., Abdel-Mageid, S., Aslan, H.: DADEM: distributed attack detection model based on big data analytics for the enhancement of the security of internet of things (IoT). Int. J. Ambient Comput. Intell. **12**(1), 114–139 (2021)
19. Thomas, M.O., Onyimbo, B.A., Logeswaran, R.: Usability evaluation criteria for internet of things. Int. J. Inf. Technol. Comput. Sci. (IJITCS) **8**(12), 10–18 (2016)

Computer Science for Education,
Medicine and Biology

Research and Practice of Applied Professional Experimental Course Construction—Takes the Automotive Service Engineering Experimental Course as an Example

Yi Wei[✉], Lu Ban, Daming Huang, GengE Zhang, and Jialiang Chen

School of Transportation, Nanning University, Guangxi 530200, People's Republic of China
1723872199@qq.com

Abstract. The transformation and development of applied higher education has become an important type and new force of higher education, and the construction of specialized experimental courses in the construction of applied courses has also received great attention and attention. This paper summarizes the characteristics and background of applied curriculum through literature review. On this basis, the connotation and significance of specialty experimental curriculum and its applied reform and construction were discussed. In view of how to realize the transformation of professional experimental courses in application-oriented universities, this paper puts forward the content of application-oriented transformation of professional experimental courses by taking eight transformation of application-oriented universities as guidance. Combined with the construction of characteristic specialty of automobile service engineering in Nanning University, this paper introduces the application reform and practice of experimental course of automobile service engineering in three aspects: curriculum system, curriculum content and curriculum implementation. The final practice results show that the automotive service engineering students of Nanning College have improved their proficiency in professional skills, and enterprises have improved their satisfaction with the automotive service engineering students of Nanning College.

Keywords: Applied universities · Professional experiments · Curriculum construction · Automotive service engineering

1 Construction Background

With the transformation and promotion of China's higher education from elite education to mass education, applied-oriented universities have emerged at the historic moment in the call of reform, developed vigorously under the guidance of policies, and become an important type and new force of China's higher education [1]. Around the transformation and development of application higher education, student-centered, application-oriented, ability-objective curriculum construction received great attention, by the Ministry of Education as one of the main tasks of application transformation of universities, clearly pointed out that "to social and economic development and industrial technology progress drive curriculum reform, integrate related professional basic courses, main course, core class, professional skills application and experimental practice class, more focus on cultivating learners' technical skills and innovation and entrepreneurial ability" [2].

The professional experimental course is an important part of the undergraduate professional course system and an important practical teaching link of undergraduate teaching [3, 4]. As the state and education, administrative departments attach great importance to university practice teaching and emphasize that in the background of application university construction and application talent training, the role of professional experimental courses in professional talent training more prominent, more prominent status, become an important carrier of college students application ability and innovation ability training and the important content of professional construction [5]. Therefore, it is of great significance to closely around the school orientation and educational concept, the industry and talent characteristics of the major, and closely promote the application-oriented reform of professional experimental courses for the construction of characteristic application-oriented majors and applied for universities.

Sun Xinghua et al. [6] from the perspective of the supply side of local undergraduate universities, guided by Thought on Socialism with Chinese Characteristics in the New Era, comprehensively promoted the construction of high-level application-oriented universities with local characteristics, implemented responsibilities, based on local areas, highlighted application and focused characteristics, and then promoted the construction of high-level application-oriented universities with local characteristics. Wang Junmei et al. [7] based on the new talent training goal of Business School of Hebei University, proposed the construction of the "134" model of" one-center, three-level and four-combination "experimental curriculum system for economics and management majors in line with students' cognitive rules and social development needs, providing reference for similar institutions. Yue Hongwei et al. [8] proposed a comprehensive experimental curriculum system based on the comprehensive experimental curriculum setting and practice of Xuchang University of New energy Materials and devices, which takes application as the main line and practices in the preparation, characterization, device assembly and performance testing of new energy materials.

Specialized experiment course is the core course of applied university and also the key point of reform and construction. In view of how to realize the transformation of professional experimental courses in applied colleges and universities, this paper puts forward the content of the application transformation of professional experimental courses by taking eight transformations of applied colleges and universities as guidance. Combined with the construction of characteristic specialty of automobile service engineering in Nanning University, this paper introduces the application reform and practice of experimental course of automobile service engineering in three aspects: curriculum system, curriculum content and curriculum implementation.

2 Connotation of the Practical Reform of Specialized Experimental Courses

The construction of application-oriented undergraduate courses is proposed to meet the transformation of application-oriented universities and the training of application-oriented talents. The application-oriented curriculum is a curriculum concept relative to theoretical courses or academic courses, and a product of the application-oriented education model corresponding to the academic education model [9]. It is well known that the curriculum plays an extremely important role in the process of education and teaching activities and talent training. The curriculum is an important carrier to implement educational activities, the basic way to achieve the training goal, and the key factor to determine the teaching effect. Therefore, in the training of applied talents, in the final analysis, must eventually fall into the construction of application-oriented courses. It is necessary to guide the curriculum construction through the concept of an application-oriented curriculum, and the training goals must be achieved through the implementation of the applied curriculum.

Regarding the concept of applied curriculum, there is no unified and clear definition in the relevant educational theory and teaching research. The interpretation and description of the application courses have different statements and different wording. Through combing and summarizing of literature evaluation [10–15], we can find that the characteristics of application courses are common in the following aspects:

1) The curriculum is oriented to the production reality, engineering practice, and work process, based on cultivating application ability.
2) The curriculum concept of structuralism, the mastery of theoretical knowledge, and practical ability cultivation of equal integration.
3) Build the teaching content according to the technical logic, application logic, and action logic.
4) The curriculum organization form and teaching method have the characteristics of open, situational, and middle school implementation.
5) The curriculum construction path of school-enterprise cooperation and integration of industry and education.

The professional experimental course mainly refers to a class of professional practice courses that separate experiment from professional theory course, formed through integration and innovation, and set up with experiment as the main content. The experimental independent course is an important exploration and practice of experimental teaching reform in the university. Experimental curriculum Settings, change is attached to the traditional concept of theory teaching and experiment teaching changes the traditional experiment repeated scattered fragments barriers, mainly the verification and demonstration experiment phenomenon, will open as the main content of the reform, the comprehensive design experiment made in the experimental teaching in engineering education concept, implementation, and fusion, the fusion possible production, and education, It fully reflects the great importance of experimental teaching and gives full play to the important role of experimental teaching in personnel training.

Experimental course setting, completely change the traditional concept of experimental teaching attached to theory teaching, change the traditional experiment fragmented, repeated barriers, mainly to validation and demonstration experiment, will open comprehensive design experiment as the main content of the reform, in the experimental teaching into engineering education concept, education integration, fully embodies the great importance to experimental teaching and give full play to the important role of experimental teaching in talent training. Through its complete experimental curriculum system and relatively independent content system, professional experimental courses achieve and strengthen students' complete, systematic and professional training of technical methods and abilities and practice of working process and methods, which is consistent with the emphasis on "practicality" and "learning by doing" in applied courses. However, there is undeniable that due to the influence of inertia thinking and coping attitude, the current experimental curriculum in curriculum objectives, curriculum model, and curriculum implementation, inevitably still has some extent of traditional course traces, can't completely get rid of the influence of academic education mode, and the requirements of application courses, which undoubtedly affects the role of experimental curriculum in applied talent training.

The application-oriented reform of professional experimental courses is based on the characteristics of application-oriented undergraduate education, the path of application-oriented university transformation, and the connotation of application-oriented curriculum construction, and carries out the reform of professional experimental courses, including curriculum concept, curriculum content, curriculum carrier, curriculum path, curriculum evaluation, and other aspects. Structural course concept to guide the course construction, with technology application logic or work action logic based on reconstruction course content, to the actual case, work situation as the key to updating the curriculum carrier, with school-enterprise cooperation, integration, collaborative innovation based on course way, achievement-oriented, output guidance based on the evaluation course effect. Through a series of reforms and construction, the application transformation of professional experimental courses is fundamentally realized.

3   The Construction of Professional Experimental Course Reflects Eight Changes

The transformation to the application-oriented colleges is not only an inevitable choice for the ordinary local undergraduate universities but also an urgent need for China's economic structure adjustment and industrial transformation and upgrading. As for the specific path and basic connotation of the transformation and development of colleges and universities, it is mainly represented by the eight transformations proposed by Hefei University, which are regarded as leading and exemplary models in the transformation and development of colleges and universities and the reform of higher education and the reform of application-oriented talents cultivation mode. The eight changes are as follows:

1) Transform the orientation of school running to application oriented.
2) Change of professional structure to demand-oriented.

3) The training program changes to knowledge output and output orientation.
4) Change of curriculum system to technical logic system.
5) The teaching process becomes student-centered.
6) The transformation of the teaching staff to dual-ability type.
7) Education mechanism should be transformed into open cooperation.
8) The quality evaluation should be changed to satisfy both students and society.

The transformation of application-oriented universities proposed by Hefei University is not only the path of application-oriented transformation, but also the result of application-oriented transformation [16]. As an important carrier of college education and teaching process and the basic unit and important foundation of professional construction, the curriculum is the key element affecting the education process and the core of realizing the training goals [17]. Therefore, the transformation of universities is bound to fall on the courses closely related to talent training and teaching quality [10]. Curriculum transformation is not only an important part of the transformation of universities, but is also considered to be the focus and difficulty in the transformation of universities. It is the key to promoting the transformation to the depth development of universities. Therefore, how to reflect the eight changes of university transformation in curriculum construction is the first key problem to be solved in the application-oriented construction of professional experimental courses [18].

Corresponding to the eight changes of university transformation, the construction of professional experimental courses should be completed in the following aspects:

1) The curriculum orientation should be transformed from academic courses to applied courses.
2) The curriculum concept has changed from the dual opposition of rationalism and utilitarianism to structuralism.
3) The curriculum goal has changed from knowledge-oriented to ability-oriented.
4) The course content should be transformed from subject logic to technology logic.
5) On the course carrier, the transformation from learning by doing to learning by doing, from opposition of knowing and doing to integration of knowing and doing.
6) Change from closed and didactic mode to open and situational mode in course teaching.
7) Change from teacher-centered to student-centered and from production-teaching separation to production-teaching integration in curriculum implementation.
8) Curriculum evaluation should be transformed from knowledge mastery to ability acquisition, from teacher satisfaction to social satisfaction, and from single standard to comprehensive standard.

4   Application-Oriented Reform Practice of Professional Experimental Courses

Nanning University is one of the first national universities of applied technology and the pilot universities for new undergraduate transformation and development in Guangxi. Under the background of the transformation and development of local undergraduate colleges and universities, Nanning University firmly strengthens the direction of application-oriented school running and talent training goals, actively explores the development path in line with local characteristics, and further promotes the reform of application-oriented talent training mode. Strive to create a local applied university distinctive characteristics, clear "local, applied ordinary undergraduate university" orientation, carry out the "application, open, new experience" school philosophy, determine "mainly multidisciplinary coordinated development, based on Nanning for Guangxi, construction of distinctive application technology university" development goals. Nanning University Automotive Service Engineering is an undergraduate major in order to meet the industrial development of Guangxi and the needs of the transportation industry and automobile service industry after it was selected as the pilot university of the National University of Applied Technology. Automotive service engineering is a broad area of technical services, non-technical services and supporting services. The automobile service industry is an organic industry composed of all automobile service providers.

According to the industry characteristics of automobile service, closely combined with the actual development of Guangxi automobile service industry, and following the principle of demand-oriented, this specialty is positioned as the training of backbone talents in the fields of automobile technology, sales and insurance services.

In order to cooperate with the school to promote the comprehensive transformation and development, fully reflect the school's application-oriented school orientation, and cultivate high-quality application-oriented talents with industrial industry, practical ability and technology application ability. Nanning school of automobile service engineering, combining the characteristic specialty construction and course construction demonstration school, around eight transformations applied universities and applied course, learn from domestic advanced colleges reform experience, the transformation of professional experimental course of applied course construction, in teaching practice and the gap between applied course construction requirements, This paper focuses on the practical reform of professional experimental courses in three aspects: curriculum system, curriculum content and curriculum implementation.

4.1 Reform of the Curriculum System

According to the requirements of the transformation of the curriculum system to the technology application logic system and the work action logic system, the automobile service Engineering major of Nanning University first carried out the reform of independent experimental courses. The traditional scattered and affiliated in professional theory courses, through integration and restructuring innovation, build independent professional experimental courses, based on summarizing experience, continuous improvement optimization, in the field of automobile service engineering technology, sales, insurance three talent demand-oriented, around technology and business two core ability training, according to the structured, modular and hierarchical curriculum concept, build two platforms four-course professional experimental curriculum system, as shown in Fig. 1:

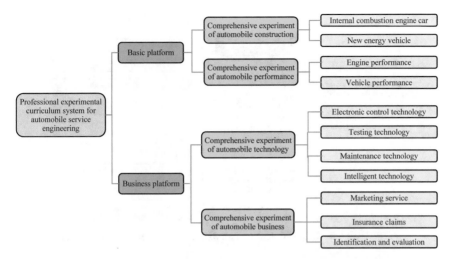

Fig. 1. Experimental course system of automobile service engineering

4.2 Reform of the Course Content

Automobile service engineering of Nanning University experiment curriculum reform is not a simple combination of traditional attachment in theory course of experiment and piled up, but in technical application ability and operation ability training as a starting point, based on technology application logic and action logic, reconstruction, reconstruction and innovation experiment content, the choice of course content and the design of the experimental unit, Emphasis should be placed on giving play to the advantages of school-enterprise cooperation and the role of the integration of industry and education, on implementing the principle of the combination of industry demand and education theory, on the integration of specialized innovation and the interaction with disciplinary and professional competitions, and on establishing a structured content system and progressive experimental paradigms at different levels [19].

The practical reform of the experimental course content of automobile service engineering major in Nanning University is mainly reflected in two aspects.

1) The experimental type is mainly comprehensive design experiment, which combines verification experiment, comprehensive design experiment and open experiment. Through the three levels of basic, application and improvement, the training process of technical application ability and work and business ability is formed from low level to high level and from simple to complex [20], as shown in Fig. 2.

2) The content of the experiment is fully in line with the actual work, production and engineering, and the professional standards are docking in the experiment, the actual cases are implanted, and the real situation is created, so that students can have the feeling and experience of doing real things and learning in the process of the experiment.

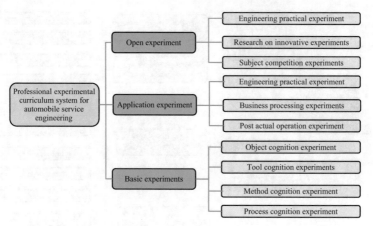

Fig. 2. Experimental content system of automobile service engineering

4.3 Reform of the Curriculum Implementation

The implementation of the curriculum should be realized through the carrier of curriculum, so the reform of curriculum implementation is the reform of the carrier of the curriculum. The reform of experimental course carrier of automobile service engineering major in Nanning University mainly revolves around "activities" and "teaching materials". The former is the important connotation of classroom teaching and the carrier of work activities. The embodiment of the reform is to carry out the integration of production and education of experimental teaching and build the "carrier of classroom activities" with comprehensive and design-based experiments. The latter is the basic element of curriculum implementation and the carrier of practical knowledge. The reform focuses on the compilation of applied experimental teaching materials based on the logic of technology application and works action to explore the "integration of knowledge and action" [21].

To implement the integration of industry and education in experimental teaching, the major of Automobile Service Engineering sets about the reform of experimental teaching content and focuses on the docking of three aspects respectively:

1) Docking with the working process, implant the actual working process into the experimental process.
2) Connect with job requirements and transform enterprise training content into experimental content.
3) Docking with enterprise development, the school-enterprise cooperation project is decomposed into experimental projects.

The implantation of a practical work processes and the transformation of enterprise training content constitute the main content of comprehensive and design experiments. To establish industry-university-research collaborative innovation cooperation with enterprises, cooperative subject research has become an important source of topic selection for independent and open experiments [20].

The technical application of experimental teaching and the logical attribute of work activities are both a teaching activity process and a work activity process. In order to fully embody the principle of "results-oriented" and the characteristics of "work task", the goal structure of the course should have both the requirements of achieving the teaching goal and the requirements of completing the work goal, with dual goals of teaching goal and work goal. Teaching objectives are the training objectives of students in knowledge, ability and emotion that are expected to be achieved through the experimental process. The work goal is the working state, effect, level or obtained data, rules, conclusions and other goals expected for students to complete the task through the experimental process.

Another aspect of teaching material reform is the reform of teaching material content structure. In the compilation of experimental units, the teacher-centered structure mode, which is written according to the purpose, content, methods and steps of the experiment, is replaced by the student-centered structure mode, which is written according to the task goal, knowledge, situation and action. Among them, the knowledge part is to provide "theoretical guidance" or "action guidance" for students, while the situation part is to build "work tasks" and "work environment" for students, to form the organic integration of theoretical knowledge and practical knowledge, and play a role in promoting the mastery of theoretical knowledge and training hands-on ability [21].

Figure 3 is the frame diagram of experimental unit content in experimental teaching materials.

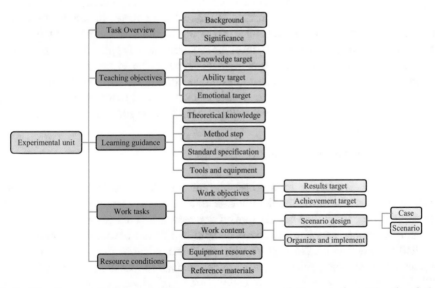

Fig. 3. The framework diagram of the content structure of the experimental unit of the experimental course of automobile service engineering

The reform of curriculum implementation is followed by the reform of the implementation path. The fundamental purpose of the application-oriented curriculum is to cultivate students' ability to solve practical production problems and complex engineering problems. The basic characteristics of the applied curriculum are authenticity, situational and openness of learning, emphasizing the unity of knowledge and action, competency-based, demand-oriented and social satisfaction. This determines that the construction of applied courses must unswervingly take the road of school-enterprise cooperation and the integration of industry and education. School-enterprise cooperation and integration of industry and education are the best strategy and effective way to realize the transformation of application courses, and the inevitable choice to realize the goal of applied courses in the course implementation path.

In terms of the reform of the implementation path of professional experimental courses, the automobile service engineering major of Nanning University makes full use of the school-enterprise cooperation and collaborative education platform, give full play to the role of integrating industry and education, and actively explore and practice the implementation path of school-enterprise cooperation and construction. Taking the comprehensive experiment of automobile business as an example, it has cooperated with the company to design the experimental project, formulation of syllabus and compilation of experimental textbooks. The experiment comes entirely from the actual production and the actual work. The experimental process implements two-teacher guidance, using the organizational form of "task-driven + case guide + context design + role-playing + group cooperation" and the combination of campus and off-class teaching methods.

In addition to the carrier and path of course implementation, Automobile Service Engineering of Nanning University also attaches great importance to the reform of the implementation conditions of experimental courses, strives to create conditions, and create a learning environment according to the real work scene. In auto business comprehensive experiment of auto insurance and claims module, for example, in the creation of working environment, in order to meet the requirements of application course authenticity and situational, laboratory environment layout as far as possible or completely according to the actual work scene to design, to difficult car insurance scene, also try to create conditions, through the sand table or using school road and experimental car accident scene simulation, try to make students have on the scene, really do experience.

5   Summary and Conclusion

Training innovative talents for engineering and application is an important development direction of the current undergraduate education talent training mode. Especially the application of technical talents applied undergraduate colleges and universities are booming, they are according to the economic and social development, training can "skilled use knowledge, solve production problems, adapt to the broad demand" the application of technical talents applied undergraduate universities, more emphasis on following engineering, oriented application, oriented engineering, application throughout the whole process of talent training, highlight the talent training specification of engineering ability characteristics and application ability characteristics.

1) This paper was guided by eight transformation of applied university, puts forward the content of applied transformation of specialized experimental courses. Combined with the construction of nanning College's characteristic specialty of automobile service engineering, the experimental course of automobile service engineering is

reformed and practiced in three aspects: curriculum system, curriculum content and curriculum. Finally, it improves the students' proficiency in professional skills and enterprises' satisfaction with the automotive service engineering students of Nanning College.

2) This paper was based on the reform and construction of nanning College automotive service engineering specialty experimental courses. Through the concept reconstruction and teaching reform, the application-oriented courses are formed in line with the school's educational orientation, close to the needs of industrial development, and highly fit the school's application-oriented talents training goals. It is the basic starting point of professional experimental curriculum construction. The application transformation of professional experimental courses is based on the action and result of this motivation.

3) Curriculum reform is a long - term and constantly advancing with The Times process. The future reform and construction will focus on the combination of on-campus, off-campus and in-class and off-class, further explore and build the course path combining online and offline, and comprehensively promote the application-oriented construction of professional experimental courses.

Acknowledgment. This project is supported by the Projects of Teaching Reform in Higher Education in Guangxi: Research on Talent Training Strategy based on the Integration of Discipline Competition and "Innovation and Entrepreneurship"—Take the Course of Foundation and Application of Microcontrofler as an example (Grant no. 2021XJJG05), the Projects of Teaching Reform in Higher Education in Guangxi: Research and Practice of Professional Experimental Course Reform in Automotive Service Engineering Based on AOE+CDIO (Grant no. 2018JGB378).

References

1. Changlin, H., Jing, L., Dan, Y.: Orientation and selection of newly-established undergraduate universities from the perspective of application-oriented universities. Educ. Res. **36**(04), 61–69 (2015)
2. Ministry of Education of the People's Republic of China: National Development and Reform Commission, The Ministry of Finance of the People's Republic of China. Guiding Opinions on guiding the transformation of some local ordinary undergraduate colleges and universities to application-oriented application. Bull. State Council People's Republic of China (06), 60–64 2016. (in Chinese)
3. Denggao, G., Jinghui, L., Jianping, L., Qiaoming, Y., Junfeng, L., Mei, Y.: Material physics & chemistry quality network curriculum construction and teaching practice. IJEME **2**(5), 24–30 (2012)
4. Zhao, C., Wang, Y.G., Ma, J.X., Yu, A.H.: Study on the examination pattern of the engineering specialized courses. IJEME **2**(2), 59–65 (2012)
5. Zhang, W., Wang, N., Yi, K., Li, J.: Exploration of talent training scheme for new energy vehicles in high-level applied universities under the background of new engineering. Auto Time (24), 45–47 (2021)
6. Sun, X., Zhou, L., Xia, G.: Research on the construction of high-level application-oriented university with local characteristics. Teacher (30), 107–108 (2021)

7. Junmei, W.: Research on the construction and implementation of experimental teaching system of economics and management major in local application-oriented universities–a case study of business school of Hebei university. Technol. Wind **35**, 105–107 (2021)

8. Yue, H., Chen, S., Tie, W., Zhu, C., Li, T.: Design and practice of integrated experimental course of new energy materials and devices in applied university. Shandong Chem. Ind. **48**(14), 199–200+202 (2019)

9. Zhu, N.: Research on the applied curriculum reform of local undergraduate universities under the background of transformation. Shaanxi Normal University (2016). (in Chinese)

10. Maoyuan, P., Qunying, Z.: See the construction of application-oriented undergraduate courses from the perspective of university classification. China Univ. Teach. **03**, 4–7 (2009). (in Chinese)

11. Wanqaing, C., Yazhou, S.: Technology applied course system construction of logic and inner structure. Educ. Teach. Forum **09**, 204–205 (2018). (in Chinese)

12. Hong, J., Ran, Z., Genhua, Z., Bin, Q., Zhu, Q.: Ideas on construction of applied course based on the university-enterprise collaborative education. Appl. Oriented High. Educ. Res. **2**(01), 40–44 (2017). (in Chinese)

13. Yanjuan, X.: On the development of curriculum models and the reform of curriculum management of china's higher engineering education. Res. High. Educ. Eng. **12**, 34–38 (2019). (in Chinese)

14. Yanmei, W.X.L.: Research on the curriculum building and reform development way of application-oriented undergraduate universities. J. Vocat. Educ. **26**(2), 58–63 (2016). (in Chinese)

15. Jianyi, Z., Yongqiang, Z., Aiguo, W.: The exploration and practice of applied curriculum construction in the background of transition. Theory Pract. Educ. **39**(15), 18–20 (2019). (in Chinese)

16. Guanglei, C., Jie, Z.: Research on the development path of the application-oriented construction in local universities. J. High. Educ. Manag. **11**(03), 66–72 (2017). (in Chinese)

17. Kerong, Z., Zhou, H., Xiaohong, W.: Characteristics, construction ways and predicaments of the application-oriented undergraduate courses. J. Vocat. Educ. **03**, 62–66 (2019). (in Chinese)

18. Guojiang, X., Modularity, C.: Exploring the path of curriculum transformation in local undergraduate universities. China High. Educ. Res. **11**, 99–102 (2014). (in Chinese)

19. Zhang, G., et al.: Planning and design of automotive service engineering based on AOE + CDIO. J. Chifeng Univ. (Nat. Sci. Edit.) **35**(09), 154–157 (2019). (in Chinese)

20. Yi, F., Daming, H.: Experimental curriculum reform on the automobile service engineering based on the AOE + CDIO. Adv. Soc. Sci. **8**(11), 8 (2019)

21. Maoyuan, P., Zhou, Q.: The training of applied talents calls for the teaching materials of the knowledge and practice system. China Educ. News **2010-04-19**(005). (in Chinese)

Transferability Evaluation of Speech Emotion Recognition Between Different Languages

Ievgen Iosifov[1,2] (ID), Olena Iosifova[1] (ID), Oleh Romanovskyi[1] (ID),
Volodymyr Sokolov[2(✉)] (ID), and Ihor Sukailo[2] (ID)

[1] Ender Turing OÜ, Tallinn, Estonia
{ei,oi,or}@enderturing.com
[2] Borys Grinchenko Kyiv University, Kyiv, Ukraine
{v.sokolov,i.sukailo.asp}@kubg.edu.ua

Abstract. Advances in automated speech recognition significantly accelerated the automation of contact centers, thus creating a need for robust Speech Emotion Recognition (SER) as an integral part of customer net promoter score measuring. However, to train a specific language, a specifically labeled dataset of emotions should be available, a significant limitation. Emotion detection datasets cover only English, German, Mandarin, and Indian. We have shown by results difference between predicting two and four emotions, which leads us to narrow down datasets to particular practical use cases rather than train the model on the whole given dataset. We identified that if emotion transfers good enough from source language to target language, it reflects the same quality of transferability in vice verse direction between languages. Hence engineers can not expect the same transferability in the mirror direction. Chinese language and datasets are the hardest to transfer to other languages for transferability purposes. English dataset transferability is one of the lowest, hence for a production environment, engineers cannot rely on a training model on English for their language. This paper conducted more than 140 experiments for seven languages to evaluate and show the transferability of speech recognition models trained on different languages to have a clear framework which starting dataset to use to achieve good accuracy for practical implementation. The novelty of this study lies in the fact that models for different languages have not yet been compared with each other.

Keywords: Speech emotion recognition · Sentiment analysis · Emotion detection · Engagement analysis

1 Introduction

Advances in automated speech recognition significantly accelerated the automation of contact centers [1, 2]. Such automation and augmentation of human agents demand speech translation to text to understand what was said and its sentiment [3]. Sentiment analysis based on text fairly often cannot recognize anger and happiness if humans do not express or articulate them through particular words. At the same time humans, articulate/encode emotions in intonations [4, 5]. It creates a need for robust SER as an integral part of net promoter score measuring.

The novelty of this study lies in the fact that models for different languages have not yet been compared with each other. To recognize dialects or lingua franca, one should understand the recognition models' limitations. If the system works with clients from different language groups, recognizing the language and its emotional coloring allows them to respond adequately.

Section 2 presents related work on the topic. Section 3 touches on the objective, problem formulation, task, limitations, and previous research. Section 4 introduced the research methodology and datasets chosen. Section 5 illustrates the experimental setup with training framework, architecture, algorithm, and model training. Section 6 describes experiment findings. Section 7 concludes conducted research.

2 Related Works

Due to the lack of labeled emotion recognition datasets and low transferability of models trained purely on English acoustic, small emotional datasets for other languages [6–12]. Despite this, only English and Chinese datasets are represented through many datasets and define diverse acoustic, lexicon, etc. Most monolingual datasets are relatively small, which is not enough to transfer even to languages of the same group. The problem of emotional data gathering through segregation from actual cases and call labeling is a sore spot and seems unlikely. Hence main research interest shifted to delivering multilingual SER models [13, 14].

In [15], research opportunities to fine-tune a pre-trained SER model with a small amount of data produces promising results but still require manual dataset preparation for the target language. [16] proposed combining acoustic features in a three-layer perceptual emotion model, which shows comparable results to monolingual SER in a new language without training. Furthermore, in [17], ensemble learning was proposed, giving promising results. Overall, assembling and layering looks profitable but not available in main speech recognition frameworks and toolkits out of the box.

In [18] and [19] two-pass classification scheme was proposed consisting of spoken language identification and SER, which also limits deployment in production to have SER models for identified languages. [20] proposed unsupervised approaches which can show great results but still require a lot of computational resources. As can be seen from the analysis of the sources, there has not been a comparison between several languages before (only one model—one language).

3 Problem Statement

3.1 Research Objective and Problem Formulation

The paper aims to address practical questions of transferability of trained SER models from one language to another and understand the limitations of such transferability for some specific languages. We tried to compare the complete language datasets with each other. This is necessary to determine which languages are closest to each other in emotional coloring.

Emotion classification and emotional engagement evaluation are some of the top demands of each business. How customers react, what triggers happiness, and what triggers sadness are endless questions to many companies. We aim to discover the possibility of robust multilingual application of emotion recognition through the evaluation of cross-lingual emotion recognition. The paper goal is to find language groups that are well comparable with each other.

3.2 Practical Limitations and Previous Research

The main limitation to the practical implementation of SER is datasets availability for each particular language as phoneme similarity varies a lot between languages. It is hard to make a dataset for each language for the SER task, and hence models are trained in one language and implemented differently.

In previous works, we have compared and selected the best techniques comparison for natural language processing [21] and automatic speech recognition methods [22]. In the following papers, unformatted text [23] was split, and the pipeline for training dataset creation [24] was assembled. In this paper, we build on our earlier findings and recognize the speaker's emotions.

4   Research Methodology

4.1 Main Approaches

The most common way and dataset for emotion recognition are multimodal. Particularly effective for emotion recognition is video data, which is a rare case for the production environment, like the contact center. In such an environment, only audio information is available to predict emotion. Emotional processes correlated with acoustic parameters (frequency, spectral energy, speech rate, shimmer, and jitter) [25].

As research progresses a lot in finding a way to represent acoustic signals and extract features of movements in the best way, the most intriguing question is where to find relevant data to train SER models.

While it is not a problem for most advanced languages in the NLP/speech area to find 200+ h of labeled data, it is still a long way for most languages. In this situation, the most obvious approach is to train a model using a dataset-rich language and apply it to a production environment with a local language. Sounds good, but in practice, it works far from good. We aim to evaluate the portability of models to different languages as an approach and to find out little places to keep in mind while doing this.

4.2 Datasets Choosing

Pre-consideration list of available speech recognition datasets for languages (Table 1) from different groups (letter designation according to ISO 639-1:2002 [26]):

- Indo-European: English (EN), German (DE), French (FR), Persian (FA), and Urdu (UR).
- Uralic: Estonian (ET).
- Sino-Tibetan: Chinese (ZN).

Table 1. Speech recognition datasets comparison

Language	Dataset	Duration, hours	Size, utterances	Modalities	Emotions
ZN	ESD [27]	29	7,000	Spoken language, voice	Neutral, anger, sadness, happiness, surprise
DE	EMODB [9]	15	800	Spoken language, voice	Neutral, anger, sadness, happiness, fear, disgust, boredom
ET	EKORPUS [7]	65	1,234	Spoken language, voice	Neutral, anger, sadness, happiness
EN	CREMA [28]	203	7,442	Multimodal, voice, and visual	Neutral, anger, sadness, happiness, fear, disgust
EN	IEMOCAP [29]	336	10,040	Multimodal, voice, and visual	Neutral, anger, sadness, happiness, fear, disgust, surprise, excitation, frustrated
EN	RAVDESS [30]	36	7,356	Multimodal, voice, and visual	Neutral, anger, sadness, happiness, fear, disgust, surprise, calm
EN	SAVEE [31]	19	480	Multimodal, voice, and visual	Neutral, anger, sadness, happiness, fear, disgust, surprise

(continued)

Table 1. (*continued*)

Language	Dataset	Duration, hours	Size, utterances	Modalities	Emotions
EN	TESS [32]	55	2,800	Spoken language, voice	Neutral, anger, sadness, happiness, fear, disgust, surprise, pleasant
FA	ShEMO [12]	196	3,000	Spoken language, voice	Neutral, anger, sadness, happiness, fear, surprise
FR	OREAU [8]	23	482	Spoken language, voice	Neutral, anger, sadness, happiness, fear, disgust, surprise
UR	URDU [13]	16	400	Spoken language, voice	Neutral, anger, sadness, happiness

From the list above, the difference in size is most striking. Datasets should be comparable in size for training and evaluation purposes. The smallest datasets consist of only 15–20 h of data, and to be similar other datasets should be truncated to the same amount of data.

After evaluation, the following datasets were chosen and truncated to comparable sizes for experiments (see Table 2).

Table 2. Dataset audio duration

Language	Dataset	Audio duration per emotion, hours			
		Neutral	Angry	Sad	Happy
ZN	ESD	4	4	4	4
DE	EMODB	3	4	4	3
ET	EKORPUS	4	4	4	4
EN	SAVEE	4	3	4	3
FA	ShEMO	4	4	4	4
FR	OREAU	4	4	4	4
UR	URDU	4	4	4	4

4.3   Model Architectures and Approach to Solve Speech Recognition Task

As comparison and evaluation of types of DNN-based architectures is not a goal of the current paper, we will rely on research in speaker-related information extraction. We will use Emphasized Channel Attention, Propagation, and Aggregation in Time Delay Neural Network (ECAPA-TDNN) [33] architecture to train the speech recognition model. The evaluations of TDNN for SER tasks can be found in work [34].

5   Experimental Setup

5.1   Training Architecture and Framework

SpeechBrain [35] toolkit was chosen as OpenSource with suitable license and supporting chosen architecture ECAPA-TDNN implementation to speed up experiments. After a few warmup evaluations, the next architecture proved to be the fastest and most accurate. Model architecture has the following form:

```
input_size: 96
channels: [512, 512, 512, 512, 1536]
kernel_sizes: [5, 3, 3, 3, 1]
dilations: [1, 2, 3, 4, 1]
attention_channels: 64
lin_neurons: 96
```

The lowest loss was at around 23 epoch. Hence we limit training for 30 epochs:

```
number_of_epochs: 30
```

Depending on experiments number of prediction classes varies between two and four (anger, neutral, happiness, sadness—starting from using only two, and up to four):

```
out_n_neurons: 4
```

5.2   Data Preparation

We prepared data by following steps. First, we conducted three experiments for the two, three, and four emotions datasets to evaluate how significantly more challenging for the model to recognize three and four emotions than just binary classification for neutral and angry. We decided to use the following sets of feelings for each stage of the experiments:

- The two-emotion dataset will consist of neutral and angry emotions.
- The three-emotion dataset will consist of neutral, angry, and sad emotions.
- The four-emotion dataset will consist of neutral, angry, sad, and happy feelings.

We decided to take these emotions for two reasons:

1) They are most practically demanded by business.
2) They are comparatively different acoustically from each other.

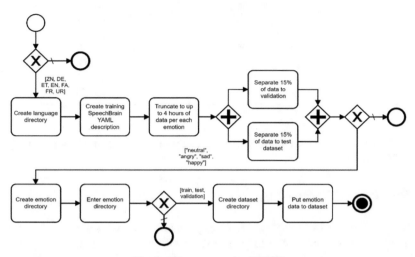

Fig. 1. Data preparation BPMN

We created a separate store (directory) for each of these sets of experiments. We made a separate store (directory)—we decided not to mix and not evaluate models trained to classify two emotions with datasets containing three and for feelings vice-versa.

We selected languages for datasets [ZN, DE, ET, EN, FA, FR, UR]. We created separate storage (directory) in the experiment set storage for each language. We truncated big datasets to comparable size to balance languages data in an equal amount of hours (see Fig. 1).

For test and validation purposes, we decided to use 15% and 15% of each language dataset, randomly selected between emotions, which means we did not take 15% of each emotion but let random choice in validation examples. Each emotion we put into storage (directory) with relevant feeling under relevant language. As of the last step, we prepare a SpeechBrain YAML description for each set of experiments to have the correct number of output neurons.

5.3 Model Training

We have trained 21 model, 7 models for each set of emotions [neutral-angry], [neutral-angry-sad], and [neutral-angry-sad-happy]. We trained models from two to four emotions (Table 3) for each language using 4 h of audio for each emotion.

5.4 Evaluation

The evaluation was conducted for pair model-language by iterating for any to any. Hence all models were evaluated by all language datasets. Evaluations were made using

Table 3. Sets of emotions

Number of emotions	Emotion set	Model trained on dataset
2	[neutral-angry]	ZN:ESD
3	[neutral-angry-sad]	DE:EMODB
		ET:EKORPUS
4	[neutral-angry-sad-happy]	EN:SAVEE
		FA:ShEMO
		FR:OREAU
		UR:URDU

a validation dataset of each model, consisting of 15% for each emotion for each language (up to 36 min).

6 Results and Discussion

We have conducted 147 experiments 49 experiments for each set of emotions. Each experiment was repeated three times to evaluate deviation inaccuracy. The standard deviation is estimated at 2% (see Tables 4, 5 and 6). Depicted numbers are the model's accuracy (0–1 scale) in predicting each particular dataset language label. As an example in Table 4, the model trained on two emotions of the English dataset shows an accuracy of 46% when predicting emotions for the Estonian language dataset.

In each column presented model language—model trained using one language data. In each row, we offer evaluation results for predicting feeling for specified in row language dataset by applying defined in column trained model. We calculated median evaluation scores and represented them in the last column and row. The model and evaluation dataset the same language is placed in the primary diagonal. For almost all languages main diagonal have the highest value. One significant outlier is the Estonian language which does not shows the best results for the Estonian evaluation dataset. It might be related to the raw intensity of emotion, which is relatively low for Estonian.

Table 4. Results of evaluation models for different languages for two emotions

Dataset language	Model language							
	ZN	DE	ET	EN	FA	FR	UR	Median
ZN	0.94	0.73	0.73	0.73	0.73	0.73	0.73	*0.73*
DE	0.39	0.97	0.76	0.82	0.94	0.73	0.73	*0.76*
ET	0.46	0.46	**0.62**	0.46	0.50	0.54	0.62	*0.50*
EN	0.46	0.54	0.42	0.96	0.58	0.69	0.73	*0.58*
FA	0.38	0.52	0.57	0.38	0.90	0.62	0.71	*0.57*

(continued)

Table 4. (*continued*)

Dataset language	Model language							
	ZN	DE	ET	EN	FA	FR	UR	Median
FR	0.29	0.54	0.66	0.71	0.77	0.74	0.54	*0.66*
UR	0.41	0.64	0.82	0.41	0.79	0.62	0.98	*0.64*
Median	*0.41*	*0.54*	***0.66***	*0.71*	*0.77*	*0.69*	*0.73*	

Table 5. Results of evaluation models for different languages for three emotions

Dataset language	Model language							
	ZN	DE	ET	EN	FA	FR	UR	Median
ZN	0.96	0.60	0.53	0.53	0.58	0.51	0.64	*0.58*
DE	0.36	0.93	0.64	0.80	0.89	0.64	0.71	*0.71*
ET	0.27	0.30	**0.46**	0.30	0.46	0.43	0.27	*0.30*
EN	0.22	0.65	0.35	0.89	0.43	0.54	0.32	*0.43*
FA	0.23	0.57	0.53	0.43	0.77	0.43	0.43	*0.43*
FR	0.35	0.52	0.40	0.52	0.48	0.62	0.40	*0.48*
UR	0.22	0.33	0.52	0.28	0.34	0.36	0.97	*0.34*
Median	*0.27*	*0.57*	***0.52***	*0.52*	*0.48*	*0.51*	*0.43*	

Table 6. Results of evaluation models for different languages for four emotions

Dataset language	Model language							
	ZN	DE	ET	EN	FA	FR	UR	Median
ZN	0.90	0.47	0.34	0.31	0.48	0.44	0.52	*0.47*
DE	0.39	0.73	0.49	0.41	0.53	0.36	0.25	*0.41*
ET	0.26	0.30	**0.34**	0.30	0.26	0.30	0.26	*0.30*
EN	0.27	0.45	0.33	0.86	0.33	0.37	0.39	*0.37*
FA	0.21	0.19	0.48	0.24	0.55	0.29	0.33	*0.29*
FR	0.28	0.27	0.30	0.23	0.25	0.66	0.23	*0.27*
UR	0.27	0.17	0.32	0.29	0.36	0.25	0.83	*0.29*
Median	*0.27*	*0.30*	***0.34***	*0.30*	*0.36*	*0.36*	*0.33*	

From the tables, we can track how much is harder to predict three emotions than two emotions, with an 18% lower median accuracy across all languages. We also can see how much is harder to predict four emotions than two emotions with 33% lower median precision across all languages.

In Figs. 2, 3 and 4, we visualize median and standard deviation for each trained model, but we have excluded evaluation by the same evaluation language as the model was introduced. This means we excluded here evaluation of the DE model by the DE evaluation dataset to have a clear view of how each model performs for non-trained language—transferability of models.

The number of emotions changes recognition accuracy both within a language and between different language pairs. As we can see, the Chinese language is not transferable to any language, even for such a simple setup as two emotions. We also can see unexpectedly how stable in transferability is Farsy trained model.

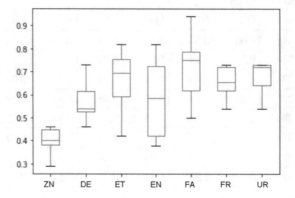

Fig. 2. Evaluation result for model language for two emotions

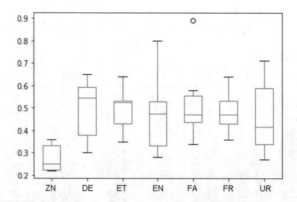

Fig. 3. Evaluation result for model language for three emotions

The accuracy of emotion recognition for different languages falls unevenly with an increase in the number of emotions. Emotions in different languages can be close.

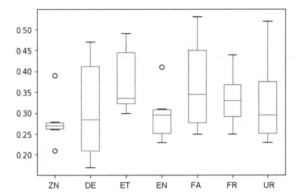

Fig. 4. Evaluation result for model language for four emotions

Therefore, with an increase in their number, some emotions are difficult to distinguish from each other. This is seen when comparing the results presented in Figs. 2, 3 and 4. When recognizing two emotions, the Estonian and Farsi models perform well. However, with an increase in the number of emotions, all models' recognition quality decreases, and the results become more uniform.

The experiment shows that an increase in emotions greatly complicates their recognition (Fig. 5). At what complexity for different languages grows non-uniformly.

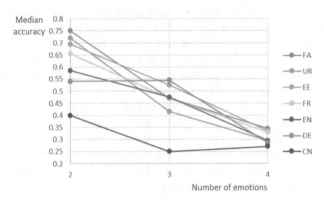

Fig. 5. Reducing the accuracy of emotion recognition with an increase in their number

From Tables 4, 5 and 6, we also track unexpected mirroring behavior (see Table 7). Let us show this on the following two pairs for DE-FA and FR-ZN emotions.

This leads us to interesting conclusions that if emotion transfers good enough from the source language to the target language, it does not mean that emotion can share as good in vice verse direction between languages.

Table 7. Forward and backward recognition accuracy for language pairs

Model language	Evaluation language	Accuracy
FR-ZN pair		
FR	ZN	0.73
ZN	FR	0.29
DE-FA pair		
DE	FA	0.52
FA	DE	0.94

7   Conclusions and Future Work

Current work aimed to evaluate speech emotion recognition transferability between seven languages. From the results of an experiment, we conclude that for transferability purposes. Chinese language and datasets are the hardest to transfer to other languages. English dataset transferability is one of the lowest, hence for a production environment, engineers cannot rely on training models in English for their language. Also, not surprisingly, predicting more emotions is much more complicated, so it is recommended to narrow down datasets to particular practical use rather than train the model on the whole given dataset.

Interestingly enough, if emotion transfers good enough from source language to target language, it does not mean that emotion can share as good in vice verse direction between languages. Hence engineers can not expect the same transferability in the mirror direction. According to the study results, it turned out that not all models are good at even recognizing emotions for their language. This can mean both the imperfection of the model and the weak expression of the emotion itself in this language. On the other hand, it can be seen that most models coped well with the recognition of two emotions. However, with an increase in emotions, recognition quality decreases linearly.

Further research will be expanded to emotion recognition based on speech and text, where text would be represented by multilingual embeddings, which pledge to be even more accurate on a significant amount of emotion and at the same time convenient by transferability of text embeddings. Future research is planned for the maximum known number of languages (15) to refine the mapping of languages (language groups).

References

1. Win, H.P.P., Khine, P.T.T.: Emotion recognition system of noisy speech in real world environment. Int. J. Image Graph. Sig. Process. **12**(2), 1–8 (2020). https://doi.org/10.5815/ijigsp.2020.02.01
2. Kumar, J.A., Balakrishnan, M., Wan Yahaya, W.A.J.: Emotional design in multimedia learning: how emotional intelligence moderates learning outcomes. Int. J. Mod. Educ. Comput. Sci. **8**(5), 54–63 (2016). https://doi.org/10.5815/ijmecs.2016.05.07
3. Dhar, P., Guha, S.: A system to predict emotion from Bengali speech. Int. J. Math. Sci. Comput. **7**(1), 26–35 (2021). https://doi.org/10.5815/ijmsc.2021.01.04

4. Shirani, A., Nilchi, A.R.N.: Speech emotion recognition based on SVM as both feature selector and classifier. Int. J. Image Graph. Sig. Process. **8**(4), 39–45 (2016). https://doi.org/10.5815/ijigsp.2016.04.05

5. Devi, J.S., Yarramalle, S., Prasad Nandyala, S.: Speaker emotion recognition based on speech features and classification techniques. Int. J. Image Graph. Sig. Process. **6**(7), 61–77 (2014). https://doi.org/10.5815/ijigsp.2014.07.08s

6. Abdel-Hamid, L.: Egyptian Arabic speech emotion recognition using prosodic, spectral and wavelet features. Speech Commun. **122**, 19–30 (2020). https://doi.org/10.1016/j.specom.2020.04.005

7. Pajupuu, H.: Estonian emotional speech corpus. Dataset V5. Center of Estonian Language Resources (2012). https://doi.org/10.15155/EKI.000A

8. Kerkeni, L., et al.: French emotional speech database—Oréau. Dataset V2 (2020). https://doi.org/10.5281/zenodo.4405783

9. Burkhardt, F., et al.: A database of German emotional speech. Interspeech (2005). https://doi.org/10.21437/interspeech.2005-446

10. Vrysas, N., et al.: Speech emotion recognition for performance interaction. J. Audio Eng. Soc. **66**(6), 457–467 (2018). https://doi.org/10.17743/jaes.2018.0036

11. Vryzas, N., et al.: Subjective evaluation of a speech emotion recognition interaction framework. In: Proceedings of the Audio Mostly 2018 on Sound in Immersion and Emotion (2018). https://doi.org/10.1145/3243274.3243294

12. Mohamad Nezami, O., Jamshid Lou, P., Karami, M.: ShEMO: a large-scale validated database for Persian speech emotion detection. Lang. Resour. Eval. **53**(1), 1–16 (2018). https://doi.org/10.1007/s10579-018-9427-x

13. Latif, S., et al.: Cross lingual speech emotion recognition: Urdu vs. Western languages. In: 2018 International Conference on Frontiers of Information Technology (FIT) (2018). https://doi.org/10.1109/fit.2018.00023

14. Roberts, F., Margutti, P., Takano, S.: Judgments concerning the valence of inter-turn silence across speakers of American English, Italian, and Japanese. Discourse Process. **48**(5), 331–354 (2011). https://doi.org/10.1080/0163853x.2011.558002

15. Neumann, M., Thang Vu, N.: Cross-lingual and multilingual speech emotion recognition on English and French. In: 2018 IEEE International Conference on Acoustics, Speech and Signal Processing (ICASSP) (2018). https://doi.org/10.1109/icassp.2018.8462162

16. Li, X., Akagi, M.: Improving multilingual speech emotion recognition by combining acoustic features in a three-layer model. Speech Commun. **110**, 1–12 (2019). https://doi.org/10.1016/j.specom.2019.04.004

17. Zehra, W., Javed, A.R., Jalil, Z., Khan, H.U., Gadekallu, T.R.: Cross corpus multi-lingual speech emotion recognition using ensemble learning. Complex Intell. Syst. **7**(4), 1845–1854 (2021). https://doi.org/10.1007/s40747-020-00250-4

18. Heracleous, P., Yoneyama, A.: A comprehensive study on bilingual and multilingual speech emotion recognition using a two-pass classification scheme. PLoS ONE **14**(8), e0220386 (2019). https://doi.org/10.1371/journal.pone.0220386

19. Sagha, H., et al.: Enhancing multilingual recognition of emotion in speech by language identification. Interspeech (2016). https://doi.org/10.21437/interspeech.2016-333

20. Scotti, V., Galati, F., Sbattella, L., Tedesco, R.: Combining deep and unsupervised features for multilingual speech emotion recognition. In: Del Bimbo, A., et al. (eds.) ICPR 2021. LNCS, vol. 12662, pp. 114–128. Springer, Cham (2021). https://doi.org/10.1007/978-3-030-68790-8_10

21. Iosifova, O., et al.: Techniques comparison for natural language processing. In: 2nd International Workshop on Modern Machine Learning Technologies and Data Science (MoMLeT&DS), vol. I(2631), pp. 57–67 (2020)

22. Iosifova, O., et al.: Analysis of automatic speech recognition methods. In: Workshop on Cybersecurity Providing in Information and Telecommunication Systems (CPITS), vol. 2923, pp. 252–257 (2021)

23. Iosifov, I., Iosifova, O., Sokolov, V.: Sentence segmentation from unformatted text using language modeling and sequence labeling approaches. In: 2020 IEEE International Conference on Problems of Infocommunications. Science and Technology (PICST), pp. 335–337 (2020). https://doi.org/10.1109/picst51311.2020.9468084

24. Romanovskyi, O., Iosifov, I., Iosifova, O., Sokolov, V., Kipchuk, F., Sukaylo, I.: Automated pipeline for training dataset creation from unlabeled audios for automatic speech recognition. In: Hu, Z., Petoukhov, S., Dychka, I., He, M. (eds.) ICCSEA 2021. LNDECT, vol. 83, pp. 25–36. Springer, Cham (2021). https://doi.org/10.1007/978-3-030-80472-5_3

25. Lech, M., et al.: Real-time speech emotion recognition using a pre-trained image classification network: effects of bandwidth reduction and companding. Frontiers Comput. Sci. 2 (2020). https://doi.org/10.3389/fcomp.2020.00014

26. ISO 639-6:2009. Codes for the representation of names of languages. Part 6. Alpha-4 code for comprehensive coverage of language variants. https://www.iso.org/standard/43380.html. Accessed 20 Nov 2021

27. Zhou, K., et al.: Seen and unseen emotional style transfer for voice conversion with a new emotional speech dataset. In: 2021 IEEE International Conference on Acoustics, Speech and Signal Processing (2021). https://doi.org/10.1109/icassp39728.2021.9413391

28. Cao, H., et al.: CREMA-D: crowd-sourced emotional multimodal actors dataset. IEEE Trans. Affect. Comput. 5(4), 377–390 (2014). https://doi.org/10.1109/taffc.2014.2336244

29. Busso, C., et al.: IEMOCAP: interactive emotional dyadic motion capture database. Lang. Resour. Eval. 42(4), 335–359 (2008). https://doi.org/10.1007/s10579-008-9076-6

30. Livingstone, S.R., Russo, F.A.: The Ryerson Audio-Visual Database of Emotional Speech and Song (RAVDESS): a dynamic, multimodal set of facial and vocal expressions in North American English. PLoS ONE 13(5), e0196391 (2018). https://doi.org/10.1371/journal.pone.0196391

31. Haq, S., Jackson, P.J.B.: Multimodal emotion recognition. Mach. Audit. 398–423 (2011). https://doi.org/10.4018/978-1-61520-919-4.ch017

32. Pichora-Fuller, M.K., Dupuis, K.: Toronto emotional speech set (TESS). Dataset 59. Scholars Portal Dataverse (2020). https://doi.org/10.5683/SP2/E8H2MF

33. Desplanques, B., Thienpondt, J., Demuynck, K.: ECAPA-TDNN: emphasized channel attention, propagation and aggregation in TDNN based speaker verification. Interspeech (2020). https://doi.org/10.21437/interspeech.2020-2650

34. Kumawat, P., Routray, A.: Applying TDNN architectures for analyzing duration dependencies on speech emotion recognition. Interspeech (2021). https://doi.org/10.21437/interspeech.2021-2168

35. Ravanelli, M., et al.: SpeechBrain: a general-purpose speech toolkit, pp. 1–34 (2020, preprint). https://arxiv.org/abs/2106.04624

Chromosome Feature Extraction and Ideogram-Powered Chromosome Categorization

Oleksii Pysarchuk[1] and Yurii Mironov[2(✉)]

[1] Faculty of Informatics and Computer Science, National Technical University of Ukraine "Igor Sikorsky Kyiv Polytechnic Institute", Kiev 03056, Ukraine
[2] Faculty of Cybersecurity, Computer and Software Engineering, National Aviation University, Kiev 03058, Ukraine
yuriymironov96@gmail.com

Abstract. Chromosomal diseases diagnostics is based on identifying chromosomes and detecting abnormalities in them. It is a sophisticated procedure that is performed manually or partially automated, therefore is prone to human error. Automation of such diagnostics is a complex multistage task that presents many challenges. One of such challenges is chromosome feature extraction and classification. Categorizing chromosomes is necessary for determining a diagnosis. This problem has been previously covered by research papers, but there is no complete solution by now. Moreover, the vast majority of proposals rely on Machine Learning (ML) and Neural Networks (NN). Usage of ML-powered solutions may pose a problem due to difficulty of dataset collection - medical data may be diverse, heterogeneous and hard to collect because of its sensitive nature. Given paper presents a proposal of chromosome feature extraction and categorization without NN. It uses Computer Vision (CV) techniques for chromosome feature extraction: medial axis is extracted from a chromosome to identify its direction, and than chromosome bands (colored segments that make up distinct chromosome pattern) are identified along the axis. Having transformed chromosome image into a discrete piece of data, it is passed to a proposed multiple-criteria decision-making algorithm. This algorithm is designed to categorize chromosomes without learning dataset, instead making use of ideogram (schematic reference chromosome) data. The algorithm is focused on processing a chromosome as a set of segments, where each segment can be recognized by a special predicate. A predefined combination of such predicates should allow recognizing chromosome type.

Keywords: Computer vision · Multiple-criteria decision-making · Edge detection · Image processing · Feature extraction

1 Problem Statement

According to WHO, approximately 295 000 newborns die before reaching the age of 28 days due to congenital anomalies [1, 2]. One of major reasons for this is chromosomal diseases. Other possible outcomes of chromosomal diseases are miscarriage and stillbirth.

© The Author(s), under exclusive license to Springer Nature Switzerland AG 2022
Z. Hu et al. (Eds.): ICCSEEA 2022, LNDECT 134, pp. 427–436, 2022.
https://doi.org/10.1007/978-3-031-04812-8_36

In order to prevent such outcomes and be aware of pathology, a karyotyping is conducted [3]. Kayotyping is a special kind of cytogenetic diagnostics. It implies analyzing chromosomal data from biological materials of potential parents or of a fetus. The process of karyotyping is time and effort consuming. Moreover, it has no production-ready means of automation. However, the correct result of this medical analysis is crucial, so it should not be prone to human error.

In order to proceed, it is necessary to consider what the basic idea behind karyotyping is. Karyotyping is a process of pairing and ordering chromosomes of an organism [4]. In order to categorize a chromosome, a special staining is used to reveal patterns. Each chromosome number, when stained, has a distinct color pattern that allows identifying it (Fig. 1).

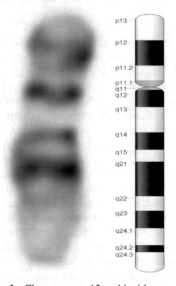

Fig. 1. Chromosome 12 and its ideogram

Staining and identifying chromosomes allows checking patient's chromosome arrangements. Such check may result in one of the following results:

- Patient has ordinary amount of chromosomes: 44 numbered chromosomes and two sex chromosomes (XX for females and XY for males), 46 in total. Each of these chromosomes is recognized and categorized;
- Structural abnormalities - patient has one or more chromosomes that have structural differences: they partially resemble the healthy version of the chromosome, but have some structural differences;
- Numerical abnormalities - patient has all the chromosomes recognized, but the amount of chromosomes is out of order. Most frequent of such cases are trisomies - three instances of a chromosome type, where usually two chromosomes are expected;
- Also patient may have both numerical and structural abnormalities simultaneously;

This problem can be resolved by providing an automated or semi-automated solution. It should include chromosome recognition and chromosomal pathologies detection. Designing such a solution is a complex task that involves image processing, extracting features to a discrete format and analyzing them to come up with a diagnosis proposal, so it should be broken down into separate steps. One of the most important steps is recognition of a single chromosome. It allows categorizing chromosomes, making it possible to detect numerical pathologies (improper count of chromosomes).

Current paper considers the problem of processing a single chromosome. Its goal is to consider a problem of chromosome feature extraction and propose a multiple-criteria decision-making model that allows categorizing a chromosome.

The scope of the paper includes related works overview, researching existing solutions for chromosome image processing and feature extraction. Application domain is overviewed, and a decision-making model is proposed. A prototype powered by this model is implemented, and tested using simple dataset. Model performance is then analyzed and conclusions are drawn.

2 Related Papers

Feature Extraction for the Classification of Human Chromosomes from G-Band Images using Wavelets [5] by **R. Nandakumar** and **K.B. Jayanthi** considers chromosome identification by their statistical features.

The mentioned paper states that the key factor to chromosome identification is its banding pattern. However, it is concluded that there is no way to directly analyze banding pattern for curved chromosomes, so a more indirect approach is proposed - Discrete Wavelet Transform. It allows to dimensionally analyze chromosome image, reducing chromosomes to a set of statistical features.

Therefore, the goal of mentioned paper is not to distinguish internal characteristics of specific chromosome as a physical object, but to come up with features of a chromosome image itself. After that, these features are reduced to discrete values and compared among chromosomes - and a certain correlation is derived. In the conclusion it is stated that algorithm enhancements and testing on larger datasets are needed.

The initial idea of mentioned paper is adapted in the research at hand - the banding pattern may be used to automatically categorize chromosomes. The major obstacle to this approach is curving chromosomes.

Another strong feature of mentioned paper is independance of neural networks and machine learning. Machine learning is a powerful tool both generally and specifically in field of medical solutions, but its major flaw is dependance on training dataset. Gathering dataset for medical neural network may be a major issue because of sensitive nature of data and heterogeneity.

New features for automatic classification of human chromosomes: A feasibility study by **Mehdi Moradi** and **S. Kamaledin Setarehdan** proposes an approach to chromosome feature extraction and classification [6]. It introduces an algorithm that is based on image transformation techniques and extracting density profile of an image.

The first step of a proposed algorithm is determining an image skeleton by means of medial axis transformation. It is needed for several reasons: firstly, it allows determining the length of a chromosome. Secondly, medial axis curve will be utilized to arrange colored segments into a list of discrete data.

The second step of an algorithm includes measurement of image density profile. The density profile will represent features of chromosomal bends, that later will be used for chromosome classification.

The third step is straightening of an image density profile. Chromosomes may have form of an arbitrary curve, so they are often more complex than a straight line. This is the main reason to make use of medial axis and density profile. After straightening, the profile will be reduced to a two-dimensional chart representing a grasycale color intensity. In order to achieve this, each pixel of a medial axis is crossed with a perpendicular line, and density profile across this line is reduced to a mean value. This turns density profile into a sequence of values that allow to estimate chromosome color across its length. Noise is removed using digital lowpass filters [7] and wavelet denoising method [8].

Neural networks are used for chromosome classification. A comparative analysis has been conducted in order to select features for optimal configuration of neural network. This results in an algorithm of chromosomal recognition with high success rate.

The mentioned paper proposes rather efficient algorithm. Neural networks can be a powerful tool for medical image processing, but using them comes with some difficulties. The main challenge is gathering proper dataset. In case of medical data, data gathering is difficult due to sensitive nature of the data and privacy issues. Speaking specifically of chromosome data, it can be very diverse. Different laboratories tend to use different equipment and chemicals during the analysis, so chromosome images may look rather varied.

Having reviewed related papers, the following may be concluded:

- The problem of chromosome classification and feature extraction is known and considered in research papers;
- There is no production-ready solution for chromosome feature extraction using image as an input;
- Existing papers heavily rely on artificial intelligence for chromosome classification, but this approach may be problematic;

The state of existing research allows making use of existing solutions for chromosome image processing, focusing on the chromosome classification.

3   Proposal

3.1   Application Domain Research

In the related papers review, some problems of utilizing neural networks in chromosome recognition have been mentioned. So it is necessary to suggest an alternative way to categorize chromosomes. The suggested alternative for using dataset of processed and tagged data is using chromosome ideograms (Fig. 1) [9].

Ideograms are schematic representations of chromosomes. They are free and publicly accessible from sources like [9]. Also, due to their visual nature, they are easy to process and convert into discrete values. Moreover, creating such a database this is a one-time activity. So, it is supposed that using ideograms may substitute dataset-powered neural networks in terms of chromosome classification.

However, there is a complication: single chromosome type is represented with multiple ideograms due to varying chromosome length and different possible resolution [10]. Moreover, ideograms represent ideal chromosome structure with fixed number of chromosome bands. As seen in [10], chromosome 12 can be represented by several ideograms, and real chromosome 12 can be in between them in terms of quality.

3.2 General Model Proposal

Having considered application domain data, it is necessary to derive a general model for ideogram-powered chromosome recognition.

The approach proposed in [6] will be used for feature extraction. The chromosome image is skeletonized, its density profile is extracted throughout the length of chromosome and turned into a set of discrete values by means of thresholding. The following step is to match length of chromosome and ideogram: since chromosomes have arbitrary length, it is necessary to match the length of chromosome and ideogram. In order not to lose data during length changes, a smaller of two data pieces should be determined, and proportionally expanded to match the larger one.

After obtaining discrete values, a decision making mechanism is used to recognize chromosome. The idea of using a decision-making mechanism is representing an ideogram as a set of values that can be compared to a real chromosome (Fig. 1). Therefore, it is possible to represent these values as a criterion for decision-making mechanism.

A single segment can be represented as follows:

$$SEG_i = \{C_i, L_i, ERR_i\} \tag{1}$$

where C_i is segment color, L_i is segment length and ERR_i is acceptable error threshold – a possible difference between "ideal" and "real" object. This single segment can be adapted to be a predicate and used as a criterion. It can be written down as follows:

$$P_i(C_{xi}, L_{xi}) \rightarrow ((C_{xi} = C_i) \wedge len_i(L_{xi})) \tag{2}$$

where xi index represents segment i of chromosome x, which is served to decision making system as an input. C_{xi} is segment x color and L_{xi} is segment x length. According to [5] the color should be binarized, so the function simply checks that at both ideogram segment and chromosome segments are black or white. The $len_i(L_{xi})$ function, designed to check matching segment length, is described like this:

$$len_i(L_{xi}) \rightarrow L_{xi} > (L_i - ERR_i) \wedge L_{xi} < (L_i - ERR_i) \tag{3}$$

The function (3) basically checks that L_{xi} is within the boundaries of acceptable error. Having described criteria p_i, it is possible to represent a chromosome as follows:

$$Ch_j = \begin{cases} SEG_1 = \{C_1, L_1, ERR_1\} \\ SEG_2 = \{C_2, L_2, ERR_2\} \\ \quad \cdots \\ SEG_n = \{C_n, L_n, ERR_n\} \end{cases} \tag{4}$$

where Ch_j is a chromosome j, consisting of n segments having their own color C_i, length L_i and error boundary ERR_i. For Ch_i, a following decision-making model P_j can be described:

$$P_j(Ch_x) \rightarrow p_1(C_{x1}, L_{x1}) \wedge p_2(C_{x2}, L_{x2}) \wedge \cdots \wedge p_n(C_{xn}, L_{xn}) \tag{5}$$

where:

- Ch_x is an input to the decision-making mechanism – a chromosome to be categorized;
- j is a chromosome type that is compared to input chromosome Ch_x. j has respective predicate P_j that is capable of determining whether chromosome Ch_x is actually of type j;

4   Prototype Overview

4.1   Used Technologies

Python is used as a programming language for implementing a prototype. It has been chosen due to a rich infrastructure of both multi-purpose and specifically scientific libraries [11]. The images and intermediate data are represented by Numpy arrays. This allows efficient data manipulations due to C++ core [12]. OpenCV is used for its wide range of image-processing utilities.

4.2   Prototype Execution Flow

Prototype covers both feature extraction and chromosome recognition. It takes single chromosome image as an input and outputs the number of the ideogram that matches provided chromosome. Figure 2 depicts the entire execution flow of the algorithm.

Feature extraction has been implemented in accordance with [6]. The feature extraction result is shown at Fig. 3. The plot-like visualization depicts color change along the chromosome length, where X is chromosome length and Y is color intensity. Correspondences between chromosome and its extracted features can be seen. The relative minimum points in the chart are situated at the same positions as darker chromosome bands.

Having extracted chromosome features like this, it is possible to express chromosome data as a sequence of segments, where each segment will have its color and length. This processed chromosome data will be compared to ideogram values Ch_j defined in formula (4). Having ideogram x, if predicate for Ch_x described in formula (5) returns "True", this means the chromosome can be categorized as chromosome x.

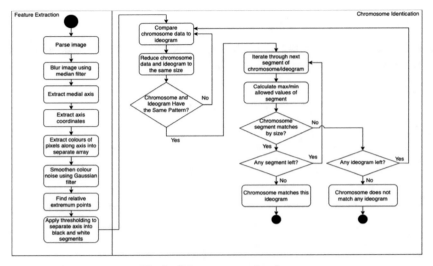

Fig. 2. Prototype execution flow

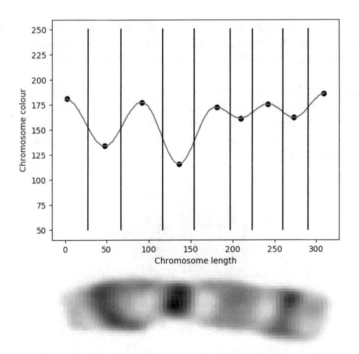

Fig. 3. Chromosome extracted features

4.3 Prototype Execution Flow

As a part of test run, a dataset of 10 chromosomes #20 have been considered, and an ideogram value Ch_{20} has been described (Table 1).

Table 1. Test Ideogram value for chromosome #20

Color	Length	Acceptable error threshold
White	47	27
Black	70	28
White	88	21
Black	38	30
White	89	24
Black	22	47
White	52	58
Black	44	82
White	52	27

The success rate of test run is 40%. 20% of errors are caused by the decision-making algorithm – to be specific, it was impossible to cover all the chromosomes out of dataset using the same Ch_{20} value. This should be addressed in the future iterations of the algorithm.

However, another significant error cause was improper medial axis from chromosome (Fig. 4). The issue is, sometimes chromosomes are too "wide" to have explicit skeleton consisting of a single continuous line along its center. This is an issue that severely affects algorithm efficiency. So, future improvements should include the revision of medial axis extraction proposed by [6].

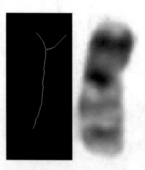

Fig. 4. Improper medial axis extraction from the chromosome

4.4   Model Performance Concerns and Corner Cases

The modal has several flaws that have to be faced in further research. Currently it is supposed that input chromosome will have matching pattern to an ideogram. The arbitrariness of real-life data should be negated by an acceptable error threshold; however, the bigger problem arises. As mentioned in application domain research, chromosomes have arbitrary length, yet ideograms are built around fixed lengths of chromosomes. So, at a current iteration of algorithm, there will still be significant error rate due to chromosomes that have size between two ideograms.

However, the main value of current mechanism is the predicate system and general architecture of decision-making model. The idea is making use of ideograms for a reference to avoid the need for dataset, since ideograms are utilized in real-life chromosome classification [9]. So, it can be possible to gain a deeper understanding of how ideograms are precepted and rearrange how predicates influence the final decision.

For example, it is possible to change iterative segments comparison into a more advanced mechanism. For example, ideograms may be recognized not for strict pattern, but for their peculiar features. So, in order to categorize a chromosome (or at least narrow the selection of options) it is possible to look for such peculiarities, for example mutual arrangement of bends. It is achievable by using proposed predicate logic and ideogram-powered approach, but it has some separate body of work to it and will be considered in the future research.

5   Summary and Conclusion

A problem of automated chromosome recognition and classification has been considered. Current state of research has been conducted and it has been found that the majority of proposals are powered by machine learning and neural networks, and no complete solution is introduced so far. Due to problems related to gathering medical datasets, it has been decided to find a way to implement chromosome recognition without machine learning.

Having researched application domain, an assumption has been made that chromosomes might be recognized by comparing them with ideograms – ideal schematic representation of chromosomes. An existing proposal [6] has been used to extract features from chromosomes, turning them into discrete values.

A predicate-powered decision-making model for chromosome categorization has been introduced. It is based on representing each chromosome segment as a predicate rule, allowing representing a chromosome as a set of such predicates. Its current version is merely a proof-of-concept intended to test the applicability of proposal, but the way it is organized allows further extension and enhancement. The predicates responsible for chromosome fragments can be reused and rearranged, creating more complicated and advanced checks and potentially resulting in more efficient categorization.

Further research should be focused on finding optimal arrangement of chromosome segment predicates. Being potentially powerful tool on their own, they should be used in the way similar to human perception of chromosome and ideogram images: for example, instead of iterative comparison of segments it may be promising to extracting "key segments" from each chromosome – the segments that bear more importance in chromosome visual recognition than others.

References

1. Tiwari, P., Gupta, M.M.: Study of lethal congenital malformations at a tertiary-care referral centre in North India. Cureus **12**(4) (2020). https://doi.org/10.7759/cureus.7502
2. Boyle, B., Addor, M.C., Arriola, L., et al.: Estimating global burden of disease due to congenital anomaly: an analysis of European data. Arch. Dis. Child. Fetal Neonatal Ed. **103**(1), F22–F28 (2018). https://doi.org/10.1136/archdischild-2016-311845
3. Moallem, P., Karimizadeh, A., Yazdchi, M.: Using shape information and dark paths for automatic recognition of touching and overlapping chromosomes in G-band images. MECS, Int. J. Image Graph. Sig. Process. **5**, 22–28 (2013). https://doi.org/10.5815/ijigsp.2013.05.03
4. O'Connor, C.: Karyotyping for chromosomal abnormalities. Nat. Educ. **1**(1), 27 (2008)
5. Nandakumar, R., Jayanthi, K.B.: Feature extraction for the classification of human chromosomes from G-band images using wavelets. Int. J. Eng. Res. Technol. (IJERT) ICEECT **8**(17), 67–72 (2020)
6. Moradi, M., Setarehdan, K.: New features for automatic classification of human chromosomes: a feasibility study. Pattern Recogn. Lett. **27**(1), 19–28 (2006). https://doi.org/10.1016/j.patrec.2005.06.011
7. Oppenheim, A.V., Schafer, R.: Image and Signal Processing and Analysis. Pearson (2015)
8. Aldroubi, A., Unser, M.: Wavelets in Medicine and Biology. CRC Press, Boca Raton (1996)
9. Huret, J.L., Ahmad, M., Arsaban, M., et al.: Atlas of genetics and cytogenetics in oncology and haematology. Nucleic Acids Res. (2013) 41(D1) (2013). https://doi.org/10.1093/nar/gks1082
10. O'Connor, C.: Chromosome mapping: idiograms. Nat. Educ. **1**(1), 107 (2008)
11. Nagpal, A., Gabrani, G.: Python for data analytics, scientific and technical applications. In: 2019 Amity International Conference on Artificial Intelligence (AICAI), pp. 140–145 (2019). https://doi.org/10.1109/AICAI.2019.8701341
12. Harris, C.R., Millman, K.J., van der Walt, S.J., et al.: Array programming with NumPy. Nature **585**(7825), 357–362 (2020). https://doi.org/10.1038/s41586-020-2649-2

Case Teaching of E-commerce Fresh Product Demand Forecasting in Logistics System Forecasting

Mengya Zhang, Zelong Zhou, Zhiping Liu, and Jinshan Dai[✉]

School of Transportation and Logistics Engineering,
Wuhan University of Technology, Wuhan 430000, China
jinshan.dai@whut.edu.cn

Abstract. In the field of Internet e-commerce, logistics system prediction is an important reference for enterprises to make appropriate decisions and achieve cost reduction and efficiency increase. This method brings case teaching method into the study of logistics system prediction method, which allows for a combination of theory and practice and enables students master it more quickly. Based on the demand of e-commerce fresh product, the appropriate logistics system forecasting model is selected. With the improvement of the basic model, a demand prediction model based on amplitude compression suitable for fresh e-commerce is proposed. The pedagogical method enables students to better understand the prediction model of logistics system based on the actual situation of deformation, and further improves the comprehensive research ability of students.

Keywords: Case teaching · Logistics system prediction · Prediction model

1 The Necessity of Case Teaching

Nowadays, with the rapid development of the logistics industry, the shortage of logistics talents has become an obstacle to the development of Chinese logistics industry. Forming a logistics talents training system which is in accord with Chinese national conditions as soon as possible can effectively promote the development of Chinese logistics industry. Actively exploring and developing the mode of case teaching method is of great significance to the development of teaching work. The exploration and development are not only beneficial to the increase of the interaction between teachers and students in teaching, enhancing students' consciousness of learning, the ability to analyze and solve problems, but also help teachers improve personal qualities, teaching quality and level. In addition, the application of case teaching is conducive to seeking a subject teaching mode for logistics majors, which assists teachers cultivate more logistics talents with good discipline literacy [1].

In the education of logistics discipline, the cultivation of students' practical skills and management ability should be emphasized. The basic research purpose of logistics case teaching is to make students master the theoretical knowledge of logistics subject,

Z. Hu et al. (Eds.): ICCSEEA 2022, LNDECT 134, pp. 437–442, 2022.
https://doi.org/10.1007/978-3-031-04812-8_37

improve their practical ability, mobilize all their knowledge and potential to solve problems, and train students' comprehensive ability to analyze and solve problems through new teaching method with case teaching [2–4]. This will undoubtedly be the crucial to improve the teaching quality of logistics discipline and train excellent logistics talents in China.

2   Methods of Logistics System Prediction

Logistics system prediction is to predict the status and trend of the logistic in a certain period using scientific methods, based on the investigation and study of the flow direction, flow, capital turnover and the supply and demand law of logistics. Whether it is the planning of logistics system, or the operation management and control of logistics system, it is necessary to estimate the demand for products and services, that is, to estimate the demand for logistics. These estimates are made using predictive methods. Through forecasting, necessary information on logistics demand could be obtained to provide reliable basis for planning, management and decision-making [5].

In logistics system forecast, historical data and market information are used to scientifically analyze, estimate and infer the future logistics demand. Commonly used demand prediction models include exponential smoothing method, regression analysis method and grey system prediction:

1) Exponential smoothing method is a very effective short-term prediction method. It is simple and easy to use, and can be used continuously with a small amount of data [6].
2) Regression analysis starts from the causal relationship between variables and finds out the internal rules between relevant variables through a large number of statistical analysis of data, so as to approximately determine the functional relationship between variables and help people to infer and predict the future value range from the past and present values of variables [7].
3) Grey system prediction is to establish a Grey Model (GM), which connects the past and the future, according to the known or uncertain information of the Grey system in the past and now. The most distinctive step in the modeling is the GM Model established for time series. Grey prediction uses generated numbers to build the different equation of the Grey Model, and gets the time response function of the different equation with integral [8].

3   E-commerce Fresh Product Demand Forecasting Case Teaching

In the learning of logistics system prediction methods, several commonly used prediction models are listed in the textbook. However, in practice, problems are often much more complex, so it is necessary to improve the basic model to be adapted to the real needs. Such improvement is also the innovation ability that students need to further improve in their research and learning. Taking the demand forecast of e-commerce fresh product as an example, students are guided to mater the deformation and practical application of prediction methods through case analysis, prediction method selection and prediction model improvement.

3.1 Case Analysis

Z company in Wuhan owns more than 40 offline stores. The retail orders of the company on the Internet are fresh products. In order to achieve lower damage and maximize customer satisfaction, the sales figures (as shown in Table 1) for the last five days are used to predict the sale figures in the future, which provides reference for next order.

Table 1. Sales volume of 4 fresh products in a certain area for 5 consecutive days

Category	Fresh beef (kg)	Shrimps (kg)	Needle mushroom (kg)	Crabs (kg)
Day1	3934	2663	3011	1686
Day2	4171	2508	3043	1650
Day3	4056	2530	3166	1695
Day4	4039	2302	2802	1455
Day5	3981	2417	2973	1380

3.2 Selection of Prediction Methods

Considering the specificity of prediction methods, the mainstream prediction models for the sales of fresh products in e-commerce include time series analysis model, multiple regression model and gray prediction model, etc. Therefore, it is necessary to guide students to choose suitable methods. Through group discussion among students, the following conclusions are drawn: The time series analysis model only uses time as the analysis factor, without considering the influence of market changes. Multiple regression model requires a large amount of data and is susceptible to external factors which leads to distortion of results [9, 10]. The grey system model has the limitation of numerical value, but it still has high accuracy [11]. Grey prediction model is the core of grey system theory. It mainly aims at the grey uncertainty prediction problem which exists in the real world, and uses a small amount of valid data and grey uncertainty data to reveal the future development trend of the system through sequence accumulation generation [12]. As grey prediction model has no strict requirements and restrictions on the change trend of data, it is suitable for the fresh food e-commerce industry with unstable demand and less historical data.

3.3 Improvement of Prediction Model

The grey prediction model has high accuracy for small sample oscillation sequence, but the final reduction formula of the traditional grey prediction model is exponential function expression, which has strict monotonicity. The simulated data predicted by this function can't well conform to the oscillation characteristics of the original sequence. Therefore, in this case teaching, it's necessary to improve the grey prediction model. To improve the prediction model, the amplitude compression method could be introduced into smoothness operator to smooth the random oscillation sequence.

1) Build smoothing operator

There are two random oscillation sequences supposed:

$$X^{(0)} = \left\{ x^{(0)}(k) \right\}_1^n \tag{1}$$

$$X^{(0)}D = \left\{ x^{(0)}(k)\mathrm{d} \right\}_1^{n-1} \tag{2}$$

$$X^{(0)}(k)d = \frac{(x^{(0)}(k) + T) + (x^{(0)}(k + 1) + T)}{4} \tag{3}$$

In formulas (1), (2) and (3), $k = 1, 2, ..., n - 1$; T is the amplitude of $X^{(0)}$; D is called the first-order smoothing operator of the sequence $X^{(0)}$; $X^{(0)}D$ is the first-order smoothing sequence of the random oscillation sequence $X^{(0)}$. By compressing the amplitude of the sequence, the random oscillation sequence is transformed into a new sequence with good smoothness.

2) Amplitude compression improvement based on DGM (1, 1) model

In this improved method, DGM (1, 1) model is established based on the first-order smooth sequence. By means of the sum formula of geometric sequence and the inverse process of smoothness operation, the stochastic oscillation sequence model is derived and established by the combination of exponential function and odd and even mutation function.

Assume the first-order smoothing sequence of $X^{(0)}$ as:

$$Y^{(0)} = (y^{(0)}(1), y^{(0)}(2), ..., y^{(0)}(n)) \tag{4}$$

Establish the DGM (1,1) model according to $Y^{(0)}$, and obtain:

$$\hat{y}^{(1)}(k + 1) = \beta_1 \hat{y}^{(1)}(k) + \beta_2 \tag{5}$$

Where

$$\hat{\beta} = (\beta_1, \beta_2)^T = (B^T B)^{-1} B^T Y \tag{6}$$

$$Y = \begin{bmatrix} y^{(1)}(2) \\ y^{(1)}(3) \\ \vdots \\ y^{(1)}(n) \end{bmatrix}, B = \begin{bmatrix} y^{(1)}(1) & 1 \\ y^{(1)}(2) & 1 \\ \vdots & 1 \\ y^{(1)}(n - 1) & 1 \end{bmatrix} \tag{7}$$

When $\hat{y}^{(1)}(1) = y^{(0)}(1)$:

$$\hat{y}^{(1)}(k + 1) = \beta_1^k y^{(0)}(1) + \frac{1 - \beta_1^k}{1 - \beta_1} \beta_2; \quad k = 1, 2, ..., n - 1 \tag{8}$$

The reducing value is:

$$\hat{y}^{(0)}(k + 1) = \alpha^1 \hat{y}^{(1)}(k + 1) = \hat{y}^{(1)}(k + 1) - \hat{y}^{(1)}(k) \tag{9}$$

Formula (9) is the DGM(1, 1) model with first-order smooth sequence $Y^{(0)}$. According to the smoothness operation:

$$\hat{y}^{(0)}(k+1) = \beta_1^{k-1} y^0(1)(\beta_1 - 1) + \beta_1^{k-1}\beta_2 \tag{10}$$

Because the parity of k is different, the expression is not the same. The expression under different odd-even conditions can be unified into a standard expression, which is composed of exponential function and odd-even mutation function. Therefore, the model can better simulate the random oscillation characteristics of the original sequence, and effectively solve the problem of low prediction accuracy of the grey model for the random oscillation sequence.

3.4 Case Result Analysis

According to the smoothing operation, the sales figures of 4 kind of fresh in the last five days are modeled as the original sequence. The simulated prediction sequence is obtained, and the average fitting relative error is calculated, as shown in Table 2.

Table 2. MATLAB calculation results

Simulation results and errors of grey model based on amplitude compression		
Modeling sequence	Simulation of sequence	Mean fitting relative error
$X_1 = (3934, 4171, 4056, 4039, 3981)$	$\widehat{X}_1 = (3934, 4117, 4046, 4066, 3943)$	0.62%
$X_2 = (2663, 2508, 2530, 2302, 2417)$	$\widehat{X}_2 = (2663, 2508, 2515, 2345, 2358)$	1.63%
$X_3 = (3011, 3043, 3166, 2802, 2973)$	$\widehat{X}_3 = (3011, 3043, 3159, 2822, 2946)$	0.617%
$X_4 = (3686, 3650, 3695, 3455, 3380)$	$\widehat{X}_4 = (3686, 3650, 3716, 3391, 3466)$	1.65%

As can be seen from the results in Table 2, the relative error between the original sequence and the simulated sequence after amplitude compression and smoothing is small. The simulated sequence after amplitude compression and smoothing can better reflect the features of the original sequence. Because the above four kinds of fresh demand changes are nonlinear and there's a certain volatility, sales forecast and the historical data is strong correlated. The amplitude compression method effectively covers the shortage that the predicted result of traditional grey model is monotonic. Therefore, the improved method is more suitable for the fresh e-commerce with large demand fluctuations and large dependence on recent data [13].

4 Conclusion

The main purpose of case teaching is to cultivate students' active learning ability, which is different from the traditional learning method. It puts emphasis on cultivating students' active learning spirit and inquiry ability. In this paper, typical cases are used to bring students into the real e-commerce scene and stimulate their interest. Through the analysis of actual product demand, students are guided to choose the appropriate prediction model and carry out independent inquiry learning, so as to improve their ability to analyze and solve practical problems. It not only cultivates students' independent and active learning ability, but also enables students to understand and master the applicability of the prediction method deeply. At the same time, based on further discussion and research, a grey prediction model based on amplitude compression is proposed, which fully demonstrates and excavates students' innovation. The validity of the model is proved through analyzing the result of the case.

References

1. Ding, Y., Huang, L.: Case teaching research of quantitative analysis method in graduate logistics major. New Finan. Econ. **1**, 114–115 (2019). (in Chinese)
2. Woschank, M., Pacher, C.: Teaching and learning methods in the context of industrial logistics engineering education. Procedia Manuf. **51**, 1709–1716 (2020)
3. Woschank, M., Pacher, C.: A holistic didactical approach for industrial logistics engineering education in the LOGILAB at the Montanuniversitaet Leoben. Procedia Manuf. 51, 1814–1818 (2020)
4. Senna, E.T.P., dos Santos Senna, L.A., da Silva, R.M.: The challenge of teaching business logistics to international students. IFAC Proc. **46**(24), 463–470 (2013)
5. Zhang, Q., Yong, G., Zhang, M.: Logistics System Engineering – Theory, Method and Case Study (3rd Edition). Publishing House of Electronics Industry, Beijing (2021). (in Chinese)
6. Mgandu, F.A., Mkandawile, M., Rashid, M.: Trend analysis and forecasting of water level in Mtera dam using exponential smoothing. Int. J. Math. Sci. Comput. (IJMSC) **6**(4), 26–34 (2020)
7. Sakpere, A.B., Oluwadebi, A.G., Ajilore, O.H., Malaka, L.E.: The impact of COVID-19 on the academic performance of students: a psychosocial study using association and regression model. Int. J. Educ. Manage. Eng. (IJEME) **11**(5), 32–45 (2021)
8. Padmaja, M., Haritha, D.: Software effort estimation using grey relational analysis. Int. J. Inf. Technol. Comput. Sci. (IJITCS) **9**(5), 52–60 (2017)
9. Chen, X., Tu, R., Li, M., Yang, X.: prediction models of air outlet states of desiccant wheels using multiple regression and artificial neural network methods based on criterion numbers. Appl. Therm. Eng. **204**, 117940 (2021)
10. Luu, Q.-H., Lau, M.F., Ng, S.P.H., Chen, T.Y.: Testing multiple linear regression systems with metamorphic testing. J. Syst. Softw. **182**, 111062 (2021)
11. Padmaja, M., Haritha, D.: software effort estimation using grey relational analysis. Int. J. Intell. Syst. Appl. **7**(2), 27–33 (2015)
12. Yichung, H.: Constructing grey prediction models using grey relational analysis and neural networks for magnesium material demand forecasting. Appl. Soft Comput. **93**, 106398 (2020)
13. Zeng, B., Liu, S.: Prediction model of random oscillation sequence based on amplitude compression. Syst. Eng. Theor. Pract. **32**(11), 2493–2497 (2012)

Training and Evaluation System of Intelligent Supply Chain Talent Based on Digital Teaching and Collaborative Innovation

Yong Gu, Di Liu(✉), and Zhiping Liu

School of Transportation and Logistics Engineering,
Wuhan University of Technology, Wuhan 430070, China
wysqh1208@163.com

Abstract. The intelligent supply chain (ISC) is a new stage of supply chain due to the integration with Internet and Internet of things, which creates new requirements for talent. We analyze the talent demand of the ISC and the main problems existing in the current talent training mode. Moreover, we explore reasonable teaching contents and methods, and an ISC talent training mode is built based on digital teaching and collaborative innovation. Furthermore, we establish a talent evaluation system of the ISC based on fuzzy analytic hierarchy process (FAHP), which contains learning ability evaluation and innovation ability evaluation. The training and evaluation system of the ISC proposed in this paper is a valuable reference for the formulation of talent training plans for supply chain specialty, and is helpful for teachers' information-based teaching ability improving, teaching effect promotion, and industry-university collaborative education development.

Keywords: ISC · Talent training mode · Talent evaluation system · Digital teaching · Collaborative innovation · FAHP

1 Introduction

In 2017, the General Office of Chinese State Council issued the Guidance on Actively Promoting the Innovation and Application of the Supply Chain. The document points out that the supply chain is an organizational form which is oriented by customer demand. It achieves efficient collaboration in the whole process of product design, procurement, production, sales and service by integrating resources, and aims to improve quality and efficiency [1]. The guidance also proposes to speed up the training of multi-level supply chain talent.

With the continuous development of technology, the supply chain has developed into a new stage of ISC which is deeply integrated with Internet and Internet of things, and the demand for talents has changed greatly compared with the traditional supply chain. All involved colleges and universities should improve the curriculum system and teaching contents according to the new requirements of the ISC, and cultivate ISC talents to meet the needs of enterprises.

Z. Hu et al. (Eds.): ICCSEEA 2022, LNDECT 134, pp. 443–451, 2022.
https://doi.org/10.1007/978-3-031-04812-8_38

2 Talent Demand of ISC

Compared with the traditional supply chain, the demand structure of talents of the ISC has changed obviously: the demand for basic posts has decreased, while the demand for technical and compound talents has increased. The changes of talent demand in the supply chain can be summarized from the following three aspects: knowledge demand, ability demand and quality demand.

2.1 Knowledge Demand

The application of new technologies such as Internet of things, big data, cloud computing, artificial intelligence makes the ISC create higher requirements for the knowledge demand of talents. Under the background of new technology, ISC talents should not only master the traditional supply chain management knowledge, but also learn the latest cutting-edge technology in the supply chain field. They must establish a compound knowledge system of ISC to match the evolving technological environment. The new demands for knowledge in the ISC are shown in Fig. 1.

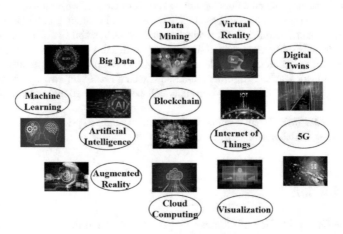

Fig. 1. New demands for knowledge in ISC

2.2 Ability Demand

The ISC talents should master the basic skills in the traditional supply chain, such as order, planning, procurement, logistics, manufacturing, etc. Moreover, they should skillfully use new technologies such as big data, cloud computing, automation and other new technologies. In addition, as the business scene becomes more complex, the ISC talents have to not only face offline problems, but also solve online problems simultaneously. Generally speaking, the ability demand of ISC for talents is reflected in the following three aspects: data operational ability, intelligent optimization analysis ability and intelligent collaborative decision-making ability [2].

2.3 Quality Demand

In addition to the demand for knowledge and ability, the demand for the quality of talents in the ISC is also increasing. The quality demand of ISC talents is mainly reflected in the following aspects: strategic thinking, international professional vision, understanding the sustainable development and social responsibility of supply chain, being good at soft skills such as negotiation and collaboration, innovative consciousness and innovative spirit, etc.

3 Main Problems of Current Talent Training Mode

At present, the training mode of ISC talent is not perfect, and the talents trained by colleges and universities are far from meeting the needs of the market, which causes the contradiction between supply and demand [3]. There are four main problems in the talent training of supply chain.

3.1 Unsuitable Curriculum System

With the wide application of big data, cloud computing and artificial intelligence in supply chain, the demand of enterprises for talents is changing from traditional supply chain management talents to compound talents. However, most of the current supply chain courses in universities lay emphasis on the teaching of classical theories while neglecting the teaching of cutting-edge technology of supply chain. As a result, the classroom teaching deviates from the actual needs of enterprises.

3.2 Rigid Teaching Method

Under the existing teaching mode, the teaching of most courses is mainly taught by teachers, supplemented by students taking notes [4]. The students can only passively accept the knowledge instilled by teachers with low classroom participation and poor classroom interaction, which cannot stimulate students' enthusiasm for learning.

3.3 Single Evaluation Method

When evaluating the students' learning effect, the most common method is the quantitative score, which is formed by the weighting of final exam results and usual scores [5]. This evaluation method emphasizes the memory of theoretical knowledge and lacks the training and evaluation of students' comprehensive literacy. It is difficult to really play the role of supervising students' learning.

3.4 Disjointed Practical Teaching with Reality

Although most universities have set up practical courses, the teaching content of these practical courses generally lags behind the development of the supply chain, and students cannot really apply what they have learned. Moreover, all the practical teaching exists in isolation and do not link up well with each other.

4   Innovation of Talent Training Mode for ISC

Digital teaching is a teaching method that uses computer technology, network and communication technology to digitally process all kinds of information such as school teaching, scientific research, management, service and so on [6]. Collaborative innovation is a talent training mode which takes knowledge increment as its core value. Its main body includes schools and enterprises. Moreover, collaborative innovation integrates the educational resources such as government, industry associations or institutions at the same time. The role of collaborative innovation in talent training is mainly embodied in the following four aspects: constructing the knowledge structure of innovation and entrepreneurship; providing the real scene of innovation and entrepreneurship; building the first-class practical innovation platform to enhance the practical skills of students; organizing projects to improve the innovative ability of students [7].

At present, the conditions of digital teaching and collaborative innovation in colleges and universities in China have been basically formed. However, they don't play their due role in the actual teaching process. This not only causes the waste of teaching resources, but also results in the talents cannot fully match the needs of enterprises, which leads to the contradiction between supply and demand of talents. Therefore, the colleges and universities should change the traditional talent training mode of supply chain, and make full use of digital teaching and collaborative innovation to rebuild the talent training mode of ISC, just as shown in Fig. 2.

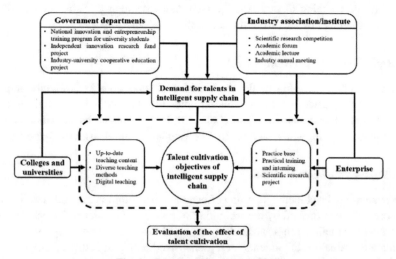

Fig. 2. Talent training mode of ISC

4.1   Updating Teaching Content

Social demand is the driving force to guide the update of teaching content. In the era of ISC, colleges and universities should combine the training of supply chain talents with professional characteristics and the latest achievements of industry development. For example, they can jointly develop course contents with enterprises and promote the achievements of scientific research into classrooms and teaching materials, which can make the teaching contents close to life, production and forefront [8]. In the curriculum system, the colleges and universities should streamline the original teaching contents and increase the contents of ISC, such as Internet of things, big data and cloud computing [9].

4.2 Adding Teaching Method

Digital teaching gives a new way to enrich the teaching method. It is necessary to make full use of the existing digital teaching resources in universities to carry out online and offline integrated teaching, which can fully mobilize the enthusiasm of students to participate in learning. When carrying out online teaching, teachers should make full use of various online teaching platforms to establish an interactive and innovative mode of education and teaching [10]. Online teaching can give full play to the open advantages of the network, and provide students with a diversified platform for learning and innovation. With its help, students can participate in cooperative research across regions and majors, which breaks the time and space restrictions in traditional learning [11]. When teaching offline, teachers can also show students the cutting-edge technologies of the ISC through digital multimedia, VR, AR and other technologies, so that students can get an immersive understanding of industry trends and the latest developments.

4.3 Training Practical Innovation Ability

Collaborative innovation mode is the support to train students' practical innovation ability. Universities and enterprises should fully share their superior resources and actively carry out cooperative teaching. With the help of technology, platform and teaching contents, they can build the collaborative innovation practice platform of ISC which can promote the practical and innovative ability of students [12]. Universities set training objectives according to the demand of the ISC talents, and utilize digital teaching conditions to carry out teaching. Meanwhile, enterprises cooperate with universities to provide students with professional practical conditions [13]. For example, they can cooperate to establish practice bases to facilitate students to carry out practical training and interning. In addition, government departments and industry associations or institutions can also provide support measures for talent training, such as innovative and entrepreneurial projects, scientific research competitions, academic forums, academic lectures, and so on [14].

5   Talent Evaluation System for ISC Based on FAHP

In order to know whether the ISC talents meet the market demand, we need to evaluate the effect of talent training. To address the problems existing in the current curriculum evaluation, this paper constructs a talent evaluation index system from the perspectives of digital teaching and collaborative innovation, and uses FAHP to evaluate talents in all directions to promote the overall growth of students.

Analytic Hierarchy Process (AHP) is a comprehensive evaluation method proposed by T.L. Saaty, an American operations planner, to solve multi-objective decision-making problems, which can effectively analyze the non-sequential relationship between the levels of the objective criteria system. FAHP is proposed based on AHP and fuzzy theory. It overcomes the problem of determining the weight in AHP, so its decision-making result is more reliable. The basic steps of using FAHP to evaluate ISC talents are as follows:

5.1   Construction of Evaluation Index System

By consulting professors with rich experience, investigating involved universities and reading literature, we finally determine the three-level evaluation index system. The first-level index includes learning ability and innovation ability. The learning ability index is subdivided into three second-level indexes: online learning, classroom learning and after-class learning, and the innovation ability is subdivided into three second-level indexes: scientific research competition results, innovative entrepreneurial achievements and simulation experimental results [15]. Each second-level index contains several third-level indexes, as shown in Table 1.

Table 1.   Talent evaluation index system of ISC

First-level index	Second-level index	Third-level index
Learning ability	Online learning	Time to watch the video
		Online practice
		Discussion forum participation
	Classroom learning	Class attendance
		Class participation
		Language expression ability
	After-class learning	Job completion
		Online test scores
		Final test scores
Innovation ability	Scientific research competition achievements	Number of participations
		Competition results
	Innovative entrepreneurial achievements	Number of participations
		Collaboration ability
	Simulation experimental achievements	Number of participations
		Software application ability

5.2 Determination of Index Weight

5.2.1 Establishment of Fuzzy Complementary Judgment Matrix

First of all, according to the scaling method in Table 2, several experts $k = (1, 2, \ldots, m)$ compare the importance of evaluation indexes at the same level to generate m fuzzy complementary judgment matrices [16].

$$A = \begin{bmatrix} a_{11} & a_{12} & \cdots & a_{1n} \\ a_{21} & a_{22} & \cdots & a_{2n} \\ \vdots & \vdots & \ddots & \vdots \\ a_{n1} & a_{n2} & \cdots & a_{nn} \end{bmatrix} \tag{1}$$

The set of m fuzzy complementary judgment matrices is marked as $A^{(k)}$:

$$A^{(k)} = \left(a_{ij}^{(k)} \right)_{m \times n} ((1, 2, \ldots, m)) \tag{2}$$

Table 2. Scaling method and interpretation

Scale	Definition	Description
0.5	Equally important	The two elements are equally important
0.6	Slightly more important	One element is slightly more important than the other
0.7	Obviously more important	One element is obviously more important than the other
0.8	Much more important	One element is much more important than the other
0.9	Extremely more important	One element is extremely more important than the other
0.1, 0.2, 0.3, 0.4	Anti-comparison	If element a_i is compared with element a_j to get r_{ij}, then element a_j is compared with element a_i to get $r_{ji} = 1 - r_{ij}$

5.2.2 Calculation of Index Weight

Firstly, the weight formula of fuzzy complementary judgment matrix is defined as follows:

$$\omega_i = \frac{\sum_{j=1}^{n} a_{ij} + \frac{n}{2} - 1}{n(n-1)}, (i = 1, 2, \cdots, n) \tag{3}$$

According to formula (3), the weight vectors of m fuzzy complementary judgment matrices are calculated respectively, and the set of m weight vectors is marked as $\omega^{(k)}$:

$$\omega^{(k)} = \left(\omega_1^{(k)}, \omega_2^{(k)}, \cdots \omega_n^{(k)} \right) (k = 1, 2, \cdots, m) \tag{4}$$

5.2.3 Consistency Test of Fuzzy Complementary Judgment Matrix

Finally, it is necessary to check the consistency of the weight vector. If it cannot pass the consistency test, the current weight is unreliable, and the fuzzy complementary judgment matrix needs to be adjusted. The characteristic matrix W of fuzzy complementary judgment matrix is calculated according to formula (5):

$$W_{ij} = \frac{\omega_i}{\omega_i + \omega_j} (\forall i, j = i = 1, 2, \cdots, n) \tag{5}$$

The set of m characteristic matrices is marked as $W^{(k)}$:

$$W^{(k)} = \left(W_{ij}^{(k)} \right)_{n \times n} (k = 1, 2, \cdots, m) \tag{6}$$

The following is a consistency test of the fuzzy complementary judgment matrix from two aspects:

(1) Check the satisfactory consistency of m fuzzy complementary judgment matrices:

$$I\left(A^{(k)}, W^{(k)} \right) = \frac{1}{n^2} \sum_{i=1}^{n} \sum_{j=1}^{n} \left| a_{ij}^{(k)} + b_{ij}^{(k)} - 1 \right| \leq \alpha (k = 1, 2, \cdots, m) \tag{7}$$

(2) Check the satisfactory compatibility of fuzzy complementary judgment matrices:

$$I\left(A^{(k)}, A^{(l)} \right) = \frac{1}{n^2} \sum_{i=1}^{n} \sum_{j=1}^{n} \left| a_{ij}^{(k)} + b_{ij}^{(l)} - 1 \right| \leq \alpha (k \neq l; k, l = 1, 2, \cdots, m) \tag{8}$$

When conditions (1) and (2) are satisfied, the mean value of weight vectors are taken as the final weight vector:

$$W^* = W_1, W_2, \cdots W_n \tag{9}$$

In the formula (9), $W_i = \frac{1}{m} \sum_{k=1}^{m} W_i^{(k)} (i = 1, 2, \cdots, n)$.

5.3 Application of Evaluation System

According to the ISC talents evaluation index system constructed above, the teacher scores the students in all directions, and calculates the comprehensive scores of the students according to the weight of the index. The score reflects the students' ability to absorb, utilize and transform the knowledge they have learned, and reflects the real learning effect of the students. Meanwhile, the score reflects the degree to which students meet the needs of enterprises, which can be used as a basis for enterprises to employ personnel and reduce the employment costs of enterprises.

6 Conclusion

In the era of ISC, it is a new challenge for colleges and universities to train ISC talents whose knowledge, ability and quality meet the needs of enterprises. Based on the existing conditions of digital teaching and collaborative innovation in universities, we propose the talent training mode and construct talent evaluation system of ISC, which is a reference for training plan formulating of ISC talent in universities.

References

1. Guiding opinions of the general office of the state council on actively promoting supply chain innovation and application [EB/OL]. http://www.gov.cn/zhengce/content/2017-10/13/content_5231524.htm. Accessed 13 Oct 2017–24 Nov 2021
2. Xiaoyan, W.: The reference and thinking of supply chain talent training in American universities. Appl.-Oriented High. Educ. Res. **3**(04), 90–95 (2018). (in Chinese)
3. Fengjiao, W.: Discussion on training mode of supply chain management talents under enterprise demand guidance. Logistics Technol. **39**(06), 134–138 (2020). (in Chinese)
4. Jiang, Q., Jin, T., Song, W., et al.: Research on cultivating undergraduates in the computer science based on students. Int. J. Eng. Manufact. **10**(6), 32–39 (2020)
5. Linli, T.: Research on teaching reform of supply chain management course based on TQM. Technol. Ind. Across Straits **06**, 19–22 (2019). (in Chinese)
6. Ni, J., Liu, L.: Research on application and development of digital teaching environment in colleges and universities. J. Liaoning Educ. Adm. Inst. **38**(2), 102–104 (2014). (in Chinese)
7. Huang Qiong, Y., Jianghua, T.T., et al.: Exploration on the training mode of college students' practical innovation ability under collaborative innovation Mode. Educ. Teach. Forum **29**, 149–152 (2018). (in Chinese)
8. Liu, D., Chen, L., Liang, H.: Teaching innovation of logistics core courses based on demand drive and integration of production and education. Logistics Technol. **37**(11), 128–132+149 (2018). (in Chinese)
9. Holcomb, M., Krul, A., Thomas, D.: Supply chain talent squeeze: how business and universities are collaborating to fill the gap. Supply Chain Manage. Rev. **19**(4), 10–18 (2015)
10. Yinhui, S., Yixuan, L.: Thinking and exploration on the education of college students' digital teaching platform in the new period. Chin. Geol. Educ. **29**(01), 19–21 (2020). (in Chinese)
11. Chen, H., Li, Z., Li, W., et al.: Discussion on teaching pattern of cultivating engineering application talent of automation specialty. Int. J. Educ. Manage. Eng. (IJEME) **2**(11), 30–34 (2012)
12. Jian, L., Yuan, Z.: The exploration and practice in innovative personnel training of computer science and technology. Int. J. Educ. Manage. Eng. (IJEME), **2**(6) (2012)
13. Birou, L., Van Hoek, R.: Supply chain management talent: the role of executives in engagement, recruitment, development and retention. Supply Chain Manage. Int. J. (2021)
14. Jiang, H., Sun, Q., Zhu, C.: The new teaching model research on supply chain management course. In: Proceedings of the 7th International Conference on Education, Management, Information and Mechanical Engineering (EMIM 2017) (2017)
15. Liu, S., Chen, P.: Research on cultivation of ethnic minorities IT talents in nationalities universities. Int. J. Mod. Educ. Comput. Sci. (IJMECS) **6**(2), 33 (2014)
16. Ji, D., Song, B., Yu, T.: The method of decision-making based on FAHP and its application. Fire ControlL Command Control **32**(11), 38–41 (2007). (in Chinese)

Effects of Students' Interaction Patterns on Cognitive Processes in Blended Learning

Zhongguo Wang[1,2] and Wenhui Peng[1(✉)]

[1] Faculty of Artificial Intelligence in Education, Central China Normal University, Wuhan 430079, China
pwh@mail.ccnu.edu.cn
[2] Center for Faculty Development, Nanyang Normal University, Nanyang 473061, China

Abstract. In the blended learning context, learners' interactive and cognitive processes are dynamically intertwined. Using content analysis, Social Network Analysis (SNA), and Epistemic Network Analysis (ENA), this study investigated the interplay of interaction patterns and cognitive processes by integrated online and offline learning traces data. The research participants were 75 undergraduate students. The results revealed that, as well as social interaction patterns in the discussion forum, the learner-content interaction affects the cognitive stages and learning performance. There were significant differences in learners' interactive and cognitive processes between high- and low-performance groups. High-performance groups with higher cognitive stages seem to engage more actively in viewing extended readings, watching lecture videos, and peer interaction. However, low-performance groups preferred to access the learning guideline and participate in answering the teacher's questions. These findings offer a fresh perspective on the interplay between the interactive and cognitive processes in BL.

Keywords: Interaction pattern · Cognitive process · Blended learning · Social network analysis · Epistemic network analysis

1 Introduction

Under the impact of the COVID-19 pandemic, online educational approaches have become mainstream. The large-scale global online teaching practice enlightens people to think about the sustainability of online learning when students return to traditional face-to-face educational environments. Many researchers believe that blended learning (BL), which combines face-to-face and online learning, will be the 'new normal' of future education [1, 2]. However, the potential influence of learners' interaction patterns on cognitive processes in BL is still not well understood.

In the BL context, according to learners' own interactive and cognitive preferences, learners participate in online and offline learning activities while interweaving individual and group cognitive processes. Integrating learners' online and offline learning traces seems to be a feasible approach to understand the potential impact of learners' interaction patterns on cognitive processes. But existing studies focus on learners' cognitive levels and interaction patterns in online learning [3, 4]. The potential association between learners' online and offline cognitive processes remains unclear. How a learner evolves within a BL community cannot be answered if the analysis is narrowed down either to online learning traces or to offline learning traces. Likewise, without the understanding of the mechanisms by which interaction and cognitive processes work together, it is unlikely to reveal the dynamics that make or hinder learning in the BL community.

Using SNA and ENA methods, this study investigated how learning interaction patterns affect cognitive processes and learning performance by integrating online- and offline-learning traces. Specifically, in the BL context, the research questions that this study aimed to address were the following:

RQ1: What learner-content interaction affects cognitive processes?
RQ2: What social interaction patterns influence cognitive processes?
RQ3: How do learners' interaction patterns influence cognitive processes?

2 Related Work

2.1 Interaction Pattern and Cognitive Process

At present, interaction patterns and cognitive processes are still two important research issues of collaborative learning. Interaction pattern means the interaction between teachers, students, and content [5]. Connectionism holds that learning relies more on the connections established by people, including the relations with content and with others [6, 7]. The cognitive process refers to the level or stage of learners' knowledge construction, which is usually identified by content analysis. Among the commonly used cognitive process coding schemes, Interaction Analysis Model (IAM) [8] has been widely used by researchers to analyze the content of the asynchronous online discussion.

A considerable volume of research has explored the interaction patterns and cognitive processes in online learning contexts and their relationships. Blended learning combines online and face-to-face learning activities. The previous research has focused on peer interaction, learner-teacher interaction, and cognitive processes. In this research, the learning interaction is considered as three interwoven categories, namely, learner-content interaction, peer interaction, and learner-teacher interaction. In addition, learner-content interaction is regarded as an important factor affecting the cognitive process.

2.2 Social Network Analysis and Epistemic Network Analysis

SNA is an approach used to visualize the network structure of interpersonal relationships and is often used to analyze the interaction patterns in collaborative learning environments [9]. Some studies have shown that the characteristic values of SNA can significantly predict the learning results. However, the relationships between the characteristic values of SNA and the cognitive process are still unclear.

Epistemic Network Analysis (ENA) is a quantitative ethnographic technique for modeling the structure of connections in data [10, 11]. The critical assumption of the method is that the structure of connections in the data is the most important in the study. ENA is thus a valuable technique for modeling the potential relationships between the interactive and cognitive processes in BL [12, 13]. Therefore, this study combined SNA and ENA approaches to visualize the network structure and relationship between students' interaction patterns and cognitive processes in blended learning.

3   Method

3.1   Participants and Context

The *Integrating Information Technology into Curriculum* course at a university in central China for educational technology students, taking during the second semester of their third year of study. Seventy-five undergraduate students (54 females, 72%) along with the instructor participated in this study. There are pre-class learning tasks, in-class group discussion, off-class individual assignments and reflection, project-based collaborative learning, mid-term and final exams in the course. Learning activities are designed to follow blended learning strategies and maintain the consistency of cognitive processes, consisting of online and offline interwoven two parts.

The online learning setting is the centerpiece of the blended course. It contains online course resources such as learning guidelines, lecture videos, slides, and extended readings. Learners can participate in diverse online learning activities, for instance, online self-directed learning, online discussion, assignments or artifacts submission, self-assessment, and peer-assessment. Each week, the instructor manages the discussion forum by posting the sections' questions of the course, each topic post as a sub-forum, and stipulating deadlines and discussion contracts. Students involved in the sub-forum by replying to the topic post and interacting with peers by commenting on others' posts. Participation in the forum is voluntary but will be considered as part of the student's final score.

The offline part is face-to-face learning in the traditional classroom, once a week (2 h) for 18 weeks, was primarily problem- or project-based collaborative learning including group discussion and negotiation, group products presentation, intra- and inter-group assessment, instructor guidance. For instance, face-to-face learning may start with a pre-class learning test or group products presentation, followed by the instructor's explanation, group discussion, or sharing between groups.

3.2   Data Collection

The online learning interactions data included counts of learners accessing learning resources (four categories: learning guidelines, lecture videos, slides, and expended readings) ranging from 1 to 237 (M = 49.04, SD = 45.42). In total, there were 807 messages posted to 11 discussion threads from 75 learners (M = 10.76, SD = 7.80) and log data (e.g., timestamps, type of post, learner ID, and reply relationship).

The offline learning activities data contained the total scores of six pre-class tests from 7.20 to 28.00 (M = 22.30, SD = 4.12). The self-assessment and peer-assessment on collaborative group learning are used to reflect the offline interaction performance of learners. The intergroup assessment of groups' learning products is based on evaluation criteria. All these assessments were considered as indicators of cognitive level and integrated into the overall cognitive process.

3.3 Data Analysis

3.3.1 Coding Scheme for Interactions

To examine interactions in the online learning process, descriptive statistical analysis was used for the learner-content interaction, aimed to explore the learners' cognitive preferences.

Social network analysis (SNA) was used for social interaction patterns analysis, including peer interaction and learner-teacher interaction in the online discussion forum. This study adopted python 3.8.3 and networkx2.5 packages. According to the posting and replying relationship, the directed weighted networks were constructed and visualized using the Kamada-Kawai path-length cost function.

3.3.2 Coding Scheme for Cognitive Processes

The cognitive processes of the online discussion forums were analyzed using the coding scheme interaction analysis developed by Gunawardena et al. (1997) [8]. Cognitive stages are categorized as i) sharing/comparing information (KC I), ii) discovery and exploration of dissonance (KC II), iii) negotiation of meaning (KC III), iv) testing and modification of proposed synthesis (KC IV), v) agreement statement (KC V). Considering that a single message posted to a topic thread may contain more than one idea, the unit of meaning was chosen as the unit of analysis. Two trained graduate students coded 807 massages with an inter-rater agreement of Cohen's kappa of 0.83, which indicated good reliability.

4 Result

4.1 Stages of Cognitive Processes

In the online discussion forum, there were 11 discussion topics consistent with the course progress. 804 task-related messages were analyzed using the coding scheme interaction analysis. Table 1 shows the distribution of the stages of the cognitive process in the discussion forum. Sharing/comparing information (F = 434) occurred the most frequently, followed by discovery and exploration of dissonance (F = 241). The high level of cognitive stages was less.

Table 1. Descriptive statistics of the stages of cognitive process

Stages of cognitive process	Frequency	Percentage
KC I	434	53.980
KC II	241	29.975
KC III	96	11.940
KC IV	6	0.746
KC V	27	3.358
Total	804	100

4.2 Learner-Content Interaction

Overall, descriptive statistics of the four types of online learning materials viewed by learners are shown in Table 2. Learners seem to prefer viewing lecture videos, followed by slides.

Table 2. Descriptive statistics of learner-content interactions

	$N_{learners=75}$		
	N	M	SD
1. Learning guideline	468	6.24	5.63
2. Lecture videos	2363	31.51	34.94
3. Slides	734	9.77	10.38
4. Extended readings	470	6.27	8.65

Note. N = Number of leaner-content interactions

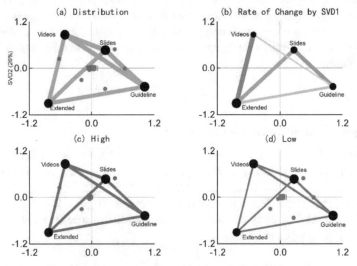

Fig. 1. Learner-content cognitive interactions preference

However, descriptive statistics data cannot describe subtle differences in learners' cognitive preferences and learning persistence. The online learning content contains 12 sections and each section with four types of learning materials. Therefore, we examined learners' engagement in every section and then reduced the dimension according to the ENA data format. According to the learning performance, learners were divided into two groups, high and low, and the connection structure of cognitive pattern was visualized as Fig. 1.

4.3 Social Interaction Patterns

Relationships of online posting and replying were converted into a directed tuple and visualized through Python 3.9 + Networkx 2.5. Figure 2 shows the social interaction patterns. The social network of the online discussion forum seems to be a sparse social network structure, as shown in Fig. 2 and the network density is 0.072 and the average clustering coefficient is 0.278.

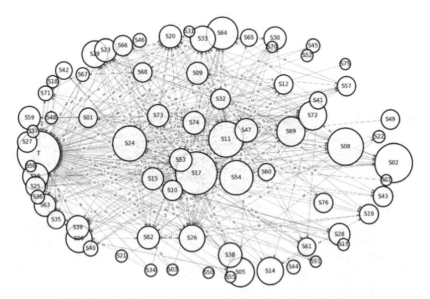

Fig. 2. Social interaction patterns in the online discussion forum

To identify the differences in the modes of learners' social interaction, the social network characteristics of each node (corresponding to each learner), such as Clustering, degree centrality, in-degree centrality, out-degree centrality, closeness centrality. In addition, peer interaction is beneficial to high-level cognitive processes. Therefore, this study considered peer interaction as a vital factor affecting the cognitive process.

4.4 Relationship of Social Interaction and Cognitive Stages

In the BL context, the online discussion forum is not the only setting for collaborative learning. In the classroom or supported by instant messengers, group collaborative learning also plays a vital role. But, the potential relationships between online interaction patterns, offline interaction patterns, and cognitive processes remain unclear. This study investigated the self-assessment and peer-assessment on group collaboration supported by scaffolding. The learners were divided into three groups according to the assessment scores: high (the top 30%), middle (the middle 40%), and low (the bottom 30%). Then, the average of cognitive process stages and social network characteristics of the three groups is shown in Fig. 3.

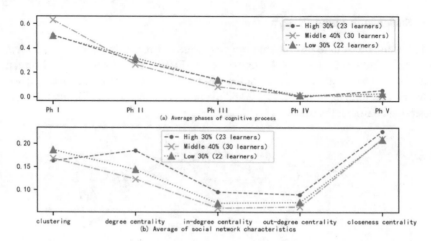

Fig. 3. The relationships between online interaction, offline interaction, and cognitive processes

Figure 3 shows that there seems to be some degree of consistent correlation between high-level cognitive stages (KC III, KC IV, and KC V), online interaction, and group collaboration. The learners with low participation in group collaboration, but are active involvement in the online discussion, can also lead to higher-level cognitive stages. However, this finding may be due to the intrinsic correlation between online cognitive stages and online discussion engagement. Therefore, a more feasible approach is to systematically explore the behavior patterns of learners participating in various online and offline activities from a broader perspective to explain the potential correlation between interaction patterns and cognitive processes.

4.5   Relationship of Interaction Patterns and Cognitive Processes in BL

To detect the differences in interaction patterns, this study considered two types of inter-action: learner-content interaction and social interaction. We integrated online and offline learning traces data, including learner-content interactions (four categories materials), pretest, SNA characteristics (out-degree centrality of nodes, SNAODC, and in-degree centrality of nodes, SNAIDC), social interaction (peer interaction and learner-teacher interaction), and cognitive stages (KC I to KC V). Then, we adopted ENA to identify the interplay of interaction patterns and cognitive processes for the high- and low learning performance. The epistemic network of learners in both groups is shown in Fig. 4.

Figure 4(a) shows a comparison of the two groups. Figure 4(b) displays the low learning performance group. And Fig. 4(c) illustrates the high learning performance group. There were significant differences in ENA characteristics between the low- and high-performance groups. The two samples t-tests results as shown in Table 3.

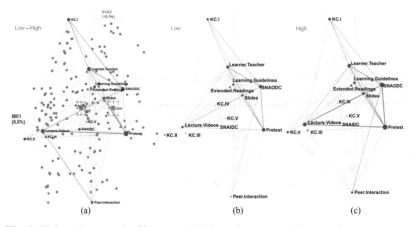

Fig. 4. Epistemic network of learners in high-performance and low-performance groups

Table 3. Results of the *t*-test for ENA characteristics between high- and low-performance groups

Dimension	Group	N	Mean	S.D	t	d
X-axis	Low-performance	198	0.31	0.60	−10.11**	0.97
	High-performance	252	−0.25	0.57		
Y-axis	Low-performance	198	0.0	0.83	0	0
	High-performance	252	0.0	0.85		

* *p < 0.01

5 Discussion

This study was to elucidate learners' interaction types in BL and how do these interaction patterns influence cognitive processes. To do this, we integrated learners' online and offline learning traces data, and then identified learner interactive and cognitive processes using content analysis and SNA. Adopted ENA to visualize the relationship structure of the cognitive processes and probed the fluctuation and interplay of interaction patterns and cognitive processes.

5.1 Learner-Content Interactions and Cognitive Processes

In the age of the global epidemic of collaborative learning, there is a risk that the role of individual interactive and cognitive processes may be obscured by an overemphasis on group learning research. This study re-examined the influence of learner-content inter-active preference on individual and group cognitive processes. The results revealed that learners with high learning performance viewed extended readings and watched lecture videos more frequently than those with low learning performance, as shown in Fig. 4 manifests that Learners with low learning performance preferred to view learning guidelines and slides. This interaction pattern is correlated with pretest and KC I. This finding elucidated one aspect of Research Question 1, the interplay of individual interactive and individual cognitive processes.

5.2   Social Interaction Patterns and Cognitive Processes

This study examined how social interaction patterns rated cognitive processes. Figure 4 shows that high-performance learners were more likely to participate in peer interaction than low-performance learners. In addition, the in-degree centrality of SNA and peer interaction seems to correlate to higher cognitive stages and high learning performance. Figure 3 Illustrates that there was a potential relationship between offline collaborative learning in a group and online learning interactive and cognitive processes. However, the interplay is unclear.

5.3   The Interplay Between Interaction Patterns and Cognitive Processes in BL

Figure 4 elucidated that there was a complex and dynamic interplay between interaction patterns and cognitive processes in BL. As mentioned above, there were potential influences of learner-content interaction and social interaction on cognitive stages. There is no denying that the interaction and cognitive processes of online and offline learning are intertwined in the BL context. Figure 4 displays the significant differences in interaction patterns and cognitive processes between high- and low-performance groups. These differences and their impact on cognitive processes should be addressed within the dynamic processes of blended learning.

6   Conclusion, Limitations, and Future Research

In BL context, individual and group different interactive and cognitive patterns lead to distinct cognitive stages and learning performance. This study used a mixed approach to elucidate the interplay of interactive and cognitive processes through integrated offline and online learning traces. These findings offer a fresh perspective on the interplay between the interactive and cognitive processes in BL. There are three limitations to this study. First, the sample analyzed is specific high education BL context and not enough to be generalized to other learning contexts. Second, although online learning traces data is large enough, it seems that obtaining offline learning processes data is still a challenge. Finally, when adopting a mixed research method to analyze multimodal data. How to calibrate and alignment data and eliminate interferential data is a challenge. Given this, consider carrying out relevant studies in different BL contexts. Future work needs to consider the interplay of individual and group interactive and cognitive processes.

Acknowledgment. This work is supported by the Project of the National Natural Science Foundation of China [number: 61977036].

References

1. Dziuban, C., Graham, C.R., Moskal, P.D., Norberg, A., Sicilia, N.: Blended learning: the new normal and emerging technologies. Int. J. Educ. Technol. High. Educ. **15**(1), 1–16 (2018)
2. Matheos, K., Cleveland-Innes, M.: Blended learning: enabling higher education reform. Revista Eletrnica deEducao **12**(1), 238–244 (2018)
3. Zhang, S., Liu, Q., Chen, W., et al.: Interactive networks and social knowledge construction behavioral patterns in primary school teachers' online collaborative learning activities. Comput. Educ. **104**(Jan), 1–17 (2017)
4. Njenga, S.T., Oboko, R.O., Omwenga, E.I., Maina, E.M.: Use of intelligent agents in collaborative m-learning: case of facilitating group learner interactions. Int. J. Mod. Educ. Comput. Sci. (IJMECS) **9**(10), 18–28 (2017)
5. Garrison, D.R., Shale, D.: Education at a distance: from issues to practice. R.E. Krieger Pub. Co. (1990)
6. Siemens, G.: Connectivism: a learning theory for the digital age. Int. J. Instr. Technol. Distance Learn. (2005)
7. Zacharis, N.Z.: Classification and regression trees (CART) for predictive modeling in blended learning. Int. J. Intell. Syst. Appl. (IJISA) **10**(3), 1–9 (2018)
8. Gunawardena, C.N., Lowe, C.A., Anderson, T.: Analysis of a global online debate and the development of an interaction analysis model for examining social construction of knowledge in computer conferencing. J. Educ. Comput. Res. **17**(4), 397–431 (1997)
9. Sharma, S., Purohit, G.N.: A new centrality measure for tracking online community in social network. Int. J. Inf. Technol. Comput. Sci. (IJITCS) **4**(4), 47–53 (2012)
10. Shaffer, D.W., Collier, W., Ruis, A.R.: A tutorial on epistemic network analysis: analyzing the structure of connections in cognitive, social, and interaction data. J. Learn. Anal. **3**(3), 9–45 (2016)
11. Shaffer, D.W., Ruis, A.R.: Epistemic network analysis: a worked example of theory-based learning analytics. In: Handbook of Learning Analytics (2017)
12. Siebert-Evenstone, A., Arastoopour Irgens, G., Collier, W., Swiecki, Z., Ruis, A.R., Williamson Shaffer, D.: In search of conversational grain size: modelling semantic structure using moving stanza windows. J. Learn. Anal. **4**(3), 123–139 (2017)
13. Hod, Y., Katz, S., Eagan, B.: Refining qualitative ethnographies using epistemic network analysis: a study of socioemotional learning dimensions in a humanistic knowledge building community. Comput. Educ. **156**, 103943 (2020)

Application of Cloud Computing in Applied Undergraduate Education and Management

Fang Huang, Jing Zuo, and GengE Zhang[✉]

School of Transportation, Nanning University, Nanning 530022, Guangxi, China
23255294@qq.com

Abstract. With the rapid development of information technology, the emergence of cloud computing promotes the rapid development of various industries, especially in the field of education. It changes the form of education. Starting from the needs of application-oriented undergraduate talent training, starting from the characteristics of cloud computing and three common service types, this paper analyzes the architecture of cloud platform in education industry. Combined with the characteristics of application-oriented talent training, this paper focuses on the application of cloud computing to education resources, teaching management, experimental training and assessment in application-oriented undergraduate education, This paper summarizes the positive impact of cloud computing on the innovation of teaching environment, the promotion of teaching equity and the unity of teaching management in applied undergraduate education, which provides a useful reference for the deep application of cloud computing in the field of education.

Keywords: Cloud computing · Applied education · Teaching management · Resource sharing

1 Introduction

Undergraduate education is the intermediate level of higher education and occupies the central and main position in the structure of Higher Education [1]. With the popularization of higher education, all kinds of undergraduate education, including general undergraduate education, vocational undergraduate education and applied undergraduate education, have attracted more and more attention [2]. Among them, the application-oriented undergraduate with the type of application technology as the school running orientation is relatively and different from the academic undergraduate, which has played a positive role in meeting China's economic and social development, the needs of high-level application-oriented talents and promoting the popularization of China's Higher Education [3]. It is imperative to strengthen the education and management of applied undergraduate.

As an emerging product of computer application technology, cloud computing has attracted the attention of all walks of life from the beginning. With the popularity of big data and cloud computing applications, more and more government agencies, enterprises and institutions and educational research structures have successively built government cloud, enterprise cloud and education cloud, such as internationally famous Amazon AWS and Microsoft azure, Alibaba cloud, Tencent cloud and Huawei cloud have emerged in China. With the increasingly prominent advantages of cloud computing, relevant education departments and scholars pay more and more attention to the practical application of cloud computing in the field of education. The outline of the national medium and long term education reform and development plan (2010–2020) proposes to fully integrate information resources, adopt cloud computing technology, form an intensive development approach of resource allocation and service, and build a stable, reliable and low-cost national education cloud service model [4]. China's ten year development plan for education informatization (2011–2020) also proposes to adopt the cloud computing service mode to form an intensive, effective and high-quality development approach for resource allocation and services, and build a stable, reliable and low-cost national education cloud service platform [5].

The main content of this paper is to apply advanced cloud computing technology to the education and management of application-oriented undergraduates.

2   Characteristics and Service Types of Cloud Computing

2.1   Cloud Computing and Its Characteristics

Cloud computing is a product of the integration of network technology and computing technology. It is a distributed computing method based on the Internet [6, 7]. It means that the huge data computing and processing program is decomposed into countless small programs through the network "cloud", and then processed and analyzed through the system composed of multiple servers. These small programs get the results and return them to users. In this way, the shared software and hardware resources and information can be provided to computers and other devices on demand. Through this technology, tens of thousands of data can be processed in a very short time (a few seconds), so as to achieve powerful network services. It is necessary to pay much attention to the security problem in cloud computing [8–10]. The protection of sensitive data in mobile cloud computing is also an important topic [11, 12].

Cloud computing has five main characteristics.

1) Self service
    Users using cloud computing services can use self-service methods to obtain services, such as independently selecting server lines, server configuration, etc.
2) Anytime, anywhere
    Users can connect to the cloud through a variety of terminals anytime, anywhere, such as mobile phones, laptops, tablets, etc., so as to use various services provided by the cloud, breaking the restrictions of time and space conditions.
3) Measurable services
    The cloud will track and record the data reading and storage capacity, CPU or other hardware usage used by users in real time, and effectively optimize the cloud background resources according to the real-time situation.

4) Rapid resource expansion and contraction

Users can apply to increase or release cloud resources according to their own needs, so as to realize the rapid expansion or reduction of software and hardware resources.

5) Resource pooling

All resources of cloud platform service providers, including software and hardware resources, are concentrated in a huge resource pool. These resources are shared by users in a multi-tenant manner, and all resources are managed and scheduled uniformly.

2.2   Cloud Computing Service Types

Cloud computing is not only a new technology integrating computer technology and network technology, but also a new service model. According to the service mode, it mainly includes the following three types: infrastructure as a service (IaaS), platform as a service (PaaS) and software as a service (SaaS). IAAs is mainly used to provide virtual machines or other resources to users as services, PaaS is mainly used to provide development platforms to users as services, and SaaS is mainly used to provide applications to customers as services.

IaaS is at the bottom of the whole cloud computing service. PaaS is at the middle layer. It is available to use various computing resources, storage resources, software and hardware resources provided by IaaS layer to build a platform. SaaS is at the top layer. It is capable to use the platform provided by PaaS layer for software development. Cloud computing service providers can focus on their own level and do not need to be fully responsible for the three levels. Upper level service providers can use the lower level cloud computing services to realize the cloud computing services of this layer.

Cloud computing and its cloud platform can be applied to different occasions, including storage cloud serving the public (or "cloud"), government cloud, medical cloud, financial cloud, education cloud, etc. for different objects [13–16].

2.3   Education Cloud and Its Platform Architecture

Different from the application scenarios of government cloud and enterprise cloud, in addition to the sharing of computing resources, storage resources and software and hardware resources, the construction of the platform should focus on the service provision of educational application resources [17–19]. As shown in Fig. 1, in the design and construction of PaaS layer and SaaS layer, it is necessary to comprehensively consider the applications related to the education industry and the development trend of the education industry in the future, so as to provide more perfect and professional data support, data analysis and prediction for teachers, education administrators, teaching researchers and other users, and support academic situation analysis and teaching quality monitoring, It provides decision-making basis for educational management and teaching reform.

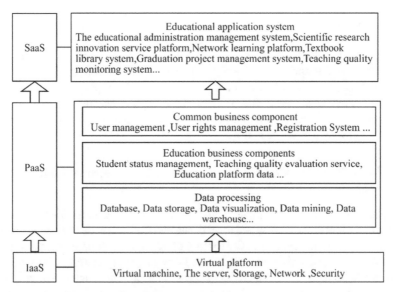

Fig. 1. Cloud platform architecture of education industry

3   Application of Cloud Computing in Applied Undergraduate Universities

With the continuous development of computer technology, cloud computing is more and more widely used in the field of education. From the sharing of digital educational resources in the initial stage to the construction of virtual simulation laboratory, it has effectively promoted the reform of teaching mode and realized the diversification of learning [20–22].

3.1   Improve the Environment of Applied Undergraduate Education

In 2015, the Ministry of education, the national development and Reform Commission and the Ministry of Finance jointly issued the opinions on guiding the transformation of some local ordinary undergraduate universities to application- oriented universities, marking that the development of Application- oriented Undergraduates in China has entered a new historical period. Applied undergraduate education is related to the supply of national professional and technical talents. The cultivation of applied undergraduate talents should change from the original knowledge accumulation to the application of skills, which requires training talents for the needs of industries and enterprises, taking practical knowledge as the starting point, adjusting the curriculum construction structure according to the existing technical practical knowledge, and enhancing the timeliness of practical knowledge, It puts forward new requirements for students' practical application ability. The traditional teaching means and methods that have been used all the time can not effectively improve students' practical ability, and it is difficult to cultivate application-oriented technical talents satisfactory to enterprises. It is necessary to reform the existing teaching methods and teaching environment [23].

Cloud computing can provide maximum sharing services for computing resources, storage, software and hardware resources, and even deeply integrate enterprise cloud and education cloud, connect education logic with industrial logic, deeply integrate schools, governments and enterprises, and jointly cultivate application-oriented technical talents [24]. Based on the above significant advantages of cloud computing, cloud computing technology provides strong technical support for applied teaching reform, especially the improvement of teaching environment, and provides platform support for improving students' engineering practice ability.

3.2 Build a Cloud Platform for Industry Education Integrated Educational Resources

The application-oriented undergraduate focuses on cultivating application- oriented talents who can adapt to the society and meet the needs of enterprises. The digital education resources are no longer limited to the sharing of education resources within schools or between colleges and universities. It is necessary to closely contact with enterprises, closely combine industry and teaching, support and promote each other, connect the education logic with the industrial logic, and make the two logic systems of industry and education cross-border, integrated restructure. Based on school enterprise cooperation and industry education integration, educational resources are deeply integrated with industrial resources and innovation resources, and curriculum related resources updated synchronously with the development of industrial technology are established. This requires the cooperation of industry, enterprises and employers to jointly build and share curriculum resources, and jointly develop curriculum resources close to the industry and production front line [25]. For the co construction and sharing of educational information resources, cloud computing has unique advantages. It can store scattered resources in cloud servers through the network, and adopt unified data standards and database systems to realize distributed storage and virtualization management of resources. Cloud computing technology is used to build an industry education integration education resource sharing platform, which can effectively integrate the teaching resources, enterprise project database, practical case database, software and hardware platforms needed in teaching practice of both schools and enterprises, and uniformly place them in the cloud for management, maintenance and use, so as to meet the requirements of teaching resources to keep up with the changes of enterprise needs and times, and constantly update and iterate Expanding demand.

3.3 Realize High-Quality Resource Cloud Sharing of School Enterprise Teaching Management

Combined with the actual situation of college teaching management, the application of cloud computing in college teaching resource management can solve the problems existing in the current teaching resource service, reduce the use cost of teaching resources, not only avoid the repeated construction of equipment, but also be in line with the international education and social needs, and greatly expand the students' professional vision. In application-oriented undergraduate colleges, most majors implement the school enterprise dual tutor system. How to let enterprise tutors participate in school education management, how to manage the teaching of enterprise teachers, and increase the in-depth integration of enterprise teaching has always been a difficult problem to solve. As shown in the architecture of the teaching management cloud platform in Fig. 2, the distributed structure is used to carry out unified and standardized management of computer hardware resources, network equipment and virtual servers, data storage, mining, user and application management of the platform, i.e. service layer, so as to realize the effective allocation of teaching resources. Enterprise tutors, school teachers and students access their required applications through web browsers, Each teaching resource system is independent of each other, and the data can also communicate with each other, so as to realize the high-quality sharing of teaching management between schools and enterprises.

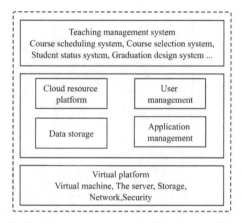

Fig. 2. Teaching management cloud platform architecture

3.4 Build a Cloud Platform for Experimental Training and Practice

For application-oriented undergraduate students, it is very important to improve their practical ability. Traditional computer laboratories are usually restricted by site, cost, curriculum and other factors, and cannot meet the teaching requirements for courses with special requirements. For example, the practical teaching of big data courses has high requirements for computer hardware, development courses need to provide a development environment similar to that of enterprises, and some courses are closely related, The phased learning results of the previous course or previous courses need to be retained for continued use in subsequent courses or professional comprehensive training. A computer laboratory needs to meet the learning needs of students of multiple majors. In order to meet different teaching needs, it is necessary to build a complex software environment, and the deployment of the experimental environment is also time-consuming and laborious. Using cloud computing technology to build the experimental training room can form an overall solution with low construction cost, short construction cycle, high operation and maintenance efficiency, easy expansion, load balance, safety and reliability, green and energy saving, and innovate the mode and architecture of training room construction [26, 27].

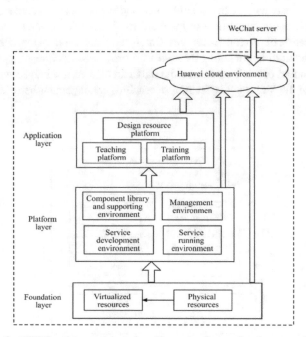

Fig. 3. HCDP architecture based on Huawei software development cloud

Taking the professional training laboratory of software engineering as an example, combined with the characteristics of Huawei cloud platform, jointly build the professional experimental training cloud platform of software engineering, as shown in Fig. 3 HCDP architecture based on Huawei software development cloud. This platform needs to meet all professional courses of students' programming practice ability, so as to comprehensively improve students' comprehensive quality and ability. The top layer of the platform architecture introduces Huawei cloud platform environment to provide various resource services for teaching, such as NodeJS based data interface service, third-party Web Service docking service, training resource platform, etc. students can access remote interfaces anytime and anywhere through the network and accept various cloud service resources, so as to provide platform guarantee for the smooth development of software training process. In the middle platform layer, Huawei soft cloud DevCloud is proposed to deploy the component library required for the training project on the software development cloud, and build the project management environment, code hosting environment and code detection environment, so as to realize the collaborative development, software detection, project management, project operation and maintenance among the members of the project team. The lowest basic layer, through the virtualization technology of Huawei cloud, deploys multiple applications in project teaching in independent spaces without affecting each other, so as to significantly improve work efficiency and serve the flexible application requirements of various projects in the teaching process.

3.5 Create an Assessment and Evaluation Cloud Platform to Promote Continuous Improvement of Talent Training

Building a cloud platform for assessment and evaluation is of great significance to promote the continuous improvement of talent training [28]. For the cultivation of applied talents, it is necessary to adopt more process assessment and evaluation or a combination of the two [29, 30]. In terms of evaluation form, it should also reflect the characteristics of application-oriented undergraduate education, adopt various methods such as cloud defense between schools and enterprises, course design display video, language course recording audio, and use cloud computing technology to build an evaluation cloud platform, which can uniformly manage all resource information, more effectively reduce the requirements of software applications on servers, operation and maintenance, and reduce risks, It is convenient to store more types of data, solve the difficulties of data sharing and interaction, break the space constraints, facilitate enterprises to participate in assessment, evaluation and feedback, provide students with opportunities to display learning achievements, and provide certain reference basis for enterprises to screen talents.

4 The Positive Impact of Cloud Computing on Applied Undergraduate Education

4.1 Innovative Education Environment

With the support of cloud computing, "Internet plus education" is a key move to promote educational informationization, promote education quality and innovate talents, and has attracted more and more attention from the domestic educational circles. Teachers can connect students with multiple projects and applications to enable students to be innovative in their learning. Cloud computing provides students with the opportunity to show their learning achievements, and gives students the opportunity to use various technologies to achieve their learning goals to meet their personalized needs. Teachers can innovate classroom structure and classroom form through cloud computing, students meet in the cloud, participate in professional learning in a high quality, whole process and depth, and accelerate the innovation of "Internet plus education".

4.2 Promoting Educational Equity

The education cloud transmits various educational resources to the cloud through the network for unified management and storage, and constructs a modern shared education model to meet the needs of educational resource sharing, students' independent choice of courses and personalized learning. Compared with traditional Internet technology, education cloud has stronger computing power and faster resource update speed. It can meet learners' access to cloud teaching resources at any time and anywhere, use any terminal, and complete the replication and mapping of the original classroom in the digital virtual space, so that every student can enjoy high-quality education, from teachers to teachers and students to students, From classroom teaching to teaching and scientific research, the East and the West jointly explore, realize new knowledge sharing, and effectively promote the complementary advantages of educational resources in Colleges and universities in the East and the West. This integrated classroom allows students to have zero distance contact with famous schools, teachers and courses, breaks the wall between colleges and universities, realizes the sharing of teaching resources between domestic Eastern and western universities, applied undergraduate universities and double first-class universities, and even applied undergraduate universities and foreign famous universities, effectively solves the problem of uneven distribution of educational resources, and is conducive to the promotion of educational equity, Accelerate the construction of modern education model.

4.3   Unified Education Management Mode

The unified management and transfer of various computing resources connected through the network is the core technology and service of cloud computing. With the help of the characteristics of cloud computing technology, education management and teaching resources are effectively integrated to realize that the scheduling of network computing resources is not limited by time and space. The construction of cloud platform can integrate and manage school level education and teaching resources, Finally, realize the linkage and sharing of education management and teaching resources, solve the problem of difficult interoperability of education management system resources at multiple levels of colleges and universities, improve the utilization rate of information resources, open up the education cloud and enterprise cloud, and realize the elastic expansion of computing resources, storage, network, software and hardware resources, which is conducive to the connection between education and industry and further improve the quality of practical teaching, Ensure the service level of digital education.

4.4   Investigation and Evaluation of Teachers and Students

In order to better check and verify the application effect of cloud computing in applied undergraduate education, questionnaires were distributed to teachers and students for investigation and analysis, 268 teacher questionnaires were distributed and recovered, and a total of 256 valid questionnaires were distributed; 320 questionnaires were distributed and collected to students, and a total of 308 valid questionnaires were issued. The questionnaire is sorted out and the statistical results in Table 1 and Table 2 are obtained.

The questionnaire for teachers and students includes 9 and 8 questions respectively. There are five options to answer, namely "fully agree", "agree", "general", "disagree" and "very disagree", which are represented by A, B, C, D and E respectively.

Table 1. Teachers' attitude feedback on using cloud teaching platform

Question	A	B	C	D	E
Rich teaching resources and improved lesson preparation efficiency	98%	2%			
Break the time and space constraints and innovate the teaching organization	96%	2.6%	1.4%		
Improve classroom efficiency, interactivity and classroom evaluation	5%	95%			
Facilitate data collection and analysis and display students' works	98%	2%			
Enhance the interaction and exchange between teachers, students and students	96%	2.3%	1.7%		
Convenient for homework release and review	92%	5.6%	2.4%		
Virtual experiment platform to improve students' practical ability	91.6%	5.3%	3.1%		
Carry out teaching and research activities, subject research and teaching reform	93%	2%	5%		
The interface and data standards are unified and easy to use	98.6%	1.4%			

Table 2. Students' feedback on using cloud teaching platform

Question	A	B	C	D	E
Improve interest in learning	11.4%	76.6%	12%		
Create personalized learning according to one self's learning situation	88.9%	8.6%	2.5%		
Carrying out pre class preview and post class review	90.3%	7.8%	1.9%		
Rich learning resources	96%	2.5%	1.5%		
Improve the ability of cooperative learning and autonomous inquiry learning	89%	6%	3.8%	1.2%	
Improve problem solving skills	90.6%	8.9%	0.5%		
Enhanced self-confidence and the courage to express themselves	97.8%	2.2%			
Convenient and smooth online interaction	98.9%	1.1%			

The data in the two tables show that the vast majority of teachers and students agree with the use of cloud platform; Only one student chose D, indicating "disagree"; The number of teachers and students with E - "very disagree" attitude is 0. It can be seen that the attitude of the respondents is very clear and generally consistent. Integrating cloud computing technology into the classroom can enable students to have more learning opportunities and promote educational equity; Cloud computing also provides opportunities for educational innovation and helps teachers innovate new teaching models. The use of cloud platform enables all users to easily access resources across multiple platforms, and promotes students' ability to effectively cooperate with others. The data of the effective questionnaire shows that cloud computing has a positive impact on the application-oriented undergraduate education and plays an obvious role in promoting the improvement of the teaching quality of the application-oriented undergraduate.

5 Conclusion

The construction of Applied Technology undergraduate course should be guided by the talent view, quality view and education view reflecting the spirit of the times and the requirements of social development, build a new discipline direction, professional structure and curriculum system to meet and meet the needs of economic and social development, comprehensively improve the teaching level, closely combine the local characteristics and professional characteristics, and pay attention to the cultivation of students' practical ability. Cloud computing and cloud platform can play an important role.

This paper introduces the characteristics, service types and education Cloud Architecture of cloud computing technology, and expounds the practical application of cloud computing technology in Applied Undergraduate Education and its positive impact on applied undergraduate education. The integration of education and teaching resources based on cloud computing not only eliminates the barriers between information and effectively improves the utilization of various resources, but also integrates high-quality resources and effectively alleviates the current situation of uneven distribution of educational resources in Colleges and universities. Combined with the current application status of cloud computing technology, this paper emphasizes to build various levels of cloud platforms in Application-oriented Undergraduate Colleges and universities, realize the linkage and sharing between clouds, actively open up the enterprise cloud of school enterprise cooperation, have the courage to explore, use innovative integrated teaching methods to open more high-quality resources, and provide students with a more personalized and more suitable learning environment. I believe that with the continuous development of cloud computing technology in the future, its value in the field of education will be more far-reaching.

Acknowledgment. The project is funded by: (1) the Basic Scientific Research Ability Improvement Project of Young and Middle-Aged Teachers in Guangxi Universities (2019ky0948), (2) the 2020 undergraduate teaching reform project of Guangxi Higher Education (2020jgz163), (3) the 2021 undergraduate teaching reform project of Guangxi Higher Education (2021jgb434) and (4) the 2020 research topic of Guangxi University Teacher Management Association (szkt202005).

References

1. Mcgrath, E.J.: Universal higher education. J. High. Educ. Manag. **2**, 184–186 (2019)

2. Tight, M.: Higher education: discipline or field of study? Tertiary Educ. Manag. **26**(prepublish), 1–14 (2020)
3. Ritzen, J.: Higher Education and Economic Development, vol. 10, pp. 1–7. Springer, Cham (2017)
4. Yuxia, D.: Continuing education policy: achievements, problems and suggestions - based on the ten-year implementation of the outline of the national medium and long-term education reform and development plan (2010–2020). Contemp. Continuing Educ. **37**(4), 23–28 (2019). (in Chinese)
5. Zhai, X., Shi, C.: Implementation status, challenges and prospects of the ten-year development plan for educational informatization (2011–2020). Mod. Educ. Technol. **30**(12), 20–27 (2020). (in Chinese)
6. Thangavelu, S., Siranjeevi, N., Shanmugam, P.K.: Fuzzy keyword search over encrypted data in cloud computing using string matching algorithm. J. Oper. Syst. Dev. Trends **2**(4), 8–12 (2017)
7. Srinivsan, S.M., Chaillah, C.: Information interpretation code for providing secure data integrity on multi-server cloud infrastructure. Int. J. Mod. Educ. Comput. Sci. (IJMECS) **12**, 26–33 (2014)
8. Almorsy, M., Grundy, J., Müller, I.: An analysis of the cloud computing security problem. Softw. Eng. **9**, 1–6 (2016)
9. Mavi, S.: Cloud computing: security issues and challenges. IITM J. Manag. IT **7**(1), 25–31 (2016)
10. Aized AminSoofi, M., Khan, I., Fazal-Amin, F.-A.: A review on data security in cloud computing. Int. J. Comput. Appl. **94**(5), 12–20 (2014)
11. Qayyum, R., Ejaz, H.: Data security in mobile cloud computing: a state of the art review. Int. J. Mod. Educ. Comput. Sci. (IJMECS) **12**(2), 30–35 (2020)
12. Wang, J., Zheng, X., Luo, D.: Sensitive data protection based on intrusion tolerance in cloud computing. Int. J. Intell. Syst. Appl. **3**(1), 58–66 (2011)
13. Seera, N.K., Jain, V.: Perspective of database services for managing large-scale data on the cloud: a comparative study. Int. J. Mod. Educ. Comput. Sci. (IJMECS) **7**(6), 50–58 (2015)
14. Khan, S.: Cloud computing: issues and risks of embracing the cloud in a business environment. Int. J. Educ. Manag. Eng. (IJEME) **9**(4), 44–56 (2019)
15. Alakbarov, R.G.: Method for effective use of cloudlet network resources. Int. J. Comput. Netw. Inf. Secur. (IJCNIS) **12**(5), 46–55 (2020)
16. Kalaiselvi, R., Kousalya, K., Varshaa, R., Suganya, M.: Scalable and secure sharing of personal health records in cloud computing. Gazi Univ. J. Sci. **3**(29), 583–591 (2016)
17. Baldassarre, M.T., Caivano, D., Dimauro, G., et al.: Cloud computing for education: a systematic mapping study. IEEE Trans. Educ. **61**(3), 234–243 (2018)
18. Tanyeri, T., Kiran, H.: Cloud computing in education. In: BIG DATA and Advanced Analytics: Collection of Materials of the Third International Scientific and Practical Conference, vol. 3, no. 5, pp. 111–133 (2017)
19. Shyshkina, M.: The general model of the cloud-based learning and research environment of educational personnel training. Comput. Soc. **6**, 1–7 (2018)
20. Cui, D.: Construction of resource sharing platform of higher vocational education based on cloud technology. In: 2018 International Conference on Information, Teaching and Applied Social Sciences (ITASS 2018), pp. 231–236 (2018)
21. Tingting, W.: Research on online and offline integrated teaching mode based on cloud platform in higher vocational logistics teaching. Logist. Eng. Manag. **40**(08), 154–155 (2018). (in Chinese)
22. Zhaodi, L.: Research on university teaching management and resource sharing based on cloud computing technology. China Manag. Informatiz. **24**(23), 197–199 (2021). (in Chinese)

23. Dimmock, C., Tan, C.Y., Nguyen, D., et al.: Implementing education system reform: local adaptation in school reform of teaching and learning. Int. J. Educ. Dev. (IJED) **80**, 102–108 (2021)
24. Zhou, Y., Shen, L., Zhang, Y.: Research on the impact of cutting-edge technology on educational governance and its countermeasures. Comput. Age **10**, 107–110+114 (2021). (in Chinese)
25. Shundi, Y.: Analysis on the impact of big data and cloud computing on the construction of educational informatization in colleges and universities. Netw. Secur. Technol. Appl. **10**, 100–102 (2021). (in Chinese)
26. Zhu, X.: Research on innovative design of big data project training room based on cloud computing. Comput. Program. Skills Maint. **11**, 70–71+96 (2021). (in Chinese)
27. Yaomin, W.: On the construction scheme of computer training room based on cloud computing technology. J. Qingdao Vocat. Tech. Coll. **34**(05), 24–27 (2021). (in Chinese)
28. Hew, K.F., Liu, S., Martinez, R., et al.: Online education evaluation: what should we evaluate? Assoc. Educ. Commun. Technol. **4**, 243–246 (2005)
29. Alkhafaji, S., Sriram, B.: Instructor's performance: a proposed model for online evaluation. Int. J. Inf. Eng. Electron. Bus. (IJIEEB) **10**(01), 34–40 (2013)
30. Harrington, C.F., Reasons, S.G.: Online student evaluation of teaching for distance education: a perfect match? J. Educ. Online **2**(1), 1–12 (2005)

Simulation Study of Multiple Ships Material Transport Planning for Replenishment at Sea

Bingbing Li$^{(\boxtimes)}$, Bing Tang, Yong Luo, Gongzhuo Xu, and Zepeng Qin

School of Logistics, Wuhan Technology and Business University, Wuhan 430065, China
767396905@qq.com

Abstract. Before a ship sets sail for a combat mission, it must supply materials that ship need, such as water, food, fuel oil. Due to different characteristics of materials and sailing time, supplying time of materials was in uncertainty. It is necessary to look for optimal decision to achieve multiple ships replenishment in time which could promote warship combat effectiveness. Materials replenishment is of vital for single ship and multiple ships material transport. To solve the problems of alongside replenishment and vertical replenishment, we build a universal model. This paper is using visual process to simulate how supply ship replenishing materials and how helicopter moving materials to combat ships. Firstly, we build mathematical model to show the influence factor about multiple ships replenishing. Secondly, we build a visual model to simulation our plan, our model is clearly to count how long the ship takes for the replenishing materials, as well as how much multiple ships' materials' quantity. In the end, we get the shortest time and its path, hoping to provide some suggestion for ships in combating at sea.

Keywords: Multiple ships · Alongside replenishment · Vertical replenishment · The shortest time · Transportation path

1 Introduction

Alongside replenishment and vertical replenishment are two most common forms of material replenishment [1, 2]. Alongside replenishment refers to one supply ship to replenish needed material for multiple ships at one time and resupply them all at once. While vertical replenishment refers to one helicopter that can only replenish one ship at a time, possibly once or more [3]. Research about shortest path becomes the hot theme. There are enough studies about ship material replenishment at sea for multi-objective programming problem, improving traveling salesman problem [4], building system dynamics to achieve shortest path. There exists much uncertainty about combat ships at sea. The ships are within unknowing position and quantity of materials. A genetic algorithm is used to seek shortest path for multiple destinations [5], shortest path could help ensure the optimal matching among user query and provider service [6]. While there is much uncertainty among combat craft's materials replenishment, for example ships' position changing in per unit time in alongside replenishment. And ships' quantity of material in vertical replenishment is also changing frequently. Many scholars focused on

Z. Hu et al. (Eds.): ICCSEA 2022, LNDECT 134, pp. 475–487, 2022.
https://doi.org/10.1007/978-3-031-04812-8_41

shortest path [7, 8], for example seeking a shortest path to arrive on time under a given condition by applying Bellman's principle of optimality, while the shortest time about the content path is also occurring in fact life, which is also very important [9]. Previous studies did not consider constructing a general model for visual simulation research to seek for the shortest time and its corresponding path. In previous studies, combination of mathematical model and simulation model is more directly to observe the process of ship replenishment [10–12]. We employ mathematical model and simulation model to build visual simulation to obtain the shortest time of materials' replenishment project, in order to provide some suggestions for our navy ships materials replenishment.

2   The Content of Model Program

The program is using visual simulation tool to research the shortest time for transport materials that were needed by ships in battle. It contains alongside replenishment and vertical replenishment. It equals to say supply ship and combat ship. First, we construct a model to restore supply activities. Secondly, we set parameters of subjects in simulation software and input data into simulation system. Thirdly, we simulated the model to look for the shortest time and its transport path. Finally, we download data from simulation software, and the methods and methods of simulation credibility evaluation are introduced. In addition, to analysis and assess which project is better.

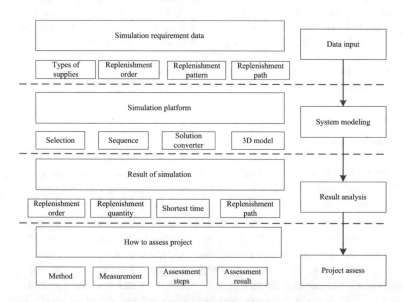

Fig. 1. Technical route of project

This program includes alongside replenishment and vertical replenishment of materials. At first, we construct the logical function model of the replenishment process. Then, we draw the flow diagram of replenishment. And then, we reconstruct the replenishment model in combination with the replenishment process. Finally, we obtain the output data by virtual simulation and evaluate the credibility of project by using the results from exported data. In the end, the results show that the credibility of this project is good and valid. The technical route of this paper is shown in Fig. 1.

3 The Project of Material Alongside Replenishment

3.1 The Content of Alongside Project

In order for naval formations to be able to carry out the assigned ship missions at all times and locations, they must rely on a strong maritime capability of material supply. Replenishment is a kind of sea replenishment method, which refers to the ship in the navigational state that both the replenishment ship and the receiving ship are at sea. To maintain a horizontal formation, replenishment ship should connect the ship in battle, keeping the materials moving to the battling ship from the replenishment ship safely.

The important thing is that supply ship should do things at the same time, one thing is to look for the best way to achieve material replenishment, the other thing is to moving itself ship to demand ship (ship in battles). There are many influence factors for supply ship to consider. For example, the speed of supply ship is one of the most important factors. It includes load time and unload time. And max travel speed, acceleration and deceleration. We should also take the quantity of replenishment into our scope of research. The aim of the project is to look for the key influence factor to explore the shortest time and its way to achieve effective material replenishment.

3.2 The Mathematical Model Construct

When the supply ship sets out to supply each battling ship, the supply ship needs to complete all the supplies required by the battle ship at once when carrying out horizontal supply. We assume that the initial inventory of all combat ships is Q_a. At time T, the Ship i send out request about material supply. According to the previous replenishment experience, the monitoring system of supply ship gives out a request to start activities of resupply supplies. Supply ship in its inner warehouse has enough needed material to be replenished. We assume that the loading of its inner warehouse of the replenishment ship is completed instantly. Then the other combat ships are also replenished, the supply capacity is B_a. Depending on the actual situation, there exists $Q_a > B_a$. At time T, the positions of each combat ship are W_a and its range is $\{(x_1, y_1), (x_2, y_2), \ldots, (x_n, y_n)\}$. We assume that the initial position of the supply ship is $(0,0)$. At time T, it is assumed that the speed at which each combat ship sails forward is S_a, and its range is $\{S_1^a, S_2^a, \ldots, S_n^a\}$. After the request for replenishment is made at time T. It is assumed that time of each ship is completing the replenishment is t, and the replenishment time is t_a, and its range is $\{t_1^a, t_2^a, \ldots, t_n^a\}$. We try to seek the relative shortest path for the shortest time. So we set the target function. Forward inference replenishment process, the first combat ship is supplied with supplies, which were L_i. Positional movement is suspended after the combat Ship i gives out a supply request, and the distance from the supply ship to the combat ship is $\sqrt[2]{x_i^{a2} + y_i^{a2}}$, and the time required is t_n^b.

In the above cases, there is no suspension, and the normal replenishment time is $t_i^a + t_i^b$. Depending on the distance of the voyage, it may exist:

$$L_i = S_i^a * \left(t_i^a + t_i^b \right) \tag{1}$$

After finishing the material replenished, the position of supply ship is $(x_i^a + x_i^b, y_i^a + y_i^b)$, the position is relational:

$$L_i = \sqrt[2]{(x_i^a + x_i^b)^2 + (y_i^a + y_i^b)^2} + S_i^a t_i^b \tag{2}$$

The values $(x_i^a + x_i^b, y_i^a + y_i^b)$ are available, it is the initial position of the second times of replenishment after one resupply of the supply ship. The material requirements for resupplying the second combat ship, assuming that the second combat ship is suspended, the distance from the supply ship to the second ship is $\sqrt[2]{(x_i^a + x_i^b - x_j^b)^2 + (y_i^a + y_i^b - y_j^b)^2}$. The time required to sail is t_j^b.

In the same way, it can be calculated that at normal replenishment time is $t_j^b + t_i^a$. Depending on the distance of the voyage, it may exist:

$$L_j = S_i^a * (t_j^b + t_i^a) \tag{3}$$

The same reason can be found after the completion of the supply location, distance, etc. In summary, the shortest time in the replenishment process can be expressed as:

$$T^* = min \sum_{i=1}^{n=5} (t_i^a + \sum_{j=1}^{n=5} \frac{\sqrt[2]{\left(x_i^a + x_i^b - x_j^b\right)^2 + \left(y_i^a + y_i^b - y_j^b\right)^2}}{S_j^a}) \tag{4}$$

3.3   Layout of Model

According to the above requirements, one supply ship and five combat ships. The process of alongside replenishment is shown in Fig. 2. Based on the simulation system, we selected 1 source, 5 queues, 1 sink to lay out the model. Source means supply ship and multiple entities are issued from the source to indicate supplies to the supply ship. Queue means combat ship and we use relative displacement. That is, the combat ship is stationary. The combat ship could store material for temporary storage, once the material arrived in the combat ship instantaneously into the combat ship warehouse, supply ships randomly left into the next combat ship for material resupply. Sink means supply ship return and assumes that the supply ship could supply material only in one time. When all combat ships are replenish, they randomly exit the combat queue, return to the starting point, and are absorbed by a sink.

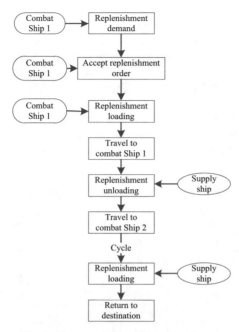

Fig. 2. Process of alongside replenishment

3.4 Model Simulation

The parameters of mathematical model are incorporated into the simulation model, and the input of the model includes the initial inventory of 5 combat ships. We assume instant replenishment is instant, that is to say, the time of replenishment is nearly to zero. In order to explore the supply path involved in the shortest time, this program uses the time setting for replenishment of the staging area.

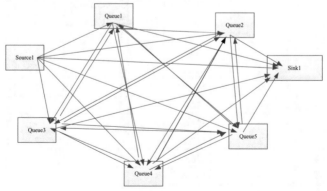

Fig. 3. Model of alongside replenishment

Because of the great uncertainty in the position of combat craft in the course of combat, it is not possible to quickly determine the specific location of combat craft. So we try to build a general model. In this stage it is necessary to try to judge the geographical location of combat craft according to real-time judgment, and then carry out material replenishment. Since there are 5 combat ships, there are 120 types of replenishment plan. Because it exists $A_5^5 = 5*4*3*2*1 = 120$. This article uses global Table to set up different starting points and destinations. The model layout is shown in Fig. 3.

By running the simulation model, we finally get results which are shown as Table 1.

The minimum replenishment time is 110.1090 s, rounding retains two decimal points, the theoretical calculation of the minimum replenishment time is 110.11 s, and the replenishment path is Path 14, that is 1-2-5-4-3. To meet the replenishment needs of the

Table 1. Path and time to alongside replenishment

Path	P1	P2	P3	P4	P5	Time	Path	P1	P2	P3	P4	P5	Time
1	2	4	5	1	3	130.86	61	1	4	3	5	2	138.61
2	2	1	5	4	3	122.10	62	4	1	3	2	5	129.09
3	3	4	2	5	1	137.85	63	4	5	2	3	1	0.00
4	1	5	3	2	4	156.50	64	3	2	5	1	4	141.06
5	2	5	3	4	1	136.52	65	2	3	5	1	4	153.69
6	3	1	4	5	2	121.33	66	4	1	3	5	2	135.15
7	5	4	2	1	3	0.00	67	1	2	5	3	4	119.85
8	4	3	2	1	5	145.09	68	1	5	4	3	2	136.69
9	3	1	2	4	5	118.17	69	2	3	5	1	4	153.69
10	5	1	3	2	4	135.65	70	5	1	4	2	3	152.69
11	1	4	2	5	3	147.27	71	2	1	5	4	3	122.10
12	5	4	1	3	2	0.00	72	2	3	4	5	1	135.18
13	4	3	5	2	1	139.26	73	1	5	4	2	3	146.00
14	1	2	5	4	3	110.11	74	4	1	5	3	2	160.77
15	1	4	2	3	5	157.35	75	2	3	5	4	1	146.82
16	3	4	1	5	2	136.45	76	1	2	5	3	4	119.85
17	3	2	4	5	1	148.11	77	1	4	2	3	5	157.35
18	5	1	2	3	4	125.28	78	5	3	4	1	2	120.71
19	4	3	2	5	1	148.88	79	2	1	5	3	4	131.85
20	1	3	2	5	4	114.97	80	5	2	3	4	1	144.72
21	1	2	4	5	3	134.87	81	3	2	4	1	5	155.94
22	3	4	1	2	5	119.29	82	4	2	5	3	1	124.21
23	3	4	1	5	2	136.45	83	5	3	4	1	2	120.71
24	4	3	2	5	1	148.88	84	4	5	1	3	2	0.00
25	5	3	4	2	1	123.59	85	1	2	4	5	3	134.87
26	1	4	3	2	5	132.55	86	3	2	1	5	4	129.52
27	1	2	5	3	4	119.85	87	4	3	1	2	5	118.94

(*continued*)

Table 1. (*continued*)

Path	P1	P2	P3	P4	P5	Time	Path	P1	P2	P3	P4	P5	Time
28	1	2	4	5	3	134.87	88	1	4	5	2	3	132.63
29	5	4	3	2	1	0.00	89	5	1	3	4	2	130.33
30	4	5	3	1	2	0.00	90	2	5	4	3	1	110.64
31	2	4	1	5	3	158.49	91	1	3	2	4	5	129.77
32	4	5	3	2	1	0.00	92	2	1	4	3	5	133.45
33	2	1	5	4	3	122.10	93	5	4	3	2	1	132.75
34	3	4	1	2	5	119.29	94	1	2	5	4	3	110.11
35	4	5	3	1	2	0.00	95	5	2	1	3	4	113.76
36	1	2	4	3	5	135.45	96	1	4	2	3	5	157.35
37	3	2	5	4	1	134.20	97	5	1	3	2	4	135.65
38	4	5	1	3	2	0.00	98	3	5	2	1	4	131.44
39	2	4	5	1	3	130.86	99	1	5	3	2	4	156.50
40	4	3	5	1	2	149.67	100	4	5	3	1	2	126.53
41	3	2	1	5	4	129.52	101	2	4	1	5	3	158.49
42	5	1	2	4	3	130.79	102	5	2	4	1	3	140.40
43	3	1	4	5	2	121.33	103	5	2	3	4	1	144.72
44	2	4	1	3	5	142.29	104	4	3	2	1	5	145.09
45	5	2	1	4	3	131.56	105	3	5	2	1	4	131.44
46	1	3	5	4	2	135.21	106	1	3	5	2	4	130.88
47	1	3	5	2	4	130.88	107	1	2	4	3	5	135.45
48	5	1	3	4	2	130.33	108	5	1	2	3	4	125.28
49	1	3	5	2	4	130.88	109	2	4	5	3	1	135.40
50	3	1	5	2	4	130.38	110	2	5	4	3	1	110.64
51	1	3	2	4	5	129.77	111	2	4	3	5	1	151.23
52	3	2	4	5	1	148.11	112	1	5	4	2	3	146.00
53	1	3	4	5	2	110.27	113	5	3	4	2	1	123.59
54	1	4	3	5	2	138.61	114	4	2	3	5	1	149.54
55	1	2	3	5	4	130.15	115	5	3	1	2	4	114.65
56	5	4	3	2	1	0.00	116	4	3	2	1	5	145.09
57	1	5	4	2	3	146.00	117	2	4	1	3	5	142.29
58	1	2	3	5	4	130.15	118	4	2	3	1	5	130.70
59	1	4	2	5	3	147.27	119	3	4	5	2	1	112.93
60	1	2	4	3	5	135.45	120	4	3	1	5	2	136.10

above-mentioned ships, the replenishment order of supply ships should be 1-2-5-4-3. In addition, Path 7, 12, 29, 30, 32, 35, 56, 84 are regarded as the most unreasonable replenishment, so their total replenishment time is zero.

4 The Project of Material Vertical Replenishment

4.1 The Content of Vertical Project

In order for naval formations to be able to carry out the assigned combat missions at all times and locations, they must rely on a strong maritime replenishment capability. Vertical replenishment is the use of helicopters to receive supplies to receiving ships activities.

Suppose that a fleet of ships consists of a supply ship and several combat ships, and a helicopter carries out a vertical replenishment mission. With the consumption of materials, the demand of materials would come out when the amount of materials decrease to the threshold. The supply demand for vertical replenishment mainly includes ammunition, medicinal materials and other dry cargo resources, as well as aviation kerosene and other barreled materials. What they have in common is that the replenishment process for these supplies is discrete, and their replenishment is carried out in batches.

In the process of vertical replenishment at sea, the materials are moving to the deck from inner warehouse of supply ship. The helicopter waits until completion of materials loading. Then helicopter brings materials flying to the combat ships. Once the helicopter close to the ships, it will unload materials vertically. Until finishing unloading all the materials, the helicopter returns back to the deck of supply ship, and waits for another transportation task. According to the actual vertical replenishment process, the complete replenishment cycle can be divided into five steps: delivery of cargo from storage, materials loading, transportation, materials unloading, return to the deck of the supply ship.

Vertical replenishment is different from alongside replenishment. Vertical replenishment only supplies one combat ship at a time, and then returns to the starting point for the next combat ship after the replenishment task is completed. However, alongside replenishment refers to supply numbers of materials to multiple ships at one time.

4.2 The Mathematical Model Construct

The project mainly relies on the mathematical model of alongside replenishment. We provide three plans for vertical replenishment. Every plan should set t_n, and its range is $\{1, 2, 3, 4, 5\}$. Time t_n is the time that combat ships send replenishment demand of materials. The quantity of materials that combat ships need is q_i. The max quantity of the helicopter can transport is q_0. So we can get the times that helicopter should transport is $m_i = q_i/q_0$. The speed of the supply ships replenishing materials is s_0.

1) The First Plan

   In the first plan, the helicopter should wait until supply ship ensuring all materials are ready for immediate shipment. The order of helicopter transporting is the earliest time that the combat ships send out message of replenishment. So the total time is sum of the time of supply ship replenishing materials and the time of helicopter transporting materials from the supply ships to combat ships. That is:

$$T_1 = \frac{\sum_{i=1}^{n=5} q_i}{s_0} + \sum_{i=1}^{n=5} \frac{m_i S_n}{s} \qquad (4)$$

2) The Second Plan

In the second plan, the helicopter should wait supply ship ensuring one unit of material ready to shipment. The order of helicopter transporting is the shortest distance from combat ship to supply ship. So the total time is the sum of the time that supply ship replenishment one unit of material and the time that supply ship replenish materials by shortest distance of supply ship and combat ship. That is:

$$T_2 = \left(1 + \sum_{i=1}^{n=5} m_i\right) \frac{q_0}{s_0} \qquad (5)$$

3) The Third Plan

In the third plan, the helicopter also should wait supply ship finishing one unit materials replenishing. The order of the helicopter transporting is the earliest time that combat ships send replenishment demand and the shortest distance from supply ship and combat ships. So that total time is:

$$T_3 = \frac{q_0}{s_0} + \sum_{i=1}^{n=5} \frac{m_i S_i}{s} \qquad (6)$$

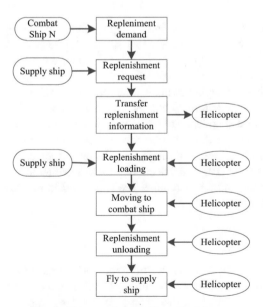

Fig. 4. Process of alongside replenishment

4.3   Layout of Model

The replenishment model is constructed based on the above model assumptions. According to the analysis of the process of vertical replenishment at sea, we select 3 types of entities, such as supply ship, receiving ship and the helicopter. These three entities are established respectively. The interaction among the three agents constitutes the maritime transport replenishment planning model. The process of vertical replenishment is as shown in Fig. 4.

The model layout is shown in Fig. 5.

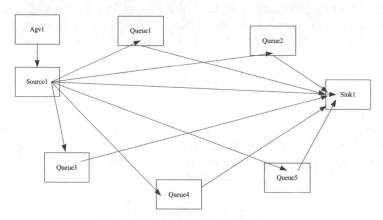

Fig. 5. Model of vertical replenishment

According to the above process, build 1 supply ship, 5 battleships and 1 helicopter. Source1 means supply ship and the Source1 supply ship responds to the ship's supply requirements and packs the supplies in the inner warehouse for shipment. Queue means combat ship and according to normal distribution statistics of inner warehouse consumption related materials, according to inner warehouse material inventory level to supply ship supplies supply demand, including the type and quantity of supplies. There are 5 ships in total that need supplies. Agv1 means helicopter and the helicopter carries out loading, flight, unloading and return operations. According to the replenishment information issued by the battleship and supply ship, it only responds quickly to one demand at a time and needs to be designated according to the queuing instruction. Sink1 means Material consumption and when supplies are consumed in the process of supply, the sink is used to achieve.

4.4   Model Simulation

With different replenishment points and different replenishment loading order, a total of 5 replenishment projects are set up. According to the actual demand of vertical replenishment materials, it is assumed that each ship's replenishment meets the following relationship, as shown in Table 2.

After we simulated 3 models about 3 related plans, we got Table 3.

Table 2. Ship replenishment and order of issuing orders

Ship and time	V_1	V_2	V_3	V_4	V_5
(Order, quantity)	(1,2)	(2,1)	(3,1)	(4,2)	(5,3)

Table 3. The time and path of vertical replenishment

Plan	Replenishment time	Replenishment path
P1	194.96	2-1-1-3-4-4-5-5-5
P2	163.96	1-1-5-5-5-4-4-2-3
P3	150.92	1-1-2-3-4-4-5-5-5

After model simulation, the shortest replenishment time and corresponding path can be obtained. The final replenishment time is 150.92 s, and the corresponding recharge path is 1-1-2-3-4-4-5-5-5.

5 The Evaluation of Project

The credibility evaluation of the visual simulation of ship material replenishment is mainly about the results of simulation could reach our material replenishment goal. The specific scenario evaluation steps are as follows: At first, we ensure that the correctness of the output data reaches our aim. In detail, we collect output data, select valid data, convert type of data and verify them. Then, the credibility of the simulation model is evaluated. The collected data is processed and expert scoring is used for model confidence assessment. Finally, the credibility evaluation of the simulation system is carried out by using the expert scoring method.

In the first step: large-scale boat supplies visual simulation system factor division, the factor set $U = \{U_1, U_2, ..., U_n\}$, $n = 3$.

In the second step: Build a fuzzy judging set, which can be represented by $V = \{V_1, V_2, ..., V_m\}$, $m = 7$. It can help us contain multiple evaluation indicators, such as very credible, better trustworthiness, basic trustworthiness, untrustworthiness, very untrustworthy. We use expert scoring method, such as the definition of 10 points to the highest score, in turn decreasing to represent the reduction of the credibility of the model.

In the third step: The weight of each factor in the calculation factor set on the visual simulation system, that is, the weight matrix $A = \{a_1, a_2, ..., a_n\}$. We select 5 associate professors and above titles to investigate, assuming that the collection of experts involved in the study is $E = \{e_1, e_2, ..., e_l\}$, $l = 5$. And follow these steps:

We take into account the weight of the influence of various factors on the overall credibility and the evaluation results of each subsystem. The corresponding judgment of the maximum value in the result set is the credibility evaluation result of the visual simulation system of ship material replenishment.

Based on the evaluation scheme, the simulation model is evaluated in step. We start with the construction of the evaluation set. Then, the calculation factor affects weight A and five associate professors are selected for evaluation. Finally, we obtain A = [0.06, 0.07, 0.07, 0.08, 0.07, 0.06, 0.07].

The ship material replenishment system contains the configuration of material handling tools (quantity of 1 ship, loading time, speed, acceleration, de-speed reduction) and whether the arrival of the supplies is consistent with the theory. By using expert scoring, and finally calculates the proportion of each expert on the evaluation set of the supply model. The result set B is obtained from the fuzzy transformation result set. And by calculating, the proportion of V_1 is the largest, exceeding 50%. The V_1 is in a very credible range, so it shows that the credibility analysis results of ship material replenishment system are very credible.

6 Conclusion

This article begins with an overview of the project's replenishment framework and what the replenishment contains. Secondly, according to the supply content to draw the replenishment process, and according to the various parameter value layout simulation model, including model conversion, model layout, model input, model simulation, data analysis, according to the simulation results scheme using the relevant model evaluation method, finally we get two better ships replenishment projects, as for alongside replenishment the route of ships order 1-2-5-4-3 and as for vertical replenishment the route of ships order 1-1-2-3-4-4-5-5-5 are the two best projects under the given conditions. It is concluded that the supply this paper provides the material Lateral supply and vertical supply scheme after evaluation can meet the current needs, and high credibility.

This paper provides two replenishment projects, under given conditions, the efficiency of materials replenishment is improved effectively. If we know every location of ships, we could calculate the shortest time about every specific path we want. However, there are still some shortcomings. The speed of ship is different, total time may introduce random variables, the exchange of ships sailing to the two projects should be studied in further.

Acknowledgment. Thanks for the support provided by Project Simulation Model of Ships Formation Alongside Replenishment based on FLXESIM from Wuhan Technology and Business University (2020H17).

References

1. Dong, P., Huang, J., Shi, H.: Research on alongside replenishment planning at sea based on two-stage optimization model. Fire Control Command Control **46**(8), 64–70 (2021). (in Chinese)
2. Dong, P., Wu, C., Wan, J., et al.: Simulation study of single ship material transport planning for underway replenishment based on multi-agent. In: 20th CCSSTA 2019 (2019). (in Chinese)
3. Yu, P., Zhang, Y., Dong, P.: Research on the vertical replenishment planning of naval vessels formation at sea. China Water Transport **19**(2), 28–30 (2019). (in Chinese)

4. Suwannarongsri, S., Bunnag, T., Klinbun, W.: Traveling transportation problem optimization by adaptive current search method. Int. J. Inf. Eng. Electron. Bus. (IJIEEB) 5(8), 33–45 (2014)
5. Moza, M., Kumar, S.: Finding K Shortest Paths in a network using genetic algorithm. Int. J. Comput. Netw. Inf. Secur. (IJCNIS) 12(5), 56–73 (2020)
6. Kumar, P., Singh, A.K.: Map reduce algorithm for single source shortest path problem. Int. J. Comput. Netw. Inf. Secur. (IJCNIS) 12(3), 11–21 (2020)
7. Fu, L., Sun, D., Rilett, L.R.: Heuristic shortest path algorithms for transportation applications: state of the art. Comput. Oper. Res. 33(11), 3324–3343 (2006)
8. Nie, Y.M., Xing, W.: Shortest path problem considering on-time arrival probability. Transp. Res. Part B 43(6), 597–613 (2009)
9. Karpenko, M., Bhatt, S., Bedrossian, N., et al.: Flight implementation of shortest-time maneuvers for imaging satellites. J. Guid. Control. Dyn. 37(4), 1069–1079 (2015)
10. Takashina, J.: Ship Maneuvering motion due to tugboats and its mathematical model. J. Jpn Soc. Naval Archit. Ocean Eng. 160, 93–102 (1986)
11. Zhang, X., Fang, Q., Zhu, W.F., et al.: Visual simulation of material supply for a large ship. Comput. Syst. Appl. 24(9), 43–48 (2015). (in Chinese)
12. Wen, C., Kai, H., Zhu, W.: Research on operation optimization problem of material handling system for large ship. Appl. Mech. Mater. 347–350, 3740–3744 (2013)

Improving U-Net Kidney Glomerulus Segmentation with Fine-Tuning, Dataset Randomization and Augmentations

Roman Statkevych[(✉)], Yuri Gordienko, and Sergii Stirenko

National Technical University of Ukraine "Igor Sikorsky Kyiv Polytechnic Institute", Kyiv, Ukraine
romanstatkevich@gmail.com

Abstract. The recent success of deep learning methods for Computer-Aided Detection (CADe) and Computer-Aided Diagnosis (CADx) provide a possibility to speed up some specific medical procedures, for example, medical image segmentation and analysis, and suggests ways to perform it in the very effective ways. In this study of the Human Kidney segmentation task, we sought to improve the efficiency of our previously proposed method based on some U-Net architectures. The investigation measures the impact of several techniques on the model performance, like test time data augmentations (TTAs) and train time data augmentations (TrTAs), shuffling, randomization, normalization, hyperparameter fine-tuning, and others. It was shown that a major improvement of the Dice Score for the test dataset - from 0.9127 as in initial research up to 0.934 (almost by 3%) - was achieved by applying dataset shuffling. The reliability of the obtained results was checked by K-Fold cross-validation for some optimization techniques. These findings can be used for different kinds of precision-focused segmentation tasks in the CADe/CADx context.

Keywords: Neural networks · Healthcare · U-Net · Image segmentation data augmentation · Optimization

1 Introduction

In recent years analyzing medical images became one of the popular areas of research in the computer vision field, especially in the context of events such as the COVID-19 pandemic, which cause a surge in rapid diagnosing of a large number of patients based on different types of visual data provided by Computer Tomography, Magnetic Resonance Imaging, microscopy, etc. Computer-Aided Detection (CADe) and Computer-Aided Diagnosis (CADx) provides a possibility to speed up diagnostic with the same or higher precision compared to a human doctor, so they can focus on more urgent tasks [1–3]. In the view of many other practical applications, it is important to constantly work on improving existing methods [4–6].

Convolution Neural Networks (CNN) are actively used as one of the approaches in CADe/CADx for medical image data processing. Multiple approaches proved to be

© The Author(s), under exclusive license to Springer Nature Switzerland AG 2022
Z. Hu et al. (Eds.): ICCSEEA 2022, LNDECT 134, pp. 488–498, 2022.
https://doi.org/10.1007/978-3-031-04812-8_42

efficient in different tasks. Most notable is U-Net [7], which is widely used in different segmentation scenarios and different data sources. In our previous research [8], we used U-Net to find glomerulus on a Kidney microscopy image from The Human BioMolecular Atlas Program [9]. In [8] we had reviewed existing methods and pointed out a lack of data and results for a various types of models, which should be covered. We have shown, that it is possible to achieve reasonable results (0.91 by Dice-Score) with a basic neural network setup. In this research we decided to try several different methods to improve a result with the same network architecture.

To our knowledge, there is not enough research regarding influence of different improvement techniques methods on the model performance for the kidney glomerulus segmentation, and we target to measure it. These measurements are important to avoid redundancy of the method and improve its resource and time consumption.

2 Methodology

2.1 Dataset

Training dataset plays a major role in defining the efficiency and precision of a CNN-based method. Mislabeled data can cause poor model conversion during evaluation, and for some scenarios, an available dataset won't be enough to train a model with good precision and also avoid overfitting. As in our previous research [8], we would use the HUBMAP Kidney dataset [9] from Kaggle, which consists of 15 high-resolution tissue images of human kidneys, split into 13 image train set and 2 image test set. For training and evaluation, images are split into tiles sized 1024×1024 pixels, excluding empty tiles (which are detected by calculating HSV (Hue-Saturation-Value) of an image, and checking, if there are more than 1000 pixels with saturation value more than 40).

2.1.1 Dataset Extension and Augmentations

In order to extend an available dataset, some augmentations could be applied that could be very useful for the relatively small datasets [3]. In this and our previous research [8] we use dataset that contains a few large high-resolution images; each image was split into smaller parts (tiles) in the way that no two tiles overlap with each other. We considered to crop tiles differently: each of two neighboring tiles overlap on 1/2 of tile size with each other on all four directions. This change increases a number of different tiles by 4 times.

Data augmentation is a deliberate modification of a training dataset (train time augmentations - TrTA) or test dataset (test time augmentations - TTA), using various image transformations, including rotating images, applying different filters, and adjusting color schemes. In the case of TrTAs, some randomized set of image transformations can be applied to the input images before forwarding pass. Test time augmentation is done differently [10] - let's assume neural network is represented by, $f(\theta, x) : X \to Y$ and there are n sets of sequential transformations represented as, $\forall i \in [1, n] \Rightarrow t_i(x) : X \to X$ where θ represents model's weights, X is input domain and Y is an output domain - than train time augmentation is represented as (2). In our previous work, we only used basic image normalization for both training time and test time augmentations, described by

(1), where M is an index of the channel (3 channels for RGB images), M is an average value for the specific channel, D is a standard deviation for a channel. It is possible, that applying additional transformation may further improve the performance of the predictions.

$$I_i^{norm} = \frac{I_i - M_i}{D_i} \tag{1}$$

$$Y_{pred} = \frac{\sum_{i=1}^{n} f(\theta, t_i(X))}{n} \tag{2}$$

For the experimental part, it was decided to use image flips as an augmentation. For TrTAs, the image could be left as is, flipped vertically, horizontally, and both vertically and horizontally with uniform probability, increasing 4 times a number of possible variations of each tile. Applying other types of augmentations (like Gaussian blur or color schema modification) led to a poor model conversion, so they were withdrawn from the experiments. For TTAs, all possible flips were applied to an input image, running model prediction on each possible combination of flips (4 in total) and calculating the pixel-wise mean of all predictions (as in formula (2)).

2.1.2 Dataset Shuffling and Randomization

Another way to modify the existing method is to use dataset shuffling. For each training epoch, dataset items are shuffled for forwarding pass. It is supposed to help overcome selection bias, which is a common issue for any studies, which use data sampling (for example, sociology [11]). Utilizing the cropping with overlapping tiles (described above) to provide more dataset entries, it is also possible to randomly sample a part of dataset for each training epoch.

2.2 Model

2.2.1 Baseline Model

As a baseline approach, our previous research is used, which uses U-Net-like model [7] with CNN architecture, weighted BCE loss function, Adam optimizer with initial learning rate 10^{-4} decreasing down to 10^{-8} when evaluation metric is on a plateau.

2.2.2 Eval Metrics

Main evaluations metric are Dice Score (DSC) [12] (3) and Intersection over Union (IoU) (4):

$$DSC = 2\frac{|GT| \wedge |PR|}{|GT| + |PR|} \tag{3}$$

$$IoU = \frac{|GT| \wedge |PR|}{|GT| \vee |PR|} \tag{4}$$

GT is a ground truth binary mask, PR is a predicted binary mask.

2.2.3 Fine-Tuning of Loss Functions

Hyperparameter tuning can often lead to improving model's performance [13]. Loss function plays a major role in prediction results, as it defines how gradients are modified during backward pass. There is a vast set of loss functions that could be applied for segmentation tasks, some of them are using thresholds, which could be fine-tuned for each specific task. In our previous research, weighted Binary Cross-Entropy loss function was used, which had one configurable threshold w, as seen in (6), responsible for determining an effect of true-positive predictions compared to true-negatives. As different datasets might have a different correlation of positive (segmented region belongs to any class) and negative (regions, which don't belong to any class) ground truth, setting a different penalty for a model for predicting false positives and false negatives might improve model's performance. As shown in previous research, prediction accuracy is close to 0.99, yet evaluation score is only 0.91, due to data imbalance (dataset consists of 90% of negative data) and consideration of both true positives and true negatives when it is far more important to predict positive segments. This makes accuracy not a representative evaluation metric in this case. The same rule applies to loss function, so it was considered to run experiments with Tversky loss (TL) [14] (7), which has parameters α and β to penalize false positives and false negatives correspondingly

$$BCE(p, y) = mean(l_1, .., l_n) \tag{5}$$

$$l_i = wy_i \cdot log(\sigma(p_i)) + (1 - y_i) \cdot log(1 - \sigma(p_i)) : p_i \in PR; y_i \in GT \tag{6}$$

$$T(GT, PR, \alpha, \beta) = \frac{\sum |GT \wedge PR|}{\sum |GT \wedge PR| + \alpha \sum |PR - GT| + \beta \sum |GT - PR|} \tag{7}$$

2.2.4 K-Fold Cross-Validation

Cross-validation [15] is another method to increase model prediction rate. The point of this method is to split the training dataset into K-folds, each containing N/K images, where N is a number of images. Then, K-models are trained, using to find best weights Θ^{max}_i for each fold, with which model has the highest evaluation on the k-th test fold. Afterward, test evaluation with trained weights can be calculated by (8), where DSC is a Dice Score by (3):

$$Score = \frac{\sum_{i=1}^{K} DSC(f(\Theta^{max}_i, X^{test}), Y^{test})}{K} \tag{8}$$

2.2.5 Environment

Experiments were written using Python with PyTorch library within Jupyter Notebook environment. Following server configuration was used:

– CPU - 12 Intel(R) Xeon(R) Silver 4214R CPU @ 2.40GHz,
– GPU - NVIDIA A100 videocard with 40 GB of video memory,
– AM - 82 GB.

3   Results

Below the results obtained by the single model training-validation method and K-Fold cross-validation method are presented.

3.1   Single Model Training-Validation Method

In this section of the experiment, the impact on the results obtained was investigated for the following different options:

- normalization,
- TrTA and TTA,
- shuffling to fight selection bias,
- various loss function hyperparameters.

In Table 1 results for different loss functions are shown. Binary Cross Entropy was used in our previous research, and it was decided to change parameter w (weight of the positive predictions) to see an influence on the results. Some measurements were omitted from Table 1, as they have shown poor results (for example, Tversky had given $DSC = 0.85$, BCE has shown $DSC = 0.77$). Results, represented in Table 1, were taken from the evaluation results of one of 30 training epochs with the highest DSC.

In Table 2 results for different dataset augmentations are shown. The baseline model is a model used in [8] (BCE with $w = 5$, with normalization, non-shuffled, no TrTA or TTA applied). In column "Parameters" the changes to the baseline model are described. "TrTA" means train time augmentations, "Shifted tiling" means that tiles are generated with step 512 (instead of 1024) pixel, increasing the number of tiles by 4x times. For shifted tiling, to make experiment consistent and to keep training time for all models the same, train sample is chosen to contain approximately the same number of tiles, as in original tiling, meaning that for each epoch only 1/4 of the extended dataset are selected, providing unique train samples for each training epoch. All models for experiments in Table 2 were run for 45 epochs.

3.2   K-Fold Cross-Validation

For experiments with K-Fold cross-validation, it was decided to split the train set into 4-folds and 6-folds. It means, that for each experimental run 4 or 6 models are trained during 45 epochs each. The best of the models are saved to be evaluated to get the final results. The following test setup options were considered:

- Shifted tiling - tiles are cropped with "shifted tiling" step (512 pixels) or original tiling step (1024 pixels); for each epoch, samples sized approximately 25% of the extended dataset is selected,
- Shuffled - tiles are shuffled randomly for each training epoch,
- TrTA - identifies if train time augmentations were applied,
- TTA - identifies if test time augmentations were applied.

Results are shown in Table 3. Each value in this table also has a fold-wise standard deviation metric, which helps to measure the stability of predictions for each model.

Table 1. Results for single model predictions depending on loss function. * - denotes a baseline configuration

Model parameters			Evaluation metrics			
Loss	Parameters	TTA	DSC	IoU	Precision	Recall
Tversky	$\alpha = 0.9, \beta = 0.1$	True	0.848	0.736	0.761	**0.957**
Tversky	$\alpha = 0.9, \beta = 0.1$	False	0.849	0.737	0.766	0.953
Tversky	$\alpha = 0.1, \beta = 0.9$	True	0.860	0.754	**0.952**	0.784
Tversky	$\alpha = 0.1, \beta = 0.9$	False	0.831	0.711	0.944	0.742
Tversky	$\alpha = 0.8, \beta = 1.2$	True	0.917	0.847	0.900	0.936
Tversky	$\alpha = 0.8, \beta = 1.2$	False	0.917	0.847	0.905	0.929
Tversky	$\alpha = 0.9, \beta = 1.1$	False	0.921	0.854	0.934	0.909
Tversky	$\alpha = 0.9, \beta = 1.1$	False	0.918	0.848	0.937	0.899
Tversky	$\alpha = 0.9, \beta = 0.7$	True	0.915	0.843	0.887	0.945
Tversky	$\alpha = 0.9, \beta = 0.7$	False	0.915	0.843	0.885	0.947
BCE*	$w = 5$	False	0.918	0.849	0.916	0.921
BCE	$w = 5$	True	**0.922**	**0.855**	0.922	0.922
BCE	$w = 7$	True	0.912	0.839	0.935	0.890
BCE	$w = 7$	False	**0.922**	**0.855**	0.936	0.888

4 Discussion

As a result of the experiment it was determined, that baseline model [8] was suffering from selection bias, which was overcame by introducing dataset shuffling. It was shown that model's performance improved from 0.914 (as in the original paper) to 0.934. Also, a small boost can be contributed to TTAs, which improved the shuffled method to 0.937. All other improvement methods have shown no additional effect in the best case and sometimes led to a decrease in the result score. Tiles shifting was expected to add additional metric improvement, but its effect seems to be minor as well. However, it seems that shifting allows the more balanced recall (0.942) and precision (0.932) values compared to the model with original tiling with precision 0.957 and recall 0.913. TrTAs have shown the worst performance out of all shuffled methods (0.929 with TTAs), however, they have shown the highest precision value (0.970) with a reasonable recall (0.89). This is a very important result for precision-focused tasks.

Table 2. Results for single model predictions depending on different dataset augmentations

Dataset augmentations		Evaluation metrics			
Augmentations	TTA	DSC	IoU	Precision	Recall
Baseline	False	0.918	0.849	0.916	0.921
Shuffled	True	0.934	0.877	0.957	0.913
Shuffled	False	0.934	0.877	0.956	0.914
Shuffled & Shifted tiling	True	**0.937**	**0.881**	0.942	**0.932**
Shuffled & Shifted tiling	False	0.933	0.875	0.941	0.926
Shuffled & No normalisation	True	0.918	0.849	0.943	0.895'
Shuffled & No normalisation	False	0.916	0.845	0.944	0.889
Shuffled & Shifted tiling & TrTA	True	0.930	0.870	0.931	0.930
Shuffled & Shifted tiling & TrTA	False	0.930	0.869	0.927	0.933
Shuffled & TrTA	True	0.929	0.867	0.970	0.890
Shuffled & TrTA	False	0.928	0.865	**0.970**	0.889

Table 3. Results with fold-wise standard deviation for K-folds cross-validation experiments.

Model parameters				Evaluation metrics			
Tiling	Shuffle	Folds	TTA	DSC	IoU	Precision	Recall
Shifted	True	4	False	0.928 ± 0.004	0.866 ± 0.007	0.915 ± 0.001	0.942 ± 0.015
Original	True	4	False	0.929 ± 0.008	0.867 ± 0.013	0.911 ± 0.024	0.947 ± 0.013
Original	False	4	False	0.901 ± 0.044	0.819 ± 0.067	0.882 ± 0.062	0.920 ± 0.037
Shifted	True	4	True	0.929 ± 0.003	0.868 ± 0.006	0.915 ± 0.007	0.944 ± 0.015
Original	True	4	True	0.931 ± 0.006	0.870 ± 0.010	0.911 ± 0.023	**0.951 ± 0.012**
Original	False	4	True	0.907 ± 0.040	0.830 ± 0.062	0.899 ± 0.056	0.924 ± 0.037
Original	True	6	True	**0.933 ± 0.007**	**0.875 ± 0.013**	**0.931 ± 0.018**	0.935 ± 0.012
Original	True	6	False	0.933 ± 0.008	0.874 ± 0.014	0.930 ± 0.02	0.936 ± 0.011

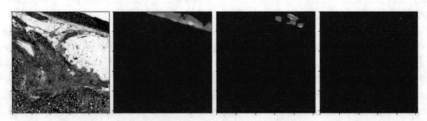

Fig. 1. Input image with problematic results with no ground truth segments, and predictions of different models.

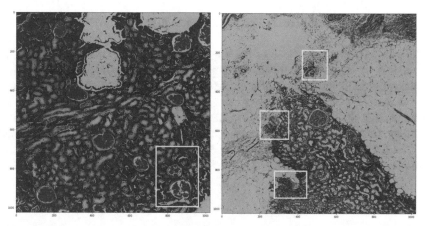

Fig. 2. Segmentation results with some artifacts (marked by hand with yellow rectangles). Blue segments show true positive results, while green are false positive.

As seen from the tables above, using new TTA (even basic) alongside image normalization almost always gives some slight improvements (0.01–0.08) compared to methods, which don't use it. This proves that TTA is an efficient way to optimize existing models, without additional training or changing model parameters. Normalization appears to be the most effective augmentation (as seen in Table 2), and the difference between normalized and non-normalized shuffled methods is almost 2%.

Interesting results were shown during experiments with different loss functions. Although Binary cross-entropy is not considered the best loss function for segmentation [16], it has shown the best and most stable performance. However, Tversky loss provides a possibility to adjust models in case-specific requirements for recall or precision values. For healthcare purposes precision (a measure of false-positive effect on model performance) could be a much more important metric than recall or Dice-Score, so in some cases model with a lesser Dice Score could be more preferable, if it has the highest precision. BCE doesn't give good control over recall/precision, while Tversky has explicit parameters α and β that allow more flexible control over these metrics.

Also, evaluation results were visually examined, to detect correlations in wrong or missed predictions. One of the problems is that there is no distinction between sclerotic and healthy glomerulus in the original dataset, which are visually different. As shown in [17], out of small biopsy samples containing 16 glomerulus on average, 32% contains at least one sclerotic glomeruli. As demonstrated in Fig. 2, on the left image there is two correctly segmented glomerulus, which looks different compared to other glomeruli, possibly because they are unhealthy, so they might introduce some bias, which could negatively affect model performance. Also, on the right image of Fig. 2 what appears to be remnants of the glomerulus are marked as false positives, because they still have a similar texture with healthy ones, so the model still considers them as glomeruli. In theory, it could be fixed by introducing additional classes, which might denote both healthy or unhealthy glomerulus, or even represent different stages of glomerulosclerosis. A similar approach was shown in [18] using a private dataset with both healthy and sclerotic classes annotated.

It was determined, that one of the major issues of predictions are false positives for sections of images, which contain no glomeruli, but some other tissue, which has similar texture. Figure 1 shows an example of this phenomenon. Leftmost image shows the test sample, which contains no ground truth, second image - prediction of K-Fold model without shuffling ($DSC = 0.907$), third image – prediction of K-Fold model with shuffling ($DSC = 0.928$), and rightmost is the prediction of the best performing model ($DSC = 0.937$). As seen from the Dice Score values, the higher the score, the lesser false-positive segment is predicted.

5 Conclusions

In this research, methods of improvement of our existing results of kidney glomerulus segmentation were investigated. It was managed to increase Dice Score for the test set from 0.9127 to 0.937, which is almost 3%. A major contribution to this increase belongs to the shuffling of the training dataset for each training epoch. Also, TTAs provided some minor increase in almost all experiments, however, it is not significant. TrTAs, shifting tiles with different steps, and using K-Fold cross-validations didn't improve Dice-Score further, although all of them increased initial results in terms of Dice Score (0.9285–0.9330), and have different effects on recall and precision metrics. For example, TTAs seem to have the best effect on the precision metric (0.97), and tiles shifting provides a more balanced precision/recall rate.

Additionally, we investigated how loss function might influence a performance of the method. Surprisingly, using standard BCE loss function had given better results, compared to a Tversky loss, despite the fact that in different researches [16] situation was opposite. This finding proves a point, that loss-functions which fit one task doesn't provide best results in other. Generalization of tasks and finding correlations between dataset's parameters and performance of different loss-function is a promising area of the research.

Most likely, there is no further possibility of improving results received during experiments. It was determined that the main reason for this is flaws of the dataset - missed labels and absence distinction between sclerotic and healthy glomerulus, etc. This proves the importance of dataset preparation and preprocessing. However, some other techniques, like pseudo-labeling [19] or Gaussian process [20], could be investigated as a part of further research, as well as different network architectures.

The finding of this research can be used as a recommendation for using shuffling and image color normalization for different kinds of segmentation tasks in the following researches in the field of Computer-Aided Detection (CADe) and Computer-Aided Diagnosis (CADx).

Acknowledgement. This work was partially supported by the KATY project which has received funding from the European Union's Horizon 2020 research and innovation program under grant agreement No 10101745.

References

1. Chen, Y.-W., Jain, L.C.: Deep Learning in Healthcare. Springer, Cham (2020). https://doi. org/10.1007/978-3-030-32606-7
2. Esteva, A., et al.: A guide to deep learning in healthcare. Nat. Med. **25**(1), 24–29 (2019)
3. Gang, P., et al.: Effect of data augmentation and lung mask segmentation for automated chest radiograph interpretation of some lung diseases. In: Gedeon, T., Wong, K.W., Lee, M. (eds.) ICONIP 2019. CCIS, vol. 1142, pp. 333–340. Springer, Cham (2019). https://doi.org/10. 1007/978-3-030-36808-1_36
4. Taran, V., et al.: Performance evaluation of deep learning networks for semantic segmenta-tion of traffic stereo-pair images. In: Proceedings of the 19th International Conference on Computer Systems and Technologies, pp. 73–80 (2018)
5. Taran, V., Gordienko, Y., Rokovyi, A., Alienin, O., Stirenko, S.: Impact of ground truth annotation quality on performance of semantic image segmentation of traffic conditions. In: Hu, Z., Petoukhov, S., Dychka, I., He, M. (eds.) ICCSEEA 2019. AISC, vol. 938, pp. 183–193. Springer, Cham (2020). https://doi.org/10.1007/978-3-030-16621-2_17
6. Vorotyntsev, P., Gordienko, Y., Alienin, O., Rokovyi, O., Stirenko, S.: Satellite image segmen-tation using deep learning for deforestation detection. In: 2021 IEEE 3rd Ukraine Conference on Electrical and Computer Engineering (UKRCON), pp. 226–231 (2021)
7. Ronneberger, O., Fischer, P., Brox, T.: U-net: convolutional networks for biomedical image segmentation. In: Navab, N., Hornegger, J., Wells, W.M., Frangi, A.F. (eds.) MICCAI 2015. LNCS, vol. 9351, pp. 234–241. Springer, Cham (2015). https://doi.org/10.1007/978-3-319-24574-4_28
8. Statkevych, R., Stirenko, S., Gordienko, Y.: Human kidney tissue image segmentation by u-net models. In: IEEE EUROCON 2021–19th International Conference on Smart Technologies, pp. 129–134 (2021)
9. Consortium, H., et al.: The human body at cellular resolution: the NIH human biomolecular atlas program. Nature **574**(7777), 187 (2019)
10. Doms, V., Gordienko, Y., Kochura, Y., Rokovyi, O., Alienin, O., Stirenko, S.: Deep learning for melanoma detection with testing time data augmentation. In: Hu, Z., Zhang, Q., Petoukhov, S., He, M. (eds.) ICAILE 2021. LNDECT, vol. 82, pp. 131–140. Springer, Cham (2021). https://doi.org/10.1007/978-3-030-80475-6_13
11. Berk, R.A.: An introduction to sample selection bias in sociological data. Am. Sociol. Rev. **48**, 386–398 (1983)
12. Dice, L.R.: Measures of the amount of ecologic association between species. Ecology **26**(3), 297–302 (1945)
13. Rizqyawan, M.I., et al.: Comparing performance of supervised learning classifiers by tuning the hyperparameter on face recognition. Int. J. Intell. Syst. Appl. (IJISA) **13**(5), 1–13 (2021)
14. Salehi, S.S.M., Erdogmus, D., Gholipour, A.: Tversky loss function for image segmentation using 3D fully convolutional deep networks. In: Wang, Q., Shi, Y., Suk, H.-I., Suzuki, K. (eds.) MLMI 2017. LNCS, vol. 10541, pp. 379–387. Springer, Cham (2017). https://doi.org/ 10.1007/978-3-319-67389-9_44
15. Refaeilzadeh, P., Tang, L., Liu, H.: Cross-validation. In: Encyclopedia of Database Systems, vol. 5, pp. 532–538 (2009)
16. Jadon, S.: A survey of loss functions for semantic segmentation. In: 2020 IEEE Conference on Computational Intelligence in Bioinformatics and Computational Biology (CIBCB), pp. 1–7 (2020)
17. Kremers, W.K., et al.: Distinguishing age-related from disease-related glomerulosclerosis on kidney biopsy: the aging kidney anatomy study. Nephrol. Dial. Transplant. **30**(12), 2034–2039 (2015)

18. Marsh, J.N., et al.: Deep learning global glomerulosclerosis in transplant kidney frozen sections. IEEE Trans. Med. Imaging **37**(12), 2718–2728 (2018)
19. Lee, D.-H., et al.: Pseudo-label: the simple and efficient semi-supervised learning method for deep neural networks. In: Workshop on Challenges in Representation Learning, ICML, vol. 3 (2013)
20. Matthews, A.G.G., Rowland, M., Hron, J., Turner, R.E., Ghahramani, Z.: Gaussian process behaviour in wide deep neural networks. arXiv preprint arXiv:1804.11271 (2018)

The Impact of Time Pressure, Emotional Intelligence and Emotion on Job Engagement: An Empirical Study in Logistics Enterprises

Ping Liu, Yunjing Zhao, Yijun Wei, and Yi Zhang[✉]

School of Business, Sichuan University, Chengdu 610065, China
657518287@qq.com

Abstract. With the deepening of organizational behavior research, the researchers are no longer committed to improve efficiency by optimizing organizational hierarchy structure, but begin to focus on positive emotions. Based on the emotional contagion theory, this paper analyzes the effects of time pressure (high/low), emotional intelligence (high/low), and emotional state (positive/negative) on employees' job engagement. A 2×2 scenario simulation experiment was conducted on 241 employees of logistics enterprises. The experiment result shows that: Time pressure has a significant impact on employees' emotions, which is negatively correlated with employees' positive emotions; Emotional intelligence has a significant impact on employees' emotions. Meanwhile, leaders with high emotional intelligence activate employees' positive emotions more than those with low emotional intelligence; The interaction impact between time pressure and emotional intelligence is significant, leaders with high emotional intelligence stimulate positive emotions more than negative emotions under high time pressure. Finally, the theoretical and practical significance and the future research direction are proposed.

Keywords: Time pressure · Emotional intelligence · Emotional contagion · Logistics enterprises

1 Background

In order to adapt to the increasingly stimulating external environment and improve the competitive advantage of enterprises, enterprises put forward higher requirements for employees. Requires employees to complete tasks with high quality and efficiency within certain working hours [1]. It has been shown that such time constraints will exert varying degrees of influence on the psychological level of employees and lead to positive or negative emotional states in the workplace [2]. In 2020, Southern Metropolis Daily released a survey report on Workplace Pressure and Overtime through research. According to statistics, as many as 72.58% of employees clearly stated that they need to work overtime. Among those who chose to work overtime, 37.31% were forced to make up for the lack of time to complete tasks in non-working hours. It can be seen that time pressure is a common problem faced by employees at work. On the one hand, time

Z. Hu et al. (Eds.): ICCSEEA 2022, LNDECT 134, pp. 499–508, 2022.
https://doi.org/10.1007/978-3-031-04812-8_43

pressure is regarded as a negative stressor, which harms the physiological and mental health of employees [3]. On the other hand, some researchers believe that time pressure will promote the work of employees and bring positive results. Schaufeli and Taris (2014) believed that time pressure, to a certain extent, would stimulate individuals' internal motivation and make them more motivated to complete their work, so that individuals would have more positive work behaviors and attitudes, such as improving work engagement and performance [4].

In team, the leader, as the direct contact of employees, can directly feel the working emotion of employees. Meanwhile, the emotional intelligence of the leader also has a direct or indirect impact on employees [5]. Studies have shown that leaders with high emotional intelligence are better at managing their own emotions, avoiding their negative emotions affecting employees, and being able to accurately identify employees' personal emotions and respond to them in time. When employees encounter difficulties and problems, leaders with high emotional intelligence will encourage employees with positive emotions, communicate with them, and solve problems together. In this way, leaders also promote employees' work enthusiasm, work engagement and a series of positive behaviors at work [6]. According to social exchange theory, when employees receive emotional trust from leaders, they are more willing to repay at work in a positive state and improve their loyalty to the organization [7]. Therefore, when the leader shows positive emotions, the employees will be affected by the positive emotions, which will also produce positive emotional experience such as happiness and joy. Positive emotional experience can improve employees' work enthusiasm, sense of identity and belonging to the company, and also help to create a positive work atmosphere and improve employees' organizational citizenship behavior [8]. Therefore, this paper introduced emotional intelligence into the analysis model, and further proposed that there is an interactive effect between time pressure and emotional intelligence, and explored how time pressure, emotional intelligence, and emotional state affect employees' job engagement.

2 Literature Review and Research Hypothesis

2.1 Time Pressure and Emotion

Time pressure is an individual's perception of the stress caused by a lack of time to do what they need to do [9]. Time pressure is a common problem faced by people in modern society. Studies have shown that time pressure is usually regarded as a negative stressor, which will bring psychological problems to employees, endanger their health and reduce their happiness at work [10]. Some researchers also found that time pressure can positively predict individual negative behaviors of employees, such as delaying work, negative coping and job burnout. When employees suffer from time pressure frequently, they will have emotional exhaustion. On the contrary, when people have enough time to complete the work, the individual subjectively has a good expectation of the time, so the time pressure will be small. Studies have shown that loose time will bring individuals higher happiness, and there is a positive correlation between the two [3]. Thus, we expect:

H1a: Time pressure is positively correlated with employees' negative emotions.
H1b: Time pressure is negatively correlated with employees' positive emotions.

2.2   Emotional Intelligence and Emotion

Emotional intelligence refers to the ability of individuals to effectively manage their own emotions and accurately identify and utilize others' emotions [11]. In an organization, leaders control their own emotions so that their negative emotions do not affect employees. At the same time, they can accurately identify employees' work emotions and give corresponding feedback and guidance according to their emotional information. Fredrickson (2003) believes that leaders with high emotional intelligence can bring positive emotional experience to employees and effectively improve their job engagement and self-efficacy [12]. Leaders with high emotional intelligence can accurately identify the emotional changes of others in the workplace, and take the best response to them. They always show positive emotional states to others in the communication process, and show their attention and respect to others [13].

In the workplace, individuals will have emotional contagion in their communication with others [14]. Among them, the emotional of the leader will affect the behavior of the employees. When the employees receive the positive emotions from the leader, they tend to have positive emotional experience. On the contrary, leaders with low emotional intelligence are more likely to show negative emotions and behave disrespectfully, coldly or rudely in the workplace, and employees will have negative emotional experience after receiving negative emotions from leaders [15]. Thus, we expect:

H2a: Emotional intelligence is positively correlated with employees' negative emotions.

H2b: Emotional intelligence is negatively correlated with employees' positive emotions.

2.3   Time Pressure, Emotional Intelligence and Emotion

In the workplace, employees will feel both time pressure and emotional messages from their leaders. McGregor (2016) believes that leadership care and team support from within the organization can reduce the negative state of employees under time pressure to a certain extent [16]. Therefore, when employees are under high time pressure, time pressure will become a negative stressor and bring negative emotional experience to employees. However, when leaders have high emotional intelligence, they can restrain the negative impact of time pressure on employees' emotions to a certain extent. Thus, we expect:

H3a: Time pressure interacts with emotional intelligence of leaders. Under high time pressure, high emotional intelligence is more likely to stimulate employees' positive emotions than low emotional intelligence.

H3b: Time pressure interacts with emotional intelligence of leaders. Under high time pressure, low emotional intelligence is more likely to stimulate employees' negative emotions than high emotional intelligence.

2.4   Emotion and Job Engagement

Relevant meta-analysis proves that employees with positive work emotions will have higher job satisfaction and performance, and thus improve organizational performance. Previous studies have shown that individual emotion is an important antecedent variable of job engagement and job burnout [17]. Emotion can also affect the individual judgment at the same time. Individuals have more positive emotions and are more likely to believe they can get the job done. They have a good expectation of the task, so they work with passion and energy and are more willing to spend time working. Thus, we expect:

H4a: The positive emotions of employees have a significant impact on job engagement. The more positive the employees' emotions are, the higher their vigor, dedication and absorption will be.

H4b: The negative emotions of employees have a significant impact on job engagement. The more negative the employees' emotions are, the lower their vigor, dedication and absorption will be.

The structural models of time pressure, emotional intelligence, emotional state and job engagement are shown as follows (Fig. 1):

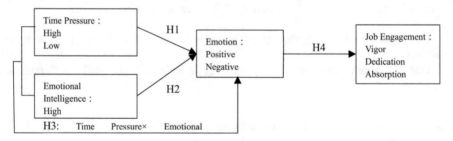

Fig. 1. Structure model

3   Method

3.1   Variable Selection and Operation

In this study, an inter-group design of time pressure (high/low) and leader emotional intelligence (high/low) was adopted, including four latent variables: time pressure, emotional intelligence, emotional state, and job engagement. Among them, time pressure (high/low) and emotional intelligence (high/low) were used as independent variables to construct four work scenarios. Emotional state and job involvement were measured by Likert scale.

First, the subjects were randomly assigned a situation material, and asked to imagine themselves as the protagonist of the material, and evaluate their own work pressure and the emotional intelligence of the leader according to the provided situation material. Then, the PANANS scale [18] was used to evaluate their emotional valence according to their subjective feelings. Finally, participants were asked to deform the constraints of the situation (do not see themselves as the protagonist) and fill in the job engagement scale.

3.2 Situational Materials Design

3.2.1 Situational Stories

Referring to the research method of Burris (2012) [19], this study designed four scenarios, and the same background information was used in each scenario: "① You are an employee of an Internet company, and your boss is Manager Wang, director of the technology Department. He is A. ② At noon, Manager arranged a task for you: finish the coding of a program before going off work tomorrow. It will take you two days to complete it according to your own ability, but B. ③ At this moment, Manager Wang suddenly asks you to go to his office. When you walk into the office C".

The four scenarios are operated as follows:

Low time pressure - high emotional intelligence: A. Cares deeply about employees and gets along with them as friends…; B. You are almost done with the task at hand…It seems there is no need to work overtime tonight. C. manager says "I believe you can do a good job. I believe in you." with smile.

Low time pressure - Low emotional intelligence: A. Very cold to employees, very demanding…; B. You are almost done with the task at hand…It seems there is no need to work overtime tonight. C. The manager is frowning and does not answer you…, angrily say "must give me as soon as possible, do not procrastinate!".

High time pressure - High emotional intelligence: A. Cares deeply about employees and gets along with them as friends…; So far there is only one day to finish…Looks like I need to work overtime tonight. C. The manager smiles and says…I believe you can do a good job. I believe in you. Come on!

High time pressure - Low emotional intelligence: A. Very cold to employees, very demanding…; So far there is only one day to finish…Looks like I need to work overtime tonight. C. The manager is frowning and does not answer you…, angrily say, must give me as soon as possible, do not procrastinate!

3.2.2 Measures

Watson et al. (1988) developed the Emotion Measurement Scale (PANAS) [18] to measure the emotional state, and finally summarized the measurement results into two dimensions: positive and negative. The scale of job engagement adopts the scale prepared by Schaufeli (2006) [20] et al. To meet the research situation, 9 items are selected for measurement. Emotional intelligence was obtained through setting up scenario experiments (classified as high/low emotional intelligence), see Sect. 3.2.1 for details.

3.3 Sample Characteristics

In this study, 241 staffs of logistics enterprises were selected as the research object. After screening, 27 invalid questionnaires were removed, and the remaining valid questionnaires were 214, with an effective rate of 88.79%. Among the effective samples,

117 (54.7%) were males and 97 (45.3%) were females. 50.47% (114 employees) have bachelor degree or above. The average length of work is 1–3 years.

4 Results

4.1 Reliability and Validity Analysis

The scale used in this study was tested for reliability and validity to ensure that each scale had good reliability and validity. Reliability refers to the reliability, stability and consistency of the measurement results of the survey scale. According to the study of Nunnally, The reliability coefficient above 0.8 is the best. The reliability test results of all scales used in this study are shown in Table 1 the reliability of the scale has passed the test.

Table 1. Reliability and validity analysis of scale

Scale	Items	Cronbach's α	KMO	Sig of Bartlett's test
PANAS-PA	9	.929	.948	.000
PANAS-NA	9	.967	.966	.000
Vigor scale	3	.910	.757	.000
Dedication scale	3	.914	.760	.000
Absorption scale	3	.927	.765	.000

According to Kaiser's research (1974) [21], KMO statistics in KMO and Bartlett's test are used to test partial correlation between variables. The closer the KMO value is to 1, the greater the correlation between variables is. KMO value greater than 0.5 indicates that the questionnaire data can be used for factor analysis. Bartlett test results showed a P value <0.05, indicating that the questionnaire was valid. As can be seen from Table 1, the validity of the scale has passed the test.

4.2 Descriptive Results

In each scenario experiment, the emotional state and job engagement of the subjects are shown in Table 2. Employees with high time pressure and low emotional intelligence had the highest negative emotions (NA $= 4.24$, SD $= 0.96$). On the contrary, the positive emotions of employees under low time pressure and high emotional intelligence were the highest (PA $= 3.96$, SD $= 0.93$).

4.3 Hypothesis Testing

The results of the influence of time pressure on employees' emotions are shown in Table 3. Time pressure has a significant influence on emotional state (F $= 5.473$, P $<$

Table 2. Descriptive results

Time pressure	Emotional intelligence	PA	NA	Vigor	Dedication	Absorption
High	High	3.34 (±0.92)	2.54 (±1.27)	3.65 (±1.13)	3.51 (±1.33)	3.58 (±1.32)
	Low	2.21 (±0.73)	4.24 (±0.96)	2.06 (±0.93)	1.96 (±0.89)	1.91 (±0.96)
Low	High	3.96 (±0.93)	2.02 (±0.98)	3.81 (±0.96)	3.83 (±1.06)	3.96 (±0.98)
	Low	1.82 (±0.67)	3.91 (±0.77)	1.59 (±0.82)	1.65 (±0.86)	1.63 (±0.79)

0.001; $F = 4.980$, $P < 0.01$), indicating that H1a and H1b are assumed to be supported. The influence of emotional intelligence on emotional state was significant ($F = 61.811$, $P < 0.001$; $F = 51.607$, $P < 0.01$), indicating that H2a and H2b are assumed to be supported. Under positive emotions, the interaction between time pressure and emotional intelligence of leadership was significant ($F = 2.329$, $P < 0.05$). Therefore, H3a was supported. However, for negative emotions, the interaction term between time pressure and emotional intelligence of leaders was not significant ($F = 1.893$, $P > 0.05$), indicating that H3b could not be supported.

Table 3. Variance analysis

Independent variable	Dependent variable	df	F	Sig.
Time pressure	Positive	3	5.473	.001
	Negative	3	4.980	.002
Emotional intelligence	Positive	3	61.811	.000
	Negative	3	51.607	.000
Time pressure × emotional intelligence	Positive	8	2.329	.021
	Negative	8	1.782	.083

It can be seen from Table 4 that the positive emotions affect job engagement (vigor, dedication, absorption) ($\beta = 0.435$, $P < 0.001$; $\beta = 0.515$, $p < 0.001$; $\beta = 0.299$, $P < 0.01$), and R^2 reached more than 70%, indicating that there is sufficient evidence that positive emotions have a positive impact on job engagement, so H4a is supported. Negative emotions to work engagement (vigor, dedication, absorption) ($\beta = -0.495$, $P < 0.001$; $\beta = -0.438$, $p < 0.001$; $\beta = -0.604$, $P < 0.001$), indicating that H4b is supported.

According to the above analysis, the research hypothesis in this paper is supported except H3b.

Table 4. Regression analysis

Variable	Vigor		Dedication		Absorption	
	M1	M2	M5	M6	M7	M9
Length of work	−0.156	0.016	−0.135	0.035	−0.221	−0.043
Education	0.072	0.007	0.056	-0.009	−0.077	−0.142
Job grade	−0.180	0.020	−0.172	0.042	−0.196	−0.16
Job category	0.071	0.066	0.142	0.140	0.001	−0.009
PA		0.435**		0.515***		0.299**
NA		−0.495***		−0.438***		−0.604***
R^2	0.279	0.757	0.083	0.802	0.013	0.738
ΔR^2	–	0.478	–	0.719	–	0.725
F	0.969	27.011***	1.042	34.777***	1.168	24.483***

5   Summary and Conclusion

5.1   Conclusion

Based on the interview study, a 2 × 2 scenario experiment was designed to explore how emotional intelligence of leaders affects employees' emotions and their job engagement under time pressure. The conclusions are as follows:

1) Time pressure has a significant impact on employees' emotions. This study result is consistent with previous research conclusions about the influence of time pressure on individuals [22]. This research explores the relationship between time pressure and individual emotion, which enriches the research on time pressure and emotion.

2) Emotional intelligence of leaders has a significant impact on employees' emotions. Leadership emotional intelligence is positively correlated with positive emotions and negatively correlated with negative emotions. This study result is consistent with previous studies on the influence of emotional intelligence on leadership on individuals, as well as the emotional contagion theory [6, 23].

3) There is a significant interaction between time pressure and emotional intelligence of leaders on employees' emotions. In the past, no studies have considered the influence of these two variables on employees' emotions. In reality, however, employees feel time pressure and are influenced by the emotional intelligence of their leaders. The results of this study indicate that the interaction between leadership time pressure and emotional intelligence has a significant effect on employees' positive emotions. The significant impact on employees' negative emotions has not been verified yet, which may be caused by the defects of the scenario simulation experiment method, and further research can be carried out in the future.

4) Employee emotion has a significant impact job engagement. The more positive employee's emotion is, the higher the employee's vigor, dedication and absorption for work are. This study verified that individuals with positive emotions can form

a kind of motivation and vitality, which can motivate employees, further verified the important role of employees' positive emotions on the organization, and helped enterprises to understand its significance and get better development.

5.2 Limitations and Directions for Future Research

Although this study has reached some meaningful conclusions, due to various subjective and objective factors, there are still some limitations and space for further improvement and improvement.

In terms of research methods, scene simulation experiment has some limitations. The scenario simulation experiment manipulated the variables through a simulated scenario story. The experiment was not carried out strictly in the laboratory or in the field, and the subjects were not in the real scene. Therefore, in this case, the experience of the subjects was generally low. At the same time, the degree of individual involvement in the simulated scene and their perception of the scene are also affected by individual characteristics. Some individuals have different perceptions of the scene, and their final cognition of the scene does not conform to our manipulation. In future studies, field experiments and questionnaire methods can be combined, or field observation and research can be carried out in the workplace.

In the operation of studying variables, the independent and dependent variables selected in this paper are not comprehensive enough. Each variable can be summarized from more dimensions, such as time pressure can be divided into accelerative and inhibitory. Employees' personal emotions are reflected in more one-way emotions, such as anger, joy and disappointment. In future studies, the study of time pressure on various one-way emotions can be expanded.

In terms of the control of variables, this study only considers demographic variables as the control variables of this study. In actual scenarios, potential variables such as stress resistance and job satisfaction of subjects will also have an impact on individual emotions. In future studies, these factors can be added into the research model to consider the impact of individual traits on individual emotions.

Acknowledgment. This research is supported by Research on the multimoding emotion recognition and mapping work-performance: based on Big Data (DSJ-KY-2021-009).

References

1. Ordonez, L., Benson, L.: Decisions under time pressure: how time constraint affects risky decision making. Organ. Behav. Hum. Decis. Process. **71**(2), 121–140 (1997)
2. Zakay, D.: Attention et judgment temporal. Psychol. Fr. **50**(1), 65–79 (2005)
3. Li, J., Huang, H.: The relationship between time and happiness: a comparison with the relationship between money and happiness. J. Southwest Univ. (Soc. Sci. Ed.) **039**(001), 76–82 (2013). (In Chinese)
4. Schaufeli, W.B., Taris, T.W.: A critical review of the job demands-resources model: implications for improving work and health. In: Bauer, G.F., Hämmig, O. (eds.) Bridging Occupational, Organizational and Public Health, pp. 43–68. Springer, Dordrecht (2014). https://doi.org/10.1007/978-94-007-5640-3_4

5. Kumar, J.A., Muniandy, B., Yahaya, W.A.J.W.: Emotional design in multimedia learning: how emotional intelligence moderates learning outcomes. Int. J. Mod. Educ. Comput. Sci. (IJMECS) **8**(5), 54–63 (2016)

6. Fredrickson, B.L., Joiner, T.E.: Positive emotions trigger upward spirals toward emotional well-being. Psychol. Sci. **13**(2), 172–175 (2002)

7. Zaied, A.N.H., Hussein, G.S., Hassan, M.M.: The role of knowledge management in enhancing organizational performance. IJIEEB **4**(5), 27–35 (2012)

8. Yukl, G.A.: Leadership in Organizations, 7th edn. Tsinghua University Press (2010)

9. Szollos, A.: Toward a psychology of chronic time pressure: conceptual and methodological review. Time Soc. **18**(2–3), 332–350 (2009)

10. Thayer, R.E.: Activation-deactivation adjective check list: current overview and structural analysis. Psychol. Rep. **58**(2), 607–614 (1986)

11. Mayer, J.D., Salovey, P.: What is emotional intelligence? (1997)

12. Fredrickson, B.L.: Positive emotions and upward spirals in organizations, pp. 163–175. Positive Organizational Scholarship (2003)

13. Tang, C., Pan, Y.: A cross-layer analysis of the impact of emotional intelligence of leadership on employees' organizational identity and organizational citizenship behavior. Nankai Manag. Rev. **13**(04), 115–124 (2010). (in Chinese)

14. Hatfield, E., Cacioppo, J.T., Rapson, R.L.: Emotional contagion. Curr. Dir. Psychol. Sci. **2**(3), 96–100 (1993)

15. Visser, V.A., Knippenberg, D.V., Kleef, G.: How leader displays of happiness and sadness influence follower performance: emotional contagion and creative versus analytical performance. Leadersh. Q. **24**(1), 172–188 (2013)

16. McGregor, A., et al.: A job demands-resources approach to presenteeism. Career Dev. Int. **21**(04), 402–418 (2016)

17. Khan, A.A., Madden, J.: Speed learning: maximizing student learning and engagement in a limited amount of time. Int. J. Mod. Educ. Comput. Sci. (IJMECS) **8**(7), 22–30 (2016)

18. Watson, D., Clark, L.A., Tellegen, A.: Development and validation of brief measures of positive and negative affect: the PANAS scales. J. Pers. Soc. Psychol. **54**(6), 1063–1070 (1988)

19. Burris, E.R.: The risks and rewards of speaking up: managerial responses to employee voice. Acad. Manag. J. **55**(4), 851–875 (2012)

20. Schaufeli, W.B., Salanova, M., González-Romá, V., et al.: The measurement of engagement and burnout: a two sample confirmatory factor analytic approach. J. Happiness Stud. **3**(1), 71–92 (2002). https://doi.org/10.1023/A:1015630930326

21. Kaiser, H.F.: An index of factorial simplicity. Psychometrika **39**(1), 31–36 (1974). https://doi.org/10.1007/BF02291575

22. Koch, S., Holland, R.W., Knippenberg, A.V.: Regulating cognitive control through approach-avoidance motor actions. Cognition **109**(1), 133–142 (2008)

23. Hofmann, D.A., Stetzer, A.: A cross-level investigation of factors influencing unsafe behaviors and accidents. Pers. Psychol. **49**, 307–339 (1996)

Application of BOPPPS Model in the Teaching Design of the Principle and Method of Micro-joining Course

Li Liu, Wei Feng, Jiaqi Li, Jian Lan, and Qilai Zhou[✉]

School of Materials Science and Engineering, Wuhan University of Technology, Wuhan 430070, China
zhou.qilai@whut.edu.cn

Abstract. The traditional classroom teaching process of engineering education majors normally faces problems that teachers dominate class, students are not active in learning, learning objectives are not clear and the learning results are not ideal. In this study, combined Chaoxing online learning system, the Principle and Method of Micro-Joining Course was taken as an example to build a new BOPPPS teaching model based on the outcomes-based education (OBE) concept, which achieve a learning result-oriented education model. Moreover, classroom participation activities were also proposed to improve student's active participation and classroom teaching effects. It proves that the classroom teaching effects can be effectively improved based on the OBE concept and BOPPPS teaching model. Moreover, it can also help transform professional knowledge into students' professional ability, making students change from passively accepting knowledge to actively acquiring knowledge. Thus, the students' ability can be improved by enhancing the teaching quality.

Keywords: BOPPPS model · Teaching design · OBE concept

1 Introduction

Recently, higher education researchers and practitioners have successively carried out active researches through the application of new teaching tools, innovative teaching model reforms and advanced teaching theory and practice, which have achieved the improvement of classroom teaching effects [1]. Higher education researchers gradually discovered that the innovation of teaching tools, the teaching designs and the educational concepts should be mutually reinforced as a whole rather than relying on single one of these areas. Therefore, exploring the comprehensive use of new teaching tools and teaching models to enhance the effectiveness of teaching under the guidance of advanced teaching concepts has become the key to higher education reform and innovation.

Thus, under the guidance of the current engineering certification OBE education concept, a teaching model based on the BOPPPS model is designed and verified in practice to compare and analyze the effective teaching characteristics of this mode. It provides ideas and references for improving curriculum teaching reform through the

Z. Hu et al. (Eds.): ICCSEA 2022, LNDECT 134, pp. 509–517, 2022.
https://doi.org/10.1007/978-3-031-04812-8_44

exploration of effective teaching designs. In this study, the Principle and Method of Micro-Joining Course was taken as an example to build a new BOPPPS teaching model based on the outcomes-based education (OBE) concept, which achieve a learning result-oriented education model combined with Chaoxing online learning system.

2 BOPPPS Model

BOPPPS model was originally created by ISW (Canada Teacher Skills Training Workshop) in 1984. It was originally mainly used for teacher skill training to improve the teaching skills and teaching effectiveness through concentrated and intensive training. Afterwards, BOPPPS model has been rapidly promoted in Canada and the United States and get wide support from teachers and students [2, 3]. BOPPPS is a closed-loop teaching process model that emphasizes student participation and feedback by six modules including Bridge in, Objective, Pre-assessment, Participatory learning, Post-assessment and Summary [4]. The modules and related task of BOPPPS model is listed in Table 1. BOPPPS teaching model can help teachers carefully analyze the entire teaching process and find teaching blind spots, which can significantly improve and enhance teaching effectiveness.

Table 1. Module and task of BOPPPS model

Symbol	Module	Task
B	Bridge in	Arouse students' interests
O	Learning objective	Know objective and emphasis well
P	Pre-assessment	Master students' prior knowledge and ability
P	Participatory learning	Core module to encourage students actively participate in teaching process
P	Post-assessment	Understand students' learning effectiveness and learning goal achievement
S	Summary	Summarize course content

Up to now, BOPPPS has been introduced and adopted by more than 33 countries around the world, and has been respected by many universities around the world. As of December in 2021, when searched with the keyword of BOPPPS in Google Scholar, a total of 277 articles related to BOPPPS model has been published from 2011 to 2021. As shown in Fig. 1, the number of published articles is on the rapid rise as a whole. In the period from 2011 to 2016, very small number of papers related to BOPPPS model has been published with the highest publishing journal number of 11. However, in the recent 5 years, the publishing paper number rapidly increases from 15 papers to 102 papers. Especially, nearly 80 articles were published per year in the past two years.

By analyzing the institutions of BOPPPS literature in Fig. 2, National University of Defense Technology and Army Engineering University of PLA were the leading

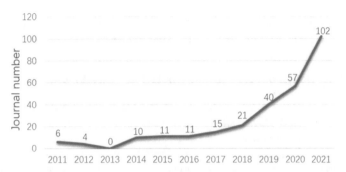

Fig. 1. The published journal number of BOPPPS literature from 2011 to 2021.

institutions for bringing BOPPPS model in their courses. For instance, National University of Defense Technology University employed BOPPPS model in various theoretical and experimental courses [5–8]. Moreover, the teachers in this university also proposed the reasonable and practical assessment methods for engineer courses embedded with BOPPPS model [9–11]. Apart from these two leading institutions, several institutions including Aga Khan University in Pakistan, Beijing Normal University in China, Nanyang Institute of Technology in Singapore, University of Bergen in Norway, and Wuhan University of Technology in China are also the backbone force of the institutions for investigation BOPPPS teaching model [12–17].

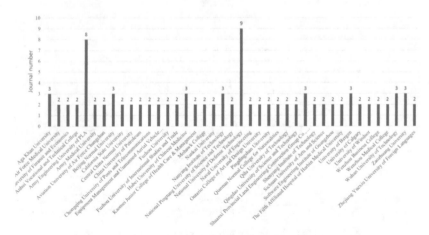

Fig. 2. Analysis of publishing institutions of BOPPPS literature.

3   Teaching Design Based on BOPPPS Model

3.1   Course Description

Principle and Method of Micro-Joining Course is an important personalized course for Materials Processing and Forming Major in Wuhan University of Technology. Since this course is a new professional direction, two classes of "Excellent Engineering Plan" with a total of 55 students were the first to attend this course in 2020 and 2021.

3.2 Teaching Design of Principle and Method of Micro-joining

Generally, the teaching design processes of Principle and Method of Micro-Joining Course based on BOPPPS model was shown in Fig. 3. The teaching design processes can be divided into three periods of before class, in class and after class. In the first period of before class, three modules of BOPPPS model were carried out, which are bridge in, objective learning, pre-assessment. Afterwards, during the second period of in class, participatory learning is the most core part of teaching design. When move to final stage of after class, post-assessment and summary modules were carried out. The main purposes of three periods of before class, in class and after class were to optimize teaching design for in class, to encourage student actively participate in class and to optimize entire teaching design for next session, respectively.

Internet technique by Chaoxing online learning system in Wuhan University of Technology was employed in teaching design of Principle and Method of Micro-Joining Course during the before class and after class periods. Chaoxing online learning system was launched by Chaoxing Group as a platform for teachers and students to lean course online [18].

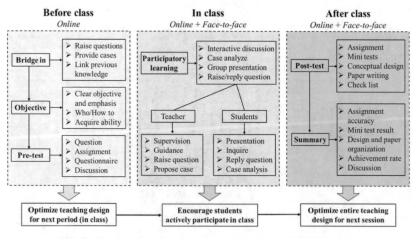

Fig. 3. The teaching design processes based on BOPPPS model.

3.2.1 Bridge In

The main function of bridge in module is to arouse students' interests and attention. Through Chaoxing online learning system, teacher can post short questions, related new cases, interesting pictures and videos to students, which enables student to think and focus on the learning content of the following class.

Moreover, as this Principle and Method of Micro-Joining Course is a professional frontier direction of materials processing and technology major. In the Materials Forming Principles course in the prior session, some related knowledge was taught. Therefore, it would be of great benefits to link the previous knowledge with the new knowledge in this new course. Thus, students can gain new knowledge by reviewing old and pay more attention to new class.

3.2.2 Learning Objective

The purpose of this module is to let students grasp the key learning objectives, emphasis, and the requiring ability and skills in the teaching course. The learning objective and emphasis of this teaching course should be very clear, specific and measurable. As a result, students can keep focused on the key knowledge in class for 45 min.

Generally, the objectives of Principle and Method of Micro-Joining Course are to provide more detailed information and theoretical understanding of important micro-joining technologies as applied to electronic devices, enable students to broaden the knowledge and to build the solid background in the new field of microelectronics. The specific objectives of this course are listed below:

1) To enable students to grasp the particularity of materials, structure and performance involved in micro-joining technology trends and relevance, and to understand the process factors that affect micro-joining performance.
2) Familiar with the solid phase bonding, joining, fusion joining and advanced joining methods of various materials, as well as the characteristics of the corresponding materials and equipment, and have the application ability to choose and design micro-joining processes for different materials.
3) Enable students to master basic literature review and engineering research capabilities, to understand background, trends and research hotspots in micro-joining field, making students able to prepare oral reports and to smoothly communicate with audiences.
4) Enable students to understand micro-joining technology, to have a certain international vision and innovative design ability to analyze and solve the practical micro-joining problems according to the specific structure and performance requirements of micro-joining process.

3.2.3 Pre-assessment

The purpose of the pre-assessment activity design is to understand the students' interest, prior knowledge and skills. Through confirming the students' specific needs from the students' feedback, teacher can specifically provide students with opportunities to express their specific needs [19]. According to the pre-assessment results, teacher should specifically analyze the actual learning situation and needs from students to adjust and modify the teaching design in next period of in class.

In this Principle and Method of Micro-Joining Course, combined methods of question, assignment, questionnaire, discussion and so on were carried out in Chaoxing online learning system. For instance, the questionnaire in pre-assessment is listed in Table 2 as an example.

Table 2. Questionnaire in pre-assessment

Number	Question
1	Write down your gender and age
2	Have your thinking ability and self-learning ability been improved by Chaoxing online learning system?
3	Do you fully understand the principle of joining at atom level in the previous Principle and Method of Micro-Joining Course?
4	Have Bridge in period arouse your interests in Principle and Method of Micro-Joining Course?
5	Have Learning objective period made you clear understand the objective, emphasis and requiring skills in Principle and Method of Micro-Joining Course?
6	Which interactive activity do you prefer to be introduced in class for improving learning effectiveness?
7	Have Chaoxing online learning system increased your learning participation rate?
8	Which communication methods do you wish to choose for communication with teacher?
9	Do you think it is necessary to enter a quiz for evaluating student's level in each course?

3.2.4 Participatory Learning

For the entire BOPPPS model, participatory learning is the most important part for students. Teacher should try their best to encourage students actively participate in class as much as possible.

For the Principle and Method of Micro-Joining Course, enormous methods such as interactive discussion, case analysis, group presentation and question raise/reply were conducted from both teacher and students' perspectives.

For instance, case-based presentation is one of the most vital participation methods in the Principle and Method of Micro-Joining Course. The teacher sets specific topics in advance to guide the students in preparing their presentation as listed in Table 3.

Table 3. Guiding topics in the student presentation

Number	Topic
1	Application of process equipment in micro-joining field
2	Advanced micro-joining process
3	Micro-joining process for dissimilar materials
4	Micro-joining process for 3D electronic package
5	Micro-joining process for flexible electronic package
6	Advanced and high-valued materials for micro-joining process

3.2.5 Post-assessment

The purpose of the post-assessment module is to test the teaching effectiveness in reaching the teaching objectives. If there are any problems in teaching goal achievement, for instance, comprehensive analysis capabilities such as basic inspections of knowledge and skills and applied analysis. The post-assessment should be able to directly detect the original teaching goal has not been achieved. Therefore, the post-assessment would be great to be consist with the results of the pre-assessment module.

The post-assessment module mainly includes classroom mini tests, assignments, conceptual design and writing papers after class and final check list in Principle and Method of Micro-Joining Course.

3.2.6 Summary

Summarize can greatly help students reflect on and integrate learning content. It would be of great benefits to emphasis on learning objective in a summary. In this Principle and Method of Micro-Joining Course, the summary is obtained by analyzing the results of the post assessment. Based on the summary results, the teaching design can be modified for the following course in next semester.

3.3 Assessment of BOPPPS Model

In School of Materials Science and Engineering based on Wuhan University of Technology, outcomes-based education (OBE) concept can be employed to analyze the achievement rate of this engineering course. The calculated achievement rate of the teaching objectively in session 3.2.2 was shown in Fig. 4, which is in a range of 0.878–0.887. In Wuhan University of Technology, when achievement rate exceeds 0.75, the teaching effective can be regarded as pass. Therefore, the teaching effect of Principle and Method of Micro-Joining Course is good.

Moreover, questionnaire was also carried out to collect students' opinions and demands.

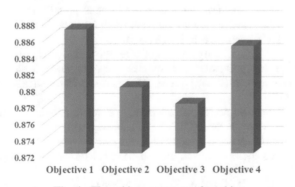

Fig. 4. The achievement rate of teaching

4 Conclusion

The main purpose of this study was to bring BOPPPS model in the Engineering course at Materials Processing and Forming major based on Wuhan University of Technology and to improve the teaching design and effectiveness on students' learning outcome. Combined BOPPPS model, OBE concept and online learning technology, the teaching design was disassembled into three periods of before class, in class and after class and six modules of bridge-in, pre-assessment, participatory learning, post-assessment and summary. Through these specific teaching periods and modules as well as online learning technique, the interactive teaching activities can be more diversity, interesting and effective. According to OBE concept, the achievement rate of this engineer course was exceeded 0.87 for all learning objectives, showing good teaching quality and design.

Acknowledgment. This work is supported by the Teaching Reform Project of Wuhan University of Technology (Grant No. w2021063 and w2020062).

References

1. Li, C., Jiang, F.: An experimental study of teaching English writing with OBE in Chinese senior high school. Theory Pract. Lang. Stud. **10**(8), 905–915 (2020)
2. Kolb, D.A.: Experiential Learning: Experience as the Source of Learning and Development. Pearson Education, New Jersey (2015)
3. Li, Q.: Design and practice of data structure teaching class based on BOPPPS model. In: Proceedings of the 8th International Conference on Information Technology in Medicine and Education, pp. 503–506 (2016)
4. Yun, X.F., Xiao, Y.R., Xiao, Y.Y.: Introducing BOPPPS computer language teaching method. Appl. Mech. Mater. **3682**(701), 1271–1274 (2015)
5. Li, C.M., Jiang, F.: An experimental study of teaching English writing with OBE in Chinese senior high school. Theory Pract. Lang. Stud. **10**, 905–915 (2020)
6. Wu, Y.F., Li, Z.G., Li, Y.: Teaching reform and research of data structure course based on BOPPPS model and rain classroom. In: Proceedings of the International Conference of Pioneering Computer Scientists, Engineers and Educators, pp. 410–418 (2021)
7. Guo, X.J., Su, S.J., Huang, Z.P.: Application of BOPPPS model in the course of embedded system design. Int. J. Learn. Teach. **4**(3), 243–246 (2018)
8. Wen, L., Zhao, Y.T., Cui, J.N.: Application of BOPPPS (bridge-in, objective, pre-assessment, participatory learning, post-assessment, summary) teaching model in electronic technology courses. In: Advances in Social Science, Education and Humanities Research, vol. 490, pp. 59–61 (2020)
9. Dominic, M., Francis, S.: An adaptable e-learning architecture based on learners' profiling. Int. J. Mod. Educ. Comput. Sci. (IJMECS) **7**(3), 26–31 (2015)
10. Strumińska-Kutra, M., Koładkiewicz, I.: Case study. In: Ciesielska, M., Jemielniak, D. (eds.) Qualitative Methodologies in Organization Studies, pp. 1–31. Springer, Cham (2018). https://doi.org/10.1007/978-3-319-65442-3_1
11. Dada, E.G., Alkali, A.H., Oyewola, D.O.: An investigation into the effectiveness of asynchronous and synchronous e-learning mode on students' academic performance in National Open University (NOUN), Maiduguri Centre. Int. J. Mod. Educ. Comput. Sci. (IJMECS) **11**(5), 54–64 (2019)

12. Jamil, Z., Naseem, A., Rashwan, E.: Blended learning: call of the day for medical education in the global South. SOTL South **3**(1), 57–76 (2019)
13. Chen, J.: Applying BOPPPS model to improve teacher's instructional design ability. In: Proceedings of the 8th International Conference on Educational Innovation through Technology, pp. 106–109 (2019)
14. Wang, Y.L., Wen, Q.T., Gao, F.: The application of BOPPPS effective teaching structure in the teaching practice of general surgery. In: Proceedings of the 3rd International Conference on Education & Education Research, pp. 225–228 (2018)
15. Reiss, M.J.: Science education in the light of COVID-19. Sci. Educ. **29**(4), 1079–1092 (2020). https://doi.org/10.1007/s11191-020-00143-5
16. Zhang, M.Y., Zhang, Q.Y., Wang, Z.Y.: Mechanical experimental platform construction based on BOPPPS model. In: Proceedings of the International Conference on Computer Science, Engineering and Education Applications, pp. 194–204 (2019)
17. Shi, J.Q., Ye, J.Y.: On the "practice-facilitation-reflection" oriented faculty development. Teach. Educ. Res. **29**(6), 81–87 (2017)
18. Yang, Z., Wang, J.: Smart classroom teaching practice under the background of information technology revolution. Curric. Educ. Res. **50**, 28–29 (2020)
19. Wang, S., Xu, X., Li, F.: Effects of modified BOPPPS-based SPOC and flipped class on 5th-year undergraduate oral histopathology learning in China during COVID-19. BMC Med. Educ. **21**, 540 (2021)

Training Mode Combined with Building Information Modeling Technology of Material Specialty in the Background of Emerging Engineering Education

Hu Yang and Qian Cao[✉]

School of Material Science and Engineering, Wuhan University of Technology, Wuhan 430070, China
caoqian@whut.edu.cn

Abstract. In the background of the construction of "Emerging Engineering Education", in order to adapt the updated material industry, a new application-oriented talent training system combined with Building Information Modeling (BIM) technology is proposed. This paper describes the training objectives and training program in detail. Teaching effects are deeply explored and analyzed. The results show that this new training program can better meet the training objectives. The general teaching effect is highly evaluated, and 82.30% of the students believe that the training is reasonable. Students' team spirit, professional ability and innovation ability have been significantly improved. The proposal of the new BIM talent training mode for material specialty has great reference value for the establishment of talent training system for related majors in other colleges and universities.

Keywords: Talent training mode · Building Information Modeling · Material specialty · Emerging engineering education

1 Introduction

The new industrial revolution represented by artificial intelligence has brought tremendous changes and varied the cultural, structural, and organizational form of the entire human society [1–4]. In order to meet the talents demand of the new industries, the Ministry of Education of China proposed the construction of "Emerging Engineering Education" in 2017 [5]. Material specialty, as one of the traditional engineering disciplines, has many cross-border and cross-integration fields. Under the background of "Emerging Engineering Education", material specialty students should not only be equipped with professional knowledge of materials, but also have interdisciplinary learning ability and engineering practice ability. College curriculum should also be modified to cultivate diversified and innovative talents to meet the needs of enterprises and to confront new challenges of the society.

Recently, most design courses in materials majors still use plane drawing, which is merely used in current industrial field. In field of building materials industry, it is replaced by an advanced technology called the Building Information Modeling (BIM) technology. BIM technology recognized as a "smart brain" of buildings. It refers to an engineering data model that integrates various relevant information of construction projects based on three-dimensional digital technology. It is a digital representation of the entity and functional characteristics of project facilities [6, 7]. BIM technology realizes information sharing in a design and construction stage of an architectural [8], and is gradually applied to the design [9], construction [10], construction management [11], simulation calculation [12, 13] etc. Using BIM technology may achieve the high efficiency, while satisfy low cost and low risk during the whole process of the project. Cultivating top-level innovative talents and industry-leading talents who have mastered the application and development capabilities of BIM technology is a major demand for adapting to the development of new material formats, meeting the upgrading and transformation of the material industry.

Up to now, BIM technology has been introduced into the classes in various universities [14–18]. Zhang Yao [19] combines BIM technology with engineering drawing course. His study shows that the introduction of BIM technology helps to cultivate students' thinking ability, hands-on ability, and practical ability. Lu Qingrui [20] introduces BIM technology into the major of geotechnical engineering and finds that the introduction of BIM technology can enhance students' interest of learning, improve their practical ability. And the new way of teaching will provide talents for the development of the industry. Song Xiaogang [21] reforms the teaching system of engineering management specialty based on BIM technology and creates a BIM collaborative innovation teaching system. This novel teaching system is proven effectiveness and progressiveness. However, there are few teaching researches focus on combining BIM technology with materials discipline.

This paper aims at the research and development of BIM talent training mode in MSE specialty. The training mode of BIM talents for material majors is proposed, and a program is constructed according to the training objectives. Then the training program has been implemented for all students majoring in materials, and a new teaching effectiveness evaluation method is used to analyze the teaching effect.

2   Training Objectives of BIM Talents Based on Material Specialty

The general training objectives of BIM talents are based on the needs of enterprises, the application status of BIM technology, and the development trend of materials engineering. They are formulated as follows:

1) Be able to design material-related equipment based on practical constraints such as environment and law;
2) Equipped with the professional ability to engage in material-related engineering modeling, data maintenance, information management, etc.;
3) Equipped with the ability to use BIM technology to support and complete the construction and management methods of engineering projects;
4) Be able to design the process flow of the integrated unit process, optimize the process design, and possess innovative consciousness;

5) Understand the cost of the material and its application in the whole cycle and process; understand the engineering management and economic issues involved, and applied to the process of design, development, and solutions.

3   BIM Talent Training Program

The new program is a combination of material major courses with BIM training projects. Under the premise of conforming to the professional personnel training plan, overall planning for the design courses of different majors is programmed to promote the implementation of joint curriculum. Through skills training, courses, competitions, graduation projects, etc., a talent training system has been constructed, as shown in Fig. 1. It can be seen from Fig. 1 that in the whole training process, different training methods and training objectives are formulated according to the different basic knowledge and abilities students mastered in different grades.

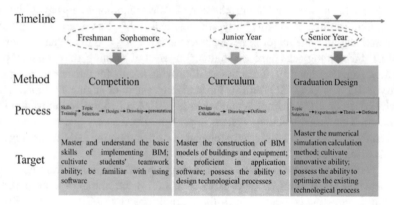

Fig. 1. Process of BIM talent training program

3.1   Freshman and Sophomore

At this stage, students possess few professional knowledge and ability, which could be mainly cultivated in the form of competition. The competition is open to all professional students of material specialty, with about 200 participants each year since 2019. Teachers use the summer vacation to train students through online and offline mixed courses and then organize competitions, so that students can understand the basic knowledge of BIM technology and the basic process related to material production. Afterwards, students choose topics, form teams, carry out basic design, and draft papers. During this period, teachers use online resources and answering questions to help students complete their works. Online resources include online courses and virtual simulation experiments of related material production processes. Finally, students submit their work and make a defense. Through the training in the freshman and sophomore stages, the teamwork ability of students is cultivated, and students will master and understand the basic skills of BIM, and be familiar with the use of basic software.

3.2 Junior Year and the First Semester of the Senior Year

Training in this stage is mainly for students majoring in materials science and engineering or inorganic non-metallic materials. Annually, there are four classes with a scale of nearly 120 students in total. After two and a-half years of college study, students are equipped with certain basic theories and knowledge of material science. This knowledge provides the basis for the following BIM learning. Students learn engineering courses of introduction to factory design, engineering design training, etc. via BIM technology, to obtain the basic methods and procedures of related material production process and to understand equipment design. This process will cultivate students' engineering awareness, improve students' engineering design ability, and may also enable students to master the ability to use BIM technology to accomplish engineering projects and solve engineering problems.

In this process, teachers need to teach offline, set task points, answer questions, and check students' progress. Students need to complete design calculations, draw drafts, and defense.

3.3 The Second Semester of the Senior Year

During the graduation design period, the graduation design topics are optimized and established. By using numerical simulation software, students' ability to solve complex engineering problems and innovative ability to confront new fields and new knowledge are cultivated. In this process, teachers need to determine graduation design topics according to the latest research progress, and guide students to conduct experiments and write graduation thesis. Students need to complete topic selection, do experiments, finish graduation thesis, and defense. Approximately 20 topics will be set up for students to choose.

To sum up, this new program is formulated for material specialty students of different majors and grades. Throughout the four academic years, it aims to improve students' ability to design and solve complex engineering problems, cultivating innovative ability and project management capability.

4   Analysis of Teaching Effect

In order to monitoring teaching quality, an evaluation system has been formed for the BIM talent training program. It aims to improve the BIM training program through evaluation of teaching links, process monitoring and teaching quality, in order to further promote the realization of training needs. In this study, the teaching quality was monitored through two forms of ability assessment and questionnaire survey.

4.1 Ability Assessment Standard

According to the various training goals at different stages, the assessment standards were separately established for competitions, curriculum, and graduation designs. As shown in Table 1, 2 and 3, respectively.

Table 1. Competition assessment standard

Target	Evaluation method	Evaluation basis	Score
Model sophistication	Correctness of the model	Conformity with drawings	70
3D display/PPT format	Presentation rating	Explanation and presentation	15
implements of interactive function	Model interaction capability	The level of model's ability of interaction	15

Table 2. Curriculum assessment standard

Design objectives	Evaluation content	Score
Topic selection and overall program design	The rationality of the topic selection, the correctness of the design, feasibility and innovation	20
Adopt correct design calculation method and have design calculation ability	Integrity of calculation process, standardization of design specification and sufficiency of references	30
Engineering expression ability	Standardization of drawings and rationality of layout; Clarity and correctness of main structure	30
Autonomy and innovation during the design process	Accuracy of answering questions, Active questioning during design process	20

Table 3. Graduation design assessment standard

Evaluation content	Score
Workload	10
Literature application ability	15
Comprehensive application ability of basic theory and professional knowledge	15
Research (Design) Capability	20
Research level and ability	20
Thesis defense	10
Normalize	10

4.2 Questionnaire

At the end of the training, the research team collected students' evaluation of the teaching effect through a questionnaire. The questionnaire used a 4-point scoring standard, with 1–4 representing "reasonable" to "unreasonable". Table 4 shows the results of questionnaire survey collected from senior students majoring in materials science and engineering and inorganic non-metallic materials of grade 2017.

Table 4. Teaching effectiveness questionnaire

Items	Reasonable	Partly reasonable	general	unreasonable
Evaluation of the rationality of the curriculum	82.30%	10.21%	5.32%	2.17%
Evaluation of curriculum content	60.20%	20.45%	10.87%	8.48%
Evaluation of the cultivation of team spirit	85.67%	8.75%	4.23%	1.35%
Evaluation of the cultivation professional and innovation ability	80.74%	14.20%	4.31%	0.75%
Evaluation of teaching conditions	70.23%	18.35%	9.71%	1.71%
Overall evaluation of teaching effect	80.50%	12.56%	6.45%	0.49%

It can be seen from Table 4 that 80.50% of the students give positive evaluations to the teaching effect and 82.30% of them believe the setting of BIM training courses is reasonable. Compared with the traditional teaching system, the introduction of BIM technology can improve students' interest in learning [20], thus good feedback could be achieved. 85.67% of the students believe their team spirit has been improved. It proves that through the BIM training, students work in groups and collaboratively [22], which can well foster team spirit. 80.74% of the students think their professional and innovation ability have been well developed. By utilizing numerical simulation methods, students learn to improve the traditional production technology and production equipment in a novel way, which may cultivate students' innovative ability [23, 24]. However, the evaluation of curriculum content and teaching conditions are lower, with a satisfaction of 60.2% and 70.23%, respectively. In order to find the reason of these results, discussions were conducted with the students. It was found that, for students, the content of the course was relatively difficult. They did not have enough knowledge about the production process of related materials, equipment structure and its connection. Meanwhile, the students needed to draw a three-dimensional model, resulting in a more difficult task and low efficiency in the process of drawing. It is difficult for some students to complete the task on time. The above problems could be solved by improving teaching conditions, perfecting the virtual simulation system, and increasing online resources. Using these methods can enhance students' understanding of the production process and equipment structure.

In general, according to the results of the questionnaire, students' evaluation of the courses combined with BIM technology is high. The new training mode is proved to be successful on cultivating students' ability of cooperation and innovation. After the training, students' understanding of material production technology and equipment structure has been significantly improved.

5 Conclusion

This paper has studied the new BIM talent training mode of materials specialty. Training objectives and program are formulated, deeply explored and analyzed. The analysis of teaching effectiveness evaluation results show that the new BIM talent training mode has achieved good results. About 82.30% of the students believe that the training is reasonable. More than 80% of the students consider their team spirit, professional ability and innovation ability have been significantly improved. However, the combination of BIM Technology may make the course more difficult for students to understand that only 60% of students think the content is reasonable. It could be ameliorated by improving teaching conditions such as perfecting virtual simulation system and introducing online resources to deepen students' understanding of structure. In any case, the new training mode has improved the quality of personnel and employment level, and will meet the major needs of industrial reform in the material industry.

References

1. Nikolay, K.: Geometrical framework application directions in identification systems. Review. Int. J. Intell. Syst. Appl. **13**(2), 1–20 (2021)
2. Khalid, S., Hadi, H.: Data analysis for the aero derivative engines bleed system failure identification and prediction. Int. J. Intell. Syst. Appl. **13**(6), 13–24 (2021)
3. Odji, E.: Influencing children: limitations of the computer-human-Interactive persuasive systems in developing societies. Int. J. Mod. Educ. Comput. Sci. **12**(5), 1–15 (2020)
4. Nor, H.Z., Shuzlina, A.R., Nor, H.U., Ismail, I.: House price prediction using a machine learning model: a survey of literature. Int. J. Mod. Educ. Comput. Sci. **12**(6), 46–54 (2020)
5. Liu, J., Zhai, Y., Xun, Z.: Analysis of the connotation of emerging engineering education and its construction. Res. High. Educ. Eng. **37**(3), 21–28 (2019). (in Chinese)
6. Wen, W., Rui, H.: A review of building information modelling. AIP Conf. Proc. **1967**(1), 1–4 (2018)
7. Vanlande, R., Nicolle, C., Cruz, C.: IFC and building lifecycle management. Autom. Constr. **18**(1), 70–78 (2008)
8. Hartmann, T., Gao, J., Fischer, M.: Areas of application for 3D and 4D models on construction projects. J. Constr. Eng. Manag. **134**(10), 776–785 (2008)
9. Linxia, Z., Yuqin, S.: Piping layout with BIM Revit MEP in gas boiler room. Sci. Technol. Ports **27**(5), 35–39 (2015). (in Chinese)
10. Sun, H., Li, M., Lu, Y.: Prospecting analysis on 3D modeling of industrial furnace based on BIM technology. Shangdong Metall. **39**(6), 33–36 (2017). (in Chinese)
11. Wang, J., Yao, P., Yi, K.: Research on the application of BIM technology in Baosteel No. 9 EAF maintenance. J. Inf. Technol. Civil Eng. Archit. **6**(5), 112–117 (2014) (in Chinese)
12. Mei, S., Xie, J., Chen, X., et al.: Numerical simulation of co-combustion of coal and refuse derived fuel in coupling with decomposition of calcium carbonate in precalciner with swirl type prechamber. CIESC J. **68**(6), 2519–2525 (2017). (in Chinese)

13. Mei, S., Xie, J., Chen, X., et al.: Numerical simulation of the complex thermal processes in a vortexing precalcine. Appl. Therm. Eng. **125**(1), 652–661 (2017)
14. Žiga, T., Andreja, I.: Toward deep impacts of BIM on education. Front. Eng. Manag. **7**(1), 1–8 (2020)
15. Wang, L., Huang, M., Zhang, X., et al.: Review of BIM adoption in the higher education of AEC disciplines. J. Civil Eng. Educ. **146**(3), 1–17 (2020)
16. Sergey, M., Aleksandrina, M., Neil, N., et al.: BIM-technologies and digital modeling in educational architectural design. IOP Conf. Ser.: Mater. Sci. Eng. **890**, 012168 (2020)
17. Casasayas, O., Hosseini, M.R., Edwards, D.J., et al.: Integrating BIM in higher education programs: barriers and remedial solutions in Australia. J. Archit. Eng. **27**(1), 1–11 (2021)
18. Taija, P., Perry, F.: Practical challenges of BIM education. Struct. Surv. **34**(4), 351–366 (2016)
19. Yao, Z.: Practice of teaching engineering drawing course based on BIM. Electron. Technol. **50**(12), 158–185 (2021). (in Chinese)
20. Lu, Q., Liu, J., Li, D., et al.: The application of BM_GeoModeler in BIM teaching of geotechnical engineering. Educ. Mod. **8**(29), 133–136 (2021). (in Chinese)
21. Song, X.: Research on BIM collaborative innovation teaching system of university engineering management. Project Manag. Technol. **19**(1), 36–40 (2021). (in Chinese)
22. Biao, Y.: Forward collaborative design of industrial furnace project based on BIM. Eng. Constr. **53**(8), 36–42 (2021). (in Chinese)
23. Abbassi, A., Khoshmanesh, K.: Numerical simulation and experimental analysis of an industrial glass melting furnace. Appl. Therm. Eng. **28**(5–6), 450–459 (2008)
24. Mette, B., Oyvind, S., Nils, E.L.H., et al.: Numerical simulations of staged biomass grate fired combustion with an emphasis on NOx emissions. Energy Proc. **75**, 156–161 (2015)

Training Mode of Compound Talents in Wuhan Shipping Enterprises

Ying Zhang[1,2], Cheng Zhang[1], Qingying Zhang[1], and Yong Zhou[1(✉)]

[1] School of Transportation and Logistics Engineering, Wuhan University of Technology, Wuhan 430063, China
zhouyong@whut.edu.cn
[2] Engineering Research Center of Port Logistics Technology and Equipment, Ministry of Education, Wuhan 430063, China

Abstract. Logistics plays a vital role in the development of society, and shipping industry, as an important part of logistics in transportation, is facing problems of compound talents. Specifically, traditional technical staff can no longer meet the needs of the current social development, which makes it increasingly necessary to cultivate multi-disciplinary talents in shipping enterprises. By understanding the development of talents in the shipping industry, analyzing the demand of compound talents in the shipping industry, combining with the development status of Wuhan shipping enterprises and the imbalance between supply and demand of compound talents in the shipping industry, the paper analyzed from multiple levels, and pointed out the shortcomings of the training of compound talents in shipping enterprises. After that it suggested the methods of training compound talents in shipping enterprises from multiple angles, established the quality evaluation system of talents training, and put forward the evaluation coefficient indexes from multiple dimensions, so as to provide scientific basis for training compound talents in shipping enterprises.

Keywords: Shipping enterprises · Compound talents · Training mode

1 Introduction

Shipping industry is a modern logistics service industry and with great development potential in the construction of global trade which is full of challenges and opportunities. It not only leads port city transportation and logistics hub, but also supports water transport economy and regional economic development. Shipping is characterized by the permanence of ships that ply traditional routes with their cargoes, linking industries and consumers. The combination of the two fundamental properties of shipping–high intensity of fixed assets, with long lead-times and lifetimes, and high exposure to volatile global flows of cargo and energy–results in complex market dynamics and high risk. The exposure of industry actors to these forces drives business models, strategies and industry developments.

From the diverse working functions and complex business types, it is typically knowledge-intensive and talents-intensive. Compound talents refer to professional technicians or high level managers with multi-disciplinary knowledge background or working experience. Pantouwakis and Karakasnaki analyzed the importance of talent factors in all quality management systems [1]. As the important metropolis in center of China, to become an international shipping center, it is a long process for Wuhan to cultivate all kinds of talents equipped with knowledge and experience. Under the background of implementing national strategies such as the Yangtze River Economic Belt and the Belt and Road, it is even more necessary for Wuhan to seize opportunities and conduct innovation for development.

There is an increasing demand for multi-disciplinary talents in shipping industry, and many scholars have done relevant research on the cultivation of shipping talents. Jiang Jun and Tan Ming analyzed the current situation of the development of the shipping talent team in Chongqing through investigation and SWOT analysis, and put forward corresponding countermeasures and suggestions [2]. Yang Dagang analyzed and compared the training curriculum system of shipping finance talents in ten universities around the world, and put forward the teaching mode of combining Industry-University-Research in China [3]. By analyzing the main training modes of foreign shipping talents and the current situation of China's shipping talents, Qu Qunzhen and Li Lixin came up with the training mode of serving Industry-University-Research [4]. Lai Yongsheng studied and analyzed the training mode of shipping talents, and put forward the training mode of objective, system and system [5]. Yin Ming and others analyzed the typical problems existing in the training of shipping professionals [6]. Wei Zhijie analyzed the importance and shipping advantages of Wuhan in the Yangtze River shipping economic belt, and put forward that Wuhan is the most ideal and promising high-end inland shipping talent training base [7]. Lei Liu and others analyzed the shortcomings in the construction of teaching staff in maritime higher vocational colleges, and put forward the countermeasures and suggestions for the construction of teaching staff in maritime higher vocational colleges [8]. Wan Jianxia and others analyzed the scientific evaluation system of the competitiveness of shipping talents, establishing a specific evaluation model of the competitiveness of shipping talents by using grey clustering and analytic hierarchy process [9]. Zhang Ying and others introduced the general idea of building a training platform for shipping compound talents, and explained in detail the construction of three bases and two resource bases that constitute the platform [10]. Zhang Ying and others discussed two interdisciplinary training modes of shipping talents and their supporting systems [11]. Pantouvakis and Karakasnaki analyzed the role of talents in the effective implementation of quality management system in shipping industry, and then evaluated the influence of talents in these enterprises [12]. Pantouvakis and Vlachos analyzed how talents and leadership influence sustainable performance, and pointed out that the influence of talents on sustainable performance is greater than that of leadership, and organizational culture [13]. In addition, many studies show some internet learning modes is not only beneficial to the dissemination of knowledge, but also can effectively promote students' learning [14–19].

In this paper, from the perspectives of shipping enterprises, it is proposed to build some training modes for compound talents, so as to promote the development of the high-level shipping talents.

2 Demand of Shipping Compound Talents in Wuhan

As shown in Table 1, according to the different working nature of enterprises related ports and shipping, the demand of shipping talents in Wuhan can be divided into five categories, including technical talents, management talents, service talents, science and education talents, and administrative talents.

Therefore, from the characteristics of the traditional shipping industry, such as transformation and development, innovation and development, sustainable development, and international development, the shipping industry in Wuhan is in urgent need of multidisciplinary professionals and managerial talents, involving shipping finance, logistics management, shipping brokerage, maritime law, shipping information, shipping consulting, shipping transactions and other related fields. To integrate liberal arts disciplines and Chinese and foreign laws, we need compound talents with rich practical experience, plan development strategies and solve practical problems, to adapt to the development of the shipping industry and meet the needs of shipping institutions.

Table 1. Classification of shipping talents in Wuhan

Shipping talent category	Composition of members
Shipping technical talents	Port technicians, Shipyard technicians
Shipping management talents	Management personnel of port and shipping authorities
Shipping service talents	Shipping traders, Shipping information consultant
Shipping education talents	Professor of shipping colleges and universities
Shipping administrative talents	Shipping customs officer, Personnel of shipping institutions

3 Situation of Talents Training in Wuhan Shipping Enterprises

In the new era of high integration of Internet economy, information technology, and shipping industry development, the contradiction between the supply and demand of shipping talents, especially high-level shipping compound talents, is restricting the rapid and sustainable development of shipping centers as the bottleneck, and the training of shipping compound talents is imperative. The growth process of shipping compound talents is as Fig. 1.

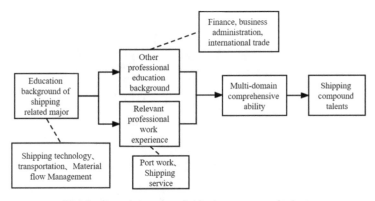

Fig. 1. Growth process of shipping compound talents

3.1 For Enterprise: Talents Growth Rate is Slow

As the main channel for cultivating compound talents in shipping, the training ability of port and shipping enterprises is uneven, and most enterprises lack high-level trainers, which slows down the process of knowledge updating and ability expansion of enterprise managers and business backbones. As the primary source of multi-disciplinary talents in shipping, seafarers and senior seafarers, due to the changes in social treatment and management environment in recent years, have led to problems such as low quality of seafarers, unstable crew, loss of seafarers, etc., so that high-quality senior seafarers are constantly added to shipping management, and the supply capacity in the new format of shipping services is weakened. Therefore, the training of shipping compound talents can only be carried out by the enterprise's own strength, and the training speed, growth quality, and knowledge structure of shipping compound talents can no longer meet the needs of the cross-development, information development, and international development of shipping industry.

Talents level is not only critical as the core competitiveness of enterprises, but also the decisive in their innovation and business development ability. The slow growth rate of talents has also become one of the main reasons for the low market competitiveness of port and shipping enterprises in Wuhan, Hubei Province. It is easy for enterprises to fall into the strange circle of "Insufficient supply of talents - Weak enterprise development - Enterprise benefit is not good - Further brain drain", as following Fig. 2.

3.2 For College: Not Match with Market Demand

As the primary source of professional and technical personnel and management cadres in port and shipping enterprises and institutions, shipping colleges and universities are currently in a state of doing their own thing at the school level. There is a lack of in-depth understanding and prediction of the talent development demand, industry development trend or leading new business trend in the future shipping center, so to some extent, there is a phenomenon that the talent training mode, professional curriculum, comprehensive quality education, and other aspects are out of touch with the actual situation of industry development and market demand. It is common for college graduates to have

Fig. 2. Enterprise vicious circle development

insufficient professional knowledge, poor hands-on ability, and poor post adaptability, which objectively limits the height of their career development and is not conducive to the development of compound talents in the future.

3.3 For Industry: Educational Resources Need to Be Integrated

Wuhan has abundant shipping science and technology education and management resources, numerous employees in ports, shipping, shipbuilding and shipping services, and a solid industrial base. However, there is no consensus and overall arrangement on the development strategy of shipping talents, including the training of compound talents. There is still a lack of urgency and crisis in the training of talents at the industry level.

There is still a lack of government policy guidance and macro guidance at the industry level in the training of shipping compound talents in enterprises. There is a lack of linkage mechanism among the relevant educational resources of government, industry, learning, and research. Generally speaking, the education and training resources are scattered, the atmosphere of talent training is not intense, and the level of talent promotion is limited, showing a situation that does not meet the needs of the rapid development of the Wuhan shipping center.

4   Training Modes of Compound Talents in Shipping Enterprises

Because of the leading talent problems and demand of shipping enterprises in Wuhan, we can focus on cultivating talents including: port technology and management, shipping management, shipping brokerage (including shipping agency and freight forwarder), shipping financial insurance, shipping brokerage, maritime safety, maritime law, ship driving and engine, intelligent port and shipping, shipping industry management and so on. It should train different types and levels of shipping talents as Fig. 3.

4.1   Non-academic Education: Professional Promotion Training Mode

According to the demand of shipping talents at different levels and in various fields, on-the-job non-academic training can be divided into four types, namely, executive training

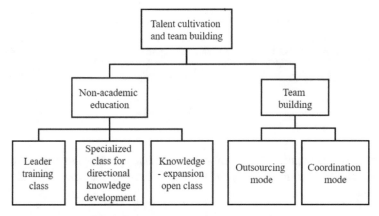

Fig. 3. Talent cultivation and team building

classes, targeted knowledge development classes, enterprise internal training classes, and knowledge development open classes.

4.1.1 Leader Training Class

(1) Training Object

It can target the training object of the executive seminar at shipping industry executives and leaders of port, shipping, and maritime authorities, including the chairman, president, general manager, decision-makers, and leaders of government departments at all levels of the above five types of shipping enterprises and institutions.

The training objective is to make students thoroughly familiar with the basic knowledge of shipping related businesses, including port, shipping, shipping finance and insurance, maritime inspection and technology, maritime law, ship and shipping brokerage, etc., and to grasp the latest shipping market trends through the course study, including the latest development trends of the essential shipping market and derivative markets such as shipbuilding, ship trading, shipping finance and insurance, maritime inspection and maritime legal consultation, to provide fundamental ideas and thinking modes for strategic decision-making of shipping industry and decision-making basis for shipping industry management and industry management.

(2) Learning Cycle

Referring to the training modes of executive seminars in Tsinghua, Peking University, and Shanghai Jiaotong University, the academic system arrangement can generally be divided into three forms as Table 2.

Combined with the working schedule of Wuhan shipping industry executives and leaders of government collection and management departments, it suggests that the study period should be one year, with two days of concentrated study every two months.

(3) Courses Contents

To facilitate the executives to formulate the strategic plan of the enterprise better, the curriculum should cover all aspects, and at the same time highlight the hot demand of the shipping industry (shipping market knowledge, maritime legal knowledge, intelligent application of port and shipping), including the professional knowledge and the

Table 2. Training modes of executive seminars

	Learning cycle	Time period arrangement	Length of study
Mode one	A year and a half	Study one day a month	Eighteen days
Mode two	A year and a half	Study for three days every two months	Eighteen days
Mode three	One year	Study two days a month	Twenty-four days

latest industry development trends of shipping economy, shipping management, port management, port technical management, introduction to maritime law and maritime inspection, etc.

(4) Source of Teachers

To meet the new needs of shipping development, it is suggested to make full use of the faculty of Wuhan universities and rely on Wuhan local universities with strong water majors to set up professional training teams. The teams can hire well-known entrepreneurs in the shipping industry, maritime legal experts, maritime bureau professionals, and senior talents of classification societies to give lectures together, so that students can grasp the characteristics of the industry from all directions and angles, fully connect the learning content with practice, and understand the latest shipping market trends.

4.1.2   Specialized Class for Directional Knowledge Development

(1) Training Object

According to the demand for specialized talents in different fields of the shipping industry, the directional knowledge development training mode is adopted, which aims at the on-the-job personnel in the fields of port, shipping, shipping finance and insurance, maritime law and maritime safety, so as to cultivate specialized talents urgently needed by the shipping industry. The types of technical talents trained are as follows:

- Port technical personnel (including port dispatching, loading and unloading technology, equipment maintenance and management)
- Shipping management talents (including shipping management, shipping agency, freight forwarder and ship brokerage talents)
- Shipping finance and insurance talents (including ship insurance, cargo insurance and shipping finance talents)
- Logistics management talents (including port collection and distribution, storage, customs declaration, logistics and freight forwarding talents)
- Maritime legal talents
- Maritime safety technical talents
- Ship inspection technical personnel
- Ship driver and marine engineer talents
- Shipping intelligence and information talents

With the emergence of unmanned ports in Yangshan Port and Qingdao Port, the application and promotion of Artificial Intelligence (AI) technology in ports and shipping fields is unstoppable. There may be unmanned ships in the future. For example, Danish Maritime Safety Administration, Google, and Beijing Ocean Economics research unmanned ships. Domestic shipbuilding enterprises have obtained the qualification of unmanned ships construction. Therefore, professionals who combine AI technology with shipbuilding, port and shipping technology and information technology should be the hot spot for future shipping. The training base of compound shipping talents in Wuhan Shipping Center should arrange the teaching plan, teachers, teaching materials, experimental subjects, and other resources in the form of training projects according to the above categories of talents to facilitate the implementation of the project.

(2) Learning Cycle

According to the needs of different fields of the shipping industry, the learning cycle can be flexibly formulated, and the length of schooling can be set to three months, one month or one week. One-week schooling can adopt an off-the-job centralized learning mode. The one-month and three-month school system can be set for two days of study on weekends.

(3) Courses Contents

In order to facilitate professional and technical personnel to better implement executive decisions and improve the industry level in their fields, it is suggested that the course content should be professional, comprehensive, and in-depth teaching should be carried out to meet the needs of professional areas, and the teaching content should be closely related to reality with rich cases.

(4) Source of Teachers

In order to meet the new needs of shipping development, it is suggested to make full use of the faculty of Wuhan universities and rely on Wuhan universities with strong water majors to set up professional training teams. The teams should be targeted to hire and invite famous entrepreneurs in the shipping industry, maritime legal experts, maritime bureau professionals, and senior talents of classification societies to give lectures together, so that students can master the specialized basic knowledge of the industry in a targeted manner and have strong ability to analyze and solve problems in professional fields.

4.1.3 Knowledge-Expanding Open Class

Knowledge-expanding open class is a flexible supplementary form to improve the professional skills of in-service personnel in shipping enterprises and institutions. For middle-level managers, technical backbones, and grass-roots managers and technicians of enterprises and institutions in various fields of the shipping industry, we broaden our horizons through online open classes, combine professional knowledge with industry skills with rich cases, and aim to extend and strengthen the professional field study of on-the-job personnel, avoid "fragmentation" and highlight the students' ability to analyze and solve problems in the core professional field of shipping. It is suggested that local colleges and universities in Wuhan or special shipping vocational education teams should organize teachers and adopt the teaching mode in enterprises (institutions). The learning cycle can be flexibly arranged for half a day, one day, three days, etc., according to the requirements of the breadth and depth of a single course.

4.2 Team Building: The Compound Professional Team Building for Business Development Needs

Facing the needs of business development, setting up a compound shipping professional team is an effective and new training mode to meet the needs of shipping enterprises and institutions for compound talents in the fields of port, shipping, shipping value-added services, maritime law and safety, ship driving and marine engineering, and shipping industry management. It is a solid guarantee to improve the overall level of shipping practitioners.

There are many ways to cultivate the compound professional team. According to the characteristics of different fields of the shipping industry, different pieces of training can be carried out to form the compound shipping professional team.

There are two modes can be considered in the training of the compound shipping professional team. One is outsourcing mode, the other is overall planning mode.

4.2.1 Outsourcing Mode

Outsourcing mode is an external cooperation mode composed of universities, vocational colleges and enterprises through making agreements, including vertical mode, and horizontal mode.

(1) Horizontal Mode

The cooperative relationship between similar colleges and universities communicates in academic and information fields through agreements to jointly cultivate a compound shipping professional team.

(2) Vertical Mode

Cooperate with shipping enterprises in colleges and universities through the Industry-University-Research project to form a vertical cooperation relationship, and cultivate a compound shipping professional team in this process. This model is characterized by complementary advantages, mainly focusing on personnel training, technology transfer, project consultation, etc.

4.2.2 Coordination Mode

The overall planning model is Industry-University-Research cooperation modeled by the government or industry associations, which have the function of comprehensive planning and coordination.

(1) Government Coordination Mode

The government coordination model has the advantages of policy, but the disadvantage is that it is too dependent on the government's relevant policies. At the same time, due to the constraints of policies, institutional mechanisms and other aspects, it lacks competitiveness, and vitality in management and market operation, so most of them can't achieve the maximum benefit.

(2) Industry Association Coordination Mode

Industry association coordination model is developed to meet the market demand. This kind of mode is characterized by that the allocation of resources is based on market regulation, and industry associations, such as ship owners' association, port association, maritime organization, navigation society, warehousing association, etc., communicate and coordinate, and form compound professional teams in the industry. This compound team mainly promotes the improvement of the overall service level of the industry, and is not responsible for solving the specific management or technical problems of specific shipping enterprises and institutions.

5   Evaluation System of Talent Training Quality

The evaluation of the training quality of shipping compound talents is a multi-objective and multi-attribute complex system, which can be quantitatively analyzed by AHP (Analytic Hierarchy Process) and fuzzy comprehensive evaluation, as shown in Fig. 4. The core of the evaluation work is to establish an evaluation index system. So a "Five-Dimensional quality evaluation index system" can be established, which includes matching of talents with regional industry, logic of training system, sharing of training resources, monitoring of training process, and evaluation of teaching quality.

5.1   Matching of Talents and Regional Industry

The evaluation of the shipping talents matching Wuhan shipping industry mainly reflects three aspects: matching industry development, meeting industrial demand, and cooperation of colleges and enterprises.

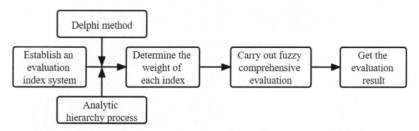

Fig. 4. Evaluation model of training quality of shipping compound talents

5.2   Logic of Training System

Find out the mapping relationship between talents training and industrial chain structure, and form a talents construction and development mechanism of "matching industrial chain, dynamic adjustment and self-improvement", which mainly includes five parts: reasonable training logic, reconstruction of curriculum system, clear development plan, integration of talent training modes, and appropriate scale of talent training.

5.3 Sharing of Training Resources

In order to strengthen the professional ability of shipping talents and cultivate sustainable quality ability, talents training can strengthen the sharing of teaching resources, which is mainly reflected in the structured teaching team, IT-based curriculum resources, sharing of experimental training, and sharing of professional practice training bases.

5.4 Monitoring of Training Process

Carry out training process monitoring, and use SWOT analysis to clarify the annual work tasks, use the "8-character spiral" operation units to carry out real-time and periodic diagnosis and modification, which can continuously improve the quality of personnel training. This factor mainly includes three parts: organization and implementation, evaluation feedback, and process diagnosis and modification.

5.5 Evaluation of Teaching Quality

Establish an evaluation committee for the training quality of shipping compound talents. The committee is mainly composed of leaders of colleges, employment offices of colleges, experts of industry and enterprises, experts of enterprise human resources, and heads of government departments, et al. It mainly evaluates the teaching quality from the employment rate, starting salary, satisfaction of employment, satisfaction of enterprises, and process of talent growth (Table 3).

Table 3. Evaluation system of personnel training quality

Main factors	Sub-index
Matching of talents and regional industry	Matching industry development
	Meeting industrial demand
	Cooperation of colleges and enterprises
Logic of training system	Reasonable training logic
	Reconstruction of curriculum system
	Clear development plan
	Integration of talent training modes
	Appropriate scale of talent training
Sharing of training resources	Structured teaching team
	IT-based curriculum resources
	Sharing of experimental training
	sharing of professional practice training bases
Monitoring of training process	Organize and implement

(continued)

Table 3. (*continued*)

Main factors	Sub-index
	Evaluation feedback
	Process diagnosis and modification
Evaluation of teaching quality	Employment rate
	Salary starting
	Satisfaction of employment
	Satisfaction of enterprises
	process of talent growth

6 Conclusion

The shortage of shipping talents, especially shipping compound talents, has become a bottleneck restricting the rapid and sustainable development of the shipping industry. Find out the problems and analyze the shortcomings from three aspects of enterprises, universities and government, and put forward the general ideas to improve the training quality of shipping enterprises. That will build a highland of shipping talents in Wuhan, improve the comprehensive quality and innovation ability of technology, management personnel and shipping college graduates of shipping enterprises, and enhance the agglomeration and competitiveness of the shipping industry.

References

1. Pantouvakis, A., Karakasnaki, M.: Role of the human talent in total quality management–performance relationship: an investigation in the transport sector. Total Qual. Manag. Bus. Excell. **28**(9–10), 959–973 (2017)
2. Jiang, J., Tan, M.: SWOT analysis and countermeasures of Chongqing shipping talent team. J. Chongqing Jiaotong Univ. (Soc. Sci. Ed.) **13**(02), 25–29 (2013). (in Chinese)
3. Yang, D.: Thoughts on the training of shipping finance talents under the national strategy of "the belt and road initiative". Shanghai Finance (06), 109–110 (2015). (in Chinese)
4. Qu, Q., Li, L.: Dynamic mechanism of innovative mode of training shipping talents. Logist. Eng. Manag. **36**(02), 166–168+162 (2014). (in Chinese)
5. Lai, Y.: Implementation guarantee and evaluation of interdisciplinary cooperation in innovative shipping talent training mode. Market Weekly (Theor. Res.) (10), 136–137 (2017). (in Chinese)
6. Yin, M., Zhang, Q., Wang, X.: Research on the training of shipping specialty talents based on supply-side structural reform thought. Navigation (04), 15–18 (2017). (in Chinese)
7. Wei, Z.: Build Wuhan into China's high-quality crew recruitment center and assignment base. China Water Transport (12), 20–22 (2016). (in Chinese)
8. Liu, L., Chen, X., Fang, L., Han, Z.: Study on the construction of teaching staff in maritime vocational colleges based on the cultivation of international shipping talents. In: Proceedings of 2020 International Conference on Education, Sport and Psychological Studies (ESPS 2020), pp. 155–159 (2020)

9. Wan, J., Zhang, S., Li, J.: Grey cluster evaluation model for competitiveness of area shipping talent. In: Proceedings of 4th International Symposium on Social Science (ISSS 2018), pp. 361–365 (2018)
10. Zhang, Y., Chen, X., Zhang, Q., Tao, D.: Construction of interdisciplinary training platform for shipping talents. In: International Conference on Economics, Law and Education Research (ELER 2021), pp. 360–367. Atlantis Press (2021)
11. Zhang, Y., Zhu, H., Tao, D., Liu, Z.: Interdisciplinary training models for shipping talents in the universities. In: Hu, Z., Zhang, Q., Petoukhov, S., He, M. (eds.) ICAILE 2021. LNDECT, vol. 82, pp. 228–237. Springer, Cham (2021). https://doi.org/10.1007/978-3-030-80475-6_23
12. Pantouvakis, A., Karakasnaki, M.: The human talent and its role in ISM code effectiveness and competitiveness in the shipping industry. Marit. Policy Manag. **45**(5), 649–664 (2018)
13. Pantouvakis, A., Vlachos, I.: Talent and leadership effects on sustainable performance in the maritime industry. Transp. Res. Part D Transp. Environ. **86**, 102440 (2020)
14. Herala, A., Knutas, A., Vanhala, E., Kasurinen, J.: Experiences from video lectures in software engineering education. Int. J. Mod. Educ. Comput. Sci. **9**(5), 17–26 (2017)
15. Fetaji, B., Fetaji, M., Ebibi, M., Kera, S.: Analyses of impacting factors of ICT in education management: case study. Int. J. Mod. Educ. Comput. Sci. **10**(2), 26–34 (2018)
16. Carmichael, D., Archibald, J.: A data analysis of the academic use of social media. Int. J. Inf. Technol. Comput. Sci. **11**(5), 1–10 (2019)
17. Jiang, Q., Jin, T., Chen, H., Song, W.: Research on cultivating undergraduates in the computer science based on students. Eng. Manuf. (6), 32–39 (2020)
18. Liu, S., Chen, P.: Research on cultivation of ethnic minorities IT talents in nationalities universities. Int. J. Mod. Educ. Comput. Sci. **6**(2), 33 (2014)
19. Chen, H., Li, Z., Li, W., Mao, H.: Discussion on teaching pattern of cultivating engineering application talent of automation specialty. Int. J. Educ. Manag. Eng. **2**(11), 30–34 (2012)

Author Index

Z. Hu et al. (Eds.): ICCSEEA 2022, LNDECT 134, pp. 539–541, 2022.
https://doi.org/10.1007/978-3-031-04812-8

Printed in the United States
by Baker & Taylor Publisher Services